Chapman & Hall/CRC

Handbooks of Modern Statistical Methods

Series Editor:

Garrett Fitzmaurice, Department of Biostatistics, Harvard School of Public Health, Boston, MA, U.S.A.

The objective of the series is to provide high-quality volumes covering the state-of-the-art in the theory and applications of statistical methodology. The books in the series are thoroughly edited and present comprehensive, coherent, and unified summaries of specific methodological topics from statistics. The chapters are written by the leading researchers in the field and present a good balance of theory and application through a synthesis of the key methodological developments and examples and case studies using real data.

Handbook of Approximate Bayesian Computation
Scott A. Sisson, Yanan Fan, and Mark Beaumont

Handbook of Graphical Models
Marloes Maathuis, Mathias Drton, Steffen Lauritzen, and Martin Wainwright

Handbook of Mixture Analysis
Sylvia Frühwirth-Schnatter, Gilles Celeux, and Christian P. Robert

Handbook of Infectious Disease Data Analysis
Leonhard Held, Niel Hens, Philip O'Neill, and Jacco Walllinga

Handbook of Meta-Analysis
Christopher H. Schmid, Theo Stijnen, and Ian White

Handbook of Forensic Statistics
David L. Banks, Karen Kafadar, David H. Kaye, and Maria Tackett

Handbook of Statistical Methods for Randomized Controlled Trials
KyungMann Kim, Frank Bretz, Ying Kuen K. Cheung, and Lisa Hampson

Handbook of Measurement Error Models
Grace Yi, Aurore Delaigle, and Paul Gustafson

Handbook of Multiple Comparisons
Xinping Cui, Thorsten Dickhaus, Ying Ding, and Jason C. Hsu

Handbook of Bayesian Variable Selection
Mahlet Tadesse and Marina Vannucci

Handbook of Matching and Weighting Adjustments for Causal Inference
José Zubizarreta, Elizabeth A. Stuart, Dylan Small, and Paul R. Rosenbaum

Handbook of Bayesian, Fiducial, and Frequentist Inference
James Berger, Xiao-Li Meng, Nancy Reid, and Min-ge Xie

For more information about this series, please visit: https://www.crcpress.com/Chapman–HallCRC-Handbooks-of-Modern-Statistical-Methods/book-series/CHHANMODSTA

Handbook of Bayesian, Fiducial, and Frequentist Inference

The emergence of data science, in recent decades, has magnified the need for efficient methodology for analyzing data and highlighted the importance of statistical inference. Despite the tremendous progress that has been made, statistical science is still a young discipline and continues to have several different and competing paths in its approaches and foundations. While the emergence of competing approaches is a natural progression of any scientific discipline, differences in the foundations of statistical inference can sometimes lead to different interpretations and conclusions from the same dataset. The increased interest in the foundations of statistical inference has led to many publications, and recent vibrant research activities in statistics, applied mathematics, philosophy, and other fields of science reflect the importance of this development. The BFF approaches not only bridge foundations and scientific learning but also facilitate objective and replicable scientific research and provide scalable computing methodologies for the analysis of big data. Most of the published work typically focuses on a single topic or theme, and the body of work is scattered in different journals. This handbook provides a comprehensive introduction and broad overview of the key developments in the BFF schools of inference. It is intended for researchers and students who wish for an overview of foundations of inference from the BFF perspective and provides a general reference for BFF inference.

Key features:

- Provides a comprehensive introduction to the key developments in the BFF schools of inference.
- Gives an overview of modern inferential methods, allowing scientists in other fields to expand their knowledge.
- Is accessible to readers with different perspectives and backgrounds.

Handbook of Bayesian, Fiducial, and Frequentist Inference

Edited by

James Berger
Xiao-Li Meng
Nancy Reid
Min-ge Xie

CRC Press
Taylor & Francis Group
Boca Raton London New York

CRC Press is an imprint of the
Taylor & Francis Group, an **informa** business
A CHAPMAN & HALL BOOK

First edition published 2024
by CRC Press
2385 NW Executive Center Drive, Suite 320, Boca Raton FL 33431

and by CRC Press
4 Park Square, Milton Park, Abingdon, Oxon, OX14 4RN

CRC Press is an imprint of Taylor & Francis Group, LLC

ISBN: 9780367321987 (hbk)
ISBN: 9781032684581 (pbk)
ISBN: 9780429341731 (ebk)

DOI: 10.1201/9780429341731

Typeset in CMR10
by codeMantra

Contents

Preface **vii**

About the editors **xi**

Contributors **xii**

1 Risky Business **1**
Stephen Stigler

2 Empirical Bayes: Concepts and Methods **8**
Bradley Efron

3 Distributions for Parameters **35**
N. Reid

4 Objective Bayesian Inference and Its Relationship to Frequentism **58**
James Berger, Jose Bernardo, and Dongchu Sun

5 Fiducial Inference, Then and Now **83**
Philip Dawid

6 Bridging Bayesian, Frequentist and Fiducial Inferences Using Confidence Distributions **106**
Suzanne Thornton and Min-ge Xie

7 Objective Bayesian Testing and Model Uncertainty **132**
James Berger, Gonzalo García-Donato, Elias Moreno, and Luis Pericchi

8 A BFFer's Exploration with Nuisance Constructs: Bayesian p-value, H-likelihood, and Cauchyanity **161**
Xiao-Li Meng

9 Bayesian Neural Networks and Dimensionality Reduction **188**
Deborshee Sen, Theodore Papamarkou, and David Dunson

10 The Tangent Exponential Model **210**
A. C. Davison and N. Reid

11 Data Integration and Model Fusion in the Bayesian and Frequentist Frameworks **238**
Emily C. Hector, Lu Tang, Ling Zhou, and Peter X.-K. Song

12 How the Game-Theoretic Foundation for Probability Resolves the Bayesian vs. Frequentist Standoff **264**
Glenn Shafer

13 Introduction to Generalized Fiducial Inference **276**
Alexander C. Murph, Jan Hannig, and Jonathan P. Williams

14 Dempster-Shafer Theory for Statistical Inference **300**
Ruobin Gong

15 Slicing and Dicing a Path Through the Fiducial Forest **323**
Joseph B. Lang

16 Inferential Models and Possibility Measures **344**
Chuanhai Liu and Ryan Martin

17 Conformal Predictive Distributions: An Approach to Nonparametric Fiducial Prediction **364**
Vladimir Vovk

18 Fiducial Inference and Decision Theory **381**
Gunnar Taraldsen and Bo Henry Lindqvist

Index **397**

Preface

When and why was the BFF community created?

The pursuit of the ideal statistical inference paradigm arguably predates the field of statistics itself, tracing back to empiricism in the 17th century Scientific Revolution. Yet, the BFF (Bayesian, Fiducial, and Frequentist) community, which champions this Handbook, only emerged a decade ago in a nearly serendipitous fashion. On 17 June 2014, at the ICSA -KISS[1] Applied Statistics Symposium in Portland, Oregon, seven statisticians found inspiration in each other's presentations at a session on distributional inference. The diverse topics included *confidence distributions* presented by Min-ge Xie of Rutgers University, *generalized fiducial inference* by Jan Hannig of the University of North Carolina and Thomas Lee of the University of California at Davis, *inferential model* by Ryan Martin of North Carolina State University, *objective Bayes* by Dongchu Sun of the University of Missouri, and *individualized inference* by Keli Liu and Xiao-Li Meng of Harvard University. The intellectual excitement occupied the after-session break, leading to the vision of an annual forum for bridging different schools of thought on statistical inference.

Dongchu Sun, who at the time was visiting East China Normal University (ECNU), which was actively pursuing international collaborations, swiftly solved the venue and funding issues. Naming the forum took a bit longer – what could be a catchy and accurate phrase to capture the group's intention? "Well, we want to connect Bayesian, Fiducial and Frequentist, so BFF?" one of us (Meng) mumbled, "but what's *that*?" "Oh, my younger son has BFF all over his (school) yearbook," another of us (Xie) exclaimed, "*That* stands for Best Friends Forever!" This wordplay captured perfectly the essence of the group's mission: to harmonize, not homogenize, statistical paradigms. After all, statistics is a study of variations, so variations in studying statistics are rather fitting, signifying the viability and vitality of statistics as a scientific discipline, forever evolving. Just as best friends are often not those who merely echo our thoughts and interests, but those who can extend our comfort zones and hence our life experiences.

The rest is history. As summarized at `www.bff-stat.org/`, after BFF1 and BFF2 took place in ECNU in 2014 and 2015, BFF3 to BFF6 were held in the US, respectively at Rutgers University, Harvard University, University of Michigan, and Duke University. Due to the COVID-19 pandemic, BFF7 at the University of Toronto was delayed until 2022, and BFF8 followed at the University of Cincinnati in 2023, when this Handbook went into production. During the pandemic, Ryan Martin and Harry Crane led a virtual BFF 6.5. in 2021; see `https://researchers.one/conferences/bff65`.

It is a very friendly history since all workshops were attended by many "BFFers," in both senses of the phrase. They were there because of their research interests and findings in BFF, often involving more than one school of thought. They were also there to seek new inspiration and make friends—in BFF spirit—across disciplines. Indeed, the attendees

[1]ICSA: International Chinese Statistical Association; KISS: Koren International Statistical Society.

came from very diverse backgrounds and expertise, ranging from philosophy to information geometry to imprecise probability, an intellectual spectrum rarely seen in other statistical conferences. That, of course, was the hope of the original seven BFFers.

What does this Handbook intend to accomplish?

The emergence of data science has magnified the need for many more efficient methodologies for collecting, analyzing, and making sense of data, for which the field of statistics has much to offer. Furthermore, the increasingly common practice of using "black-box" methods calls for significantly more research and understanding of the theory behind these methods both for their implementation efficiency (e.g., by reducing brute-force trial and error) and for their safe application (e.g., by preemptively avoiding their hazardous zones). To make meaningful progress, foundational underpinning is essential. However, the different and sometimes competing paradigms of statistical inference can be quite confusing to the broad data science community, especially because these paradigmatic differences, "unlike most philosophical disputes, have immediate practical consequences." (Efron, 2013*a*). Fraser (2011) stressed that "as a modern discipline, statistics has inherited two prominent approaches (i.e., Bayesian and frequentist methods) to the analysis of models with data ... How can a discipline, central to science and to critical thinking, have two methodologies, two logics, two approaches that could frequently give different answers to the same problems?"

Indeed, whereas competing approaches are a natural progression of scientific enquiries, the different inferential paradigms can sometimes lead to qualitatively different interpretations and conclusions from the same data. Statisticians have long been aware of the need to at least properly frame and explain such differences, especially when building modern statistical inference that "matches contemporary attitudes" (Kass, 2011) and hence there has always been a desire to find a common roof for at least the major inferential paradigms, especially BFF. However, the search for such a common roof remains "perhaps the most important unresolved problem in statistical inference" (Efron, 2013*b*).

To continue the search and research, many scholars have been working to link and compare approaches to inference. For instance, objective Bayesian methods, oriented towards the development of default prior distributions, have led to major advances in the last two decades, which include establishing solid links to frequentist and fiducial methods and computing algorithms. In addition, effective nonparametric Bayesian approaches and Bayesian computing methods have promoted the use of Bayesian methods in increasingly complex settings. On the side of frequentist inference, the development of confidence distributions, frequentist analogues of Bayesian posteriors, has provided a platform to associate, compare, and combine inferences across the aisles of Bayesian, fiducial, and frequentist schools. On the fiducial inference front, there is renewed interest, including Fraser's structural model, Dempster-Shafer inference, Hannig's generalized fiducial inference and Liu and Martin's inferential models. These new developments provide a "useful statistical tool for problems where frequentist methods with good properties were previously unavailable" (Hannig, 2009). Relating to these BFF developments, Reid and Cox (2015) highlighted several principles of statistical inference and outlined how these principles "are, or could be, used to inform the interpretation of results, and to provide a greater degree of coherence for the foundations of statistics." The annual BFF workshop series represents a growing collective interest in communicating among different inferential perspectives and in seeking common ground that will help to build an inferential foundation for data science.

Recent vibrant cross-disciplinary research activities in statistics, applied mathematics, computer science, philosophy, and other fields have been accompanied by increased awareness of the importance of foundational research. The BFF development bridges the foundations for scientific learning, facilitates objective and replicable scientific research, and provides directions for searching for scalable computing methodologies for handling many types of complex data. However, most of the published work typically focuses on a single topic or theme, and the body of work is scattered in many journals, which makes it a challenging task for anyone, especially for students and newcomers, who wish for a holistic picture of BFF and understand their respective strengths and limitations. This handbook, therefore, is meant to be a coherent introduction to and overview of the key ideas underlying the synergistic BFF movement and a showcase of its developments in its first decade.

Who are the target readers and how they should read the Handbook?

As we wrote in the book proposal, this book is aimed at

I. Researchers who already work on the foundations of statistical inference but desire a broader, more structured, and stimulating overview;

II. Researchers and students entering the field of statistical science, as a reference book for general BFF inference;

III. Scientists in other fields who work on or apply cutting-edge statistical inference tools and are interested in acquiring a more thorough knowledge of modern inferential methods.

To effectively cater to our varied readership, we have curated a collection of 18 articles. Nearly all of them feature multi-layered content, enriched with a plethora of examples, designed to cultivate intuition, offer insights, and enable readers to draw connections across various concepts. Categorizing these articles into three distinct schools of thought was not trivial due to their overlapping perspectives. Nevertheless, we managed to arrange the 18 chapters into four parts. Part I, encompassing the first six chapters, primarily offers big-picture, overview-style articles.

Chapter 1 is based on Stephen Stigler's banquet speech at BFF4, a rather risky business because an effective academic banquet speech needs to deliver a trinity: tasteful humor to captivate the audience, substantial content to sustain their interest, and a surprise element to make it memorable. Stigler's "Risky Business" exemplifies how to accomplish this trifecta seamlessly. However, we feature it to lead the Handbook for an additional reason. As the preeminent statistical historian of our time, Stigler's delightful journey into the BFF's past situates the ongoing movement within a vast historical framework, alerting us to both the exhilarating opportunities and significant challenges that await.

Chapter 2, Brad Efron's overview of one of the most surprising inferential methods, empirical Bayes, is delightful because of its rich historical development and connections, from the much celebrated James-Stein estimator to the exceedingly impactful framework of controlling false discovery rates. In Chapter 3, Nancy Reid connects and compares the three schools of thought through the shared goal of creating an inferential distribution for parameters of interest and interlaces the other chapters in this Handbook into its rich exposition.

Chapter 4 by James Berger, Jose Bernardo, and Dongchu Sun brings readers to another rich exposition on a challenging topic, Objective Bayes, a seemingly paradoxical notion. Indeed, as the authors stated, the desire to define objectivity objectively is generally misguided. Nevertheless, it is useful to search for prior that would impose as little procedural information as possible, that is, information imposed purely for permitting us to use the Bayesian procedures, not because of scientific knowledge or informed judgment.

Searching for such "objective priors" has occupied much of the BFF history both because of its deep connections with frequentisim and because fiducial inference was precisely motivated by the desire to avoid prior specification altogether, yet still produce a post-data distribution for our inferential targets. However, there is no free lunch. The fiducial approach comes with a host of challenges, including logical inconsistency, as reviewed and explained in Chapter 5 by Philip Dawid.

Chapter 6, by Suzanne Thornton and Min-ge Xie, concludes Part 1 with an overview of the concept of confidence distribution, an inferential construct that can be harmonized with any of the BFF perspectives, at least from a procedural point of view.

Part II (Chapters 7-9) and Part III (Chapters 10-12) are devoted to topics with more Bayesian and frequentist emphasis, respectively, albeit this distinction is often more a matter of angle of exposition. For example, Chapter 8 by Xiao-Li Meng documented the author's learning journey from being taught as a frequentist to become a fiducial sympathizer. But the chapter is placed in Part II because these learning stories were told from a pragmatic Bayesian perspective. In contrast, Part IV (Chapters 13-18) is almost exclusively devoted to fiducial inference and its extensions and general formulations. This devotion reflects the fact that fiducial inference, despite its long history starting with R.A. Fisher, is much less studied in general compared to Bayesian and frequentist.

We hope the handbook will help facilitate new interest and developments that will in turn lead to advances in statistical theory, facilitate new methodologies and applications, and establish a solid theoretical foundation for statistical inference in data science.

Bibliography

Efron, B. (2013*a*), 'Bayes' theorem in the 21st century', *Science* **340**(6137), 1177–1178.

Efron, B. (2013*b*), 'Discussion on "Confidence distribution, the frequentist distribution estimator of a parameter" by Xie and Singh', *International Statistical Review* **81**(1), 41–42.

Fraser, D. A. (2011), 'Is Bayes posterior just quick and dirty confidence?', *Statistical Science* **26**(3), 299–0316.

Hannig, J. (2009), 'On generalized fiducial inference', *Statistica Sinica* **19**(2), 491–544.

Kass, R. E. (2011), 'Statistical inference: The big picture', *Statistical Science* **26**(1), 1–9.

Reid, N. and Cox, D. R. (2015), 'On some principles of statistical inference', *International Statistical Review* **83**(2), 293–308.

About the editors

James Berger, PhD, is the Arts and Sciences Distinguished Professor Emeritus of Statistics at Duke University. Dr. Berger received his PhD in Mathematics from Cornell University in 1974. Among the awards and honors, Dr. Berger has received Guggenheim and Sloan Fellowships, the COPSS President's Award in 1985, the Sigma Xi Research Award at Purdue University for the contribution of the year to science in 1993, the COPSS Fisher Lecturer in 2001, the Wald Lecturer of the IMS in 2007 and the Wilks Award from the ASA in 2015. He was elected as a foreign member of the Spanish Real Academia de Ciencias in 2002, elected to the USA National Academy of Sciences in 2003, was awarded an honorary Doctor of Science degree from Purdue University in 2004, and became an Honorary Professor at East China Normal University in 2011.

Xiao-Li Meng, PhD, is the Whipple V. N. Jones Professor of Statistics at Harvard University. Dr. Meng received his PhD in Statistics from Harvard University. He is the Founding Editor-in-Chief of Harvard Data Science Review. In 2020, he was elected to the American Academy of Arts and Sciences. His interests range from the theoretical foundations of statistical inferences to statistical methods and computation.

Nancy Reid, PhD, is a University Professor of Statistical Sciences at the University of Toronto. Dr. Reid received her PhD in Statistics from Stanford University and is a Fellow of the Royal Society, the Royal Society of Canada, the Royal Society of Edinburgh, and a Foreign Associate of the National Academy of Sciences. In 2015, she was appointed Officer of the Order of Canada. Her research interests include the foundations and theory of statistical inference.

Min-ge Xie, PhD, is a Distinguished Professor at Rutgers, The State University of New Jersey. Dr. Xie received his PhD in Statistics from the University of Illinois at Urbana-Champaign (UIUC). He is the current Editor of The American Statistician and a co-founding Editor-in-Chief of The New England Journal of Statistics in Data Science. His research work on confidence distributions was described as a "grounding process with energy and insight." His research interests include statistical inference, foundations of data science, fusion learning, and interdisciplinary research.

Contributors

Jose Bernado
Universidad de Valencia
Valencia, Spain

Anthony C. Davison
Ecole Polytechnique Fédérale de Lausanne
Lausanne, Switzerland

Philip Dawid
University of Cambridge
Cambridge, UK

David Dunson
Duke University
Durham, NC, USA

Bradley Efron
Stanford University
Stanford, CA, USA

Gonzalo Garcıa-Donato
Universidad de Castilla-La Mancha
Ciudad Real, Spain

Ruobin Gong
Rutgers University
New Brunswick, NJ, USA

Jan Hannig
University of North Carolina at Chapel Hill
Chapel Hill, NC, USA

Emily C. Hector
North Carolina State University
Raleigh, NC, USA

Joseph B. Lang
University of Iowa
Iowa City, Iowa, USA

Bo Henry Lindqvist
Norwegian University of Science and Technology
Trondheim, Norway

Chuanhai Liu
Purdue University
West Lafayette, IN, USA

Ryan Martin
North Carolina State University
Raleigh, NC, USA

Elias Moreno
Universidad de Granada
Granada, Spain

Alexander Murph
University of North Carolina at Chapel Hill
Chapel Hill, NC, USA

Theodore Papamarkou
The University of Manchester
Manchester, UK

Luis Pericchi
University of Puerto Rico Rio Piedras
San Juan, Puerto Rico

Deborshee Sen
University of Bath
Bath, UK

Glenn Shafer
Rutgers University
New Brunswick, NJ, USA

Peter X.K. Song
University of Michigan
Ann Arbor, MI, USA

Stephen Stigler
Retired
Chicago, IL, USA

Dongchu Sun
East China Normal University
Shanghai, China

Lu Tang
University of Pittsburgh
Pittsburgh, PA, USA

Gunnar Taraldsen
Norwegian University of Science and Technology
Trondheim, Norway

Suzanne Thornton
Swarthmore College
Swarthmore, PA, USA

Vladimir Vovk
Royal Holloway, University of London
Egham, UK

Jonathan P. Williams
North Carolina State University
Raleigh, NC, USA

Ling Zhou
Southwestern University of Finance and Economics
Chengdu, China

1

Risky Business

Stephen Stigler

University of Chicago

The BFF consortium brings together three approaches to statistical inference. This is a very difficult task, and historically, only the uncommonly courageous have attempted it. One of the three approaches is a radical approach; one is enjoying a spurt of recent popularity; and one has had steady support for many decades. Can you guess which is which? The data that come from the use of Google's NGram Viewer (which tracks the use of the relevant terms over time in a large database from Google Books) will surprise you (Figure 1.1).

The radical approach is "Frequentist" – it barely exists! The use of the term is vanishingly small compared with the other two. "Bayesian" is enjoying greatly rising popularity, and "Fiducial" is steady-as-she-goes.

I will review all three, starting by giving each a suggested 'visualization', as is now the vogue.

1.1 First, Frequentist Inference

This picture (Figure 1.2) comes from an official WWII Army Airforce manual, giving survival advice to pilots whose planes were disabled – should they attempt to nurse the

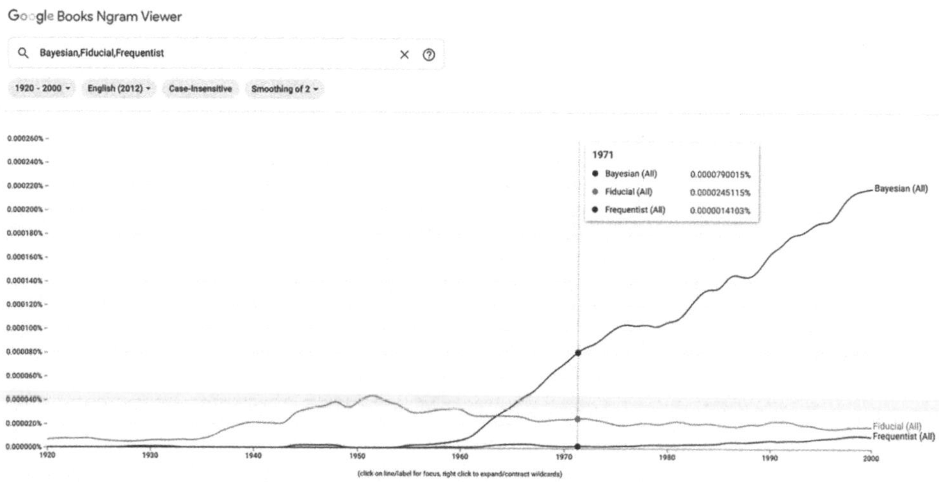

FIGURE 1.1
Relative frequency of use of the terms 'Bayesian', 'fiducial', and 'frequentist', in Google Books from 1920 to 2000. From Ngram Viewer.

DOI: 10.1201/9780429341731-1

FIGURE 1.2
Airforce manual advice on whether or not to abandon ship: randomize! From Army Airforce WWII manual.

plane to a safe landing or should they bail out? The advice was frequentist: Randomize! This was a clear case of de-personalized medicine. Despite its Google non-existence, Frequentist inference is taught in every university, and it serves as a convenient whipping boy when things go wrong: this branch of inference is often blamed for concentrating on "straw man" hypotheses, and Frequentist is itself often the strawman of the inference wars. It is criticized for uncritically advocating mindless significance tests and as the well-spring of irreproducible results. And many of these charges can be true, at least in some examples cited as "typically frequentist" in the attacks. It is surely risky to be a Frequentist these days.

Apparently one of the most pressing problems of the day has arisen because "we" –meaning most statistics instructors and everyone who takes a course in statistics – pay insufficient attention to what we say on the very first day of the first class. That is the class where we say probability is assigned to subsets of possible outcomes in the set of all possible outcomes; that is, in the Sample Space. Sometimes I put a question on the final exam: True or False: The Sample Space is the place we store the sample. The results are predictably appalling. I remember when I first became aware of the problems that can arise with an incompletely specified sample space or reference set. It was listening to a talk by Glenn Shafer, who was then insisting on the importance of understanding what he called the protocol – the rules that led to the observed outcome. For example, the outcome from a tossed coin that led to one head and one tail would have a different probability if the protocol said the coin would be tossed until the first tail than if it were simply to be tossed twice. Or to get to a slightly more deceptive example, when answering a question such as "Given that a family includes at least one boy, what is the probability that there are two or more?" Was the questioner planning all along to use that male-centric phrasing? And if so, what would the questioner have asked if the children were all girls? The way a question is posed generally carries information, and dealing correctly with that information may require facts that are simply not available.

I remember bringing this up at the lunch table of the University of Chicago's Quadrangle Club one day, and one colleague, probabilist Pat Billingsley, granted the point, but he didn't think it was that important. For example, he said, his major professor at Princeton, Willy Feller, would never have made such an error. Of course we went right back to check if Feller had indeed avoided this, and of course we found that Feller had included exactly that question in all three editions of his famous Volume I.

Feller erred, and so does everyone else, at least occasionally. And the question is a basic one for the analysis of data. Where did the data come from? What choices were made in framing the question asked, and how did those choices depend upon the data themselves? Identifying the sample space is never as easy in practice as it seems on that first day in class.

Andy Gelman has been beating this drum loudly – maybe too loudly – with regard to significance tests. Even in that form the problem has been recognized for a long time – for at least 250 years the problem has been pointed out, often by people who have then gone on to make the same error themselves. The American Statistical Association (ASA) made a cause célèbre of this in 2016, asking 20 statisticians to write pieces on the value and interpretation of P-values. Several of the 20 pieces were excellent, and many of the other pieces were pretty good too. It will not surprise the reader that about 5% of these failed to reach significance. Thus far, all was well. But then, the ASA muffed their chance. Instead of publishing all (or almost all) of these as splendid examples of the ways good statisticians with differing viewpoints would address this problem, they boiled the message down to the almost-empty intersection of the results and published that, while burying the pieces themselves in a distant website. It was a supreme irony that the ASA, faced with a rich trove of variable scientific data bearing on a broad and important problem in science, would summarize the data with the verbal equivalent of an off-the-shelf 5% level test.

1.2 Second, Bayesian Inference

The picture I present is of Nora Bayes instead of the usual portrait of Thomas Bayes; she is more appealing, and the picture is at least equally authentic (Figure 1.3). Nora and Thomas are alike, in that neither is responsible for modern Bayesian Inference, and unlike, in that Thomas was largely ignored during his lifetime and for a century later, while Nora had several Number One Hits, including the first recording of "Over There," the famous WWI song written by George M. Cohan. (The "Normandie" song whose sheet music is pictured seems a clear reference to Laplace's birthplace.) Nora was probably a relative of Thomas, but it was a distant relation. She was born in Joliet, Illinois, in 1880 with the name Eleanora Goldberg. And yes, I have taken a look for important early 1700 works by Thomas Goldberg, in case this name change was not the first in the family. No major finds yet.

Bayesian inference is a risky business, being attacked from all sides, even from inside, by other Bayesians. Fisher characterized Bayesian inference as "fundamentally false and devoid of foundation." Its recent popularity has come in part from spreading the Bayesian tent widely: there are objective Bayesians, subjective Bayesians, empirical Bayesians, computational Bayesians, improper Bayesians, least favorable Bayesians, maximum entropy Bayesians, noninformative Bayesians, preposterous Bayesians, savage Bayesians, and good Bayesians. Nora Bayes had five husbands before she died in 1928 at age 47, and her followers have also spread their affections widely.

FIGURE 1.3
The singer, Nora Bayes. From author's collection.

1.3 Finally, Fiducial Inference

Figure 1.4 is a picture I took a few years ago in Orléans, France. I call your attention to the obvious symbolism recalling Eve and the apple, offering a clear temptation to sin.

FIGURE 1.4
Fiducial, Eve and the apple. Author's photograph.

Fiducial inference may be the riskiest of all. In his 1998 paper, "R. A. Fisher in the 21st Century," Brad Efron [5] provided a map of different forms of statistical inference. It consisted of a triangle with his favorite, Fisherian inference, at one corner, and Bayesian and Frequentist at other corners. Conditional inference and the bootstrap were on the side connecting Fisherian and Frequentist; empirical Bayes was on the side connecting Bayesian and Frequentist. When this paper was selected to appear in a volume of Brad's papers a few years ago, I was asked to provide commentary, and here is what I said about this diagram:

> Eager young statistical adventurers heading out into research might take caution from Efron's Statistical Triangle, a barycentric picture of modern research with Fisherian, Bayesian, and Frequentist as the three vertices. It bears some resemblance to the dreaded Bermuda Triangle: relative safety near the edges, and Fiducial probability lurking in the center.

The implication was clear: those who bite of this forbidden fruit risk the fate of being lost at sea without a trace.

Fiducial inference has received vicious criticism from many quarters over the years; how has it managed to survive so well? It may help to take a look at its history, which Phil Dawid and Mervyn Stone [4] called "the chequered history of fiducial inference." Some of this history predates Ronald A. Fisher. The earliest versions were of course not so well articulated as those of Fisher and later, but with the clarity of hindsight, the basic idea can be seen there, without in the least undermining the originality of later work. With some generosity, we might say Bayes had a taste of the fiducial apple (that's Thomas, not Nora;

Nora served the apple). More clearly, Laplace was more a fiducialist than a Bayesian, and just so with several others, such as Gauss and Poisson, all the way to Edgeworth. Closer to the present, of course, we come in 1930 to Fisher who was the modern originator and the first to insist it was a separate form of inference. Interestingly, his first inspiration seems to have come from thinking about estimating the correlation coefficient rather than the means of normal distributions [1].

Following Fisher, we have various contributions by Neyman offering different views, and by several others: the more familiar names being E. B. Wilson, Maurice Fréchet, Corrado Gini, Ed Deming, Sam Wilks, and Serge Bernstein. Not all of these were favorable to the idea, but all were seriously engaged. The modern era may be dated from 1951, the year when John Tukey made his first presentation on the subject to a September IMS Meeting in Minneapolis. In the mid-50s, Tukey engaged Fisher in correspondence about fiducial inference [2, pp. 220ff] Instead of leading to a meeting of minds, it led to a parting of the ways. In 1995, Tukey recalled his final visit to see Fisher in 1956:

> Well, I was talking with Sir Ronald in his [home] office. And I think roughly what I said was that I didn't see the logical strength of the fiducial argument but I gave a lot of weight to the fact that he thought it was a good idea. At which point he did his best to show me the door. Since Elizabeth [Tukey's wife] was out in the garden in the other direction, talking to one of his daughters, I didn't get shown the door. So he grabbed his hat and his cane and went toodling out the door himself. [3, page 40]

Tukey took up fiducial inference seriously for the last time in his four Wald Lectures in 1958. In the 1960s, there were seminal works by Don Fraser, including eventually two books, and important articles by Art Dempster and by George Barnard; all adding remarkable clarity and precision to a subject not previously known for these virtues. And so it continued: many works looked carefully at the subject and passed judgment on it, but any final verdict, such as through widespread acceptance, was to prove elusive. I, of course, stop short of the recent BFF period in this review.

This history is superficial, and to really understand the matter, we must look more closely. It turns out that major interest in the subject has been episodic, and there is a telling pattern to be seen by close inspection. Of course Fisher's introduction was itself one such point. But in 1940, there was a flurry of activity when Fréchet wrote to dozens of statisticians, including Fisher, seeking a clear understanding. The responses came from far and wide and included E. S. Pearson and no less than eight letters from Fisher. A decade later, Tukey's 1951 talk led to further interest, including by Jimmie Savage. And then in the early 60s, there was the deep and lasting work of Fraser, Dempster, and Barnard, punctuated by a session at the 1963 ISI Session at Ottawa. If we take all of these and plot the results, we find a remarkable pattern (Figure 1.5).

Of course, the series in the background is the famous slightly smoothed series of Sunspot numbers. It would be hard to avoid concluding a causal relationship. The direction of the causality requires more study. Since Stanley Jevons in the 1870s linked sunspots to both business cycles and to monsoons in India, the problem is complicated. I trust that modern causal inference is equal to the challenge.

This is truly a risky business, and the brave modern warriors for better inference show exemplary courage. The task is fraught with danger, but the reward, if successful, would be great.

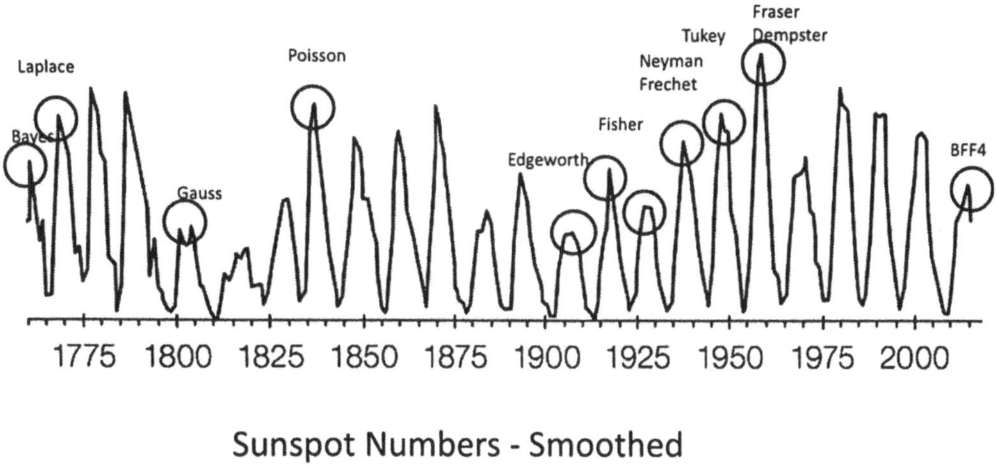

FIGURE 1.5
Fiducial activity and sunspot occurrence. Author's chart

Bibliography

[1] Aldrich, J., 2000. Fisher's "Inverse Probability" of 1930. *International Statistical Review, 68*, pp. 155–172.

[2] Bennett, J. H., 1990. *Statistical Inference and Analysis: Selected Correspondence of R. A. Fisher*. Oxford: Clarendon Press.

[3] Brillinger, D. R., Fernholz, L. T. and Morgenthaler, S., 1997. *The Practice of Data Analysis: Essays in Honor of John W. Tukey*. Princeton: Princeton University Press.

[4] Dawid, A. P. and Stone, M., 1982. The functional-model basis of fiducial inference. *Annals of Statistics, 10*, pp. 1054–1067.

[5] Efron, B., 1998. R. A. Fisher in the 21st century. *Statistical Science, 13*, pp. 95–114.

2

Empirical Bayes: Concepts and Methods

Bradley Efron
Stanford University

2.1 Introduction

Robbins (1956) coined the name "empirical Bayes" for his method of dealing with the following kind of inferential situation: an unknown probability density $g(\theta)$ ("density" here including the possibility of discrete atoms) has produced a random sample of realizations $\theta_1, \theta_2, \ldots, \theta_N$; the θ_i are unobserved, but each one has yielded an observed random variable x_i according to a known family of densities $f(x_i \mid \theta_i)$:

$$\theta_i \sim g(\theta) \quad \text{and} \quad x_i \sim f(x_i \mid \theta_i) \tag{2.1}$$

independently for $i = 1, 2, \ldots, N$. From the observed sample $\boldsymbol{x} = (x_1, x_2, \ldots, x_n)$ we wish to estimate the parameters $\theta_1, \theta_2, \ldots, \theta_N$.

Robbins' monumental insight was that, in addition to x_i, the other observations x_j, $j \neq i$, could assist in the estimation of θ_i. Table 2.1 concerns one of his examples. An automobile insurance company has tabulated the number of claims made by each of its 9461 customers during the past year, with x_i the number of claims made by customer i. The table shows that 7840 of the customers made $x = 0$ claims, 1317 made $x = 1$ claim, etc., up to the one customer, perhaps not the most dependable driver, who made seven claims. Now the insurance company wants to set premiums for the succeeding insurance year. Of course it would be helpful to know how many claims each customer could be expected to make.

TABLE 2.1
Counts y_x of the Number of Insurance Claims Made in a Single Year by 9461 Policy Holders; Robbins' Formula (2.3) Estimates the Number of Claims in the Following Year. For Instance, a Customer Who Had Made $x = 2$ Claims was Expected to Make 0.527 Claims the Next Year. *Gamma MLE* is from Fisher's Gamma Model, Section 2.2

Claims x	0	1	2	3	4	5	6	7
Counts y_x	7840	1317	239	42	14	4	4	1
Formula (2.3)	0.168	0.363	0.527	1.33	1.43	6.00	1.25	–
Gamma MLE	0.165	0.376	0.669	0.94	1.16	1.32	1.45	1.55

 DOI: 10.1201/9780429341731-2

Let y_x be the number of customers making x claims in the past year, $y_0 = 7840$, $y_1 = 1317$, etc. Robbins supposed that $f(x_i \mid \theta_i)$ in equation (2.1) was Poisson,

$$f(x_i \mid \theta_i) = e^{-\theta_i}\theta_i^{x_i}/x_i!, \tag{2.2}$$

$x_i = 0, 1, 2, \ldots$, θ_i being the Poisson expectation for customer i. (Bad drivers have high θ_i values.) Without any assumptions about the prior density $g(\theta)$ in equation (2.1), he showed that the Bayes posterior expected value for θ_i given x_i, $E\{\theta_i \mid x_i\}$, could be estimated by

$$\widehat{E}\{\theta_i \mid x_i\} = (x_i + 1)y_{x_i+1}/y_{x_i}. \tag{2.3}$$

So, for instance, a driver who had had no accidents could be expected to have $\widehat{E}\{\theta_i \mid 0\} = y_1/y_0 = 0.168$ accidents in the coming year. Robbins' derivation of formula (2.3) appears at the end of this section.

Robbins' compact little formula has a certain magical quality. It raises some intriguing ideas that we'll be dealing with in what follows:

- The insurance company's data set, the line of counts in Table 2.1, by itself provides the information for estimating $E\{\theta_i \mid x_i\}$, a Bayesian quantity, without appeal to prior knowledge. In other words, the estimate $\widehat{E}\{\theta_i \mid x_i\}$ is purely frequentistic (doesn't involve prior assumptions), justifying the name "empirical Bayes". Frequentist or not, somehow we've produced a Bayesian result without any Bayesians. The idea that a large data set could imply its own prior distribution was revolutionary.

- The estimate $\widehat{E}\{\theta_i \mid x_i\}$ for driver i depends on his or her past experience x_i, but also on the experience of the other drivers. A driver with a clean record, $x_i = 0$, has estimate $\widehat{E}\{\theta_i \mid x_i = 0\} = y_1/y_0$, partly depending on less-clean drivers who had had one accident each. There is a tacit assumption being made on the relevance of others' experiences to any one individual's insurance premium; see Section 2.6.

- Robbins' formula is nonparametric in the sense that it doesn't depend on any parametric form for $g(\theta)$. In fact, $g(\theta)$ is never estimated in the proof of (2.3). Robbins' derivation, at the end of this section, finesses the role of $g(\theta)$.

- Formula (2.3) goes astray at the right side of Table 2.1, where the counts y_x get small. The more stable "Gamma MLE" estimates begin by parametrically modeling $g(\theta)$ as a gamma distribution, as explained in Section 2.2 and discussed in Section 2.5.

- The insurance company doesn't care if formula (2.3) is individually accurate–only if it is close enough overall to give a good picture of forthcoming claims. Any one driver, though, might hope for individual applicability. This distinction becomes more salient if, say, we discuss individual versus overall prognosis of a serious medical condition.

This chapter is not intended to be a systematic review of the empirical Bayes literature. The basic idea of empirically estimating part of a Bayesian procedure can take on forms other than (2.1). A great amount of effort, starting with Robbins' original papers, has gone into verifying the asymptotic accuracy of formulas like (2.3), but none of that is part of what follows. Here I will focus on some notable empirical Bayes success stories and the conceptual and methodological issues they raise.

Derivation of Robbins' formula If $\theta \sim g(\theta)$, where $g(\theta)$ is the prior density, and $x \mid \theta \sim f(x \mid \theta)$, then according to Bayes rule, the posterior density of θ given x is

$$g(\theta \mid x) = g(\theta)f(x \mid \theta)/f(x), \tag{2.4}$$

where $f(x)$ is the marginal density of x,

$$f(x) = \int_\Theta f(x \mid \theta) g(\theta) \, d\theta, \tag{2.5}$$

with Θ the set of possible θ values.

For the Poisson density $f(x \mid \theta) = \theta^x e^{-\theta}/x!$ we get

$$\begin{aligned} f(x) &= \int_0^\infty (e^{-\theta}\theta^x/x!) g(\theta) \, d\theta \\ \text{and} \qquad g(\theta \mid x) &= g(\theta) \frac{e^{-\theta}\theta^x/x!}{f(x)}. \end{aligned} \tag{2.6}$$

This yields the posterior expectation

$$\begin{aligned} E\{\theta \mid x\} = \int_0^\infty \theta g(\theta \mid x) \, d\theta &= \frac{\int_0^\infty \left(e^{-\theta}\theta^{x+1}/x!\right) g(\theta) \, d\theta}{f(x)} \\ &= (x+1) \frac{\int_0^\infty \left(e^{-\theta}\theta^{x+1}/(x+1)!\right) g(\theta) \, d\theta}{f(x)} \\ &= (x+1) f(x+1)/f(x). \end{aligned} \tag{2.7}$$

In the empirical Bayes situation (2.1), we don't know $f(x)$ or $f(x+1)$, but they are easy to estimate. The expected value of y_x/N is $f(x)$. Substituting y_x/N for $f(x)$ and y_{x+1}/N for $f(x+1)$ in the bottom line of (2.7) gives Robbins' formula (2.3). We will see this same tactic—expressing a posterior quantity of interest directly in terms of the marginal density $f(x)$—show up again in Section 2.2 through Section 2.4, as a way of avoiding estimating the prior $g(\theta)$. Section 2.5 discusses methods that *do* estimate $g(\theta)$.

2.2 The Missing Species Problem

Robbins named empirical Bayes and made the concept explicit, but an earlier paper can be recognized as a clear empirical Bayes triumph: Fisher, Corbet, and Williams (1943). Corbet, a leading naturalist, had been trapping butterflies in Malaysia (then Malaya) for two years. Table 2.2 shows his trapping data, in the same format as Table 2.1. Let

$$y_x = \text{number of species trapped } x \text{ times} \tag{2.8}$$

during the two years: $y_1 = 118$ species were so rare they had been trapped just one time each, $y_2 = 74$ trapped twice each, up to the three species trapped 24 times each. (We will take Table 2.2 as comprising all the data though in fact Corbet also had counts for the more common, less interesting species.)

There is a crucial difference between Table 2.2 and Table 2.1: in the latter case, there is no $x = 0$ entry. These are the "missing species". Corbet asked Fisher a seemingly unanswerable missing species question: if he continued trapping for one additional year, how many new species could he expect to see? Fisher answered Corbet using what we would now call a parametric empirical Bayes model. We will return to Fisher's solution, but first present a nonparametric method due to Good and Toulmin (1956).

TABLE 2.2
Butterfly Data; Number y_x of Species Seen x Times Each in Two Years of Trapping; 118 Species Trapped Just Once, 74 Trapped Twice Each, etc.

x	1	2	3	4	5	6	7	8	9	10	11	12
y_x	118	74	44	24	29	22	20	19	20	15	12	14
x	13	14	15	16	17	18	19	20	21	22	23	24
y_x	6	12	6	9	9	6	10	10	11	5	3	3

Let i index the species, say

$$i = 1, 2, \dots, S, \tag{2.9}$$

S being the total number of species both observed and missing. The basic assumption of both the parametric and nonparametric analyses is that species i is being trapped according to a Poisson process with intensity parameter θ_i.

Let x_i be the number of times species i is trapped in the original one unit of time (the unit being two years for Corbet), and $x_i(t)$ the number of times trapped in a hypothetical future period of t units. The Poisson process assumption says that

$$\Pr\{x_i = x\} = e^{-\theta_i}\theta_i^x/x! \tag{2.10}$$

and

$$\Pr\left\{x_i = 0 \text{ and } x_i(t) > 0\right\} = e^{-\theta_i}\left[1 - e^{-\theta_i t}\right]. \tag{2.11}$$

This last is the probability that species i is *not* seen in the initial trapping period but *is* seen in the future period, i.e., it is one of Corbet's hoped-for "new" species.

The prior density $g(\theta)$ can be thought of as an empirical density putting probability $1/S$ in each of the S parameters θ_i. We can write $E\{t\}$, the expected number of new species seen in an additional trapping period of t units, as

$$E\{t\} = S\int_0^\infty e^{-\theta}[1 - e^{-\theta t}]g(\theta)\, d\theta \tag{2.12}$$

(ignoring the discreteness in $g(\theta)$). Expanding $[1 - e^{-\theta t}]$ in a Taylor series gives

$$E\{t\} = S\int_0^\infty e^{-\theta}\left[\theta t - \frac{(\theta t)^2}{2} + \frac{(\theta t)^3}{3!} - \cdots\right] g(\theta)\, d\theta. \tag{2.13}$$

Summing (2.10) over the S species gives an expression for e_x, the expected value of y_x,

$$e_x = E\{y_x\} = S\int_0^\infty \left[e^{-\theta}\theta^x/x!\right] g(\theta)\, d\theta. \tag{2.14}$$

Comparing (2.13) and (2.14), we can write

$$E\{t\} = e_1 t - e_2 t^2 + e_3 t^3 - \cdots; \tag{2.15}$$

e_x is unbiasedly estimated by the count y_x, leading to Good and Toulmin's intriguing estimator[1]

$$\widehat{E}\{t\} = y_1 t - y_2 t^2 + y_3 t^3 - \cdots. \tag{2.16}$$

Corbet asked Fisher about one more year of trapping; that is 1/2 of a 2-year time unit, giving an estimated number of newly trapped butterfly species $\widehat{E}\{0.5\}$ equaling

$$\widehat{E}\{0.5\} = 118 \cdot 0.5 - 74 \cdot 0.5^2 + 44 \cdot 0.5^3 - \cdots = 45.2. \tag{2.17}$$

For two more trapping years, $t = 1$, formula (2.16) gives $\widehat{E}\{1\} = 75$. The heavy curve in Figure 2.1 graphs $\widehat{E}\{t\}$ for t between 0 and 1. Formula (2.16) diverges for t greater than 1, a defect considered next.

Fisher's answer to Corbet's missing species question began with the assumption that the prior density $g(\theta)$ for the Poisson process parameters θ_i had a scaled gamma form,

$$g_{\alpha,\beta}(\theta) = \frac{\theta^{\alpha-1}e^{-\theta/\beta}}{\beta^\alpha \Gamma(\alpha)} \qquad (\theta > 0). \tag{2.18}$$

That is, $\theta \sim \beta G_\alpha$ for G_α a standard gamma variate with shape parameter α. In this case, changing parameters from (α, β) to (α, γ) with $\gamma = \beta/(1+\beta)$, the marginal density $f(x)$ (2.5) is

$$f_{\alpha,\gamma}(x) = c_{\alpha,\gamma} \frac{\Gamma(x+\alpha)}{x!\Gamma(\alpha)} \gamma^x (1-\gamma)^\alpha \qquad [\gamma = \beta/(1+\beta)], \tag{2.19}$$

proportional to a negative binomial density, namely the probability of getting x tails before α heads, flipping a coin with probability γ of tails. The proportionality constant $c_{\alpha,\gamma}$ is necessary in the missing species problem because the data is truncated; for Corbet's data, $c_{\alpha,\gamma}$ equals one over the sum for x from 1 to 24 of the untruncated negative binomial density.

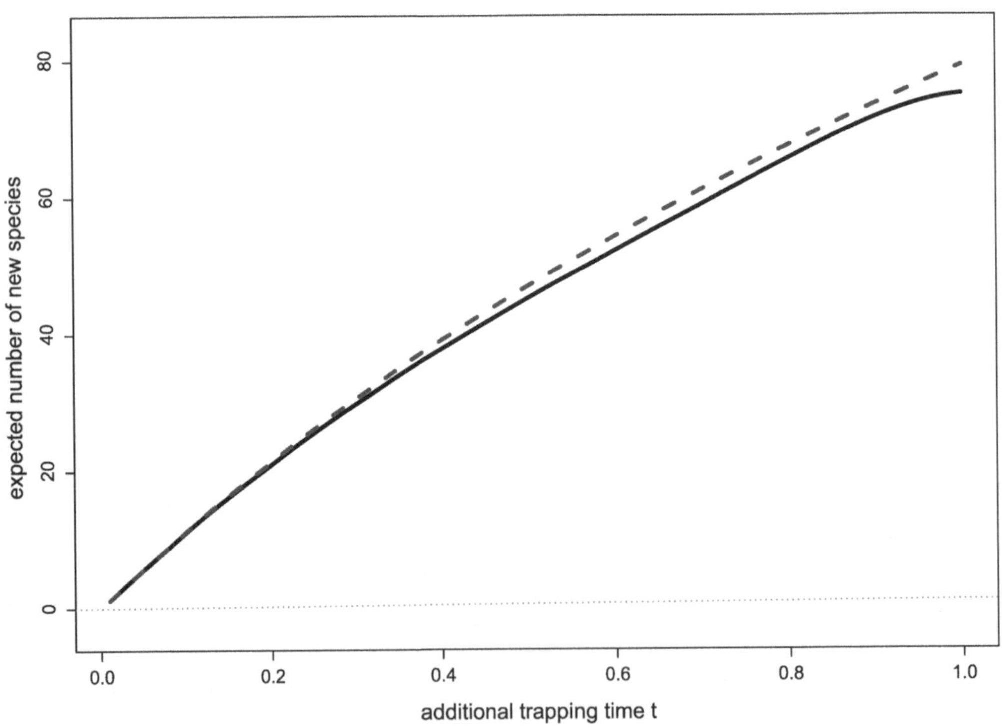

FIGURE 2.1
Expected number of new butterfly species seen in t units additional trapping time; nonparametric (solid), parametric (dashed).

Thinking of the counts y_x in Table 2.2 as a multinomial sample, maximization of the log likelihood

$$\sum_{x=1}^{24} y_x \log\left(f_{\alpha,\gamma}(x)\right) \tag{2.20}$$

gave maximum likelihood estimates

$$\hat{\alpha} = 0.104 \quad \text{and} \quad \hat{\gamma} = 0.989, \qquad \text{so } \hat{\beta} = \frac{\hat{\gamma}}{(1-\hat{\gamma})} = 89.79. \tag{2.21}$$

Fisher then used the prior density $g_{\hat{\alpha},\hat{\beta}}(\theta)$ to estimate $E\{t\}$ by substitution in (2.12).

There is no particular reason to think that the gamma prior (2.18) applies to Corbet's data, but in this case, it nicely agrees with Good and Toulmin's nonparametric solution, as seen in Figure 2.1. For $t = 1/2$, one year of additional trapping, it gives $\widehat{E}\{1/2\} = 46.95$ compared with 45.2 before.

Fisher's gamma model provides a neat formula for $\widehat{E}\{t\}$,

$$\widehat{E}\{t\} = \frac{y_1\left[1 - \frac{1}{(1+\hat{\gamma}t)^{\alpha}}\right]}{\hat{\gamma}\hat{\alpha}}. \tag{2.22}$$

This gives reasonable-looking predictions for t greater than 1, for instance

$$\widehat{E}\{2\} = 123, \qquad \widehat{E}\{4\} = 176. \tag{2.23}$$

"Reasonable" does *not* necessarily mean convincing. The MLE value $\hat{\alpha} = 0.104$ endows $g_{\hat{\alpha},\hat{\beta}}(\theta)$ with an enormous spike near zero, representing very rare species. As t increases, more and more of the estimate $\widehat{E}\{t\}$ depends on the exact shape of the spike. Letting t go to infinity in (2.22) gives

$$\widehat{E}\{\infty\} = 11422, \tag{2.24}$$

an estimate of the total number of species S: not something one wants to bet on. The problem of very rare species becomes more acute in the next example, where the species are words in Shakespeare's vocabulary.

Table 2.3 provides a statistical view of the Shakespearean canon.[2] Altogether the canon comprises 884,647 words, of which there are 31,534 *distinct* words. (Henceforth, "words"

TABLE 2.3
Shakespeare's Word Counts; 14,376 Distinct Words Appeared Once Each in the Canon, 4343 Distinct Words Twice Each, etc.; the Canon Contains 31,534 Distinct Words, with 884,647 Words in Total, Counting Repeats

	1	2	3	4	5	6	7	8	9	10
0+	14376	4343	2292	1463	1043	837	638	519	430	364
10+	305	259	242	223	187	181	179	130	127	128
20+	104	105	99	112	93	74	83	76	72	63
30+	73	47	56	59	53	45	34	49	45	52
40+	49	41	30	35	37	21	41	30	28	19
50+	25	19	28	27	31	19	19	22	23	14
60+	30	19	21	18	15	10	15	14	11	16
70+	13	12	10	16	18	11	8	15	12	7
80+	13	12	11	8	10	11	7	12	9	8
90+	4	7	6	7	10	10	15	7	7	5

will mean "distinct words".) The upper left entry shows that 14,376 words were so rare; they appeared just once each in the canon; 4343 appeared twice each, 2292 three times each, up to 5 words that appeared 100 times each; 846 words appeared more than 100 times each, but these play no role in what follows.

We can ask Corbet's question for Shakespeare: suppose one found some previously unknown Shakespeare amounting to proportion t of the canon, that is, $884,647 \cdot t$ total words, counting repeats. (This would be the equivalent of a new trapping period of length t.) How many distinct new words, not among the original 31,534, would we expect to find? Letting t go to infinity, the question becomes, "How many words did Shakespeare know but not use?"

For $t = 1$, i.e., a find of new Shakespeare equal in length to the canon, Good and Toulmin's nonparametric estimator (2.16) gives

$$\widehat{E}\{1\} = 11430. \tag{2.25}$$

Fisher's Gamma model is again in good agreement, giving

$$\widehat{E}\{1\} = 11483.^3 \tag{2.26}$$

The maximum likelihood estimates for the gamma prior $g_{\alpha,\beta}(\theta)$ were

$$\hat{\alpha} = -0.3954 \quad \text{and} \quad \hat{\beta} = 104.26 \tag{2.27}$$

($\hat{\gamma} = 0.9905$). Now the prior $g_{\hat{\alpha},\hat{\beta}}(\theta)$ is improper, the spike near zero being infinite–this is allowable since we don't have to deal with the missing species count y_0. As $t \to \infty$, Fisher's estimate $\widehat{E}\{t\}$ goes to infinity as more and more hypothetical cases are pulled out of the infinite spike.

One sees that there can't be a way to pin down $E\{\infty\}$, the number of words Shakespeare knew but didn't use. However, Efron and Thisted (1976) used linear programming methods to put a nonparametric lower bound on $E\{\infty\}$,

$$E\{\infty\} > 35,554. \tag{2.28}$$

Adding in the 31,534 distinct words in the canon, this implies a Shakespearean vocabulary exceeding 67,000 words.[4]

In November of 1985, noted Shakespearean scholar Gary Taylor found a short love poem of 429 words, "Shall I fly?", which he attributed to Shakespeare. This proved immediately controversial, no new Shakespearean work having been discovered since the 1600s (and the poem not being very good).

Thisted and Efron (1987) analyzed the poem's rare word counts using Good and Toulmin's approach. Here the new trapping period t is quite small, $t = 429/884,647$, a little less than 0.0005. Formula (2.16) predicts $\widehat{E}\{t\} = 6.97$ while nine distinct words in the poem had actually never appeared in the canon. Good and Toulmin's method also provides predictions for the number of words in the poem that appeared once each in the canon, twice each, etc., as shown in Table 2.4. The predictions are consistently lower than the observations, but other than that they seem reasonably agreeable.

Thisted and Efron concluded that Shakespearean authorship could not be ruled out on the basis of rare word counts. "Shall I fly?" has never been accepted into the canon, but Taylor and others have gone on to use statistical methods in a variety of the Bard's attribution questions such as a possible collaboration with Marlowe on Parts I, II, and III of *Henry VI*. The authoritative *New Oxford Shakespeare* does indeed list Marlowe as a collaborator, to the fury of scholars less swayed by statistics.

TABLE 2.4
Counts y_x and Their Estimate $\widehat{E}_x$ (2.16) for the Poem "Shall I fly?"; 9 Distinct Words in the Poem Never Appeared in the Canon ($x = 0$), with (2.16) Predicting 6.97

x	0	1	2	3	4	5	6	7	8	9
y_x	9	7	5	4	4	2	4	0	2	3
$\widehat{E}_x$	6.97	4.21	3.33	2.84	2.53	2.43	2.16	2.01	1.87	1.76

TABLE 2.5
A Sequence Analysis Combining Data from 10 Mice Gave a Library of 237,863 Distinct Clones, of Which 142,500 Occurred Only Once Each, 28,524 Occurred Twice Each, etc. If Another 10 Mice Were Sequenced, Good and Toulmin's Nonparametric Method Estimates Seeing about 123,000 Previously Unseen Distinct Clones

x	1	2	3	4	5	6	7	8	9	10
y_x	142,500	28,524	13,296	7699	5161	3709	2785	2217	1822	1510
x	11	12	13	14	15	16	17	18	19	20
y_x	1235	1151	1067	942	874	785	724	655	659	542

The missing species problem shows up in biogenetical investigations. Table 2.5 concerns a sequence analysis of mouse DNA. Ten mice had their genomes sequenced and split into "clones", i.e., short fragments. Some of the clones were common but 142,500, others were so rare they only showed up once in the combined pool of ten mice; 28,524 appeared twice each, 13,296 three times each, etc. Application of Good and Toulmin's formula (2.16) showed that the observed library of 287,863 clones was by no means complete: a new set of ten mice could be expected to yield $\widehat{E}\{1\} = 122,900$ previously unseen clones.

The differences between Good and Toulmin's and Fisher's approaches go further than nonparametric versus parametric. Fisher's method actually estimates the prior $g(\theta)$, whereas Good and Toulmin's method doesn't. With $\hat{g}(\theta)$ in hand, the statistician can assess Bayesian quantities other than $E\{t\}$, for instance $\Pr\{\theta \leq 1 \mid x\}$. More on the two empirical Bayes approaches, *f-modeling* and *g-modeling*, will be seen in Section 2.6.

2.3 The James–Stein Estimator

Early developments of empirical Bayes methodology featured Poisson distributions for observations x_i in model (2.1). The James–Stein estimator (James and Stein, 1961) focused instead on normally distributed observations,

$$x_i \overset{\text{ind}}{\sim} \mathcal{N}(\theta_i, \sigma_0^2) \qquad (i = 1, 2, \ldots, N), \tag{2.29}$$

though the empirical Bayes connection took a while to develop.

Faced with situation (2.29), and without any prior knowledge of how differently the unknown parameters θ_i might arrange themselves, the obvious estimators are

$$\hat{\theta}_i^{\text{MLE}} = x_i \qquad \text{for } i = 1, 2, \ldots, N, \tag{2.30}$$

the "MLE" superscript noting that these are maximum likelihood estimates. Their expected total squared error is

$$E\left\{\sum_{i=1}^{N}\left(\hat{\theta}_i^{\mathrm{MLE}}-\theta_i\right)^2\right\}=N\sigma_0^2. \tag{2.31}$$

(We will take σ_0^2 as known.)

James and Stein's 1961 paper, building on Stein's earlier work, proved the following result:

Theorem 2.1. *Let*

$$\hat{\theta}_i^{\mathrm{JS}}=\overline{x}+\left(1-\frac{(N-3)\sigma_0^2}{S}\right)(x_i-\overline{x}) \qquad \left(S=\sum_{i=1}^{N}(x_i-\overline{x})^2\right). \tag{2.32}$$

Then, assuming model (2.29) *with* $N \geq 4$,

$$E\left\{\sum_{i=1}^{N}\left(\hat{\theta}_i^{\mathrm{JS}}-\theta_i\right)^2\right\}<N\sigma_0^2 \tag{2.33}$$

for all possible choices of the θ_i*'s.*

The James–Stein theorem has a good claim to being the most striking and disruptive statistics result of the post-war era: striking in its simplicity and disruptive of conventional wisdom. A century of experience with normal theory OLS regression models, all of which employ estimates like (2.30), seemed suddenly to be on shaky ground. One could do better, at least if "better" is defined by expected square error loss. (That's a crucial caveat, as we'll see in Section 2.6.)

If the improvement offered by the James–Stein estimates was infinitesimal, only theorists would be interested. In favorable situations, the gains can, in fact, be enormous. Table 2.6 (Efron and Morris, 1972) concerns one such situation. The batting averages of 18 major

TABLE 2.6
Batting Averages of 18 Baseball Players. *Column 1*, Observed Batting Average after 45 Attempts. *Column 2*, James–Stein Estimate (2.32), $\sigma_0^2 = 0.066$. *Column 3*, Average for Remainder of the 1970 Season. *Column 4*, Number of Attempts in Remainder of the Season

	MLE	JS	final avg	# final AB
Clemente	.400	.294	.346	367
F. Robinson	.378	.289	.298	426
F. Howard	.356	.284	.276	521
Johnstone	.333	.279	.222	275
Berry	.311	.275	.273	418
Spencer	.311	.275	.270	466
Kessinger	.289	.270	.263	586
L. Alvarado	.267	.265	.210	138
Santo	.244	.261	.269	510
Swoboda	.244	.261	.230	200
Unser	.222	.256	.264	277
Williams	.222	.256	.256	270
Scott	.222	.256	.303	435
Petrocelli	.222	.256	.264	538
E. Rodriguez	.222	.256	.226	186
Campaneris	.200	.251	.285	558
Munson	.178	.246	.316	408
Alvis	.156	.242	.200	70

league baseball players were recorded over the 1970 season.[5] Column 1 of the table shows their batting averages after 45 attempts, about one-tenth of the full season. Player i's observed average x_i can be thought of as the maximum likelihood estimate $\hat{\theta}_i^{\text{MLE}}$ of θ_i, his true batting average. The grand average of the 18 observed averages was

$$\bar{x} = \sum_{i=1}^{18} x_i/18 = 0.265. \tag{2.34}$$

Column 2 gives the James–Stein estimates $\hat{\theta}_i^{\text{JS}}$ (2.32), taking

$$\sigma_0 = \left[\frac{\bar{x}\,(1-\bar{x})}{45}\right]^{1/2} = 0.066, \tag{2.35}$$

the binomial standard deviation based on 45 coin flips with probability of success $\bar{x}$. We don't know the true averages, but as a surrogate, Column 3 lists the players' final averages over the remaining nine-tenths of the 1970 regular season, θ_i^{final}.

The prediction errors for $\hat{\theta}^{\text{MLE}}$ and $\hat{\theta}^{\text{JS}}$ were

$$\sum_{i=1}^{18} \left(\hat{\theta}_i^{\text{MLE}} - \theta_i^{\text{final}}\right)^2 = 0.0754 \quad \text{and} \quad \sum_{i=1}^{18} \left(\hat{\theta}_i^{\text{JS}} - \theta_i^{\text{final}}\right)^2 = 0.0213. \tag{2.36}$$

In this case using the James–Stein rule reduces prediction error by a factor of more than 3. Nobody cares much about 50-year-old baseball statistics, but if we were predicting cancer cure rates at 18 hospitals, three times less error might be life-saving.

James and Stein's proof of Theorem 2.1 is an elegant exercise in frequentist inference. Efron and Morris (1972) motivated formula (2.32) from a less elegant empirical Bayes viewpoint. They envisioned a "normal-normal" version of setup (2.1),

$$\theta_i \sim \mathcal{N}(M, A) \quad \text{and} \quad x_i \sim \mathcal{N}(\theta_i, \sigma_0^2), \tag{2.37}$$

independently for $i = 1, 2, \ldots, N$. That is, the prior density $g(\theta)$ is assumed to be normal with mean M and variance A, both M and A unknown to the statistician.

The Bayes estimate of θ_i under model (2.37) is

$$\hat{\theta}_i^{\text{Bayes}} = M + B(x_i - M) \qquad \left[B = A/(A + \sigma_0^2)\right]. \tag{2.38}$$

The marginal distribution of the x_i under (2.37) is

$$x_i \overset{\text{ind}}{\sim} \mathcal{N}(M, V) \qquad (V = A + \sigma_0^2), \tag{2.39}$$

so that

$$\bar{x} \sim \mathcal{N}(M, V/N) \qquad \text{independent of } S \sim V\chi^2_{N-1}, \tag{2.40}$$

with S as in (2.32). Parameters M and B in (2.38) are estimated unbiasedly by

$$\widehat{M} = \bar{x} \quad \text{and} \quad \widehat{B} = \left[1 - \frac{(N-3)\sigma_0^2}{S}\right], \tag{2.41}$$

the last following from

$$E\left\{\left(\chi^2_{N-1}\right)^{-1}\right\} = (N-3)^{-1}.$$

Using (2.41) in (2.32), we can rewrite the James–Stein estimator as

$$\hat{\theta}_i^{\text{JS}} = \widehat{M} + \widehat{B}\left(x_i - \widehat{M}\right). \tag{2.42}$$

The term "empirical Bayes" for the James–Stein rule seems entirely appropriate, $\hat{\theta}_i^{\text{JS}}$ being $\hat{\theta}_i^{\text{Bayes}}$ (2.38), with the prior hyperparameters B and M frequentistically estimated from the observed sample $x_1, x_2, \ldots, x_n$.

The James–Stein theorem is a wonderfully exact result: the simple rule (2.32) *always* bests the MLE, with no exceptions or further assumptions. Exact results exert a powerful attraction in statistics, a sprawling and usually inexact science. The immediate history of Theorem 2.1 involved searching for other estimators that dominated maximum likelihood, particularly admissible ones (which (2.32) isn't; see Strawderman, 1971).

And yet it is an *inexact* aspect of (2.32) that perhaps has had the greatest long-term effect: that *shrinkage* is a powerful tool for the simultaneous estimation of multiple parameters. The baseball example violates the assumptions of Theorem 2.1—the batting averages are binomial rather than normal—but the savings accrue nonetheless.

Shrinkage is quite extreme for the baseball players, each estimate $\hat{\theta}_i^{\text{JS}}$ being shrunk 78% $(= 1 - \widehat{B})$ of the way from x_i to the grand mean $\bar{x}$. This is illustrated in Figure 2.2, where we can see that the James–Stein rule is an *over-shrinker*, a favorable tactic for total squared error risk it inherits from Bayes rule (2.38). The tactic works because the unbiased estimates x_i are, as a group, overdispersed,

$$\begin{aligned} &\frac{E\left\{\sum_{i=1}^{N}(x_i - \bar{x})^2\right\}}{N-1} = A + \sigma_0^2, \\ \text{compared to} \quad &\frac{E\left\{\sum_{i=1}^{N}(\theta_i - \bar{\theta})^2\right\}}{N-1} = A. \end{aligned} \tag{2.43}$$

Regularization, the big brother of shrinkage, is used in modern estimation algorithms such as the lasso. Empirical Bayes methods tend toward strong regularization.

A good question: what if the normal prior $\theta_i \sim \mathcal{N}(M, A)$ in equation (2.37) isn't a good match to the problem at hand? Is there a version of the James–Stein rule applicable more generally? The answer is yes. Efficient empirical Bayes methods, supported by modern

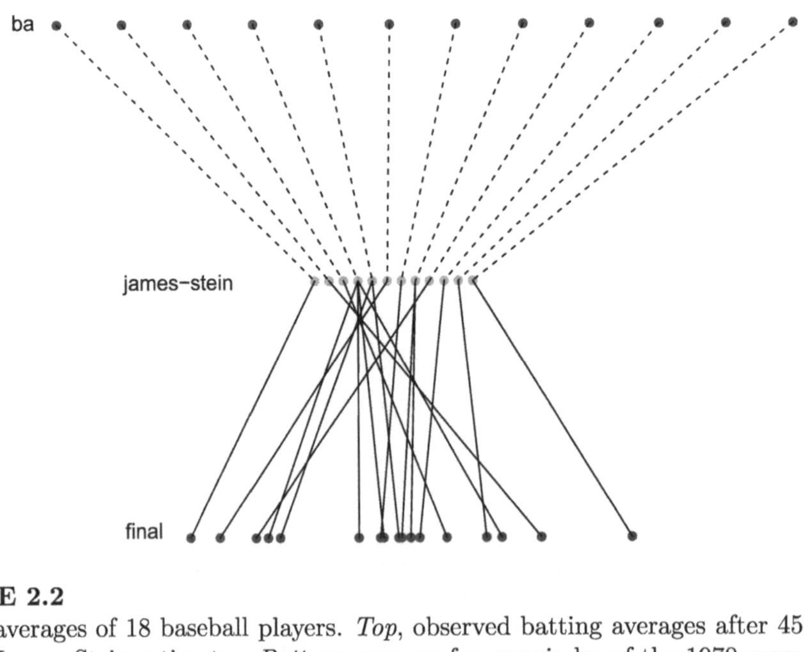

FIGURE 2.2
Batting averages of 18 baseball players. *Top*, observed batting averages after 45 attempts. *Middle*, James–Stein estimates. *Bottom*, average for remainder of the 1970 season.

computational capabilities, do the job (though without the frequentist guarantees of the James–Stein theorem). An example, artificial but informative, follows next.

Figure 2.3 shows the "two towers" example: $N = 1500$ x_i values have been observed,

$$x_i \sim \mathcal{N}(\theta_i, 1) \qquad (i = 1, 2, \ldots, N = 1500), \tag{2.44}$$

where 500 of the θ_i's were uniformly selected in the interval $[-1.7, -0.7]$ and 1000 uniformly selected in $[0.7, 2.7]$. The solid histogram shows the two towers of θ_i values, while the 1500 x_i values from (2.44) appear in the dashed histogram.

The statistician wishes to estimate the θ_i's using some estimation rule $e(x)$,

$$\hat{\theta}_i = e(x_i) \qquad (i = 1, 2, \ldots, N), \tag{2.45}$$

with loss function the expected average squared error (EASE),

$$\text{EASE} = E\left\{\sum_{i=1}^{N}\left(\hat{\theta}_i - \theta_i\right)^2 / N\right\}, \tag{2.46}$$

the expectation being over model (2.44). The MLE rule $e(x_i) = x_i$ has

$$\text{EASE}_{\text{MLE}} = 1. \tag{2.47}$$

Suppose though a friendly Oracle has told us the order statistic of the θ_i's

$$\theta_{(1)} \leq \theta_{(2)} \leq \cdots \leq \theta_{(N)},$$

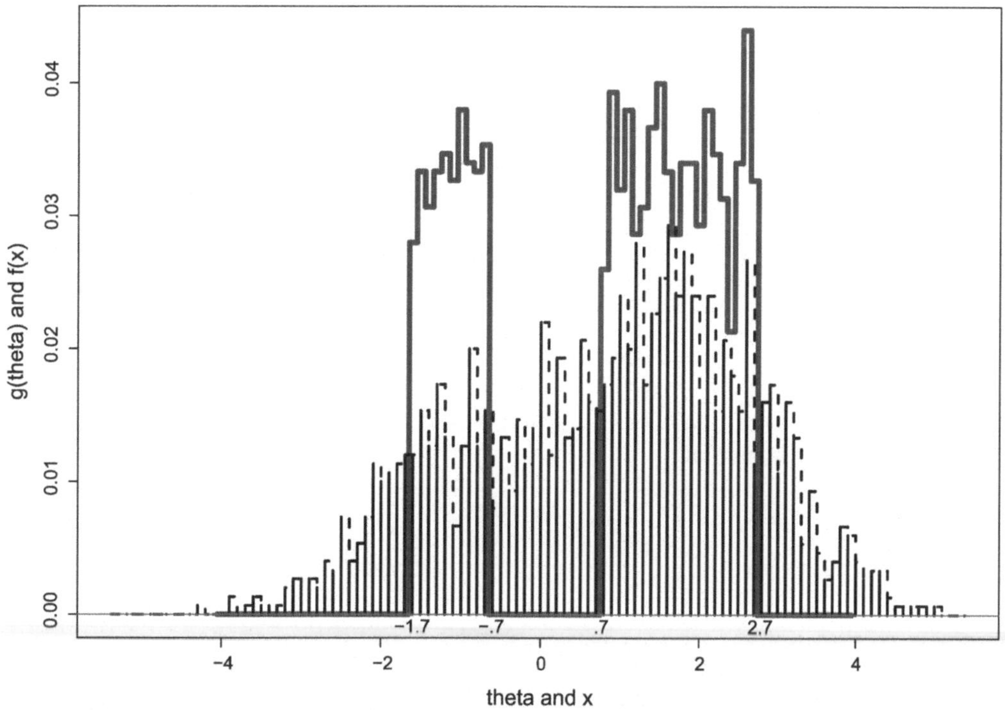

FIGURE 2.3
Two towers example. *Solid histogram,* $N = 1500$ *parameters* θ_i*. Dashed histogram,* observations $x_i \sim \mathcal{N}(\theta_i, 1)$.

without letting on which $\theta_{(j)}$ goes with which θ_i. (Equivalently, the Oracle tells us the histogram of the θ_i's. Oracle calculations, as in Jiang and Zhang, 2009 and Brown and Greenshtein, 2009, go back to Robbins' original papers.) Let $\bar{g}(\theta)$ denote the discrete density putting probability $1/N$ on each $\theta_{(j)}$. The Oracle's information allows us to compute the Bayes posterior expectation of θ_i given x_i assuming prior $\bar{g}(\theta)$,

$$e_{\bar{g}}(x) = \frac{\sum_{j=1}^{N} \theta_{(j)} \varphi(x - \theta_{(j)})}{\sum_{j=1}^{N} \varphi(x - \theta_{(j)})}, \tag{2.48}$$

with $\varphi(x)$ the standard normal density $\exp(-x^2/2)/\sqrt{2\pi}$.

For the two tower prior in Figure 2.3, numerical calculations show that rule (2.48) has

$$\text{EASE}_{\bar{g}} = 0.563, \tag{2.49}$$

more than 40% less than EASE_{MLE}. There aren't any Oracles of course, but this is where modern empirical Bayes comes in, allowing us to do nearly as well as (2.49) using a fully data-based estimator. This amounts to extending the James–Stein rule to non-normal priors $g(\theta)$, though without Theorem 2.1's guaranteed superiority.

The Bayesian model underlying our empirical Bayes estimates is

$$\theta \sim g(\cdot) \quad \text{and} \quad x \mid \theta \sim \mathcal{N}(\theta, 1). \tag{2.50}$$

Given x, the conditional distribution has mean and variance denoted as

$$\theta \mid x \sim (e_g(x), v_g(x)). \tag{2.51}$$

Let $f(x)$ indicate the marginal density of x (2.5),

$$f(x) = \int_{\mathcal{T}} \varphi(x - \theta) g(\theta) \, d\theta, \tag{2.52}$$

where $\mathcal{T}$ is the domain of θ. *Tweedie's formulas* provide convenient expressions for $e_g(x)$ and $v_g(x)$ in terms of $l(x) = \log f(x)$,

$$e_g(x) = x + \dot{l}(x) \quad \text{and} \quad v_g(x) = 1 + \ddot{l}(x), \tag{2.53}$$

$\dot{l}(x) = dl(x)/dx$ and $\ddot{l}(x) = d^2 l(x)/dx^2$; see Efron (2010).

If $\hat{\theta} = e(x)$ is some estimation rule for θ in model (2.50), its overall squared error risk $R(g, e)$ is

$$R(g, e) = E\left\{\left(\hat{\theta} - \theta\right)^2\right\}, \tag{2.54}$$

the theoretical version of EASE (2.46). The Bayes conditional expectation $e_g(x)$ minimizes (2.54), the Bayes risk equaling

$$\begin{aligned} R(g, e_g) = \int_{-\infty}^{\infty} v_g(s) f(x) \, dx &= \int_{-\infty}^{\infty} \left(1 + \ddot{l}(x)\right) f(x) \, dx \\ &= 1 - \int_{-\infty}^{\infty} \dot{l}(x)^2 f(x) \, dx, \end{aligned} \tag{2.55}$$

the last following from the general result $E\{\dot{l}^2\} = -E\{\ddot{l}\}$. Using $\hat{\theta} = e(x)$ not equal to $e_g(x)$ must increase $R(g, e)$, the *regret* being

$$R(g, e) - R(g, e_g) = \int_{-\infty}^{\infty} \left(e(x) - e_g(x)\right)^2 f(x) \, dx. \tag{2.56}$$

Section 2 of Efron (2019) verifies (2.55)–(2.56).

Figure 2.3 suggests an empirical Bayes estimation approach:

1. Estimate the marginal density $f(x)$ by a smooth curve $\hat{f}(x)$ drawn through the bar tops of the histogram for $x_1, x_2, \ldots, x_{1500}$.
2. Estimate e_g by

$$\hat{e}(x) = x + \frac{d}{dx} \log \hat{f}(x), \tag{2.57}$$

as in equation (2.53). Section 2.5 discusses a convenient way to fit $\hat{f}(x)$ based on Poisson regression to the histogram bin counts.

The solid curve in Figure 2.4 shows $e_g(x)$ (2.48), the Bayes posterior expectation based on the Oracle's empirical prior $\bar{g}(\theta)$. $R(\bar{g}, e_{\bar{g}}) = 0.563$, the value of $\text{EASE}_{\bar{g}}$ given previously. (It isn't difficult to show that this is the minimum possible value of EASE for any rule of the form $\hat{\theta} = e(x)$.) The dashed curve is $\hat{e}(x)$, an empirical Bayes estimate (2.57), with $\hat{f}(x)$ obtained from a natural spline fit, five degrees of freedom, to the histogram bin counts of the 1500 x_i's in Figure 2.3.

The empirical Bayes rule $\hat{e}(x)$ does a reasonably good job of approximating $e_{\bar{g}}$, missing badly only in the lower 1% tail of the x_i values. Its empirical Bayes regret (2.56) was

$$\int_{-\infty}^{\infty} \left(\hat{e}(x) - e_{\bar{g}}(x)\right)^2 \bar{f}(x)\, dx = 0.026, \tag{2.58}$$

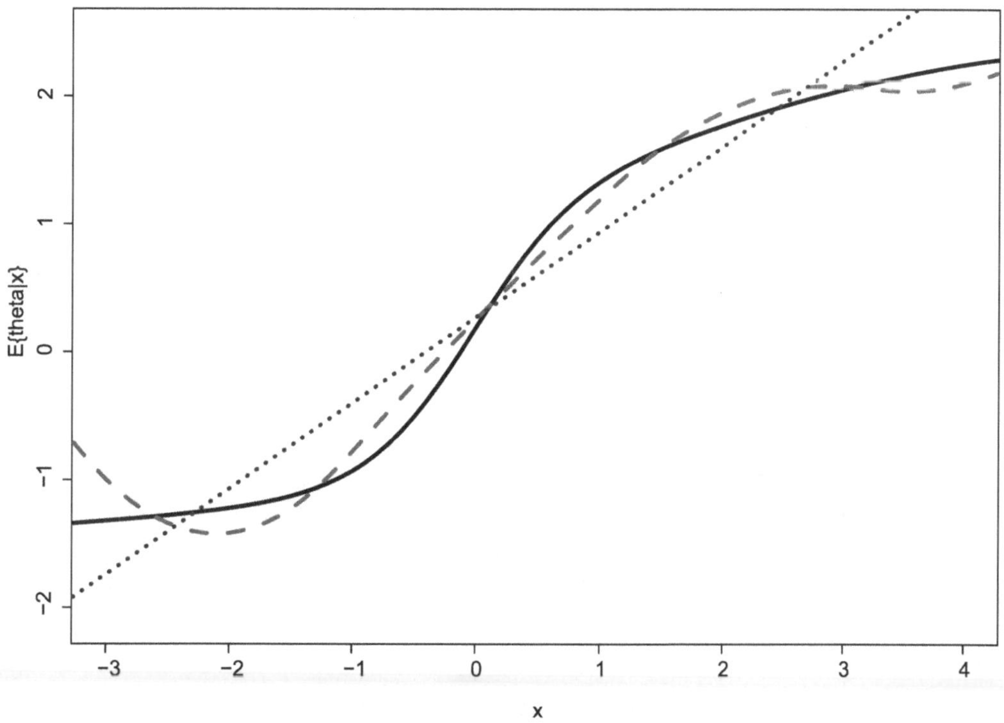

FIGURE 2.4
Posterior expectations $E\{\theta \mid x\}$: Oracle (solid); estimated (dashed) from a natural spline fit with 5 degrees of freedom (i.e., with 6 free parameters counting the intercept); James–Stein (dotted).

using the Oracle marginal density

$$\bar{f}(x) = \sum_{1}^{N} \frac{\varphi(x - \theta_{(i)})}{N}.$$

One can do better, but the estimated risk $0.563 + 0.026 = 0.589$ is still more than 40% less than the MLE risk 1.

The James-Stein rule (2.42) is the straight line graphed in dots for Figure 2.4 and has empirical Bayes regret 0.117. The two towers prior is emphatically non-normal so it isn't surprising to see James–Stein erring here.

2.4 False Discovery Rates

The false discovery rate story is strikingly similar to that of the James–Stein estimator: a frequentist theorem that provides surprising and appealing data analysis possibilities, disruptive of prevailing wisdom, is later reinterpreted in empirical Bayes terms. The disruptive theorem in this case is the FDR control algorithm of Benjamini and Hochberg (1995), summarized next.

Suppose we have N independent testing situations, each of which provides a p-value p_i against its null hypothesis H_{0i}. Let $r(\boldsymbol{p})$ be some rule that inputs the vector $\boldsymbol{p}$ of N p-values, and outputs a decision for or against each H_{0i}, $i = 1, 2, \ldots, N$. The rule is observed to reject H_{0i} for R of the N tests. The *false discovery proportion* is then

$$\text{Fdp} = R_0/R, \tag{2.59}$$

where R_0 is the number of the R rejections where H_{0i} was actually true, i.e., the number of false discoveries. (R_0 and Fdp are not observable statistics.) The *false discovery rate* of rule $r(\cdot)$ is defined to be the expected value of Fdp,

$$\text{Fdr} = E\{\text{Fdp}\}, \tag{2.60}$$

with Fdp set equal to zero if $R = 0$.

The Benjamini–Hochberg algorithm is a rule $r_q(\boldsymbol{p})$ guaranteed to have Fdr $\leq q$ where q is some small pre-chosen constant like 0.1. The algorithm begins by ordering the p-values,

$$p_{(1)} \leq p_{(2)} \leq \cdots \leq p_{(N)}. \tag{2.61}$$

Let i_{max} be the largest index i for which

$$p_{(i)} \leq (i/N)q; \tag{2.62}$$

rule r_q rejects $H_{0(i)}$ for $i \leq i_{\text{max}}$, and accepts $H_{0(i)}$ for $i > i_{\text{max}}$.

Theorem 2.2. *The rule r_q has*

$$\text{Fdr} = \hat{\pi}_0 q \text{ where } \hat{\pi}_0 = \text{N}_0/\text{N}, \tag{2.63}$$

N_0 the number of true null hypotheses.

N_0 is unobservable but in the kind of "fishing expedition" where mass testing is often used, N_0/N is near 1. In any case, (2.63) is usually taken to say that r_q controls Fdr at level q.

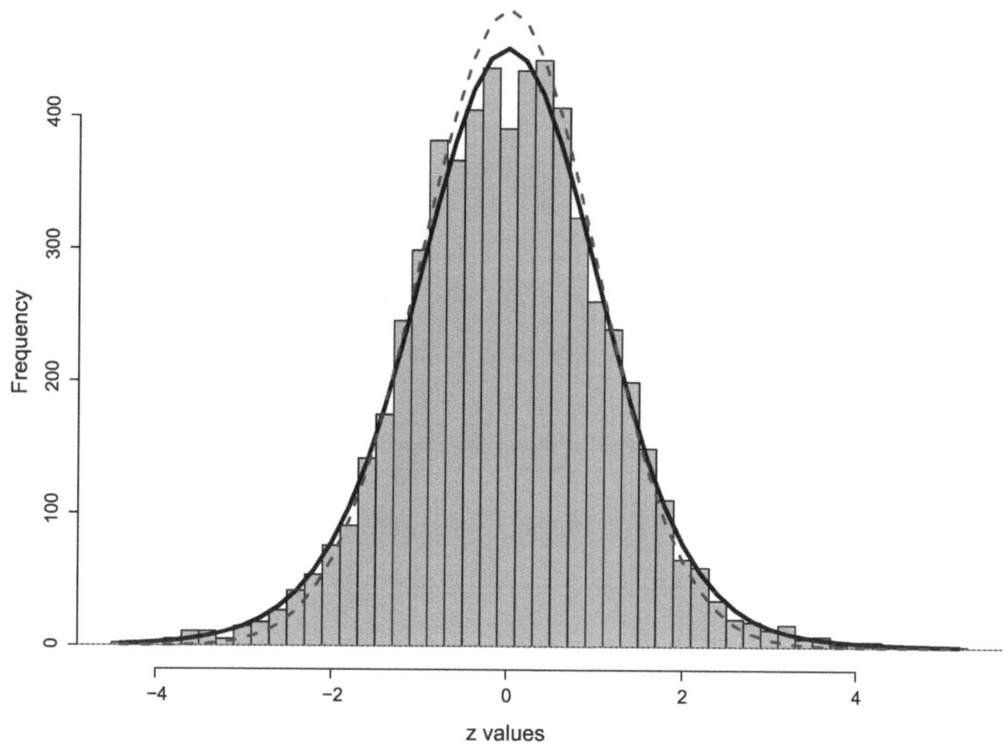

FIGURE 2.5
Frequencies for 6033 z-values in prostate microarray study; heavy curve for natural spline fit with 5 degrees of freedom; light dashed curve counts proportional to $\mathcal{N}(0,1)$ fit.

As an example, the histogram in Figure 2.5 shows z-values from a microarray study involving 50 prostate cancer patients and 52 healthy controls. Genetic activities for 6033 genes were compared between cancer and control subjects, giving a z-value z_i for each of the $N = 6033$ genes, the null hypothesis of no activity difference between cancer patients and controls being

$$H_{0i} : z_i \sim \mathcal{N}(0,1), \qquad i = 1, 2, \ldots, N. \tag{2.64}$$

If all the H_{0i} were true, the histogram would follow a $\mathcal{N}(0,1)$ density, the dashed curve, which is seen to be too high in the center and too low in the tails. (Section 2.1 of Efron, 2010 gives a fuller description of the prostate cancer data.)

Figure 2.6 illustrates the Benjamini–Hochberg rule r_q, $q = 0.1$, applied to the right side of the histogram; that is, using right-sided p-values

$$p_i = 1 - \Phi(z_i), \tag{2.65}$$

Φ the standard normal cdf. The ordered values $p_{(i)}$ are plotted versus index i. They cross the control line $(i/N) \cdot q$ just after $i = 28$, so r_q rejects $H_{0(i)}$ for $i = 1, 2, \ldots, 28$ and accepts for $i \geq 29$. The rejection threshold corresponds to $z_i > 3.33$.[6]

As with the James–Stein estimator, the Benjamini–Hochberg rule (2.62) can be motivated by a simple Bayesian model. We suppose that each z-value is randomly selected to be either "null" with prior probability π_0 or "non-null" with probability $\pi_1 = 1 - \pi_0$; nulls follow density $f_0(x)$—the $\mathcal{N}(0,1)$ density $\varphi(z)$ in (2.64)—while non-nulls follow some other density $f_1(z)$,

$$\pi_0 \longrightarrow f_0 \longrightarrow z; \qquad \pi_1 \longrightarrow f_1 \longrightarrow z. \tag{2.66}$$

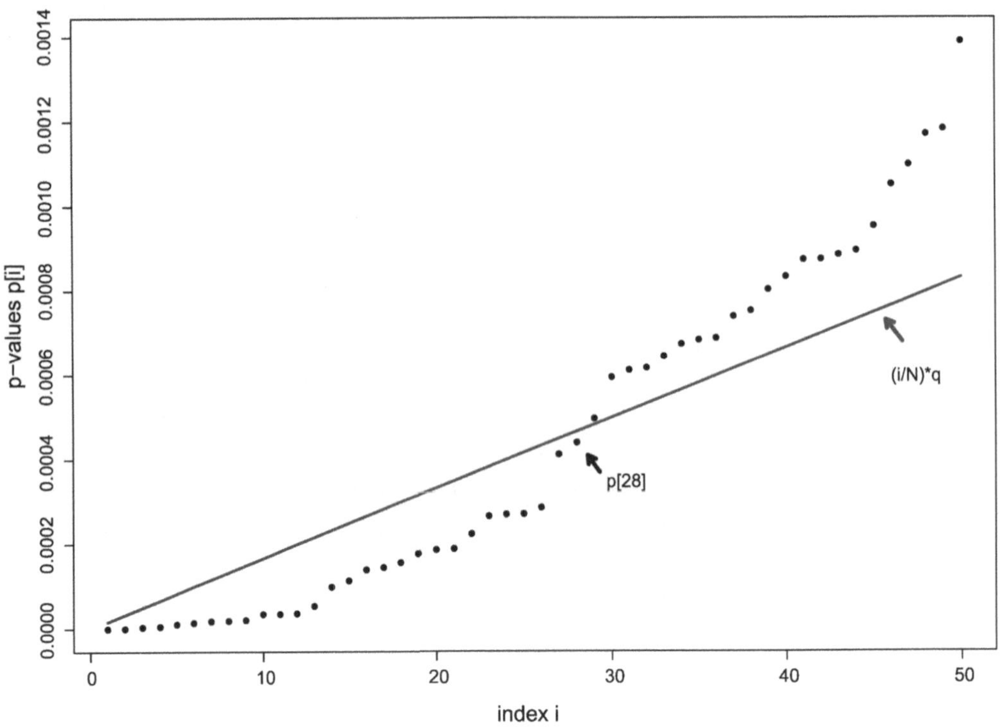

FIGURE 2.6
Benjamini–Hochberg rule r_q with $q = 0.1$ applied to right side of prostate data; rejects largest 28 z-values $z_i > 3.33$.

Let $F_0(z)$ and $F_1(z)$ be the right-sided[7] cdfs,

$$F_0(z) = \int_z^\infty f_0(x)\, dx \quad \text{and} \quad F_1(z) = \int_z^\infty f_1(x)\, dx \tag{2.67}$$

and

$$F(z) = \pi_0 F_0(z) + \pi_1 F_1(z); \tag{2.68}$$

define $\text{Fdr}(z)$ as the posterior probability of nullness given an observed z-value exceeding z,

$$\text{Fdr}(z) = \Pr\{\text{null} \mid \text{observed } z\text{-value } \geq z\} = \pi_0 F_0(z)/F(z) \tag{2.69}$$

according to Bayes rule. Even if $F_0(z)$ is known (as is usually assumed in hypothesis testing situations) and π_0 is taken to be nearly 1, still the marginal cdf $F(z)$ will most often be unknown. However, there is an obvious nonparametric estimate of $F(z)$, the empirical cdf of the N z-values,

$$\widehat{F}(z) = \#\{z_i \geq z\}/N. \tag{2.70}$$

Plugging this into (2.69) gives the *empirical Bayes Fdr estimate*

$$\widehat{\text{Fdr}}(z) = \hat{\pi}_0 F_0(z)/\widehat{F}(z), \tag{2.71}$$

where $\hat{\pi}_0$ could be 1 in practice. We might decide to reject H_{0i} if the ratio $\widehat{\text{Fdr}}(z_i)/\hat{\pi}_0$ of posterior- to prior-nullness is smaller than some threshold q. In terms of the ordered values $z_{(i)}$, we reject H_{0i} if

$$F_0(z_{(i)})/\widehat{F}(z_{(i)}) \leq q. \tag{2.72}$$

But $F_0(z_{(i)})$ is the p-value $p_{(i)}$, and $i/n = \widehat{F}(z_{(i)})$. Looking at (2.62), we see that the empirical Bayes rejection rule (2.72) is the same as the Benjamini–Hochberg rule r_q!

From a Bayesian point of view, we should prefer the *local false discovery rate*

$$\text{fdr}(z) = \Pr\{\text{null} \mid z\} = \pi_0 f_0(x)/f(z), \tag{2.73}$$

to the tail area rate Fdr(z) (2.69). Estimating fdr(z) requires a parametric estimate of the marginal density,

$$f(z) = \pi_0 f_0(z) + \pi_1 f_1(z),$$

as contrasted with the nonparametric $\widehat{F}(z)$ (2.70).

Lindsey's method (Lindsey, 1974) offers a convenient way to construct such estimates. The method begins by assuming an exponential family form for the density $f(z)$,

$$\log f_\beta(z) = \beta^\top s(z) - \psi(\beta), \tag{2.74}$$

where $s(z)$ is a vector of sufficient statistics; for example,

$$s(z) = (z, z^2, z^3, z^4, z^5) \tag{2.75}$$

would allow $\log f_\beta(z)$ to be any fifth-degree polynomial; $\psi(\beta)$ is whatever constant is needed to make $f_\beta(z)$ integrate to 1.

It looks like special software would be needed to calculate the MLE for β, but Lindsey's method finesses the problem by discretization. The histogram in Figure 2.5 has $K = 49$ bins, with bin k having, say centerpoint x_k and count y_k. Count y_k has expected value

$$\mu_k(\beta) = \exp\{\beta_0 + \beta^\top s_k\}, \tag{2.76}$$

$s_k = s(x_k)$, where β_0 is a proportionality constant chosen to make

$$\sum_1^N \mu_k(\beta) = N.$$

We assume that the counts y_k are independently Poisson with expectations (2.76). This is a multiparameter exponential family, for which standard GLM software gives the MLE $\hat{\beta}$, nearly identical to the MLE in family (2.74).

Model (2.74) was applied to the $N = 6033$ z-values of Figure 2.5, taking $s(z)$ in (2.74) to be a fifth-degree natural spline rather than polynomial (2.75). It gave MLE density $\hat{f}(z)$ and estimated local false discovery rate

$$\widehat{\text{fdr}}(z) = f_0(z)/\hat{f}(z), \tag{2.77}$$

with $f_0(z) = f(z)$ and $\pi_0 = 1$. This is shown in the left panel of Figure 2.7.[8]

Fifty-one of the genes had fdr$(z_i) \leq 0.2$, a conventional threshold for non-nullness. How non-null were they? Suppose

$$z_i \sim \mathcal{N}(\theta_i, 1), \qquad i = 1, 2, \ldots, N, \tag{2.78}$$

θ_i the *effect size* for gene i. As in (2.50)–(2.53),

$$\begin{aligned} &\theta \mid z \mathrel{\dot\sim} (\hat{e}(z), \hat{v}(z)), \\ \text{with} \quad &\hat{e}(z) = z + \frac{d}{dz}\log \hat{f}(z) \quad \text{and} \quad \hat{v}(z) = 1 + \frac{d^2}{dz^2}\log \hat{f}(z). \end{aligned} \tag{2.79}$$

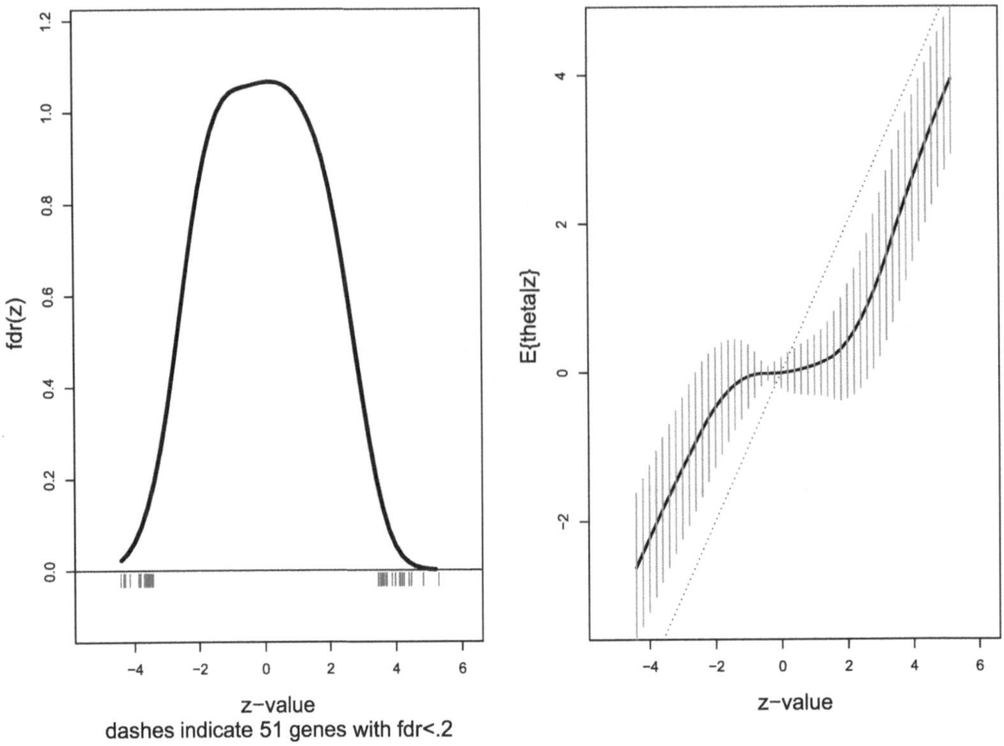

FIGURE 2.7
Left panel, estimated local false discovery rate fdr(z) for prostate cancer study (ns fit/5 df). *Right panel*, expected effect size $E\{\theta \mid z\}$; bars are +/− posterior standard deviation.

The right panel of Figure 2.7 shows $\hat{e}(z) \pm \hat{v}(z)^{1/2}$. At $z = 4$ for instance, the distribution of θ given z is estimated to have posterior expectation 2.60 and standard deviation 1.11.

The precise nature of Theorem 2.2 elicited a substantial literature devoted to exact frequentist control of FDR. Empirical Bayes methods are less exact, but, as with the James–Stein rule, they open up a wider range of possibilities such as the local false discovery rate. An important difference is that Theorem 2.2 requires independence among the observations z_i while empirical Bayes expectation results like $\hat{e}(z)$ (2.79) do not.

2.5 *f*-modeling and *g*-modeling

Our four empirical Bayes success stories—Robbins' formula (2.3), the missing species estimate (2.16), the James–Stein rule (2.42), and false discovery rates (2.70)—all share one important characteristic: they avoid estimating the prior density $g(\theta)$ in the basic empirical Bayes setup (2.1),

$$g(\theta) \longrightarrow \theta_i \quad \text{and} \quad f(x \mid \theta_i) \longrightarrow x_i, \tag{2.80}$$

$i = 1, 2, \ldots, N$. Instead, they focus on directly estimating the marginal density $f(x)$ from the observations

$$x_i \sim f(x), \qquad i = 1, 2, \ldots, N. \tag{2.81}$$

The reason is simple enough: estimating $g(\theta)$ can be hard work. In the classic situation where $f(x_i \mid \theta_i)$ is the normal kernel $\varphi(x_i - \theta_i)$, finding $g(\theta)$ is the famously difficult *deconvolution* problem. Estimating $g(\theta)$ in the general context (2.80) can be called the *empirical Bayes deconvolution* problem, as in Efron (2016).

And yet there are good reasons one might want an estimate of the prior $g(\theta)$, involving questions that can't be answered directly in terms of the marginal density $f(x)$. In the prostate cancer effect-size model $z_i \sim \mathcal{N}(\theta_i, 1)$ (2.74), what is $\Pr_g\{\theta_i \geq 2\}$ say, or $\Pr_g\{\theta_i \geq 2 \mid z_i = 3\}$?

The term *g-modeling* refers to methods that *do* begin with direct estimation of $g(\theta)$, as opposed to *f-modeling* which focuses on $f(x)$. Fisher's parametric gamma prior for the missing species problem (2.18) was the initial g-modeling example. Nonparametric g-modeling, in which $g(\theta)$ is discretely represented by a small number of point masses, goes back to Robbins' original publication (Kiefer and Wolfowitz, 1956; Laird, 1978) and has enjoyed renewed interest in an age of computational advances (Koenker and Mizera, 2014). Here we will only discuss recent developments in parametric g-modeling (Efron, 2016).

The mechanics of g-modeling are easiest to explain in a discrete setting where θ can take on m possible values, say $\theta \in \mathcal{T}$,

$$\mathcal{T} = \left\{\theta^{(1)}, \theta^{(2)}, \ldots, \theta^{(m)}\right\}, \tag{2.82}$$

and x only n possible values, $x \in \mathcal{X}$,

$$\mathcal{X} = \left\{x^{(1)}, x^{(2)}, \ldots, x^{(n)}\right\}. \tag{2.83}$$

Now $g(\theta)$ can be expressed as an m-vector $\boldsymbol{g}$,

$$\boldsymbol{g} = (g_1, g_2, \ldots, g_m)^\top, \tag{2.84}$$

where $g_j = \Pr\{\theta = \theta^{(j)}\}$. Define

$$f_{ij} = \Pr\left\{x = x^{(i)} \mid \theta = \theta^{(j)}\right\}, \tag{2.85}$$

and let $\boldsymbol{M}$ be the $n \times m$ matrix $\{f_{ij}\}$. The marginal density $f(x)$ is an n-vector

$$\boldsymbol{f} = (f_1, f_2, \ldots, f_n)^\top, \qquad \text{where } f_k = \Pr\left\{x = x^{(k)}\right\},$$

with $\boldsymbol{f}$ given by the matrix product

$$\boldsymbol{f} = \boldsymbol{M}\boldsymbol{g}. \tag{2.86}$$

The observed data $x_1, x_2, \ldots, x_N$ is a random sample from $\boldsymbol{f}$. Sample space $\mathcal{X}$ being discrete, the counts

$$y_k = \#\left\{x_i = x^{(k)}\right\}, \qquad k = 1, 2, \ldots, n, \tag{2.87}$$

are sufficient statistics, with $\boldsymbol{y} = (y_1, y_2, \ldots, y_n)^\top$ having multinomial distribution

$$\boldsymbol{y} \sim \text{Mult}_n(N, \boldsymbol{f}) \tag{2.88}$$

(N draws on n categories with probability vector $\boldsymbol{f}$).

Parametric g-modeling assumes that $\boldsymbol{g}$ is a smooth function of a p-dimensional parametric vector α,

$$\{\boldsymbol{g}(\alpha), \alpha \in \mathcal{A}\}, \tag{2.89}$$

$\mathcal{A}$ a subset of $\mathcal{R}^p$; α produces the count vector $\boldsymbol{y}$ according to

$$\alpha \longrightarrow \boldsymbol{g}(\alpha) \longrightarrow \boldsymbol{f}(\alpha) = \boldsymbol{M}\boldsymbol{g}(\alpha) \longrightarrow \boldsymbol{y} \sim \text{Mult}_n\left(N, \boldsymbol{f}(\alpha)\right). \tag{2.90}$$

Maximizing the multinomial log likelihood

$$\text{loglik}(\alpha) = \sum_{k=1}^{n} y_k \log\left(f_k(\alpha)\right) \tag{2.91}$$

gives MLE $\hat{\alpha}$ and $\hat{\boldsymbol{g}} = \boldsymbol{g}(\hat{\alpha})$; an estimate of the prior density.

Efron (2016) proposed taking $\boldsymbol{g}(\alpha)$ to be a p-parameter exponential family,

$$\log(gj(\alpha)) = Q_j^\top \alpha - \phi(\alpha) \qquad (j = 1, 2, \ldots, m). \tag{2.92}$$

Here the vectors $Q_j^\top$ are the rows of an $m \times p$ structure matrix $\boldsymbol{Q}$ chosen by the g-modeler. An example is given next. Function $\phi(\alpha)$ normalizes $\boldsymbol{g}(\alpha)$ to sum to 1,

$$\phi(\alpha) = \log \sum_{j=1}^{n} \exp\left(Q_j^\top \alpha\right). \tag{2.93}$$

Model (2.89)–(2.92) was applied to the prostate cancer data of Section 2.4, now with $z_1, z_2, \ldots, z_N$, $N = 6033$, playing the role of the x_i (2.80) and $z_i \sim \mathcal{N}(\theta_i, 1)$ as in (2.74). $\mathcal{X}$ equaled $\{-4.4, -4.2, \ldots, 5.2\}$ ($n = 49$), the bin centers in Figure 2.5, with the counts y_k the heights of the histogram bars. $\mathcal{T}$ was $\{-4.2, -4.0, \ldots, 4.2\}$, $m = 43$. The structure matrix $\boldsymbol{Q}$ was chosen to be

$$\underset{49\times 43}{\boldsymbol{Q}} = \left(I_0, \text{ns}(\mathcal{T}, 5)\right), \tag{2.94}$$

$I_0 = (0, 0, \ldots, 1, 0, 0, \ldots, 0)^\top$, the 1 at $\theta^{(22)} = 0$, and $\text{ns}(\mathcal{T}, 5)$ the R language natural spline function having five degrees of freedom (from the R package `splines`).

With this choice of $\boldsymbol{Q}$, (2.92) represents a "spike and slab" prior, with the spike at $\theta = 0$ accommodating the large number of null genes (zero effect size), while $\text{ns}(\mathcal{T}, 5)$ allows for smooth modeling of the non-null effects. Figure 2.8 shows the results: $\hat{\boldsymbol{g}}$ put probability 0.83 on $\theta = 0$, with most of the 0.17 slab not very far from 0. For example, $\widehat{\Pr}\{\theta \geq 2\}$ was only 0.013. Taking $\hat{\boldsymbol{g}}$ literally allows for Bayesian estimates, e.g.,

$$\widehat{\Pr}\{\theta \geq 2 \mid z = 3\} = 0.35. \tag{2.95}$$

Section 2 of Efron (2016) provides the details for maximum likelihood estimation in model (2.92). Some regularization is necessary for stability; the regularization constant c_0 of Efron (2016, Section 3) was 0.5 in Figure 2.8.

g-modeling is inherently less straightforward than f-modeling. A low-dimensional parametric model, like that used for Figure 2.8, amounts to a bet on the form of $g(\theta)$. In (2.94), the bet is that the non-null prior distribution can be adequately approximated by a fifth-degree natural spline. The contribution by Koenker and Giu to the discussion of Efron (2019) describes nonparametric g-modeling in its modern computational form, which requires less betting but doesn't easily permit estimates like (2.95).

f-modeling has dominated empirical Bayes applications. Ingenious techniques like those of Robbins, Good and Toulmin, and Tweedie have allowed certain special problems to be solved directly in terms of $f(x)$, rendering g-modeling unnecessary. Venturing beyond the class of "special problems" may lie in empirical Bayes' future, in which case g-modeling will have its day.

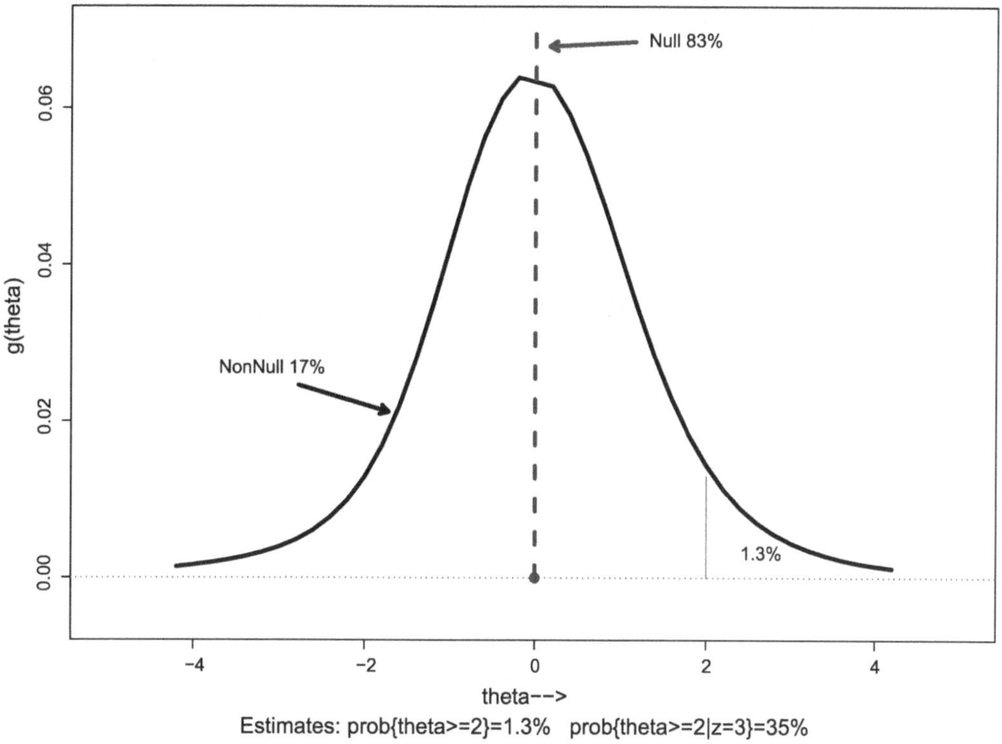

FIGURE 2.8
Prostate study: estimated prior $g(\theta)$; g-modeling $\boldsymbol{Q} = \text{ns}(\theta, 5)+$ atom at 0; estimated null probability 83%.

Whether or not that day arrives, g-models are the only game in town for analyses like those in Figure 2.8 or (2.95). Even on f-modeling's home ground, g-model estimates can be more efficient; see Tables 2.1 and 2.2 in Section 2.5 of Efron (2019). The computations in Figure 2.8 used the R package `deconvolveR`, fully described in Narasimhan and Efron (2020). g-modeling can be thought of as the computer-intensive child of Fisher's gamma model, just as the f-modeling methods of Section 2.4 are descendents of the James–Stein rule.

2.6 Relevance

Previously we have discussed empirical Bayes methods in terms of two dichotomies:

1. parametric or non-parametric
2. f-modeling or g-modeling

A third dichotomy concerns the use to which the methods are aimed,

3. ensemble or individual

(where "ensemble" would be the equivalent of "population" in a survey sampling framework). Ensemble methods hope to do well according to some overall measure of accuracy, as for instance EASE $= E\{\sum(\hat{\theta}_i - \theta_i)^2/N\}$ (2.46). "Individual" refers to inferences for a specific case, for example interpreting (2.95) to say that $\widehat{\Pr}\{\theta_i > 2 \mid z_i = 3\}$ applies to a specific gene having $z_i = 3$ in the prostate cancer study.

A fourth dichotomy is:

4. frequentist or Bayesian

Can empirical Bayes procedures be frequentist? The answer is yes, and in fact most of the literature has been approached from a frequentist point of view. The minimization of EASE in Section 2.3 is a frequentist criterion, with the prior $g(\theta)$ playing no role in the analysis except to suggest the form of the rule $\bar{g}(\theta)$ (2.48). Robbins used the term *compound Bayes* to distinguish the Oracle Bayes situation, where $\boldsymbol{\theta} = (\theta_1, \theta_2, \ldots, \theta_N)$ is fixed, from *empirical Bayes*, where we truly believe $\theta_i \sim g(\cdot)$—but without much difference in his methodology.

Ensemble loss functions like $\sum(\hat{\theta}_i - \theta_i)^2$ are conducive to frequentism. Bayesian considerations become crucial when empirical Bayes methods are applied to individual inferences. Suppose the two towers in Figure 2.3 concern an insurance company where the right tower represents automobile claims and the left tower flood damages. The 40% savings of the empirical Bayes rule $\hat{e}(x)$ would still be relevant to the company's bottom line, but would be inappropriate if applied to any one individual's automobile or flood premium.

Similar considerations apply to the 18 baseball players in Table 2.6, Section 2.3. Baseball fans know that Clemente was the best hitter of his generation. Putting him in a James–Stein analysis with 17 less accomplished hitters almost guarantees an underestimate of his true ability. James–Stein shrinkage, Figure 2.3, is a winning tactic for minimizing total squared error loss, but risks "sacrificing the extremes for the sake of the center".

The question here is one of *relevance*. How relevant are the other 17 players to any one player's estimation? Frequentist estimates, like the MLE column in Table 2.6, base inference for each case on just its own data. Empirical Bayes methods share data across individuals, for the benefit of overall performance but at possible peril to unusual individuals. Bayes estimators have the same property—though usually concealed by a tacit assumption that all θ_i's obtained from the prior $g(\theta)$ are relevant to each other—but relevance is easier to worry about in an empirical Bayes setting, where we can see the "prior" being constructed from individual cases.

Figure 2.9 illustrates an example of individual inferences in an empirical Bayes setting: $n = 51$ values $x_0, x_1, x_2, \ldots, x_5$ have been observed from the empirical Bayes model

$$\theta_i \sim g(\theta) \quad \text{and} \quad x_i \mid \theta_i \sim \mathcal{N}(\theta_i, 1), \tag{2.96}$$

$i = 0, 1, 2, \ldots, 50$, with $g(\cdot)$ unknown and the θ_i unobserved. We wish to assess the posterior distribution of θ_0 given $x_0 = 5$ and $\boldsymbol{x} = (x_1, x_2, \ldots, x_{50})$. We are specifically interested in θ_0, and not some omnibus criterion such as EASE.

g-modeling[9] was used to produce an estimate $\hat{g}(\theta)$ for $g(\theta)$, the dotted curve in Figure 2.10. The likelihood function for θ_0, $\varphi(\theta_0 - 5)$, is shown by the dashed curve. Bayes rule then gives the posterior density

$$\hat{g}(\theta_0 \mid x_0, \boldsymbol{x}) = c\varphi(\theta_0 - 5)\hat{g}(\theta_0), \tag{2.97}$$

shown as a solid curve. The Bayesian effects are quite strong here: $\hat{g}(\theta_0 \mid x_0, \boldsymbol{x})$ reaches its maximum at $\theta_0 = 3.8$, compared to the standalone frequentist estimate $\hat{\theta}_0^{\text{MLE}} = 5$; (2.97) puts only 18% of its mass above 5, compared to 50% for $\varphi(\theta_0 - 5)$. All of this requires us to take the Bayesian assumption $\theta_i \sim g(\theta)$ literally and not just as the hint it was in the EASE calculations of Section 2.3.

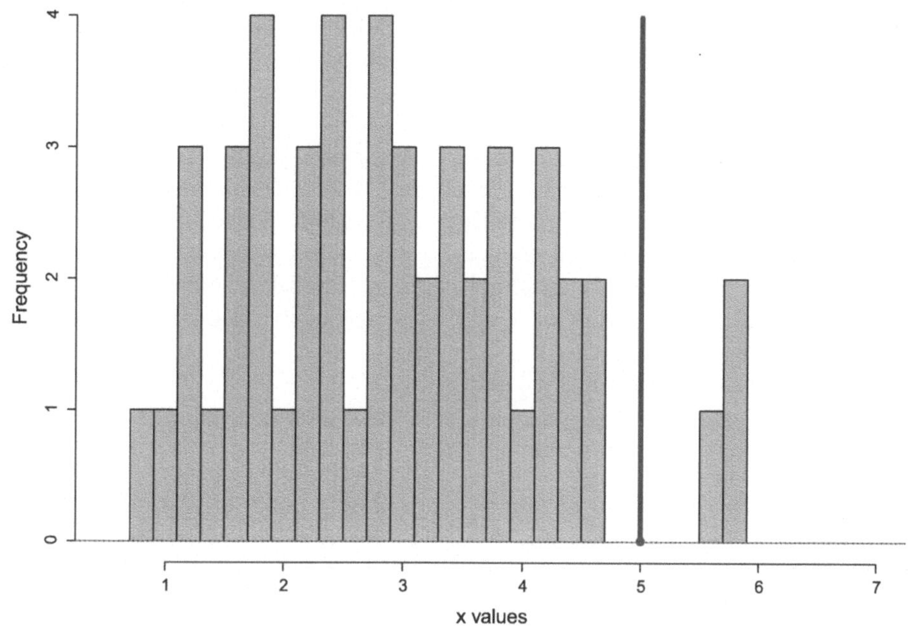

FIGURE 2.9
$x_0 = 5$ and 50 other values x_i are observed from model (2.96), shown as shaded histograms. What can we say about θ_0?

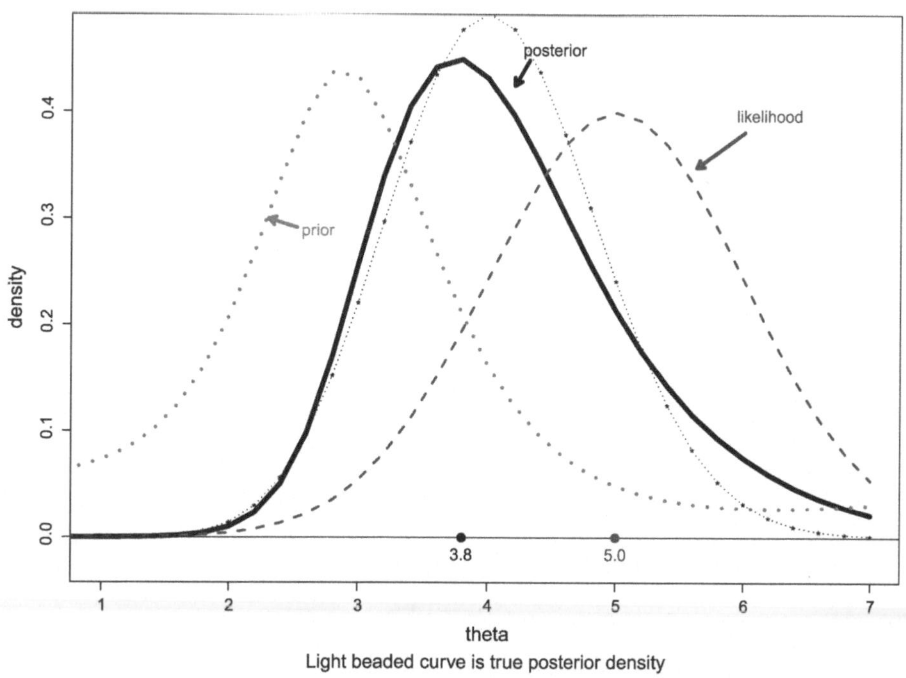

FIGURE 2.10
Empirical Bayes posterior inference for θ_0: likelihood (dashed), g-model prior (dotted), posterior density (solid).

With only $N = 50$ values x_i to work with, one might worry about the accuracy of $\hat{g}(\theta_0 \mid x_0, \boldsymbol{x})$. The true prior $g(\theta)$ used to generate the x_i was $\text{Gamma}_9/3$ (gamma with 9 degrees of freedom, divided by 3) and at least in this case, $\hat{g}(\theta_0 \mid x_0, \boldsymbol{x})$ is a reasonable approximation to the true posterior density $g(\theta_0 \mid x_0, \boldsymbol{x})$ (shown by the faint sparsely dotted curve).

In a genuine (unsimulated) setting, a more pressing worry might concern the relevance of $x_1, x_2, \ldots, x_{50}$ to the estimation of θ_0. Empirical Bayes considerations have moved the point estimate from 5.0 down to 3.8, more than a full standard error from the MLE. Do we trust the empirical Bayes connection enough for such a drastic change? In a James–Stein context, Efron and Morris (1972) proposed capping the distance of the empirical Bayes estimate from the MLE, sacrificing some of the savings of $\hat{\theta}^{\text{JS}}$ in order to protect outliers like Clemente from overshrinkage.

A standard inference problem might begin with data

$$x_{01}, x_{02}, \ldots, x_{0n} \overset{\text{iid}}{\sim} f(x \mid \theta_0), \tag{2.98}$$

in which case there is no question of the relevance of each observation to the estimation of θ_0. In the empirical Bayes model (2.96), the "other" observations x_i are less directly linked to θ_0; they have the same "grandparent" as $x_0 \sim f(x \mid \theta_0)$ through the common prior $g(\theta)$ but not the same parent θ_0, which is to say they are cousins to x_0 rather than siblings.

Specific relevance applications appear in Efron and Morris (1972), Efron (2010, Section 10.3), and Efron (2019, Section 7). A different methodology is explored in Mukhopadhyay and Wang (2020). As a topic area, relevance is underdeveloped in the empirical Bayes literature, even though it can be an obvious concern in situations like that of the baseball players where the individual cases have evocative identities.

Notes

1. Evidently, the missing species problem also arose in World War II code-breaking efforts: Good credited his fellow code-breaker Alan Turing for some of the ideas here. The argument leading to (2.16)—substituting observed counts for their expectations—is similar to that for Robbins' formula, though the authors worked separately.
2. These numbers are based on the Concordance in Spevack (1968); differences in what is considered "Shakespeare" have altered the counts a little since then.
3. All this might seem like pure speculation. However, cross-validation tests were run: one of the plays was removed from the canon, Table 2.3 recomputed for the remainder, and predictions $\widehat{E}\{t\}$ made for the omitted play. Accuracy of both parametric and nonparametric predictions was impressive.
4. "Words" here means any distinct spelling, and "persons" a different word than "person". Contemporaries such as Marlowe and Jonson appear to have used larger vocabularies than Shakespeare.
5. *Batting average* is the number of successful hits per attempt, or at-bat. Clemente, for instance, hit successfully in 18 of his first 45 at-bats, for an average of $18/45 = 0.400$ ("batting .400"). Not knowing anything about baseball is an advantage for getting the full force of this example.
6. The familiar Bonferroni bound $p_i \leq 0.05/N$ rejects for $z_i \geq 4.40$, only the top 4 genes making the cutoff. The popularity of FDR has much to do with its relative liberality.
7. Right-sided here to agree with the example in Figure 2.6 where we rejected for large positive values of z.
8. The fact that $\widehat{\text{fdr}}(z)$ slightly exceeds 1 near $z = 0$ demonstrates that π_0 is actually a little less than 1. This permits an estimate of π_0, ignored here.
9. Using a natural spline with five degrees of freedom; a correction for the small sample size, $N = 50$, was made; see Section 6 of Efron (2019).

Bibliography

Benjamini, Y. and Hochberg, Y., 1995. Controlling the false discovery rate: A practical and powerful approach to multiple testing. *Journal of the Royal Statistical Society, Series B*, *57*(1), pp. 289–300.

Brown, L. D. and Greenshtein, E., 2009. Nonparametric empirical Bayes and compound decision approaches to estimation of a high-dimensional vector of normal means. *Annals of Statistics*, *37*(4), pp. 1685–1704.

Efron, B., 2010. *Large-Scale Inference: Empirical Bayes Methods for Estimation, Testing, and Prediction*, Volume 1 of Institute of Mathematical Statistics Monographs. Cambridge: Cambridge University Press.

Efron, B., 2016. Empirical Bayes deconvolution estimates. *Biometrika*, *103*(1), pp. 1–20.

Efron, B., 2019. Bayes, oracle Bayes and empirical Bayes. *Statistical Science*, *34*(2), pp. 177–235. with commentary and a rejoinder by the author.

Efron, B. and Morris, C., 1972. Empirical Bayes on vector observations: An extension of Stein's method. *Biometrika*, *59*, pp. 335–347.

Efron, B. and Thisted, R., 1976. Estimating the number of unseen species: How many words did Shakespeare know? *Biometrika*, *63*(3), pp. 435–447.

Fisher, R., Corbet, A. and Williams, C., 1943. The relation between the number of species and the number of individuals in a random sample of an animal population. *Journal of Animal Ecology*, *12*, pp. 42–58.

Good, I. and Toulmin, G., 1956. The number of new species, and the increase in population coverage, when a sample is increased. *Biometrika*, *43*, pp. 45–63.

James, W. and Stein, C., 1961. Estimation with quadratic loss. In *Proceedings of the Fourth Berkeley Symposium on Mathematical Statistics and Probability, Vol. I*, pp. 361–379. Berkeley: University of California Press.

Jiang, W. and Zhang, C.-H., 2009. General maximum likelihood empirical Bayes estimation of normal means. *Annals of Statistics*, *37*(4), pp. 1647–1684.

Kiefer, J. and Wolfowitz, J., 1956. Consistency of the maximum likelihood estimator in the presence of infinitely many incidental parameters. *Annals of Mathematical Statistics*, *27*, pp. 887–906.

Koenker, R. and Mizera, I., 2014. Convex optimization, shape constraints, compound decisions, and empirical Bayes rules. *Journal of the American Statistical Association*, *109*(506), pp. 674–685.

Laird, N., 1978. Nonparametric maximum likelihood estimation of a mixed distribution. *Journal of the American Statistical Association*, *73*(364), pp. 805–811.

Lindsey, J., 1974. Construction and comparison of statistical models. *Journal of the Royal Statistical Society, Series B*, *36*, pp. 418–425.

Mukhopadhyay, S. and Wang, K., 2020. On the problem of relevance in statistical inference. *arXiv e-prints*, arXiv:2004.09588.

Narasimhan, B. and Efron, B., 2020. deconvolveR: A g-modeling program for deconvolution and empirical Bayes estimation. *Journal of Statistical Software*, *94*(11), pp. 1–20.

Robbins, H., 1956. An empirical Bayes approach to statistics. In *Proceedings of the Third Berkeley Symposium on Mathematical Statistics and Probability, 1954–1955, Vol. I*, pp. 157–163. Berkeley and Los Angeles: University of California Press.

Spevack, M., 1968. *A Complete and Systematic Concordance to the Works of Shakespeare, Vols 1–6.* Hildesheim: George Olms.

Strawderman, W. E., 1971. Proper Bayes minimax estimators of the multivariate normal mean. *Annals of Mathematical Statistics*, *42*(1), pp. 385–388.

Thisted, R. and Efron, B., 1987. Did Shakespeare write a newly-discovered poem? *Biometrika*, *74*(3), pp. 445–455.

3

Distributions for Parameters

N. Reid

University of Toronto

3.1 Introduction

This handbook is the outcome of a series of workshops, as described in the preface. In this introductory chapter, I attempt to give a high-level overview of different inferential approaches. I have addressed this through what I view as a common goal among Bayesian, fiducial and frequentist approaches — finding a distribution for a parameter or parameters of interest. Confidence distributions, generalized fiducial distributions, inferential models, and belief functions are some of the terms associated with these approaches. In this overview, I discuss both historical developments and modern versions. For a more detailed comparison of Bayesian and frequentist approaches, see Berger (2022). A more eclectic discussion of Bayes, Fiducial and Frequentist is given in Meng (2024) though a series of compelling examples. An exploration of frequentist properties of Bayesian methods in a quite different context, of testing fixed hypotheses, is presented in Berger *et al.* (2024).

The history of statistical inference, while young, is surprisingly rancorous. The hostility between Fisher and Neyman may have contributed to a divergence of developments in the United Kingdom and in the United States that hampered the maturing of the subject (Zabell, 1992; Lehmann, 1993). About Bayesian inference, Fisher (1930) wrote:

> I know of only one case in mathematics of a doctrine which has been accepted and developed by the most eminent men of their time, and is now perhaps accepted by men now living, which at the same time has seemed to a succession of sound writers to be fundamentally false and devoid of foundation.

The discussions at the BFF workshops are much more congenial. One theme of various approaches discussed there is the desire to obtain a distribution for parameters of a statistical model that enables us to make inferential statements. In this chapter, I give an introduction to BFF from this point of view. The topics are treated in more depth in other chapters in this volume and, true to our history, present divergent viewpoints.

3.2 An Historical Sketch

The most familiar way to describe a distribution for parameters is to use the calculus of conditional probability embodied in Bayes' theorem. We assume a model for the observation of a random process, described by a family of probability density functions for possible

DOI: 10.1201/9780429341731-3

outcomes, denoted $\{f(y;\theta), \theta \in \Theta\}$; the form of the density function is assumed known. We further assume that y takes values in $\mathcal{Y}$ and that the sets $\mathcal{Y}$ and Θ are known. Given a probability distribution for θ with density function $\pi(\theta)$, and interpreting $f(y;\theta)$ as the conditional density for y, given θ, standard probability calculus gives the conditional density for θ, given y:

$$\pi(\theta \mid y) = \frac{f(y;\theta)\pi(\theta)}{\int f(y;\theta)\pi(\theta)d\theta} = \frac{f(y;\theta)\pi(\theta)}{m(y)}. \tag{3.1}$$

We assume that $m(y)$, the marginal density of y, is positive in this definition; a more rigorous formulation is given in, for example, the introductory chapters of Schervish (1995). In statistical applications y is the data, often a sample of size n, $\pi(\theta)$ is called the prior density for θ, and $\pi(\theta \mid y)$ is called the posterior density. The informal notion is that $\pi(\theta)$ is available to the experimenter before any data are collected, and the data modify the prior to give a probability density for θ suitable for inference once y has been observed. Stigler (2013) noted that the original title of Bayes' paper was "A method of calculating the exact probability of all conclusions based on induction", but the published title was somewhat more tentative (Bayes, 1763).

The phrase "all conclusions" is widely relied upon in modern applications of Bayesian inference. Armed with the posterior density, the statistician can compute probabilities of any events of interest, such as $\Pr(\theta > 0 \mid y)$, or $\Pr(\theta_1 > \theta_2 \mid y)$. Thus a potential conclusion from a Bayesian analysis of data from an experiment to compare treatments A and B might take the form "from this data and model we see that the (posterior) probability that μ_A is larger than μ_B is 0.09"; μ_A and μ_B here being parameters in the statistical model for the responses. Kline *et al.* (2021) reported on a Bayesian analysis of a household survey in Ohio, undertaken in 2020 to assess the sero-prevalence of antibodies to the SARS-COV-2 virus. One of their conclusions is

> Based on our model, the posterior mean prevalence is 1.3% with a 95% credible interval of (0.2%, 2.7%).

The data are fixed in this conclusion, so the probability referred to essentially comes from the prior distribution. As a result, much turns on the choice of the prior and its meaning. In Bayes (1763), the parameter space Θ is $[0, 1]$ and $\pi(\theta)$ is a uniform density on Θ, a choice that was supported by Laplace; see Stigler (1986). It was quickly realized that a uniform prior is not invariant to the parametrization of the model, so many attempts were made to find priors by other means. Jeffreys (1961) used parametrization invariance to motivate priors based on Fisher information; see Section 3.4.1 and Kass and Wasserman (1996, Section 2).

In subjective Bayesian inference (Savage, 1954), emphasis is placed on formulating a proper prior distribution from consideration of probabilities of events about θ in the absence of data. This can be useful as an aid to personal decision-making, but is less helpful for establishing a scientific consensus.

Often inference is labelled "Bayesian" in settings where there is considerable empirical evidence for the distribution of the parameters in a model. For example, in Kline *et al.* (2021), the model parameters include some components related to the sensitivity and specificity of the antibody tests, information that was externally available from manufacturers as part of their validation studies. Strictly speaking, the results then follow directly from probability calculus, and it is debatable whether this should be described as a Bayesian analysis. In this chapter, and in most discussions of BFF, we consider that such concrete prior information is not available.

The fiducial approach to inference proposed in Fisher (1930) was an attempt to deduce a probability distribution from the data without using a prior distribution. Although Cox

(2016) described this as "almost certainly a major error", there have been many attempts to both make Fisher's ideas more precise and to put fiducial inference on a more solid footing. Dawid (2024) and Lang (2024, Section 15.4) give more detailed summaries of Fisher's original development. Taraldsen and Lindqvist (2024) make a distinction between the fiducial model and the statistical model. Vovk (2024, Section 17.2) discusses Fisher's approach to fiducial prediction.

Suppose there is some one-dimensional summary of the data, which Fisher denoted by T, and a one-dimensional parameter of interest, θ. As a very simple example, suppose the model for T is a location model with known density,

$$T = \theta + e, \quad e \sim f_0. \tag{3.2}$$

Once $T = t$ is observed, we can write $\theta = t - e$, and can think of translating to θ the probability model assigned to the error variable, e. More generally, if we fix a percentile of the distribution function for T, $F(t;\theta)$, we create a link between T and θ. Fisher (1930) expressed this as

$$\mathrm{d}f = -\frac{\partial}{\partial\theta}F(t;\theta)\mathrm{d}\theta \tag{3.3}$$

and called this the fiducial probability density function for θ. The location model explication of fiducial inference is due to Fraser (1966), who called it a structural density, to distinguish it from Fisher's construction. The structural density was defined for models based on groups, as the link between observation and parameter is direct in such models. Fisher (1930) wrote "this then is a definite probability statement about the unknown parameter". Fraser (1968, p. 20) states "A probability statement concerning e is *ipso facto* a probability statement concerning θ".

Confidence distributions, described in Cox (1958), are closely related to fiducial distributions, but are perhaps more accessible. Neyman's original theory (Neyman, 1937) described the construction of an interval of parameter values at a given confidence level, often taken to be 0.95. The interpretation of the confidence level is indirect: in repeated sampling from the model, 95% of such confidence intervals will include the true value of the parameter. This is a statement about the statistical procedure, not the particular interval that has just been constructed; it is like a warranty from the statistician. Berger (2022) calls this "procedural frequentism". Cox (1958) suggested:

> ... in simple cases ... there seems no reason why we should not work with confidence distributions for the unknown parameter ... These can either be defined directly, or ... introduced in terms of the set of all confidence intervals.

While confidence intervals of the form $[a, b]$ are nearly always used in practice, it is easier to define a confidence distribution using a set of one-sided intervals of the form $(-\infty, a]$, following Efron (1993):

> Let $\theta_x(\alpha)$ be the upper endpoint of an exact or approximate one-sided level-α confidence interval for θ. ... The confidence distribution for θ is defined to be the distribution having density
>
> $$\pi_x^\dagger(\theta) = \mathrm{d}\alpha_x(\theta)/\mathrm{d}\theta; \tag{3.4}$$

here $\alpha_x(\theta)$ is the inverse function to $\theta_x(\alpha)$. Efron added:

> ... assigns probability 0.05 to θ lying between the upper endpoints of the 0.90 and 0.95 confidence intervals, etc. ... Of course this is logically incorrect, but it has powerful intuitive appeal. ... *in the case of no nuisance parameters, [this] is exactly Fisher's fiducial distribution* [emphasis added].

Confidence distributions were called significance functions in Fraser (1991), to reflect his construction based on the computation of p-values:

> ... significance [p-value] records the probability left of the observed data point... the significance function is a plot of this probability as a function of θ ... the full spectrum of confidence intervals is obtained ... suggesting the alternate name confidence distribution function.

I added the emphasis in the quote from Efron (1993) because accommodating nuisance parameters is the rock on which all these constructions have foundered, and one reason there are active investigations and extensions underway today. A distribution function is most useful if it is available for a single parameter of interest, but statistical models useful for the analysis of data have many unknown parameters. Even if our data are assumed to come from a single normal distribution, we may in different circumstances be interested in the expected value of that distribution, with the variance parameter treated as a "nuisance", or vice versa. In a fully Bayesian setting, we can in principle construct a joint posterior distribution for all the parameters in the model, and then integrate over the nuisance parameters to obtain a marginal posterior distribution for the parameter of interest. It is not possible to apply the same reasoning to fiducial distributions, confidence distributions, or significance functions, because the probability that underlies their interpretation is fundamentally different.

Marginalization of probabilities for vector parameters to probabilities for scalar parameters is one version of the combination of probabilities, but the probability calculus for computing the combined probabilities of several events also does not apply to fiducial and confidence distributions. This is addressed in Dawid (2024, Section 5.6) and in Taraldsen and Lindqvist (2024). Example 5.8 in Dawid (2024), originally due to Stein (1959), is a simple but informative illustration of this. As a result, recent work on fiducial, confidence and significance functions has needed to address the combination problem separately. In Section 3.4, I will look at the modern approaches being developed as part of the BFF discussions. Before turning to this, it is helpful to consider the meaning of probability.

3.3 The Nature of Probability

Reid and Cox (2015), following many other writers on the subject, including Fisher (1956, p. 31), describe two meanings of probability. One meaning, sometimes called *epistemic* probability, describes uncertainty of knowledge. This is most often associated with Bayesian approaches, and the uncertainty may be, as Jeffreys (1961) proposed, a rational or impersonal degree of belief, or, as in Savage (1954), a personal degree of belief. Personal degrees of belief are sometimes described through a willingness to bet with different odds on the outcome of an event, and underpin subjective Bayesian inference.

The second meaning, usually called *empirical* or *aleatory* probability, encapsulates properties of a "real" world, either through the postulated existence of an infinite population, or a process of repetition under constant conditions. These will of necessity be hypothetical or highly idealized, but may still be suitable for the development and reporting of conclusions based on data. A textbook example of aleatory probability is the meaning of the statement "the probability of tossing heads with a fair coin is 0.5"; usually explained to students as a convergence to 0.5 of the observed fraction of heads in an endlessly repeated sequence. Berger (2022) calls this "empirical frequentism". Lang (2024, Section 15.3), gives a different view on the meaning of probability; for example, he describes both epistemic and aleatory probabilities as subjective; the subjectivity in the latter coming from the uncertain knowledge of the modeller.

It is not always very clear in the literature which meaning of probability is intended to be ascribed to fiducial, structural, or confidence, distributions, but in general it seems to be aleatory probability. That is, we expect that an inferential interval for a parameter θ which is ascribed a probability of 95%, should, in hypothetical repeated sampling from the model, include the true value of the parameter 95% of the time. Cox (1958) introduced the confidence distribution as simply a concise way to present a continuum of such (nested) intervals, instead of specifying a single level of confidence. On the other hand, in their book, Schweder and Hjort (2016, Chapter 1.6) refer to "objective epistemic probability". As I understand their discussion, they suggest that the interpretation of the confidence distribution is epistemic, but its properties in repeated sampling can be verified empirically.

In contrast, the probability that is ascribed to a Bayesian statement, based on a posterior distribution, is an epistemic probability describing the uncertainty in our knowledge. This uncertainty is derived from our uncertainty about the value of the parameters that govern the model, i.e., from the prior distribution. Taraldsen and Lindqvist (2024, Section 18.1) state that the interpretation of fiducial probability "is exactly as for a Bayesian posterior", which suggests an epistemic interpretation.

The calculus of probability, following axioms set out by Kolmogorov, is indifferent to the meaning ascribed to probabilistic statements. This calculus specifies in particular, that marginal probabilities can be obtained by summing or integrating joint probabilities, and that the total probability of all possible outcomes is 1, so that the probability of a set in the sample space is one minus the probability of its complement, and so on. Lang (2024) state that only aleatory probability satisfies Kolmogorov's axioms; Schweder and Hjort (2016, p. 7) write "No axiomatic theory has been worked out for epistemic probability to be used in science, except for Bayesian probabilities".

It is proposed in Reid and Cox (2015, Section 2.2) that by requiring Bayesian probability statements to have a calibration property in repeated sampling from the model, we avoid the need for two potential interpretations of probability. In particular, if

$$\Pr\{\theta \geq \theta^{(1-\alpha)}(y) \mid y\} = 1 - \alpha \tag{3.5}$$

then we should require as well that

$$\Pr\{\theta \geq \theta^{(1-\alpha)}(Y) \mid \theta\} = 1 - \alpha, \tag{3.6}$$

where the first probability statement comes from a Bayesian posterior distribution and is thus conditional on the observed data, and the second comes from the statistical model, and is thus conditional on the parameter. Berger (2022) argued that "even [subjective] Bayesians should accept the empirical frequentist principle"; see also Berger (2006).

If it were possible to choose a single prior distribution that ensures this calibration, then the discussion of inferential approaches would have ended many years ago. However, the current position is that this is not possible, using the notions of probability described so far.

Approaches to inference using an expanded notion of probability, such as Dempster–Shafer belief functions (Gong, 2024), or imprecise probabilities (Walley, 1990), have been argued to be the only way to ensure such calibration. While important for the foundations, these methods have not yet found widespread application, perhaps because the conclusions are more difficult to understand, and thus to explain to non-experts.

The search for distributions for parameters has re-emerged in recent literature, with a view to extending the ideas beyond the fairly limited range of models treated in the past. In the next section, I describe some of the progress that has been made, and the gaps that remain.

3.4 Recent Developments

3.4.1 Objective Bayes

"Objective Bayes" generally means any method proposed for deriving a posterior distribution that is not based on personal probability or subjective prior distributions. Classes of objective priors include the reference priors of Berger and Bernardo (1992) and probability matching priors, reviewed in Datta and Mukerjee (2004), but there are many others. The fact that so many variants exist suggests there is a problem in making the notion of objectivity precise. Kass and Wasserman (1996) provided an excellent overview of various approaches to what they call the formal selection of priors.

As discussed in Section 3.3, one way to avoid using epistemic probability in Bayesian inference is to consider this as one inferential method among many, and test it against its properties in sampling from the model. Thus, the probability referred to in the posterior conclusion becomes an empirical probability about the method. Reid and Cox (2015) write

> ... as with other measuring devices, within this scheme of repetition probability is defined as a hypothetical frequency... it is unacceptable if a procedure yielding high-probability regions in some non-frequency sense are poorly calibrated ... such procedures, used repeatedly, give misleading conclusions.

This suggests that for a given prior distribution on θ, if we compute a posterior quantile, say $\theta_\pi^{(1-\alpha)}(y)$, as at (3.5), now with the dependence on the prior explicitly noted, then we need to investigate the distribution of $\theta_\pi^{(1-\alpha)}(Y)$ under the model $f(y;\theta)$ for the observed random variable. It turns out that if $\theta \in \mathbb{R}$ is a scalar parameter, then Jeffreys' prior provides this type of calibration, at least approximately. Writing $\ell(\theta; y) = \log f(y;\theta)$ for the log-likelihood function of the model, the expected Fisher information is

$$i(\theta) = \mathrm{E}\{-\partial^2\ell(\theta;Y)/\partial\theta^2\}. \tag{3.7}$$

The prior density

$$\pi_J(\theta) \propto i^{1/2}(\theta) \tag{3.8}$$

is the unique prior for which

$$\Pr\{\theta_\pi^{(1-\alpha)}(Y) \le \theta \mid \theta\} = 1 - \alpha + O(n^{-1}), \tag{3.9}$$

when $Y = (Y_1, \ldots, Y_n)$ is an independent sample of size n from the model $f(y;\theta)$. In (3.9) the term $O(n^{-1})$ is a remainder that roughly speaking looks like A/n, where $A < \infty$ may depend on θ, but not on n.

It is necessary to consider a remainder term of $O(n^{-1})$ in equation (3.9), as any proper prior will lead to posterior limits that are calibrated to $O(n^{-1/2})$; sometimes described as the "prior is washed out by the data". This follows from the central limit theorem, usually in this context called the Bernstein-von Mises theorem, for the posterior distribution. This first-order equivalence prompted the title of Fraser (2011). Kleijn and van der Vaart (2012) is a helpful reference on the Bernstein-von Mises theorem, comparing the usual case to that under model misspecification.

The practical appeal of Bayesian approaches is in multi-parameter problems, where we may write, for example,

$$\pi(\psi \mid y) = \int_{\{\theta:\psi(\theta)=\psi\}} \pi(\theta \mid y)d\theta. \tag{3.10}$$

If $\theta \in \mathbb{R}^d$, the expected Fisher information is a $d \times d$ matrix, and (3.8) becomes

$$\pi_J(\theta) \propto |i(\theta)|^{1/2}, \tag{3.11}$$

based on the determinant of the information matrix. Unfortunately, except in very special settings, posterior probability bounds computed using this prior, and marginalized as at (3.10), are not calibrated, even approximately:

$$\Pr\{\psi^{(1-\alpha)}(Y) \leq \psi \mid \theta\} \neq 1 - \alpha + O(n^{-1}). \tag{3.12}$$

There is nothing special about Jeffreys' prior in this disappointment: there is rarely a single prior for a vector parameter that provides calibration for arbitrary derived scalar parameters.

An illuminating and well-studied example due to Stein (1959) is the "many-normal-means" problem: $y_i \sim N(\theta_i, 1/n), i = 1, \ldots, d$, with the parameter of interest $\psi = \Sigma_i \theta_i^2$. Jeffreys' prior for θ is proportional to 1, and the posterior for θ, given y, is multivariate normal with expected value y and covariance matrix I/n, where I is the $d \times d$ identity matrix. The posterior density for ψ is a non-central χ_d^2, and posterior probability bounds based on this diverge from their nominal value in the sampling model quite dramatically. Stein (1959) showed, in the context of fiducial inference, that the expected value under the model of the mean of the marginal posterior density is $\psi + 2d/n$, whereas the expected value of the maximum likelihood estimator is $\psi + d/n$, and emphasized the divergence of approximate confidence intervals that follows from that. The divergence of the entire marginal posterior density as d increases with n fixed was noted in Fraser and Reid (1992) and studied in more detail in Reid and Sun (2010).

That flat priors are not well-calibrated for components or scalar functions of θ is a version of the marginalization paradox of Dawid *et al.* (1973); see also Dawid (2024, Sections 5.6 and 5.7.2), but was also noted in earlier literature in the context of particular examples. It was re-discovered in the particle physics literature in the 1990s, when many simulation experiments were being developed in preparation for data from the Large Hadron Collider (Heinrich, 2006). In spite of this lack of calibration, there are many applications of Bayesian inference in complex models where the authors say something along the lines of "we used standard vague priors for the parameters of our model", also assume independence of these priors, and proceed without attention to the properties of the resulting inference. There are somewhat fewer applied papers where checks on the stability of the inference to the form of the prior are included as part of the model-checking. There is an important theoretical literature on robustness of Bayesian inference (e.g., Berger *et al.*, 2000) and on prior-data conflict (e.g., Evans and Moshonov, 2006).

A "flat" prior for an unbounded parameter, while not a proper probability density function, can often lead to a proper posterior density, i.e., a density that is finite and integrates to 1. Many objective priors are defined this way, including Jeffreys' prior above, the uniform prior $\pi_U(\theta) \propto 1$, and most matching and reference priors. There are examples in the literature, however, where improper priors lead to improper posterior densities. If these densities are estimated by simulation schemes such as Markov chain Monte Carlo, the results are meaningless as the algorithm cannot converge.

Since it is not possible in general to obtain calibrated posterior distributions for arbitrary parameters of interest from a single prior, objective priors leading to calibration need to be targetted on the parameter of interest. This is explicitly recognized in the work on reference priors (Berger and Bernardo, 1992), and matching priors, but the theoretical results have not percolated very widely into practice. Gelman *et al.* (2008) suggested the use of weakly informative priors, with the expressed goal of finding a "somewhat informative prior distribution that can nonetheless be used in a wide range of applications". The weakly informative prior distribution is chosen to be proper and to put positive probability on all "plausible" values of the unknown parameter or parameters. The general calibration properties of the posteriors from weakly informative priors seems not to have been studied; numerical work for models widely used in ecology is provided in Lemoine (2019). If such priors are meant to give calibrated inferences, this would need to be checked in each instance, and, as usual, for each parameter of interest.

Another feature of priors leading to calibrated posterior inference for a parameter is that the priors necessarily depend on the form of the model, for example, Jeffreys' prior depends on the expected Fisher information. As another example, the probability matching priors developed by explicitly requiring (3.6) to hold to $O(n^{-1})$ (Peers, 1965; Mukerjee and Ghosh, 1997) must satisfy an equation of the form

$$\sum_{a=1}^{d} \frac{\partial}{\partial \theta_a} [\{i^{11}(\theta)\}^{-1/2} i^{a1}(\theta)\pi(\theta)] = 0, \tag{3.13}$$

where $i^{ab}(\theta)$ is the (a, b) entry of $i^{-1}(\theta)$ and we assume that the first component of θ is the parameter of interest. If the Fisher information matrix has $i_{1a}(\theta) = 0$, for $a = 2, \ldots, d$, in which case θ_1 is orthogonal to the remaining parameters, then the solutions of (3.13) are

$$\pi(\theta) \propto i_{11}^{1/2}(\theta) g(\theta_2, \ldots, \theta_d), \tag{3.14}$$

where $g(\cdot)$ is an arbitrary function of its arguments (Tibshirani, 1989; Nicolaou, 1993). Although $\pi(\theta)$ is expressed as a prior for the whole vector, its property of calibration of posterior intervals applies *only* to θ_1. It is not in general possible to find a single prior that satisfies (3.13) simultaneously for all components. Unfortunately, imposing the next-order requirement, $\Pr\{\theta_1^{(1-\alpha)}(Y) \leq \theta_1 \mid \theta\} = 1 - \alpha + O(n^{-3/2})$ does not help, as the dominant term of this expansion is a function of derivatives of the log-likelihood function (Mukerjee and Ghosh, 1997).

By matching higher-order frequentist confidence limits to posterior probability limits, Fraser *et al.* (2010b) showed that the only way to have calibration to order $O(n^{-3/2})$ is to let the prior depend on the data. Data-dependent priors also featured in Box and Cox (1964) for a transformed regression model, and in Wasserman (2000) for mixture models. The use of these "priors" makes clear that the probability statements obtained from a Bayesian argument have no probabilistic meaning aside from their frequentist behaviour as confidence limits. This suggests that it may be more fruitful to directly target these intervals without using prior probabilities, which is the goal of fiducial and confidence approaches.

This discussion has assumed a (finite-dimensional) parametric model; nonparametric models raise many more difficulties. Diaconis and Freedman (1986) showed that consistency of Bayes estimates in nonparametric models can depend crucially on "arbitrary details of the prior"; see also Wasserman and Robins (2000). Mukerjee (2008) described data-dependent priors for empirical likelihoods that have probability-matching properties. Consonni *et al.* (2018) reviewed objective Bayes procedures with an emphasis on estimation and on model selection. The many-normal-means problem mentioned above is often used as a prototype for nonparametric methods. Frequentist properties of Bayesian inference for the many-normal-means problem are discussed, for example, in Castillo *et al.* (2015), Bhadra *et al.* (2016), van der Pas *et al.* (2017), and Martin and Ning (2020), although not always with an emphasis on calibration.

3.4.2 Generalized Fiducial

The fiducial argument develops a statement of uncertainty about the parameter from the structure of the model. Fisher assumed that there was a function $T(Y, \theta)$ with a known distribution. Once Y is observed, he argued that the probability that had been assigned to the unknown quantity Y could be transferred to the unknown parameter θ. This probability does not obey the usual probability calculus, which led to some confusion in the literature. Stein (1959) used the normal means example in the previous section to point out a problem with the fiducial distribution.

In spite of these reservations, Hannig and co-authors have developed generalized fiducial inference intended to be useful for complex models and high-dimensional parameters. The basis of the approach is the observation that Fisher's construction of $T(Y, \theta)$ with a known distribution is well-suited to today's method of simulating random variables. Hannig (2009) takes as a starting point a data-generating equation of the form

$$Y = G(U, \theta), \tag{3.15}$$

where U has a known distribution. If this equation can be inverted as a function of Y, then for a fixed value y, we can write

$$\theta = Q_y(U). \tag{3.16}$$

The generalized fiducial distribution for θ is defined to be $Q_y(U^*)$, where U^* is an independent copy of U. The data-generating equation is also the starting point for the discussion of fiducial decision theory by Taraldsen and Lindqvist (2024).

The function G will typically not be invertible, however, as there is no *a priori* requirement that Y and θ are structurally related, as in Fraser (1966), i.e., Y and θ take values in different spaces, usually of different dimensions. In independent sampling and continuous models, where $Y \in \mathbb{R}^n$, and $\theta \in \mathbb{R}^d$, inversion would only be possible if the particular model could be reduced, for example by sufficiency, to a d-dimensional sufficient statistic. More generally, Hannig's generalized fiducial approach introduces an auxiliary function $J(\theta; y)$ and defines the fiducial density

$$r(\theta; y) \propto f(y; \theta) J(\theta; y). \tag{3.17}$$

Here $J(\theta; y)$ functions as a kind of data-dependent prior, modifying the likelihood function $f(y; \theta)$. The construction of the fiducial density is summarized in Hannig *et al.* (2016), and (3.17) is given as a "user-friendly" formula for its construction. They note

> ... $r(\theta \mid y)$ is not a conditional density in the usual sense. Instead, we are using this notation to emphasize that the fiducial density depends on the fixed observed data.

Hannig *et al.* (2016) gave conditions under which intervals for θ computed from $r(\theta; y)$ have, in the limit, the stated coverage probabilities. (Hannig *et al.* (2016) use conditional notation as in the quote above; Murph *et al.* (2024) use the notation $r_y(\theta)$.)

But, as is the case with confidence distributions and Fisher's (1930) original fiducial distribution (Fisher, 1930), the calculus of probability does not apply, and marginal fiducial densities, generalized or not, have no probability interpretation. They may lead to confidence intervals or sets of confidence intervals that can be verified in simulations to have the stated coverage, but this needs to be checked in each problem.

A strength of the generalized fiducial approach is that by emphasizing the data-generating equation, it forces users to think about the structure of the model being fitted. It also lends itself to simulation. Hannig and co-authors have applied the method to a number of modern problems, including non-parametric estimation of a survival function (Cui and Hannig, 2019), high-dimensional regression (Lai *et al.*, 2015), linear mixed models (Cisewski and Hannig, 2012; Lidong *et al.*, 2008), and wavelet regression (Hannig and Lee, 2009). Hannig *et al.* (2016) is a good entry into this literature; see also Murph *et al.* (2024). The approach has also been used for models with discrete observations, although the inversion at (3.16) involves additional randomness (Hannig, 2009); the resulting confidence intervals are similar to continuity corrected intervals, at least in simple cases.

A weakness of the generalized fiducial approach is that additional devices seem to be needed in many complex problems, and the theoretical basis of these is not always clear, although each device may seem sensible in the context of the problem. Dawid's Example 5.17 (Dawid, 2024), is called there "non-partitionable", which in particular means that the fiducial solution must necessarily involve some *ad hoc* choice. He concludes "it could be argued ... that the model simply does not support fiducial inference".

Hannig *et al.* (2016) write:

> From a practical point of view, GFI is used in a way similar to the use of a posterior computed using a default (objective) prior, such as probability matching, reference, or flat prior. The main technical difference is that the objective prior is replaced by a data-dependent Jacobian.

As with objective Bayesian approaches, the resulting construct is not a probability statement about the parameter, but rather a statement whose empirical properties must be investigated.

3.4.3 Confidence Distributions

A general approach to the construction of confidence distributions was reviewed in Xie and Singh (2013), building on Cox (1958), Efron (1993), Schweder and Hjort (2002), and Xie *et al.* (2011), and reviewed in Thornton and Xie (2024)

Xie and Singh (2013, Definition 1) used the definition of a confidence distribution of Schweder and Hjort (2002): any function on the parameter space and sample space that is monotone increasing between 0 and 1 as a function of the parameter, and has a uniform $(0, 1)$ distribution as a function of the data. More precisely, they write

> A function $H_n(\cdot) = H_n(\mathbf{x}, \cdot)$ on $\mathcal{X} \times \Theta \to [0, 1]$ is called a confidence distribution for a parameter θ, if R1) For each given $\mathbf{x} \in \mathcal{X}$, $H_n(\cdot)$ is a cumulative distribution function on Θ; R2) At the true parameter value $\theta = \theta_0$, $H_n(\theta_0) \equiv H_n(\mathbf{x}, \theta_0)$, as a function of the sample $\mathbf{x}$, follows the uniform distribution $U[0, 1]$.

In the notation of Xie and Singh (2013), $\mathbf{x} = (x_1, \ldots, x_n)$ is a sample, and the dependence on n is introduced in order to extend the exact requirement of uniformity to an approximate one, with the approximation error (hopefully) decreasing as n increases.

This definition is agnostic about the source of the function H_n, allowing as well that there may be more than one such function. As several discussants of Xie and Singh (2013) point out, this vagueness in definition makes it difficult to distinguish a confidence distribution from a fiducial distribution, beyond the almost tautological point that confidence distributions are meant to be built up from sets of nested confidence intervals, derived by some other means. The parameter of the confidence distribution in Xie and Singh (2013) is one-dimensional, although even with this restriction confidence distributions cannot be manipulated like probability distributions. Schweder and Hjort (2013) noted that a confidence distribution for μ in a $N(\mu, 1)$ model cannot be transformed into a confidence distribution for $|\mu|$, for example.

Schweder and Hjort (2002; 2016, Chapter 6) discussed the close connection between confidence distributions and Fisher's fiducial approach. They proposed, in most cases, to seek confidence distributions for single parameters of interest, which they call focus parameters, often using as the basic quantity the profile log-likelihood ratio: in the notation of Section 3.4.1, with a model $f(y; \theta)$ and focus parameter $\psi = \psi(\theta)$, they proposed to base confidence curves for ψ on an identified pivotal quantity, i.e., a function of ψ and the data with a known distribution, and in regular statistical models they emphasize the pivotal

$$D(\psi) = 2\{\ell(\hat{\theta}) - \ell(\hat{\theta}_\psi)\}, \quad \hat{\theta}_\psi = \arg\sup \ell(\theta)_{\psi(\theta)=\psi}, \tag{3.18}$$

where $\ell(\theta) = \ell(\theta; y) \propto f(y;\theta)$ is the log-likelihood function. This pivotal quantity has, at least approximately, a χ^2_1 distribution, when θ is the data-generating value, subject to some regularity conditions on the model.

To extend the range of problems that can be considered, Schweder and Hjort (2016, Chapters 3 and 4) use pivotal quantities other than (3.18) where the problem demands it. Examples include problems where the maximum likelihood estimate might be on the boundary of the parameter space, or problems with increasing numbers of nuisance parameters that can be handled by finding an appropriate marginal or conditional likelihood from which to construct a pivotal quantity. For example in the many-normal-means problem, if the target parameter is $\Sigma\theta_i^2$, then a preliminary reduction to Σy_i^2 can be used to construct a confidence distribution. While the range of problems treated is impressively large, the approach does not in general avoid an arbitrariness similar to that embodied in the definition in Xie and Singh (2013) quoted above.

When θ is a d-dimensional parameter, the χ^2_d approximation to the distribution of $D(\theta) = 2\{\ell(\hat{\theta}) - \ell(\theta)\}$ can be used to define a d-dimensional confidence distribution. Schweder and Hjort (2016, Chapter 9) note that "reducing the full confidence distribution to marginal confidence distributions for the parameters of interest ... cannot be done by integrating the confidence density", but go on to note that this might work approximately – "Whether the approximation is satisfactory must be checked, usually by simulation". This is consistent with the approach taken in generalized fiducial inference (Cisewski and Hannig, 2012; Hannig *et al.*, 2016).

A special issue on confidence distributions was published in 2018 by the *Journal of Statistical Planning and Inference* and includes many extensions of the methodology. See also Nadarajah *et al.* (2015) for an overview and several examples.

3.4.4 Significance Functions

Significance functions, or p-value functions, are essentially confidence distributions by a different name, but with an explicit focus on a single parameter. Fraser (1991) sets out the basic idea and emphasizes the link to the likelihood function. Just as the likelihood function is not a density for a parameter, so the significance function is not a distribution for the parameter, but rather a way of describing what values of the parameter are consistent with the data.

Fraser and Reid (1993), Fraser and Reid (1995) and Fraser (2017) combined the notion of significance function with a particular construction of that function. In a model for data $y \in \mathbb{R}^n$, with parameter $\theta \in \mathbb{R}^d$, and a parameter of interest $\psi \in \mathbb{R}$, as with confidence distributions and fiducial distributions, a pivotal quantity forms the basis for the significance function. In Fraser et al. (1999), the construction of the pivotal quantity first requires a dimension reduction from n to d, which is based on a local location model, generalizing an argument in Fraser (1964). This reduction is discussed in Davison and Reid (2024).

To motivate this reduction, consider Fisher's fiducial equation (3.3), but now using y in place of T:

$$\mathrm{d}f = -\frac{\partial}{\partial\theta}F(y;\theta)\mathrm{d}\theta = f(y;\theta)\frac{-\partial F(y;\theta)/\partial\theta}{f(y;\theta)} = L(\theta;y)\left|\frac{\mathrm{d}y}{\mathrm{d}\theta}\right|, \tag{3.19}$$

where $\mathrm{d}y/\mathrm{d}\theta = -F_{;\theta}(y;\theta)/F_{y;}(y;\theta)$ is the change in y with respect to θ, when the value of the distribution function is held fixed (and the abbreviated subscript notation is used for the derivative with respect to y or θ). This is generalized to the (n, d) setting as

$$\frac{\mathrm{d}y}{\mathrm{d}\theta} = \begin{pmatrix} V_1(\theta) \\ \vdots \\ V_n(\theta) \end{pmatrix} = V(\theta), \tag{3.20}$$

with each d-vector $V_i(\theta) = \mathrm{d}y_i/d\theta^{\mathrm{T}}$, obtained from the cumulative distribution function of the component y_i. (The determinant in equation (3.19) is then $|V^{\mathrm{T}}(\theta)V(\theta)|^{1/2}$.) The generalized fiducial density of Murph *et al.* (2024, Theorem 13.1) has a similar form, although they allow the "determinant-like object", here denoted $|dy/d\theta|$, to be defined in different ways.

The reduction to a d-dimensional statistical model uses the matrix $V(\hat{\theta})$ to define a tangent exponential model with canonical parameter $\varphi(\theta)$ (Reid and Fraser, 2010; Davison and Reid, 2024):

$$f_{\mathrm{TEM}}(s;\theta) = \exp\{\varphi^{\mathrm{T}}(\theta)s(y) + \ell(\theta; y^0)\}h(s). \tag{3.21}$$

In (3.21) $\ell(\theta; y^0)$ is the log-likelihood function at the observed data point y^0, and

$$\varphi(\theta) = \varphi(\theta; y^0) = \frac{\partial \ell(\theta; y^0)}{\partial V} = \sum_{i=1}^{n} \frac{\partial \ell(\theta; y^0)}{\partial y_i} V_i, \tag{3.22}$$

where the last equality assumes the components $y_1, \ldots, y_n$ are independently distributed. The model (3.21) is the basis for the development of an approximation to a significance function for ψ. The saddlepoint approximation to the density of s in an exponential family model can be applied to the density in equation (3.21), following arguments very similar to those developed in Barndorff-Nielsen (1990), and the resulting conclusion is that a modified version of the log-likelihood ratio statistic is a suitable one-dimensional pivotal quantity.

Let $r = r(\psi; y) = \pm\{2D(\psi)\}^{1/2}$, where $D(\psi)$ is defined at (3.18), and the sign on r is the same as the sign of $\hat{\psi} - \psi$. As the limiting distribution of D is χ^2_1, the limiting distribution of r is standard normal. Thus, an approximate significance function for ψ is $\Phi\{r(\psi; y^0)\}$. This approximation is accurate to $O(n^{-1/2})$, and is identical to the deviance-based approximate confidence distribution (3.18) of Schweder and Hjort (2016). A more accurate approximate significance function is $\Phi\{(r^*(\psi; y^0)\}$, where

$$r^*(\psi; y) = r + \frac{1}{r} \log \frac{Q}{r}, \tag{3.23}$$

and $Q = Q(\psi; y^0)$ is computed from $\ell(\theta; y^0)$ and $\varphi(\theta; y^0)$ in (3.21), see Davison and Reid (2024) for an explicit expression.

As a function of y with ψ fixed, the distribution of $\Phi(r^*)$ is $U(0, 1)$ under the model, to $O(n^{-3/2})$, so provides a confidence distribution in the sense of Schweder and Hjort (2002) and Xie and Singh (2013). The details of the construction of the tangent exponential model, and the normal approximation to the distribution of $r^*(\psi; y)$ under the model $f(y; \theta)$ are given in Davison and Reid (2024), building on Brazzale *et al.* (2007, Section 8.6) and a series of papers by Fraser and others.

The inferential argument that $r^*(\psi; y)$ is an appropriate pivotal quantity is more difficult and relies on showing that it emerges from conditioning on approximately ancillary statistics, first for θ in the full model, and then for ψ in the reduced model and under the constraint that ψ is fixed but θ is otherwise unconstrained. The argument is sketched in

Reid and Fraser (2010) and Fraser et al. (1999), and questions of uniqueness are considered in Fraser et al. (2010a).

3.4.5 Belief Functions and Inferential Models

A different approach to describing probability for unknown quantities is to change the formulation of probability itself. An early version of this is the development of upper and lower probabilities in Dempster (1966), extended to belief functions in the Dempster–Shafer theory (Shafer, 1976). Helpful overviews of this theory are provided in Wasserman (1990) and Dempster (2008); see also Shafer (2024) and Gong (2024). By assigning probabilities to sets of events, and defining upper and lower probabilities, the theory permits a conclusion of uncertainty for the truth of an event. This is related to the theory of imprecise probabilities developed in Walley (1990). While this theory is philosophically appealing from some points of view, it has not had broad impact on statistical practice.

One modern version of this theory is the development of inferential models (Martin and Liu, 2016). These models are developed in a manner reminiscent of Hannig's generalized fiducial distribution (Hannig, 2009). The latter is computed using a data-generating equation $Y = G(U, \theta)$ and an independently generated copy of U^*, as in Section 3.4.2. The inferential model approach proposes using a random set to predict U; this random set is converted to a belief function. Dawid (2024, Section 5.9) describes how Hannig's generalized fiducial inference and Martin and Liu's inferential models build on the theory of belief functions.

The derivation of the inferential model is guided by the requirement that the resulting inference is both valid and efficient, properties which are analogous to coverage and length of a confidence interval; definitions are given in Martin and Liu (2016, Chapter 4). Martin (2019) argued that it is not possible to obtain valid statistical inference in general without using a theory of non-additive probability, such as that provided by inferential models. Balch *et al.* (2019) argued that Bayesian and fiducial methods of inference are necessarily at risk of being invalid, because they are based on the usual probability calculus, rather than a system of belief functions.

Liu and Martin (2024) link their development of inferential models to the theory of possibility measures, which involve a particular type of belief function.

3.5 Comparisons

As developments in statistical science turned increasingly to implementation of methods, the historical tension among Bayesian, frequentist and fiducial approaches was not resolved so much as ignored. The current study of the theory described above is ongoing, and the test of time may be which versions are most flexible and computationally feasible in complex models.

Some points of comparison of the approaches include the nature of the probability statements sought, the adjustment or allowance for nuisance parameters, and the potential for calibration.

Even the nature of the probability statement intended is not always clear from what is written. There seems to be a presumption in the application of Bayesian methods in practice that the posterior probability has an epistemic interpretation, but the marginal posterior probability for a quantity of interest has no such interpretation unless the prior does, and various proposed noninformative or objective priors have no such meaning. Berger (2006) provided a series of philosophical viewpoints around Bayesian analysis using objective priors which emphasizes their practical importance over their interpretation. The weakest of these is his fourth: "Objective Bayesian analysis is simply a collection of *ad hoc* but useful methodologies for learning from data"; this is close to the viewpoint outlined by Fraser (2011). Berger *et al.* (2010) discussed the formal probability interpretation of a posterior distribution obtained using the reference prior of Bernardo (1979).

Empirical Bayes inference (Efron, 2010; 2024) is a seemingly firmly frequentist approach to inference, estimating from the sample either the marginal density $m(y)$ or the prior density $\pi(\theta)$: Efron's f-estimates and g-estimates, respectively. Much of the literature on empirical Bayes methodology emphasizes its use in nonparametric settings and establishing concentration rates for posteriors (see, e.g. Donnet *et al.* (2018); Petrone *et al.* (2014); Zhang and Gao (2020)). Efron (2024, Sections 2.5, 2.6) treats the estimated posterior as a distribution for θ; see, e.g., his equation (2.95), but it is not clear to me how much is assumed in his phrase "Taking $\hat{g}$ literally".

Generalized fiducial distributions are described as a pragmatic way to generate (sets of) confidence intervals, for example Hannig *et al.* (2016, p. 1348) write "GFD ... should be viewed as a distribution estimator ... of the fixed true parameter. To validate this distribution estimator in a specific example, we then typically demonstrate good small sample performance by simulation and prove good large sample properties by asymptotic theorems". This suggests the operational definition of a method of inference that is assessed by its calibration properties, as described in Reid and Cox (2015). Significance functions are used exactly in the same way in Fraser (1991), as a summary of the information about a parameter available from the data, with properties of the summary to be assessed by calibration.

Schweder and Hjort (2016, p. 4) described a confidence distribution as a "gold standard for epistemic probability distributions in science". They emphasized that confidence densities cannot be treated by the usual probability calculus, but they ascribed this to the inapplicability of the usual calculus to epistemic probability. In their construction of confidence distributions, they ensured that its properties in repeated sampling, i.e. its aleatory or empirical properties, are guaranteed. The epistemic nature of the confidence distribution is then simply a statement about the uncertainty of any conclusions we may draw from the data at hand. It is not clear what is gained by describing the confidence distribution as an epistemic probability, which seems to put it on the same foundational footing as a posterior probability. But their approach does satisfy the requirement described in Reid and Cox (2015) that any inferential procedure be treated as a measuring device, and its properties assessed empirically.

The approach to inference based on confidence distributions outlined by Xie and Singh (2013) is similarly proposed as a method for generating reasonable inferences, "a redefinition of the confidence distribution as a *purely frequentist concept*", although in the same section they describe a confidence distribution as a "probability distribution on the parameter space", while making clear in later sections that this distribution does not obey the usual rules of probability calculus. More recent work emphasizes techniques for combining confidence distributions; see, e.g., Shen *et al.* (2020).

Thornton and Xie (2024) argue that the theory of confidence distributions can serve as a bridge between Bayes, frequentist and fiducial approaches.

> Regardless of whether the setting is parametric or nonparametric, normal or non-normal, exact or asymptotic, CD inference is possible as long as one can create confidence intervals (or regions) of all levels for the parameter of interest.

This is essentially the description of a confidence distribution as a summary statistic, or function, and is agnostic about the method used to derive it. From that point of view, the difficulties inherent in eliminating nuisance parameters, for example, have been solved by the method used to create the confidence distribution, but there remains no general theory for ensuring that this is possible, and the interpretation of the result is firmly aleatory.

The conclusions from inferential models are intended to be epistemic: "The inferential output should be interpreted as a belief probability, or a range of belief probabilities. That is, there is no notion of frequencies in its interpretation, it is simply a measure of evidence in data" (Martin and Liu, 2016, p. 13). In this sense, inferential models and belief functions provide a different axiomatic theory, seemingly contradicting the statement of Schweder and Hjort above. However, Martin and Liu (2016) emphasized what they call validity as a necessary component of inferential models, which seems to me to put priority on the calibration of the resulting inference.

Indeed all the methods outlined in Section 3.4 are generally described by nearly all their proponents as needing to be assessed by their frequentist performance. The only method for which such assessment is seen as irrelevant is a completely subjective Bayes approach, although even there the possibility may be allowed that for any particular application of the method, the inference obtained is not very much influenced by the choice of prior distribution (Goldstein, 2006).

There is also general agreement among the proponents of approaches of Section 3.4 that inference for a parameter of interest cannot be obtained by marginalizing over a joint posterior, or confidence, or fiducial, distribution. The theory of reference priors, for example, is very clear about this (Berger and Bernardo, 1992). Schweder and Hjort (2016, Chapter 2.4) emphasized the construction of confidence distributions for a parameter of interest, which they call a focus parameter, by reduction of the data in advance of the construction. This typically proceeds by finding an exact or approximate pivotal quantity.

In more general settings, but with smooth models, approximate pivotal quantities can be obtained by using the classical summary statistics from likelihood inference. Xie and Singh (2013) often used the Wald statistic $(\hat{\psi} - \psi)/\hat{\sigma}_{\psi}$, which under repeated sampling from the model follows an approximate standard normal distribution, subject to regularity conditions. Schweder and Hjort (2016) usually use the asymptotically equivalent pivotal $D(\psi) = 2\{\ell(\hat{\psi}, \hat{\lambda}) - \ell(\psi, \hat{\lambda}_{\psi})\}$; its approximate sampling distribution is χ^2_1. In non-regular models special arguments are needed to find a suitable pivotal quantity. Hannig's approach to generalized fiducial distributions sometimes reduces the problem to a pivotal quantity first, and sometimes marginalizes a multi-parameter generalized fiducial distribution, but then checks that the resulting inference is approximately valid.

As described in Davison and Reid (2024), the theory of significance functions identifies a particular pivotal quantity, $r^*(\psi; y)$, based on arguments derived from higher-order asymptotic expansions of likelihood-based quantities. An advantage of insisting on this, beyond the possibly improved distributional approximation, is that approximate elimination of nuisance parameters is built into its construction. In effect, nuisance parameters are integrated out of the model using Laplace approximation, and the validity of this integration step also rests on higher-order approximations to likelihood-based quantities. In this way, the significance function approach removes the *ad hoc* choices involved in developing confidence distributions or generalized fiducial distributions. Fraser was convinced that the pivotal quantity $r^*(\psi; y)$ was essentially unique, but the arguments in support of this still seem to me to be somewhat elusive.

Higher-order asymptotic theory assumes a high degree of regularity in the models, whereas with non-regular problems there is often a clever construction that can be exploited. For example, Cunen *et al.* (2018) proposed two methods for constructing confidence distributions for change-points, one based on a test for no change, and another based on the simulated distribution of the profile log-likelihood ratio statistic. Cui and Hannig (2019) showed how to invert a pair of data-generating equations for censored survival data to obtain a non-parametric generalized fiducial distribution for the survival function. I am not sure if either of these problems can be readily handled using the r^* construction.

The theory of inferential models has both a method for conditioning on auxiliary variables, to reduce the dimension of the data to that of the parameter, and a method for marginalizing over auxiliary variables to eliminate nuisance parameters; see Martin and Liu (2016, Chapters 6 and 7).

3.6 Conclusion

This is necessarily a very selective overview, and the quotes from other papers in this volume and in the literature can give only a limited view. The emphasis on distributions for parameters means that nonparametric models are not addressed. Vovk (2024) describes nonparametric fiducial methods for predicting future observations, and shows how a generalization of this is connected to the theory of conformal prediction. Taraldsen and Lindqvist (2024) discuss fiducial decision theory; the emphasis is on an action, or decision, made on the basis of some data, model and loss function.

Sen *et al.* (2024) study an enormously complex parametric model, which could be viewed as practically non-parametric, arguing that a Bayesian approach to inference is more likely to be reproducible. Implicitly, the high-dimensional posterior distribution is integrated to obtain marginal posterior distributions; it seems likely that this is subject to the Stein phenomenon described in Section 3.4.1. Similarly, the massive data sets envisaged in Hector *et al.* (2024) require more emphasis on computation than interpretation; the goal in that paper is to construct efficient estimators of parameters that borrow strength from datasets of similar structure, which requires first assessing which datasets these might be. The focus on individual parameters and their potential distribution is of less interest in these complex models.

Berger (2022) categorized several types of frequentists, or several flavours of frequentism, and Berger (2006) described four versions of objective Bayes. This introduction, and volume, confirms that there are many approaches to find distributions for parameters, and perhaps also confirms that the collective noun for statisticians is *a disagreement.*

It is perhaps natural, given the subtlety of the concepts involved, that there continues to be disagreement, or at least not convergence, among various approaches to the interpretation of distributions for parameters. The true test of statistical theory, though, is its utility and reliability for applications, and often the disagreement isn't as stark, at least in problems with relatively well-understood models and modest amounts of data.

A recent series of papers on the estimation of the probability of collisions among orbiting satellites illuminates some of these points. The calculations are essentially those for inference on the length of a multivariate normal vector, (Stein, 1959), discussed in Section 3.4.1 and in Dawid (2024, Section 5.7.2). Balch *et al.* (2019) note the mis-calibration of the Bayes posterior, and state a general principle which they call the false confidence theorem. Cunen *et al.* (2021) pointed out that the problem is solved by first reducing to the obvious pivotal quantity, the length of the response vector, before constructing a confidence distribution. Martin *et al.* (2021) responded that their theorem continues to be valid, as confidence distributions cannot be manipulated as if they were probability distributions. As far as practical solution of the problem of satellite collisions, though, the approach of Cunen *et al.* (2021) or the improvement developed in Elkantassi and Davison (2022) based on Section 3.4.4 seem quite satisfactory.

We rely on foundations and theory to provide a basis for developing solutions to new practical problems; without this each application would require some *ad hoc* solution anew. The papers in this volume illustrate the subtleties in foundational concepts and are at the same time motivated by the desire to improve statistical practice.

Bibliography

Balch, M. S., Martin, R. and Ferson, S., 2019. Satellite conjunction analysis and the false confidence theorem. *Proceedings of the Royal Society A*, *475*, p. 120180565. https://doi.org/10.1098/rspa.2018.0565.

Barndorff-Nielsen, O. E., 1990. Approximate interval probabilities. *Journal of the Royal Statistical Society: Series B*, *52*, pp. 485–496.

Bayes, T., 1763. An essay towards solving a problem in the doctrine of chances. *Philosophical Transactions of the Royal Society*, *53* and *54*, pp. 370–418; 296–325. Reprinted in *Biometrika*, *45*, 1958, pp. 293–315.

Berger, J. O., 2006. The case for objective Bayes analysis. *Bayesian Analysis*, *1*, pp. 385–402.

Berger, J. O., 2022. Four types of frequentism and their interplay with Bayesianism (with discussion). *New England Journal of Statistics in Data Science*, *1*, pp. 126–137.

Berger, J. O. and Bernardo, J. M., 1992. On the development of reference priors (with discussion). In *Bayesian Statistics 4*, eds J. M. Bernardo, J. O. Berger, A. P. Dawid and A. F. M. Smith, pp. 35–60. Oxford: Oxford University Press.

Berger, J. O., Bernardo, J. M. and Sun, D., 2010. The formal definition of reference priors. *Annals of Statistics*, *37*, pp. 905–938.

Berger, J. O., Garcia-Donato, G., Moreno, E. and Pericchi, L., 2024. Objective Bayesian testing and model uncertainty. In *A Handbook of Bayesian, Fiducial, and Frequentist Inference*, eds J. Berger, X-L. Meng, N. Reid and M. Xie, pp. 132–160. Boca Raton: Chapman and Hall/CRC.

Berger, J. O., Insua, D. R. and Ruggeri, F., 2000. Bayesian robustness. In *Robust Bayesian Analysis*, eds D. R. Insua and F. Ruggeri, Lecture Notes in Statistics, vol. 152, pp. 1–32. New York: Springer.

Bernardo, J. M., 1979. Reference posterior distributions for Bayesian inference (with discussion). *Journal of the Royal Statistical Society: Series B*, *41*, pp. 113–147.

Bhadra, A., Datta, J., Polson, N. G. and Willard, B., 2016. Default Bayesian analysis with global-local shrinkage priors. *Biometrika*, *103*, pp. 955–969.

Box, G. E. P. and Cox, D. R., 1964. An analysis of transformations (with discussion). *Journal of the Royal Statistical Society: Series B*, *26*, pp. 211–246.

Brazzale, A. R., Davison, A. C. and Reid, N., 2007. *Applied Asymptotics*. Cambridge: Cambridge University Press.

Castillo, I., Schmidt-Hieber, J. and van der Vaart, A., 2015. Bayesian linear regression with sparse priors. *Annals of Statistics, 43*, pp. 1986–2018.

Cisewski, J. and Hannig, J., 2012. Generalized fiducial inference for normal linear mixed models. *Annals of Statistics, 40*, pp. 2101–2127.

Consonni, G., Fouskakis, D., Liseo, B. and Ntzoufras, I., 2018. Prior distributions for objective Bayesian analysis. *Bayesian Analysis, 13*, pp. 627–679.

Cox, D. R., 1958. Some problems connected with statistical inference. *Annals of Mathematical Statistics, 29*, pp. 375–372.

Cox, D. R., 2016. Some pioneers of modern statistical theory: A personal reflection. *Biometrika, 103*, pp. 747–759.

Cui, Y. and Hannig, J., 2019. Nonparametric generalized fiducial inference for survival functions under censoring. *Biometrika, 106*, pp. 501–518.

Cunen, C., Hermansen, G. and Hjort, N. L., 2018. Confidence distributions for change-points and regime shifts. *Journal of Statistical Planning and Inference, 195*, pp. 14–34.

Cunen, C., Hjort, N. L. and Schweder, T., 2021. Confidence in confidence distributions! *Proceedings of the Royal Society A, 476*, p. 20190781.

Datta, G. S. and Mukerjee, R., 2004. *Probability Matching Priors: Higher Order Asymptotics.* New York: Springer.

Davison, A. C. and Reid, N., 2024. The tangent exponential model. In *Handbook of Bayesian, Fiducial, and Frequentist Inference*, eds J. Berger, X-L. Meng, N. Reid and M. Xie, pp. 210–237. Boca Raton: Chapman and Hall/CRC.

Dawid, A. P., 2024. Fiducial inference: Then and now. In *Handbook of Bayesian, Fiducial, and Frequentist Inference*, eds J. Berger, X-L. Meng, N. Reid and M. Xie, pp. 83–105. Boca Raton: Chapman and Hall/CRC.

Dawid, A. P., Stone, M. and Zidek, J. V., 1973. Marginalization paradoxes in Bayesian and structural inference. *Journal of the Royal Statistical Society: Series B, 35*, pp. 189–233.

Dempster, A. P., 1966. New methods for reasoning towards posterior distributions based on sample data. *Annals of Mathematical Statistics, 37*, pp. 355–374.

Dempster, A. P., 2008. The Dempster–Shafer calculus for statisticians. *International Journal of Approximate Reasoning, 48*, pp. 365–377.

Diaconis, P. and Freedman, D., 1986. On the consistency of Bayes estimates. *Annals of Statistics, 14*, pp. 1–26.

Donnet, S., Rivoirard, V., Rousseau, J. and Scricciolo, C., 2018. Posterior concentration rates for empirical Bayes procedures with applications to Dirichlet process mixtures. *Bernoulli, 24*, pp. 231–256.

Efron, B., 1993. Bayes and likelihood calculations from confidence intervals. *Biometrika, 80*, pp. 3–26.

Efron, B., 2010. *Large-Scale Inference: Empirical Bayes Methods for Estimation, Testing, and Prediction.* Cambridge: Cambridge University Press.

Efron, B., 2024. Empirical Bayes: Concepts and methods. In *Handbook of Bayesian, Fiducial, and Frequentist Inference*, eds J. Berger, X-L. Meng, N. Reid and M. Xie, pp. 8–34. Boca Raton: Chapman and Hall/CRC.

Elkantassi, S. and Davison, A. C., 2022. Space oddity? A statistical formulation of conjunction assessment. *Journal of Guidance, Control, and Dynamics, 45*, pp. 2258–2274.

Evans, M. and Moshonov, H., 2006. Checking for prior-data conflict. *Bayesian Analysis, 1*, pp. 893–914.

Fisher, R. A., 1930. Inverse probability. *Proceedings of the Cambridge Philosophical Society, xxvi*, pp. 528–535.

Fisher, R. A., 1956. *Statistical Methods and Scientific Inference.* Edinburgh: Oliver & Boyd.

Fraser, A. M., Fraser, D. A. S. and Staicu, A.-M., 2010a. Second order ancillary: A differential view with continuity. *Bernoulli, 16*, pp. 1208–1223.

Fraser, D. A. S., 1964. Local conditional sufficiency. *Journal of the Royal Statistical Society: Series B, 26*, pp. 52–62.

Fraser, D. A. S., 1966. Structural probability and a generalization. *Biometrika, 53*, pp. 1–9.

Fraser, D. A. S., 1968. *The Structure of Inference.* New York: Wiley.

Fraser, D. A. S., 1991. Statistical inference: Likelihood to significance. *Journal of the American Statistical Association, 86*, pp. 258–265.

Fraser, D. A. S., 2011. Is Bayes posterior just quick and dirty confidence? *Statistical Science, 26*, pp. 299–316.

Fraser, D. A. S., 2017. *p*-values: The insight to modern statistical inference. *Annual Review of Statistics and Its Application, 4*, pp. 1–14.

Fraser, D. A. S. and Reid, N., 1992. Strong matching of frequentist and Bayesian parametric inference. *Journal of Statistical Planning and Inference, 103*, pp. 263–285.

Fraser, D. A. S. and Reid, N., 1993. Third order asymptotic models: Likelihood functions leading to accurate approximation to distribution functions. *Statistica Sinica, 3*, pp. 67–82.

Fraser, D. A. S. and Reid, N., 1995. Ancillaries and third order significance. *Utilitas Mathematica, 47*, pp. 33–53.

Fraser, D. A. S., Reid, N., Marras, E. and Yi, G. Y., 2010b. Default priors for Bayesian and frequentist inference. *Journal of the Royal Statistical Society: Series B, 72*, pp. 631–654.

Fraser, D. A. S., Reid, N. and Wu, J., 1999. A simple general formula for tail probabilities for frequentist and Bayesian inference. *Biometrika, 86*, pp. 249–264.

Gelman, A., Jakulin, A., Pittau, M. G. and Su, Y. S., 2008. A weakly informative default prior distribution for logistic regression models. *Annals of Applied Statistics, 2*, pp. 1360–1383.

Goldstein, M., 2006. Subjective Bayesian analysis: Principles and practice. *Bayesian Analysis, 1*, pp. 403–420.

Gong, R., 2024. Dempster-Shafer theory for statistical inference. In *Handbook of Bayesian Fiducial and Frequentist Inference*, eds J. Berger, X-L. Meng, N. Reid and M. Xie. Boca Raton: Chapman and Hall/CRC. pp. 300–322.

Hannig, J., 2009. On generalized fiducial inference. *Statistica Sinica*, *19*, pp. 491–544.

Hannig, J., Iyer, H., Lai, R. C. S. and Lee, T. C. M., 2016. Generalized fiducial inference: A review and new results. *Journal of the American Statistical Association*, *111*, pp. 1346–1361.

Hannig, J. and Lee, T. C. M., 2009. Generalized fiducial inference for wavelet regression. *Biometrika*, *96*, pp. 847–860.

Hector, E.C., Tang, L., Zhou, L. and Song, P. X.-K., 2024. Data integration and model fusion in the Bayesian, fiducial and frequentist framerwork. In *Handbook of Bayesian, Fiducial, and Frequentist Inference*, eds J. Berger, X-L. Meng, N. Reid and M. Xie, pp. 238–263. Boca Raton: Chapman and Hall/CRC.

Heinrich, J., 2006. The Bayesian approach to setting limits – what to avoid. In *Statistical Problems in Particle Physics, Astrophysics and Cosmology*, eds L. Lyons and M. K. Ünel, pp. 98–101. London: Imperial College Press.

Jeffreys, H., 1961. *Theory of Probability* (3rd ed.). Oxford: Oxford University Press.

Kass, R. E. and Wasserman, L., 1996. The selection of prior distributions by formal rules. *Journal of the American Statistical Association*, *91*, pp. 1343–1370.

Kleijn, B. J. K. and van der Vaart, A. W., 2012. The Bernstein-Von-Mises theorem under model misspecification. *Electronic Journal of Statistics*, *6*, pp. 354–381.

Kline, D., Li, Z., Chu, Y., Wakefield, J., Miller, W. C., Turner, A. N. and Clark, S. J., 2021. Estimating seroprevalence of SARS-CoV-2 in Ohio: A Bayesian multilevel poststratification approach with multiple diagnostic tests. *Proceedings of the National Academy of Sciences*, *118*. https://doi.org/10.1073/pnas.2023947118.

Lai, R. C. S., Hannig, J. and Lee, T. C. M., 2015. Generalized fiducial inference in ultrahigh-dimensional regression. *Journal of the American Statistical Association*, *110*, pp. 760–772.

Lang, J. B., 2024. Slicing and dicing a path through the fiducial forest. In *Handbook of Bayesian, Fiducial, and Frequentist Inference*, eds J. Berger, X-L. Meng, N. Reid and M. Xie, pp. 323–343. Boca Raton: Chapman and Hall/CRC.

Lehmann, E. L., 1993. The Fisher, Neyman-Pearson theories of testing hypotheses: One test or two? *Journal of the American Statistical Association*, *88*, pp. 1242–1249.

Lemoine, N., 2019. Moving beyond noninformative priors: Why and how to choose weakly informative priors in Bayesian analysis. *Oikos*, *128*, pp. 912–928.

Lidong, E., Hannig, J. and Iyer, H. K., 2008. Fiducial intervals for variance components in an unbalanced two-component normal mixed linear model. *Journal of the American Statistical Association*, *103*, pp. 854–864.

Liu, C. and Martin, R., 2024. Inferential models and possibility measures. In *Handbook of Bayesian, Fiducial, and Frequentist Inference*, eds J. Berger, X-L. Meng, N. Reid and M. Xie, pp. 344–363. Boca Raton: Chapman and Hall/CRC.

Martin, R., 2019. False confidence, non-additive beliefs, and valid statistical inference. *International Journal of Approximate Reasoning, 113*, pp. 39–73.

Martin, R., Balch, M. S. and Ferson, S., 2021. Response to the comment confidence in confidence distributions! *Proceedings of the Royal Society A, 477*, p. 20200579.

Martin, R. and Liu, C., 2016. *Inferential Models: Reasoning with Uncertainty.* Boca Raton: Chapman & Hall/CRC.

Martin, R. and Ning, B., 2020. Empirical priors and coverage of posterior credible sets in a sparse normal mean model. *Sankhya A, 82*, pp. 477–498.

Meng, X.-L., 2024. A BFFer's exploration with nuisance constructs: Bayesian p-value, H-likelihood, and Cauchyanity. In *Handbook of Bayesian, Fiducial, and Frequentist Inference*, eds J. Berger, X-L. Meng, N. Reid and M. Xie, pp. 161–187. Boca Raton: Chapman and Hall/CRC.

Mukerjee, R., 2008. Data-dependent probability matching priors for empirical and related likelihoods. In *Pushing the Limits of Contemporary Statistics: Contributions in Honor of Jayanta K. Ghosh*, eds B. Clarke and S. Ghosal, pp. 60–70. Hayward: IMS Collections.

Mukerjee, R. and Ghosh, M., 1997. Second-order probability matching priors. *Biometrika, 84*, pp. 970–975.

Murph, A., Hannig, J. and Williams, J. P., 2024. Introduction to generalized fiducial inference. In *Handbook of Bayesian, Fiducial, and Frequentist Inference*, eds J. Berger, X-L. Meng, N. Reid and M. Xie, pp. 276–299. Boca Raton: Chapman and Hall/CRC.

Nadarajah, S., Bityukov, S. and Krasnikov, N., 2015. Confidence distributions: A review. *Statistical Methodology, 22*, pp. 23–46.

Neyman, J., 1937. Outline of a theory of statistical estimation based on the classical theory of probability. *Philosophical Transactions of the Royal Society A, 236*, pp. 333–380.

Nicolaou, A., 1993. Bayesian intervals with good frequentist properties in the presence of nuisance parameters. *Journal of the Royal Statistical Society: Series B, 55*, pp. 377–390.

van der Pas, S., Szabo, B. and van der Vaart, A., 2017. Uncertainty quantification for the horseshoe (with discussion). *Bayesian Analysis, 12*, pp. 1221–1274.

Peers, H. W., 1965. On confidence points and Bayesian probability points in the case of several parameters. *Journal of the Royal Statistical Society: Series B, 27*, pp. 9–16.

Petrone, S., Rousseau, J. and Scricciolo, C., 2014. Bayes and empirical Bayes: Do they merge? *Biometrika, 101*, pp. 285–302.

Reid, N. and Cox, D. R., 2015. On some principles of statistical inference. *International Statistical Review, 83*, pp. 293–308.

Reid, N. and Fraser, D. A. S., 2010. Mean loglikelihood and higher-order approximations. *Biometrika, 97*, pp. 159–170.

Reid, N. and Sun, Y., 2010. Assessing sensitivity to priors using higher order approximations. *Communications in Statistics—Theory and Methods, 39*, pp. 1373–1386.

Savage, L. J., 1954. *The Foundations of Statistics.* New York: Wiley.

Schervish, M. J., 1995. *Theory of Statistics.* New York: Springer.

Schweder, T. and Hjort, N. L., 2002. Confidence and likelihood. *Scandinavian Journal of Statistics, 29*, pp. 309–332.

Schweder, T. and Hjort, N. L., 2013. Discussion of "Confidence distribution, the frequentist distribution estimator of a parameter". *International Statistical Review, 81*, pp. 56–68.

Schweder, T. and Hjort, N. L., 2016. *Confidence, Likelihood, Probability: Inference with Confidence Distributions.* Cambridge: Cambridge University Press.

Sen, D., Papamarkou, T. and Dunson, D., 2024. Bayesian neural networks and dimensionality reduction. In *Handbook of Bayesian, Fiducial, and Frequentist Inference*, eds J. Berger, X-L. Meng, N. Reid and M. Xie, pp. 188–209. Boca Raton: Chapman and Hall/CRC.

Shafer, G., 1976. *A Mathematical Theory of Evidence.* Princeton: Princeton University Press.

Shafer, G., 2024. How the game-theoretic foundation for probability resolves the Bayesian vs. frequentist standoff. In *Handbook of Bayesian, Fiducial, and Frequentist Inference*, eds J. Berger, X-L. Meng, N. Reid and M. Xie, pp. 264–275. Boca Raton: Chapman and Hall/CRC.

Shen, J., Liu, R. Y. and Xie, M., 2020. iFusion: Individualized fusion learning. *Journal of the American Statistical Association*, 115, pp. 1251–1267. https://doi.org/10.1080/01621459.2019.1672557.

Stein, C., 1959. An example of wide discrepancy between fiducial and confidence intervals. *Annals of Mathematical Statistics, 30*, pp. 877–880.

Stigler, S., 1986. Laplace's 1774 memoir on inverse probability. *Statistical Science, 1*, pp. 359–363.

Stigler, S., 2013. The true title of Bayes's essay. *Statistical Science, 28*, pp. 283–288.

Taraldsen, G. and Lindqvist, B. H., 2024. Fiducial inference and decision theory. In *Handbook of Bayesian, Fiducial, and Frequentist Inference*, eds J. Berger, X-L. Meng, N. Reid and M. Xie, pp. 381–395. Boca Raton: Chapman and Hall/CRC.

Thornton, S. and Xie, M., 2024. Bridging Bayesian, frequentist and fiducial inferences using confidence distributions. In *Handbook of Bayesian, Fiducial, and Frequentist Inference*, eds J. Berger, X-L. Meng, N. Reid and M. Xie, pp. 106–131. Boca Raton: Chapman and Hall/CRC.

Tibshirani, R., 1989. Noninformative priors for one parameter of many. *Biometrika, 76*, pp. 604–608.

Vovk, V., 2024. Conformal predictive distributions: An approach to nonparametric fiducial prediction. In *Handbook of Bayesian, Fiducial, and Frequentist Inference*, eds J. Berger, X-L. Meng, N. Reid and M. Xie, pp. 364–380. Boca Raton: Chapman and Hall/CRC.

Walley, P., 1990. *Statistical Reasoning with Imprecise Probabilities.* London: Chapman & Hall.

Wasserman, L. A., 1990. Belief functions and statistical inference. *Canadian Journal of Statistics, 18*, pp. 183–196.

Wasserman, L. A., 2000. Asymptotic inference for mixture models using data-dependent priors. *Journal of the Royal Statistical Society: Series B*, *62*, pp. 159–180.

Wasserman, L. A. and Robins, J., 2000. Conditioning, likelihood and coherence: A review of some foundational concepts. *Journal of the American Statistical Association*, *95*, pp. 1340–1346.

Xie, M. and Singh, K., 2013. Confidence distributions: The frequentist distribution estimator of a parameter. *International Statistical Review*, *81*, pp. 3–39.

Xie, M., Singh, K. and Strawderman, W. E., 2011. Confidence distributions and a unifying framework for meta-analysis. *Journal of the American Statistical Association*, *106*, pp. 320–333.

Zabell, S. L., 1992. R. A. Fisher and the fiducial argument. *Statistical Science*, *7*, pp. 369–387.

Zhang, F. and Gao, C., 2020. Convergence rates of empirical Bayes posterior distributions: A variational perspective. *arXiv e-prints*, arXiv:2009.03969.

4

Objective Bayesian Inference and Its Relationship to Frequentism

James Berger
Duke University

Jose Bernardo
University of Valencia

Dongchu Sun
East China Normal University

4.1 Introduction

In objective Bayesian analysis, the prior distribution is chosen to represent 'minimal information,' the idea being that this allows Bayesian inferences to be made based only on the model (and data). This is thus the part of Bayesian analysis that is compatible with the BFF agenda since fiducial and frequentist analyses are likewise based (primarily) on the model (and data).

There have been many efforts to formally define what it means to be objective and then to develop objective prior distributions. Our belief is that this is generally misguided; there is no unambiguous objectivity. The best one can do is to find prior distributions that represent minimal information, in some sense, and which have desirable properties, such as resulting in statistical procedures that have good frequentist performance. Once found, such priors can be the *conventional priors* to use in objective Bayesian analysis, in the sense that we professionally agree to use them as the objective priors in specified inference problems.

The rest of this section presents background issues that are useful for the development and understanding of objective Bayesian analysis. Section 4.2 reviews some of the most prominent approaches to objective Bayesian analysis that have been developed. Section 4.3 presents our preferred approach, called the *reference prior* approach. The reference prior approach can result in different prior distributions for different unknown parameters in a model, which can be unwieldy in practice. Hence, in Section 4.4, approaches to developing overall objective priors are discussed, with hierarchical normal modeling being highlighted.

4.1.1 Conditioning and Frequentist Performance

Objective Bayesian inference is closely tied with frequentism in various ways, as will be seen throughout this chapter. Here we discuss two general issues needed to understand the relationship.

DOI: 10.1201/9780429341731-4

4.1.1.1 Conditioning

Conditional inference is a crucial concept in statistics, but is often neglected. This is partly because conditioning occurs automatically in Bayesian analysis and is, hence, not taught there. On the other hand, in the frequentist paradigm, conditioning is very difficult, and there is no established theory as to how to choose conditional procedures. It is useful to begin with a simple example, taken from [13], which discusses conditioning in general.

Example 4.1 There are two observations, x_1 and x_2, having distribution

$$x_i = \begin{cases} \theta + 1 & \text{with probability } \frac{1}{2} \\ \theta - 1 & \text{with probability } \frac{1}{2} \end{cases} .$$

Consider the confidence set, for the unknown θ, given by

$$C(x_1, x_2) = \begin{cases} \text{the point } \{\frac{1}{2}(x_1 + x_2)\} & \text{if } x_1 \neq x_2 \\ \text{the point } \{x_1 - 1\} & \text{if } x_1 = x_2. \end{cases}$$

The frequentist coverage of this confidence set can be shown to be

$$\Pr(C(x_1, x_2) \text{ contains } \theta \mid \theta) = 0.75.$$

This is a very inadequate report once the data is observed. Indeed, observe that, if $x_1 \neq x_2$, then it is certain that their average is equal to θ, so the confidence should then be given as 100%. Conversely, if $x_1 = x_2$, θ could be the data's common value plus one or the common value minus one, each of which is equally likely to have led to this result.

A common way to obtain sensible frequentist answers here is to define the conditioning statistic $s = |x_1 - x_2|$ chosen to measure the 'strength of evidence' in the data: $s = 2$ reflects data with maximal evidential content, whereas $s = 0$ reflects data of minimal evidential content. One then defines frequentist coverage in the usual way, but does so conditional on the strength of evidence s. An easy computation gives this conditional confidence, for the two distinct cases, as

$$\begin{aligned} \Pr(C(x_1, x_2) \text{ contains } \theta \mid s = 2, \theta) &= 1 \\ \Pr(C(x_1, x_2) \text{ contains } \theta \mid s = 0, \theta) &= \tfrac{1}{2}. \end{aligned}$$

Note that the unconditional coverage is still 75% (the report of 100% occurs half the time and the report of 50% occurs half the time), so these conditional reports are still proper frequentist reports. But clearly, the conditional reports are much more informative.

Finding good conditioning statistics in complex problems is very difficult, so that the conditional frequentist theory of statistics is underdeveloped. In contrast, the objective Bayesian approach automatically conditions properly.

Example 4.2 *Example 4.1 continued.* The objective prior for this example is $\pi(\theta) = 1$ (since θ is a location parameter — see Section 4.2.4). Application of Bayes theorem shows that, if $x_1 \neq x_2$, the posterior distribution for θ gives probability one to the point $(x_1 + x_2)/2$ while, if $x_1 = x_2$, the posterior distribution gives probability 1/2 each to the common value of the data plus 1 and the common value minus 1. Thus, the objective Bayesian confidence statements for $C(x_1, x_2)$ are 1 and 0.5 for the two cases, respectively, which is the correct answer.

In the aforementioned example, the objective Bayesian analysis produced the same stated confidences as did the optimal conditional frequentist procedure, automatically conditioning correctly. This is not a coincidence. Quite often, objective Bayesian procedures yield results that are (nearly) optimal (conditional) frequentist answers, and often the conditional frequentist results could not be easily obtained in any other way.

4.1.1.2 Frequentist Performance

Objective Bayesian priors are often chosen to achieve (unconditional) frequentist goals. For instance, 'frequentist-matching' priors (see Section 4.2.3) are the ones that produce Bayesian credible sets with good frequentist properties. Reference priors will also be seen to yield procedures that have excellent frequentist performance.

Objective Bayesian inference can also overcome some problems that face frequentists. Here is an example.

Example 4.3 *A Variance Components Problem:* For $i = 1, \ldots, m$,

$$x_i \sim \text{Normal}(x_i \mid \mu_i, 1) \quad \text{and} \quad \mu_i \sim \text{Normal}(\mu_i \mid \xi, \tau^2),$$

the μ_i and ξ being means and 1 and τ^2 being variances. The marginal density of x_i, given ξ and τ^2, is

$$m(x_i \mid \xi, \tau^2) = \int \text{N}(x_i \mid \mu_i, 1)\text{N}(\mu_i \mid \xi, \tau^2)\, d\mu_i = \text{N}(x_i \mid \xi, 1 + \tau^2)\,.$$

The marginal likelihood for the full data $\boldsymbol{x} = (x_1, \ldots, x_m)$, and with $s^2 = \sum(x_i - \bar{x})^2$, is then

$$m(\boldsymbol{x} \mid \xi, \tau^2) = \prod_{i=1}^{m} \text{N}(x_i \mid \xi, 1 + \tau^2) \propto \frac{1}{(1 + \tau^2)^{m/2}} \exp\left\{ -\frac{n(\bar{x} - \xi)^2 + s^2}{2(1 + \tau^2)} \right\}. \tag{4.1}$$

The standard maximum liklihood estimate (mle) $\hat{\xi}_{\text{mle}} = \bar{x}$ is fine, but $\hat{\tau}^2_{\text{mle}} = \max\{0, \frac{s^2}{m} - 1\}$, the marginal maximum likelihood estimate of τ^2, is problematical. This is particularly so if $s^2/m < 1$, in which case the mle would be $\hat{\tau}^2_{\text{mle}} = 0$. Setting τ^2 to 0 is equivalent to declaring that $\mu_1 = \cdots = \mu_p$ exactly (since $\mu_i \sim \text{Normal}(\cdot \mid \xi, \tau^2)$); this would be silly.

Indeed, in this situation, there is actually a great deal of uncertainty about τ^2. This can be seen by looking at the marginal likelihood of τ^2 found by integrating (4.1) over ξ, resulting in

$$m(\boldsymbol{x} \mid \tau^2) \propto (\tau^2 + 1)^{-(m-1)/2} \exp\left\{ -\frac{s^2}{2(\tau^2 + 1)} \right\}. \tag{4.2}$$

For illustration, Figure 4.1 shows this marginal likelihood when $m = 5$ and $s^2 = 5$. It is mostly decreasing away from 0, but not very quickly, indicating that there is

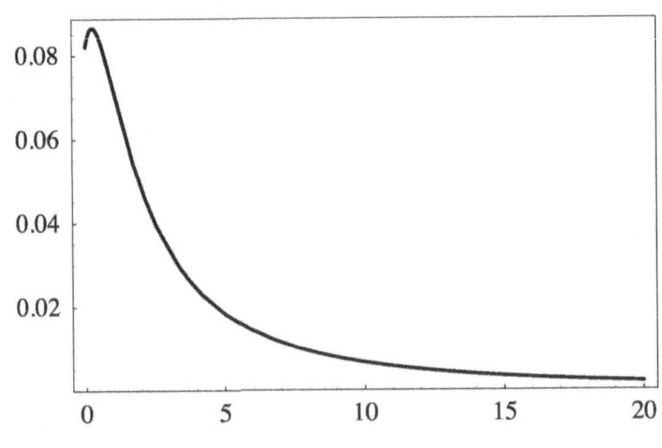

FIGURE 4.1
Marginal likelihood function of τ^2 when $p = 5$ and $s^2 = 5$ is observed.

considerable uncertainty about τ^2 even though $\hat{\tau}^2_{\text{mle}} = 0$. Frequentist methods have difficulty incorporating the uncertainty in τ^2 when the maximum of the likelihood is achieved at a boundary, whereas objective Bayesian analysis results in a posterior distribution for τ^2 that correctly reflects the uncertainty in the likelihood.

While a value of $s^2/m \leq 1$ is somewhat unusual here (if, for instance, $m = 5$, $n = 1$ and $\tau^2 = 1$, then $\Pr(s^2/5 < 1) = 0.264$), it is common in variance component models with multiple variances to have at least one mle variance estimate equal to 0.

Another frequentist use of objective Bayesian procedures is an alternative to asymptotics. Asymptotics is often used to develop statistical procedures, but the performance of the procedures with small or moderate sample sizes may well be uncertain. The objective Bayesian approach is an attractive alternative to asympotics for two reasons.

First, objective Bayesian procedures typically automatically have the desired asymptotic properties. For instance, the central limit theorem was first developed by [27] as a normal approximation to the posterior distribution; this approximation is identical to the frequentist central limit theorem commonly taught today. Objective Bayesian procedures will typically even have the correct asymptotics when the problem is irregular (i.e., the usual central limit theorem does not hold); Example 4.17 provides an illustration. Note that objective Bayesians need not do any asymptotic computations; the Bayesian procedures will automatically yield the correct asymptotic answer if the sample size is large enough for the asymptotics to apply. The second motivation for using objective Bayesian analysis is that the objective Bayesian answers are still likely to be sensible when the sample size is too small for asymptotics to apply.

4.1.2 Informal Objective Bayesian Solutions

A variety of *ad hoc* methods have been proposed for choosing a default prior distribution, and it is useful to briefly discuss some of them so as to see the problems that can be encountered. The use of a constant prior distribution is the most common objective Bayesian method and will be discussed in Section 4.2.1. While a constant prior is often fine, examples will be seen in which the constant prior is inadequate.

4.1.2.1 Truncation of the Parameter Space

If the parameter space is unbounded, there is often worry about using, say, a constant prior, because the prior then has infinite mass and there is no guarantee that the resulting posterior distribution will be proper. A common 'solution' is to, instead, choose the prior to be constant over some (large) bounded region of the parameter space, since it will then be proper (after normalization). This does not really solve the problem, however, because, if the posterior resulting from the constant prior on the entire parameter space is improper, then inferences arising from the truncated prior will strongly depend on the (arbitrary) choice of the truncation point. Thus, to use this approach one would need to perform careful sensitivity studies to determine the effect of the truncation points.

4.1.2.2 Vague Proper Priors

The use of vague proper priors is popular; the following is a typical example.

Example 4.4 *Normal mean.* Suppose the problem is to estimate a normal mean θ, with known variance. The standard objective prior is $\pi(\theta) = 1$. A vague proper prior alternative is to use the $\mathrm{N}(\theta \mid 0, K)$ prior distribution, with a large value of K. In estimation problems this will yield essentially the same answer as the constant prior and is thus reasonable to use but, in hypothesis testing, such vague proper priors can be very bad (see the chapter "Objective Bayesian Testing and Model Uncertainty" in this book).

It is not uncommon to see vague proper priors used that are an approximation to an inferior default prior. The use of the vague proper prior then is also inferior and its use can hide the problem. Here is a commonly encountered example.

Example 4.5 *Example 4.3 continued.* In the variance components example, consider priors of the form $\pi(\xi, \tau^2) = \pi^*(\tau^2)$ (constant in ξ). The resulting posterior distribution for τ^2 is

$$\begin{aligned} \pi(\tau^2 \mid \mathbf{x}) &\propto \int \frac{1}{(1+\tau^2)^{m/2}} \exp\left\{-\frac{n(\bar{x}-\xi)^2+s^2}{2(1+\tau^2)}\right\}\pi^*(\tau^2)d\xi \\ &\propto \frac{1}{(1+\tau^2)^{(m-1)/2}} \exp\left\{-\frac{s^2}{2(1+\tau^2)}\right\}\pi^*(\tau^2). \end{aligned}$$

A commonly used prior for a normal variance σ^2 is $1/\sigma^2$, and so it is tempting to choose $\pi^*(\tau^2) = 1/\tau^2$. This results in an improper posterior, however, because there is a non-integrable singularity at zero,

$$\int_0^{\epsilon} \pi(\tau^2 \mid \mathbf{x})d\tau^2 \approx \exp\left\{-\frac{s^2}{2}\right\} \int_0^{\epsilon} \frac{1}{\tau^2} d\tau^2 = \infty\,.$$

A commonly used vague proper alternative is $\pi^*(\tau^2) \propto \tau^{-2(1+\epsilon)} e^{-\epsilon/\tau^2}$, and the resulting posterior will be proper. However, this posterior gives almost all of its mass to a small region near 0, and will essentially ignore the data. Thus the vague proper prior does not fix the problem, and may even cause the problem to be hidden. Instead, one should simply use a good objective prior for the problem. The simplest choice is $\pi^*(\tau^2) = 1/\sqrt{\tau^2}$, which results in an excellent objective posterior distribution.

Another commonly used ad hoc vague proper prior is to choose a flat prior (e.g., a constant) that 'spans the range of the likelihood function.' The rationale for doing this is typically stated to be a desire to reduce the influence of the prior, by choosing a prior that is compatible with the likelihood function. It is not really possible to evaluate this strategy. On the one hand, it may often be better than using an arbitrarily chosen objective prior (e.g. a constant prior), but it corresponds to a disturbing double use of the data.

4.1.3 Justifying Posteriors from Improper Priors

When $\pi(\theta)$ is improper, Bayes theorem no longer applies, and so it is not obvious how to justify the use of the posterior density

$$\pi(\theta \,|\, \boldsymbol{x}) = \frac{p(\boldsymbol{x} \,|\, \theta)\, \pi(\theta)}{\int_{\Theta} p(\boldsymbol{x} \,|\, \theta)\, \pi(\theta)\, d\theta}\,. \tag{4.3}$$

For this expression to define a probability density, it is clearly necessary that

$$p(\boldsymbol{x}) = \int_{\Theta} p(\boldsymbol{x} \,|\, \theta)\, \pi(\theta)\, d\theta < \infty\,, \tag{4.4}$$

and we take this as a minimal requisite condition for the use of the improper prior to make sense (but see [28] for counterarguments).

This section presents a strong justification for the use of (4.3) with improper priors, showing that $\pi(\theta \,|\, \boldsymbol{x})$ is the limit of proper posteriors arising from proper priors that approximate the improper $\pi(\theta)$. To this end, for an improper prior $\pi(\theta)$, consider an increasing sequence of subsets of the parameter space $\{\Theta_1, \Theta_2, \ldots\}$, such that

$$\Theta_i \subset \Theta_{i+1}, \int_{\Theta_i} \pi(\theta) < \infty, \ \forall\, i \geq 1, \ \text{ and } \lim_{i\to\infty} \Theta_i = \Theta. \tag{4.5}$$

Then, $\pi(\theta)$ can be normalized within each of the Θ_i's to produce the sequence of *proper* priors $\{\pi_1(\theta), \pi_2(\theta), \ldots\}$ defined by

$$\pi_i(\theta) = \begin{cases} \dfrac{\pi(\theta)}{\int_{\Theta_i} \pi(\theta)\, d\theta} & \text{if } \theta \in \Theta_i \\ 0 & \text{otherwise.} \end{cases}$$

Since the $\pi_i(\theta)$'s are all proper probability distributions, Bayes theorem can be applied, given the data $\boldsymbol{x}$, to obtain the corresponding sequence of posterior densities $\{\pi_1(\theta \mid \boldsymbol{x}), \pi_2(\theta \mid \boldsymbol{x}), \ldots\}$ defined by

$$\pi_i(\theta \mid \boldsymbol{x}) = \begin{cases} \dfrac{p(\boldsymbol{x} \mid \theta)\, \pi_i(\theta)}{\int_{\Theta_i} p(\boldsymbol{x} \mid \theta)\, \pi_i(\theta)\, d\theta} & \text{if } \theta \in \Theta_i \\ 0 & \text{otherwise.} \end{cases}$$

Here is the type of convergence of distributions that we consider.

Definition 4.1 Reverse KL Convergence. A sequence of probability distributions, defined by their density functions $\{p_i(\boldsymbol{x})\}_{i=1}^{\infty}$, is said to reverse KL converge to a probability distribution with density $p(\boldsymbol{x})$ whenever

$$\lim_{i\to\infty} \int_{\mathcal{X}} p_i(\boldsymbol{x}) \log \frac{p_i(\boldsymbol{x})}{p(\boldsymbol{x})}\, d\boldsymbol{x} = 0\,.$$

Theorem 4.1 *Let $\pi(\theta)$ be an improper prior on Θ, having a sequence of subsets satisfying (4.5). If $\int_\Theta p(\boldsymbol{x} \mid \theta)\, \pi(\theta)\, d\theta < \infty$ at $\boldsymbol{x}$, then $\{\pi_i(\theta \mid \boldsymbol{x})\}$ is reverse KL convergent to $\pi(\theta \mid \boldsymbol{x})$ at $\boldsymbol{x}$.*

This theorem, from [7], is quite powerful in that it says that only the minimal condition (4.4) is needed for reverse KL convergence; thus if the formal posterior is proper, its use is justified.

4.2 Approaches to Objective Bayesian Analysis

There have been many efforts to develop and implement objective Bayesian analysis. Some of the more prominent approaches are discussed in this section. Discussion of our favorite approach, called *reference analysis*, is delayed until Section 4.3. Two other approaches, the confidence distribution approach and the fiducial approach, can also be considered to be attempts at implementing objective Bayesian analysis, but we do not present them here because they are extensively discussed in other chapters of the handbook.

4.2.1 The Constant Prior and Inverse Probability

The first clear Bayesian was the Reverend Thomas Bayes. His paper (published posthumously) *An essay towards solving a problem in the doctrine of chances* [1] was the first unambiguous statement and use of what is now known as Bayes theorem.

The paper was also a significant work in objective Bayesian analysis. In today's language, the paper considered the basic binomial problem of observing x from $p(x \mid \theta) = \text{Bi}(x \mid n, \theta)$, with θ unknown. While inference concerning θ given an observed x (via Bayes theorem) was one focus of the paper, the other focus was that of coming up with a reasonable objective prior distribution, $\pi(\theta)$, for θ. Bayes argued that the marginal distribution of x should be uniform and showed that $\pi(\theta) = 1$ achieved this; this was thus the recommended objective prior.

The real inventor of objective Bayesian analysis, as a paradigm for statistical analysis, was Simon Laplace. In a series of papers culminating in *Théorie Analytique des Probabilités* [27], he invented a statistical paradigm that was to dominate statistics for well over a century. The approach of Laplace to statistical inference was essentially objective Bayesian analysis with the utilization of a constant prior density for all unknown parameters of statistical models. Laplace did not do this blindly, but rather argued that one should choose a parameterization of the problem in which different values of the unknown were equally likely.

To distinguish Laplace's method from ordinary probability reasoning (computing probabilities of various data arising from the model) Augustus de Morgan [20] called reasoning from the data to the model *inverse probability*. This name reigned until roughly 1950, when it became replaced by 'Bayesian analysis' for rather unclear reasons, as extensively discussed in [22].

Inverse probability (*i.e.*, objective Bayesian analysis using a constant prior density) was arguably the dominant mode of formal statistical inference until the 1920s [17], [24]. While dominating statistical practice, inverse probability was not without its critics. Much of the criticism was ill-placed, but at least some of the philosophical criticisms were legitimate. Foremost among these criticisms was the fact that the statistical answers arising from the use of inverse probability will depend on the choice of the parameterization that is used for the statistical model.

Example 4.6 *Inference on a binomial parameter.* Suppose x is distributed according to the binomial $\text{Bi}(x \mid n, \theta)$ distribution. Inverse probability says use the objective prior $\pi(\theta) = 1$. Suppose someone chose $\psi(\theta) = \frac{2}{\pi}\text{ArcSin}(\sqrt{\theta})$ as the unknown parameter, rather than θ. Inverse probability then says use the prior $\pi(\psi) = 1$ (note that $0 < \psi < 1$).

To see that this will typically result in different Bayesian answers, make the change of variables from ψ back to θ. With this change of variables, the objective prior $\pi(\psi) = 1$ transforms to

$$\pi(\theta) = 1 \times \psi'(\theta) = \pi^{-1}\, \theta^{-1/2}(1-\theta)^{-1/2},$$

which is the $\text{Be}(\theta \mid \frac{1}{2}, \frac{1}{2})$ density. The posterior mean for θ corresponding to the uniform prior can be shown to be $(x+1)/(n+1)$, while that corresponding to the $\text{Be}(\theta \mid \frac{1}{2}, \frac{1}{2})$ prior will be $(x+\frac{1}{2})/(n+\frac{1}{2})$, which are clearly different. Of course, these are typically very close; if $x = 5$ and $n = 30$, for instance, the two posterior means are 0.188 and 0.177 respectively, and the difference is negligible from a practical perspective (since the posterior standard errors are on the order of 0.07). Still, it is unappealing to have different answers depending on the rather arbitrary choice of parameterization.

4.2.2 The Jeffreys-rule Prior

In the 1930s, Harold Jeffreys, being aware of the inconsistencies that could arise if one always used the constant prior, sought a 'rule' to specify an objective prior which would give the same results, no matter which parameterization is chosen. In [26] he developed such a rule, for any continuous parameter vector $\theta = (\theta_1, \ldots, \theta_m) \in \Theta \subset \mathbb{R}^m$, based on the *Fisher information matrix* $I(\theta)$, namely the $m \times m$ matrix having elements (assuming they exist)

$$I(\theta)_{ij} = \mathrm{E}^{\boldsymbol{x}\,|\,\theta}\left[-\frac{\partial^2}{\partial\theta_i\partial\theta_j}\log p(\boldsymbol{x}\,|\,\theta)\right] = \int_{\mathcal{X}}\left[-\frac{\partial^2}{\partial\theta_i\partial\theta_j}\log p(\boldsymbol{x}\,|\,\theta)\right]p(\boldsymbol{x}\,|\,\theta)d\boldsymbol{x}. \tag{4.6}$$

It can be easily shown that the prior distribution

$$\pi(\theta) \propto |I(\theta)|^{1/2}, \tag{4.7}$$

called the *Jeffreys-rule* prior, is invariant under one-to-one reparameterization and so will result in coherent answers across different parameterizations.

The Jeffreys-rule prior can be proper (when normalized) or improper. When improper, it has the surprising property of virtually always resulting in a proper posterior $\pi(\theta\,|\,\boldsymbol{x}) \propto p(\boldsymbol{x}\,|\,\theta)\,|I(\theta)|^{1/2}$, *i.e.*, $\int_\Theta p(\boldsymbol{x}\,|\,\theta)|I(\theta)|^{1/2}\,d\theta < \infty$.

When one has n iid observations, $I(\theta)$ is just n times the Fisher information for a single observation. Since proportionality constants like n are irrelevant if the prior is improper and are renormalized anyway if the prior is proper, one only needs to compute the Fisher information for one observation.

4.2.2.1 The Single Parameter Case

In one parameter *regular* continuous models, $\{p(x\,|\,\theta), \theta \in \Theta \subset \mathbb{R}, x \in \mathcal{X}\}$, where the sampling space $\mathcal{X}$ does not depend on the parameter θ, and the model probability density is twice continuously differentiable with respect to θ, Jeffreys rule (4.7) reduces to

$$\pi(\theta) \propto \sqrt{I(\theta)}, \quad I(\theta) = \mathrm{E}^{x\,|\,\theta}\left[-\frac{\partial^2}{(\partial\theta)^2}\log p(x\,|\,\theta)\right]. \tag{4.8}$$

Example 4.7 *Jeffreys-rule prior for the parameter of a Poisson model.* Let $\boldsymbol{x} = (x_1, \ldots, x_n)$ be a random sample of size n from a Poisson probability model $\mathrm{Pn}(x\,|\,\theta) = e^{-\theta}\theta^x/x!$. Recalling that, in the iid case, it suffices to compute the Fisher information for a single observation, and noting that $\mathrm{E}[x\,|\,\theta] = \theta$, the Fisher information is

$$I(\theta) \; = \mathrm{E}^{x\,|\,\theta}\left[-\frac{\partial^2}{(\partial\theta)^2}\log\frac{e^{-\theta}\theta^x}{x!}\right] = \mathrm{E}^{x\,|\,\theta}\left[\frac{x}{\theta^2}\right] = \frac{1}{\theta}\,.$$

Thus, using (4.8), the Jeffreys-rule prior is the improper prior

$$\pi(\theta) \propto \sqrt{I(\theta)} = \theta^{-1/2}.$$

Example 4.8 *Location models.* Let $\boldsymbol{x} = (x_1, \ldots, x_n)$ be a random sample from a location model $p(x\,|\,\theta) = h(x-\theta)$, $x \in \mathbb{R}$, $\theta \in \mathbb{R}$. Equation (4.8) gives

$$\pi(\theta) \propto \left[\int_{-\infty}^{\infty}\frac{h'(x-\theta)^2 - h(x-\theta)h''(x-\theta)}{h(x-\theta)}\,dx\right]^{1/2},$$

and changing the integration variable to $y = x - \theta$ shows that the integral does not depend on θ and, hence, the resulting prior is proportional to some constant.

Example 4.9 *Scale models.* Consider the model $p(x\,|\,\theta) = \theta^{-1}g(x/\theta)$. This is a general scale model and can be transformed to the location model $p(y\,|\,\phi) = g(\exp\{y-\phi\})$, with the change of variables $y = \log x$ and $\phi = \log\theta$. For ϕ, the Jeffreys-rule prior is constant and changing variables back to θ results in $\pi(\theta) \propto \pi(\phi)|d\phi/d\theta| \propto \theta^{-1}$, which is thus the Jeffreys-rule prior for scale models.

As stated earlier, Jeffreys rule cannot be applied to non-regular problems. This includes those situations where the sampling space $\mathcal{X}$ depends on the parameter.

Example 4.10 *Uniform data on* $[0, \theta]$. Let x be a sample from the uniform distribution on the real interval $(0, \theta)$, so that $p(x \mid \theta) = \theta^{-1}$, if $0 < x < \theta$, and zero otherwise. The sample space $\mathcal{X} = (0, \theta)$ depends on the parameter θ and, therefore, Jeffreys rule (4.8) is not applicable. (Blind use of the formula yields a negative information function.) This will be considered, again, in Section 4.3.2, where the reference prior approach will be shown to yield the sensible objective prior $\pi(\theta) = 1/\theta$.

4.2.2.2 The Multi-parameter Case

Although Jeffreys rule (4.7) does provide a multivariate invariant objective prior, the results are very often unappealing.

Example 4.11 *Inference on a normal mean, with variance unknown.* Let $\boldsymbol{x} = (x_1, \ldots, x_n)$ be a random sample of size n from a normal distribution $\mathrm{N}(x \mid \mu, \sigma^2)$, both parameters unknown. Using (4.6), the corresponding Fisher information matrix for a single observation is

$$I(\mu, \sigma^2) = \begin{pmatrix} 1/\sigma^2 & 0 \\ 0 & 1/[2\sigma^4] \end{pmatrix}, \tag{4.9}$$

and, hence, by (4.7), the Jeffreys-rule prior is

$$\pi(\mu, \sigma^2) \propto |I(\mu, \sigma^2)|^{1/2} \propto \sigma^{-3}. \tag{4.10}$$

If the prior is of the form $\pi(\mu, \sigma^2) \propto \sigma^{-\alpha}$, for some $\alpha > 0$, Bayes theorem yields a joint posterior

$$\pi(\mu, \sigma^2 \mid \boldsymbol{x}, \alpha) \propto \sigma^{-(n+\alpha)} \exp\left\{-\frac{n}{2\sigma^2}[s^2 + (\overline{x} - \mu)^2]\right\}, \tag{4.11}$$

where $\overline{x}$ is the sample mean and $s^2 = \sum(x_i - \overline{x})^2/n$. Integrating out σ^2, the corresponding marginal posterior for μ is (provided $n + \alpha > 3$), the Student t density

$$\pi(\mu \mid \boldsymbol{x}, \alpha) = \mathrm{St}(\mu \mid \overline{x},\ s/\sqrt{n + \alpha - 3}, n + \alpha - 3). \tag{4.12}$$

The use of (4.10) ($\alpha = 3$) thus leads to $\pi(\mu \mid \boldsymbol{x}) = \mathrm{St}(\mu \mid \overline{x},\ s/\sqrt{n}, n)$.

Jeffreys was unhappy with this result; he mentions that this is against the 'standard practice' of removing one degree of freedom from the sample size per estimated parameter (which is the result for $\alpha = 2$). This led Jeffreys to recommend the use of $\alpha = 2$, *i.e.*, of $\pi(\mu, \sigma^2) \propto \sigma^{-2}$, even though it contradicted his multivariate rule. His argument to rationalize this choice was to treat μ and σ independently. This Jeffreys-recommended prior has now become known as the *Jeffreys independence prior.*

Example 4.12 *Multinomial distribution.* Suppose $\boldsymbol{x} = (x_1, \ldots, x_m)$ is multinomial $\mathrm{Mu}(\boldsymbol{x} \mid n, \theta_1, \ldots, \theta_m)$ (suppressing the $(m+1)$st cell count, $x_{m+1} = n - \sum_{j=1}^m x_j$, and probability, $\theta_{m+1} = 1 - \sum_{j=1}^m \theta_j$, since they are determined by the others) so that

$$p(\boldsymbol{x} \mid n, \theta_1, \ldots, \theta_m) = \frac{n!}{\prod_{j=1}^m x_j!(n - \Sigma x_j)!} \prod_{j=1}^m \theta_j^{x_j} (1 - \Sigma\theta_j)^{n - \Sigma x_j}.$$

Using (4.6), the computation of the Fisher information matrix yields

$$I(\theta_1, \ldots, \theta_m) = \frac{n}{1 - \Sigma\theta_j} \begin{bmatrix} \frac{1+\theta_1-\Sigma\theta_j}{\theta_1} & 1 & \ldots & 1 \\ 1 & \frac{1+\theta_2-\Sigma\theta_j}{\theta_2} & \ldots & 1 \\ \ldots & \ldots & \ldots & \ldots \\ 1 & 1 & \ldots & \frac{1+\theta_m-\Sigma\theta_j}{\theta_m} \end{bmatrix}$$

with

$$|I(\theta_1, \ldots, \theta_m)| = n^m \left[\left(1 - \sum\nolimits_{j=1}^m \theta_j\right) \prod\nolimits_{j=1}^m \theta_j\right]^{-1}.$$

Thus, the Jeffreys-rule prior, proportional to $|I(\theta_1, \ldots, \theta_m)|^{1/2}$, is the proper Dirichlet prior

$$\pi^J(\theta_1, \ldots, \theta_m) \propto \left(1 - \sum\nolimits_{j=1}^m \theta_j\right)^{-1/2} \prod_{j=1}^m \theta_j^{-1/2}, \tag{4.13}$$

i.e., the $\text{Di}((\theta_1, \ldots, \theta_m) \,|\, (\frac{1}{2}, \ldots, \frac{1}{2}))$ distribution. Multiplying this by the multinomial likelihood immediately yields that the corresponding posterior distribution is $\text{Di}((\theta_1, \ldots, \theta_m) \,|\, (x_1 + \frac{1}{2}, \ldots, x_m + \frac{1}{2}))$.

Again, (4.13) does *not* have appropriate behavior for an objective prior. As a numerical illustration of this, consider the case where the sample size n is small relative to the number of classes $m+1$; thus we have a large sparse table. For instance, suppose $n = 3$ and $m = 1000$, with $x_{240} = 2$, $x_{876} = 1$, and all the other $x_i = 0$. The posterior means resulting from using the Jeffreys-rule prior can be shown to be

$$\text{E}[\theta_i \,|\, \boldsymbol{x}] = \frac{x_i + 1/2}{\sum_{j=1}^m [x_j + 1/2]} = \frac{x_i + 1/2}{n + m/2} = \frac{x_i + 1/2}{503},$$

so that $\text{E}[\theta_{240} \,|\, \boldsymbol{x}] = 2.5/503$, $\text{E}[\theta_{876} \,|\, \boldsymbol{x}] = 1.5/503$, and $\text{E}[\theta_i \,|\, \boldsymbol{x}] = 0.5/503$. So, cells 240 and 876 only have total posterior probability of $4/503 = 0.008$, even though all 3 observations are in these cells.

The problem is that the Jeffreys-rule prior effectively added 1/2 to the 998 zero cells, making them more important than the cells with data! That the Jeffreys-rule prior can encode much more information than is contained in the data is hardly desirable for an objective analysis.

In this section, we have seen two important examples — the normal model with both parameters unknown and the multinomial model — where the multiparameter form of the Jeffreys-rule prior fails to provide appropriate objective posteriors. We actually know of no multiparameter example in which the Jeffreys-rule prior has been verified to be satisfactory. In higher dimensions, that prior always seems to be either 'too diffuse' as in the normal examples, or 'too concentrated' as in the multinomial example.

4.2.3 Frequentist Matching

We have encountered situations where Bayesian credible sets are identical to frequentist confidence sets and have the same stated confidence or coverage. Ensuring that this holds, at least approximately, is the goal of the frequentist matching approach to developing objective priors.

4.2.3.1 Definition of Matching Priors

It is customary to consider one-sided credible sets $(-\infty, \theta_\alpha(\boldsymbol{x})]$, for a real parameter θ, where $\theta_\alpha(\boldsymbol{x})$ denotes the $\alpha-$posterior quantile of θ given $\boldsymbol{x}$, i.e.,

$$\Pr(\theta \leq \theta_\alpha(\boldsymbol{x}) \,|\, \boldsymbol{x}) = \alpha, \tag{4.14}$$

for $\alpha \in (0, 1)$. Now, consider $C(\boldsymbol{x}) \equiv (-\infty, \theta_\alpha(\boldsymbol{x})]$ to be a frequentist confidence set, with $\boldsymbol{x}$ random and arising from $p(\boldsymbol{x} \,|\, \theta)$, for given θ. The frequentist coverage probability of this confidence set is clearly

$$\Pr(\theta \leq \theta_\alpha(\boldsymbol{x}) \,|\, \theta)\,. \tag{4.15}$$

Note that, in (4.14), θ is random while $\boldsymbol{x}$ is fixed but, in (4.15), $\boldsymbol{x}$ is random while θ is fixed. The idea is now to study how close the frequentist coverage probability $\Pr(\theta \leq \theta_\alpha(\boldsymbol{x}) \,|\, \theta)$ is to α, the stated Bayesian coverage.

Definition 4.2 If the frequentist coverage (4.15), of the one-sided Bayesian credible sets $(-\infty, \theta_\alpha(\boldsymbol{x})]$, is exactly equal to the Bayesian credible probability α, then the Bayesian credible sets are said to be *exact frequentist matching* (or 'exact matching' for short).

Example 4.13 *Normal mean with variance known.* Let $\boldsymbol{x} = (x_1, \ldots, x_n)$ be a random sample of size n from a normal distribution $\mathrm{N}(x \,|\, \mu, \sigma^2)$, with μ unknown but σ^2 known. Under the Jeffreys prior, $\pi(\mu) \propto 1$, Bayes theorem yields $\pi(\mu \,|\, \boldsymbol{x}) = \mathrm{N}(\mu \,|\, \overline{x}, \sigma^2/n)$. In this case, the posterior α-quantile of μ given $\boldsymbol{x}$ is $\mu_\alpha(\boldsymbol{x}) = \overline{x} + Z_\alpha \sigma/\sqrt{n}$, where Z_α is the α-quantile of $\mathrm{N}(0,1)$. Algebra yields that the frequentist coverage is

$$\Pr(\mu \leq \mu_\alpha(\boldsymbol{x}) \,|\, \mu) = \Pr\left(\frac{\sqrt{n}(\overline{x} - \mu)}{\sigma} \geq -Z_\alpha \,\Big|\, \mu\right) = \Pr\left(\frac{\sqrt{n}(\overline{x} - \mu)}{\sigma} \leq Z_\alpha \,\Big|\, \mu\right) = \alpha,$$

$\forall \mu \in \mathbb{R}$ and $\alpha \in (0,1)$. Consequently, the constant prior for μ is an exact matching prior for μ.

Having exact matching credible sets is actually a fairly common phenomenon. Indeed, this occurs for a wide class of models that are 'invariant' to certain transformations; see Section 4.2.4 for development. But more generally, one must settle for a weaker form of matching.

Definition 4.3 Asymptotic Matching Prior. A prior $\pi(\theta)$ is an i-th order matching prior for $\eta = g(\theta)$ if

$$P(\eta \leq \eta_\alpha(\boldsymbol{x}) \,|\, \theta) = \alpha + o(n^{-i/2}), \quad \forall \theta \in \Theta \text{ and } \alpha \in (0,1). \tag{4.16}$$

Note that virtually any prior with full support is matching up to an error term of order $O(n^{-1/2})$. What is typically thus sought is an error term of order $O(n^{-1})$, which would be a 'first order' matching prior. Higher order matching is rare (except for the exact matching scenario), but sometimes possible.

4.2.3.2 Single Parameter Case

When there is a single parameter, [34] proved that a one-sided posterior credible interval arising from the use of the Jeffreys-rule prior (proportional to the square root of the Fisher information $I(\theta)$), has the desired frequentist coverage probability up to $O(n^{-1})$. In fact, they showed that a prior π is a first-order matching prior if, and only if, the differential equation

$$\frac{\partial}{\partial \theta}\left(\frac{\pi(\theta)}{\sqrt{I(\theta)}}\right) = 0 \tag{4.17}$$

is satisfied. Consequently, under the regularity conditions where the Fisher information exists, the Jeffreys prior is the only first-order matching prior.

4.2.3.3 Multiparameters

[30] was among the first authors to study matching priors for the two-parameter case, where one is a parameter of interest and the other is a nuisance parameter. [31] extended the results in [34] and [30] and introduced a method to find a prior that is matching for a parameter to order $O(n^{-1})$, in the presence of nuisance parameters. [33] generalized this method by using a one-to-one transformation of the parameter vector into a parameter of interest and

a nuisance parameter vector orthogonal in the sense of [16]. [18] derived a matching prior for a smooth function $\eta = \eta(\theta)$ for $\theta = (\theta_1, \ldots, \theta_m)$, their result is stated in the following theorem.

Theorem 4.2 *Let $I(\theta)$ be the Fisher information for one observation, and define the gradient vector*

$$\nabla\eta = \{\partial\eta/\partial\theta_1, \ldots, \partial\eta/\partial\theta_m\}^t \,. \tag{4.18}$$

Also, define

$$\xi(\theta) \equiv (\xi_1, \ldots, \xi_m)^t = \frac{I(\theta)^{-1}\nabla\eta}{\sqrt{(\nabla\eta)^t I(\theta)^{-1}\nabla\eta}}. \tag{4.19}$$

Then a first-order matching prior $\pi(\theta)$ for η must satisfy

$$\sum_{i=1}^{m} \frac{\partial}{\partial\theta_i}[\pi(\theta)\xi_i(\theta)] = 0. \tag{4.20}$$

A matching prior is not unique in the multiparameter case. Indeed, the class of solutions of (4.20) is often quite large, growing with the dimension m. Also, note that the equation to be solved depends on the parameter of interest, so that there could be a different matching prior for each parameter of interest. If the interest is in all the coordinates of θ, one might hope that there is a simultaneous solution to all m matching equations. Sometimes this is so, but not always (cf. [19]).

Example 4.14 *Inference on a normal variance with unknown mean.* Consider again Example 4.11 and define $s^2_{n-1} = \sum(x_i - \overline{x})^2/(n-1)$. Consider the class of priors $\pi_a(\mu, \sigma^2) \propto 1/\sigma^a$.

When $\eta = \sigma^2$, $\nabla\eta = (0,1)^t$, and $\xi \equiv (0, \sqrt{2}\sigma^2)^t$. Thus (4.20) becomes

$$\sqrt{2}\,\frac{\partial\sigma^2\pi(\mu,\sigma^2)}{\partial\sigma^2} = 0.$$

Clearly the independent Jeffreys prior for (μ, σ^2) is a first-order matching prior for σ^2, but the Jeffreys-rule prior ($a = 3$) is not. *Thus, contrary to the one-parameter case (under regularity), the Jeffreys-rule prior need not be first-order matching in multiparameter problems.*

A comprehensive study of higher order matching priors can be found in [19]. A different version of matching, based on higher order local likelihood approximations, is given in [23]. Interestingly, these priors can be data-dependent.

4.2.4 Invariance Priors

The big advance of Jeffreys-rule priors was in providing inferences that were consistent no matter what parameterization was chosen for the model; this is often called 'invariance to parameterization.' Here we discuss an additional type of invariance, namely invariance to other transformations (of both the data and the parameters) that seem to leave the problem unchanged. There are many illustrations of this in the literature but the most extensively studied and reliable invariance theory is 'invariance to a group operation on the model'; here is an illustration.

Example 4.15 Consider a *scale parameter* density

$$p(x \mid \theta) = \frac{1}{\theta} f\Big(\frac{x}{\theta}\Big), \quad x > 0, \quad \theta > 0.$$

- Consider a change of scale for the problem (*e.g.*, measure in terms of meters, rather than feet) to $x^* = c\,x > 0$ and $\theta^* = c\,\theta > 0$.
- The density of x^* can be shown to be $p(x^* \mid \theta^*) = (1/\theta^*) f(x^*/\theta^*)$, which is of exactly the same mathematical form as the original density. Another density of exactly the same mathematical form is $(1/\sigma) f(y/\sigma)$.
- Objective priors are determined by the mathematical form of the density so it would seem that the objective prior for θ should be the same function as that for the objective prior for θ^* (or σ).

This logic, when combined with 'invariance to parameterization,' leads to what are called 'invariance priors.'

The development of invariance priors is achieved through the mathematical tools of *group theory*. In the above example, the transformation 'multiplication by positive constants' is called the *scale group* or *multiplication group*. We do not delve deeply into group theory here and, hence, only present invariance priors in simple cases. For an introduction to the use of group theory in developing invariance priors, see [2].

Curiously, the development of invariance priors initially mirrors the success and failure of the Jeffreys-rule prior. For one-parameter problems having a group-invariance structure, the invariance prior is the Jeffreys-rule prior (also the reference prior and first-order matching prior), so all is well. In multi-parameter problems, however, the natural invariance prior is what is called the *left Haar prior*[1], and is problematical in much the same ways that the Jeffreys-rule prior is problematical in multi-parameter problems.

Unlike the situation with the Jeffreys-rule prior, however, a 'fix' was developed for the invariance prior in multi-parameter problems; instead of using the left Haar prior, use what is called the *right Haar prior*. This development was actually a frequentist development, finding the prior distribution in invariant problems that is exact frequentist matching; it is the right-Haar prior that achieves this.

Example 4.16 A model of the form $p(x \mid \mu, \sigma) = (1/\sigma) f((x-\mu)/\sigma)$ is called a location-scale model. (Were this the normal model, σ would be the standard deviation, which is differently parameterized than using the variance.) It can be shown for such a model that the left-Haar prior is $\pi^l(\mu, \sigma) = \sigma^{-2}$ (the Jeffreys-rule prior), whereas the right-Haar prior is $\pi^r(\mu, \sigma) = \sigma^{-1}$ (the independence-Jeffreys prior); it is the latter that results in Bayesian credible sets with correct frequentist coverage (e.g., the left-Haar prior gave the wrong degrees of freedom in the normal case).

As a specific example, suppose the $(x_1, \ldots, x_n)$ are iid observations from the Cauchy density $p(x \mid \mu, \sigma) = 1/(\pi\sigma[1 + (x-\mu)^2/\sigma^2])$ and that it is desired to find a 90% frequentist confidence set for $\xi = \mu + \sigma$, which is the third quartile of the distribution. The natural Bayesian credible set for ξ is the equal-tailed credible set, given by $C(\boldsymbol{x}) = (l(\boldsymbol{x}), u(\boldsymbol{x}))$, where $\pi^r(\mu, \sigma \mid \boldsymbol{x})$ is the posterior distribution arising from $\pi^r(\mu, \sigma)$,

$$\int_{\{(\mu,\sigma):\mu+\sigma<l(\boldsymbol{x})\}} \pi^r(\mu, \sigma \mid \boldsymbol{x})\, d\mu d\sigma = \int_{\{(\mu,\sigma):\mu+\sigma>u(\boldsymbol{x})\}} \pi^r(\mu, \sigma \mid \boldsymbol{x})\, d\mu d\sigma = 0.05\,.$$

$C(x)$ is guaranteed to be a 90% frequentist confidence set. Furthermore, it is a 90% frequentist confidence set conditional on the (ancillary) statistic $s = (\frac{x_2 - x_1}{x_1}, \ldots, \frac{x_n - x_1}{x_1})$; as discussed in Section 4.1.1, it is important to find confidence sets that condition properly on relevant statistics, such as s. Trying to develop conditional frequentist confidence procedures directly from frequentist reasoning would be virtually impossible here.

A simple version (without conditioning on $\boldsymbol{s}$) of the exact matching property of the right-Haar prior can be found in [2]. A more general version can be found in [15]. The whole class of such theorems is sometimes called the Hunt-Stein theorem, which was developed in [25] for testing.

4.3 Reference Priors

We have not yet presented a general method for developing good objective priors in multi-parameter problems. The Jeffreys-rule prior in multi-parameter problems simply seems to be unsuitable. Matching priors and invariance priors in multiparameter problems are typically suitable, but they are often unavailable (e.g., the solutions to the system of differential equations may not be available, or the model may not have a suitable group invariance structure). In this section, we describe a method for deriving an objective prior in multiparameter problems which seems to be essentially always effective.

4.3.1 Introduction to the Reference Prior Approach

In the matching prior approach for multi-parameter models, we saw that different parameters in the model could have different matching priors. This is also true of the reference prior approach. A reference prior will be a prior distribution on the full parameter space, but it will be developed to focus on the parameter of interest. If there is more than one parameter of interest in a problem, there could thus be more than one reference prior.

The information to be expected from one observation from model $\mathcal{M} \equiv \{p(\boldsymbol{x} \mid \theta), \boldsymbol{x} \in \boldsymbol{\mathcal{X}}, \theta \in \Theta\}$, when the proper prior for θ is $\pi(\theta)$, is commonly taken to be the Kullback-Liebler divergence between the prior distribution and the posterior distribution, which is given by

$$I\{\pi \mid \mathcal{M}\} = \int\int_{\boldsymbol{\mathcal{X}}\times\Theta} p(\boldsymbol{x} \mid \theta)\pi(\theta) \log \frac{\pi(\theta \mid \boldsymbol{x})}{\pi(\theta)}\, d\boldsymbol{x}\, d\theta\,. \tag{4.21}$$

The sharper the prior, the more information it contains, reducing the information expected from the data; in the extreme, a point mass prior provides complete information about θ, so the data can add no information. The prior that maximizes $I\{\pi \mid \mathcal{M}\}$ can thus claim to be that prior which maximizes the amount of information provided by the data and is, hence, a natural candidate for an objective prior.

Unfortunately, the resulting prior turns out to be problematical, in that it will typically be a discrete prior that depends on the sample size, even when θ is a continuous parameter ([6]). This makes it unappealing in practice.

An alternative, proposed in [14], is to apply this idea asymptotically, considering k independent replications of $\mathcal{M}$, resulting in the sequence of realizations $\boldsymbol{x}^{(k)} = \{\boldsymbol{x}_1, \ldots, \boldsymbol{x}_k\}$ and the ensuing replicated model $\mathcal{M}^k$. Then $I\{\pi \mid \mathcal{M}^k\}$ is the amount of information in $\boldsymbol{x}^{(k)}$. As $k \to \infty$, the posterior distribution will typically converge to a point mass, providing complete information about θ. Thus, intuitively, $\lim_{k\to\infty} I\{\pi \mid \mathcal{M}^k\}$ provides a measure of the missing information about θ associated to the prior $\pi(\theta)$. Maximizing this missing information is thus another natural way to define an objective prior and is the basis of the *reference prior* approach that we will be discussing.

4.3.2 The Reference Prior for a Real Parameter

Consider first the situation in which the model has only one real continuous parameter θ. The following theorem, from [7] in which the proof can be found, provides an explicit expression for the reference prior for θ (i.e., that prior which maximizes the asymptotic missing information), under mild conditions. Recall that $\boldsymbol{x}$ refers to the entire vector of observations from the model and $\boldsymbol{x}^{(k)} = \{\boldsymbol{x}_1, \ldots, \boldsymbol{x}_k\}$ refers to a vector of (artificial) independent replicates of these vector observations from the model. It is often more convenient (and equivalent) to work with sufficient statistics $\boldsymbol{t}_k = \boldsymbol{t}_k(\boldsymbol{x}^{(k)}) \in \boldsymbol{\mathcal{T}}_k$ for the replicated observations, so the theorem will be stated in terms of such sufficient statistics.

Proposition 4.1 Explicit form of the reference prior. *Consider a standard model $\mathcal{M} \equiv \{p(\boldsymbol{x}\,|\,\theta), \boldsymbol{x} \in \boldsymbol{\mathcal{X}}, \theta \in \Theta \subset \mathbb{R}\}$. Let $\pi^*(\theta)$ be a continuous strictly positive function such that the corresponding formal posterior*

$$\pi^*(\theta\,|\,\boldsymbol{t}_k) = \frac{p(\boldsymbol{t}_k\,|\,\theta)\,\pi^*(\theta)}{\int_\Theta p(\boldsymbol{t}_k\,|\,\theta)\,\pi^*(\theta)\,d\theta} \tag{4.22}$$

is proper and asymptotically consistent and, for any interior point $\theta_0 \in \Theta$, define

$$f_k(\theta) = \exp\left\{\int_{\boldsymbol{\mathcal{T}}_k} p(\boldsymbol{t}_k\,|\,\theta)\,\log\left[\pi^*(\theta\,|\,\boldsymbol{t}_k)\right]d\boldsymbol{t}_k\right\} \quad \text{and} \tag{4.23}$$

$$f(\theta) = \lim_{k\to\infty} \frac{f_k(\theta)}{f_k(\theta_0)}\,. \tag{4.24}$$

Suppose $f_k(\theta)$ is continuous and, for any fixed θ and sufficiently large k, $\{f_k^0(\theta)/f_k^0(\theta_0)\}$ is either monotonic in k or bounded above by some $h(\theta)$ which is integrable on any compact set. Then $\pi^R(\theta) = f(\theta)$ is the reference prior.

The choice of π^* is essentially arbitrary and can be chosen for computational convenience, and the choice of θ_0 is immaterial.

Example 4.17 *(Uniform data, continued.)* Suppose x is one observation from a uniform distribution $\{p(x\,|\,\theta) = \theta^{-1}, 0 < x < \theta, \theta > 0\}$. Here the sampling space is $\mathcal{X} = [0, \theta]$, which depends on the parameter θ, so that Fisher information is not defined and, hence, there is no Jeffreys-rule prior.

The i.i.d. replication data is the vector $\boldsymbol{x}^{(k)} = \{x_1, \ldots, x_k\}$ and using (for mathematical convenience) $\pi^*(\theta) = \theta^{-1}$, the corresponding posterior distribution is the Pareto density

$$\pi^*(\theta\,|\,t_k) = \mathrm{Pa}(\theta\,|\,k, t_k) = k\,\frac{t_k^k}{\theta^{k+1}}\,, \quad \theta > t_k, \tag{4.25}$$

where $t_k = \max\{x_1, \ldots, x_k\}$ is a sufficient statistic. The sampling distribution of t_k is the inverted pareto density

$$p(t_k\,|\,\theta) = \mathrm{Ip}(t_k\,|\,k, \theta^{-1}) = k\,\frac{t^{k-1}}{\theta^k}\,, \quad 0 < t_k < \theta. \tag{4.26}$$

Computation yields

$$f_k(\theta) = \exp\left[\int_0^\theta \mathrm{Ip}(t_k\,|\,k, \theta)\,\log\left(\mathrm{Pa}(\theta\,|\,k, t_k)\right)dt_k\right] = \frac{c(k)}{\theta}, \tag{4.27}$$

where $c(k) = k\,e^{-(1+1/k)}$. It follows that, for all k (not just in the limit), $f_k^0(\theta)/f_k^0(\theta_0) = \theta_0/\theta$. Thus $\pi^R(\theta) = \theta^{-1}$ is the reference prior.

This is an excellent prior; for instance, it is exact frequentist matching. Indeed, using Bayes theorem, the corresponding reference posterior for an i.i.d. sample $\boldsymbol{x} = (x_1, \ldots, x_n)$ is the Pareto distribution $\text{Pa}(\theta \mid n, t_n)$, having density

$$\pi(\theta \mid \boldsymbol{x}) = \pi(\theta \mid t) = \begin{cases} n\, t^n \theta^{-(n+1)} & \text{if } \theta \geq t \\ 0 & \text{otherwise} \end{cases} .$$

A one-sided posterior credible set, at level $1-\alpha$, can be shown to be the interval $(t_n, t_n\, \alpha^{-1/n})$ and, using (4.26), the frequentist coverage of this, as a confidence procedure, can easily be shown to be $1-\alpha$. Interestingly, this interval has width $t_n(\alpha^{-1/n}-1) = t_n(\exp\{-\frac{1}{n}\log(\alpha)\}-1) \approx \frac{-\log(\alpha)}{n}$ for large n, showing that this model has non-standard asymptotics. (Regular models have confidence intervals of width proportional to $C/\sqrt{n}$ for some constant.) This is, thus, an example of how objective Bayesian analysis automatically produces answers with the correct non-regular asymptotics.

4.3.3 Multiple Continuous Parameters

Applying the idea of maximizing missing information works wonderfully if θ is one-dimensional but fails – if utilized directly – when θ is multi-dimensional; indeed, the result will often be the multivariate Jeffreys-rule prior, which was seen in Section 4.2.2.2 to be highly problematical. The third key component of the reference prior approach is thus to apply the idea of maximizing missing information sequentially.

To see the idea, suppose that θ_1 is the one-dimensional parameter of interest and that the remaining parameters, $\theta^2 = (\theta_2, \ldots, \theta_m)$, are viewed as nuisance parameters. Suppose we were told that the 'correct' prior for the nuisance parameter is $\pi(\theta^2 \mid \theta_1)$ (the prior being allowed to depend on θ_1 and assumed here to be proper). Then θ^2 could be integrated out of the model, obtaining the (proper) marginal model

$$\mathcal{M}^* \equiv \{p(\boldsymbol{x} \mid \theta_1),\ \boldsymbol{x} \in \boldsymbol{\mathcal{X}},\ \theta_1 \in \Theta_1\},\ p(\boldsymbol{x} \mid \theta_1) = \int_{\Theta^2} p(\boldsymbol{x} \mid \theta_1, \theta^2)\, \pi(\theta^2 \mid \theta_1)\, d\theta^2 .$$

As this marginal model only depends on θ_1, the method of maximizing missing information can be applied (using, say, the implementation in the previous section) to obtain the (marginal) reference prior $\pi^R(\theta_1)$. The final overall reference prior would then simply be $\pi^R(\theta_1, \theta^2) = \pi(\theta^2 \mid \theta_1)\pi^R(\theta_1)$.

If the conditional prior $\pi(\theta^2 \mid \theta_1)$ is not known, one can condition on θ_1, and then find the reference priors $\pi^R(\theta^2 \mid \theta_1)$ for θ^2, given each θ_1. If θ^2 is one-dimensional, these can be determined as in the previous section. In the general case, these can be determined in the regular situation where posterior distributions are asymptotically normal; the result is given in Proposition 4.2. If the $\pi^R(\theta^2 \mid \theta_1)$ are proper, one can proceed as above to form the marginal model and find the reference prior for θ_1 in the marginal model.

Unfortunately, the conditional reference priors $\pi^R(\theta^2 \mid \theta_1)$ are typically not proper, in which case $p(\boldsymbol{x} \mid \theta_1)$ is not proper and so cannot be used to determine $\pi^R(\theta_1)$. To overcome this difficulty, one must operate, as in Section 4.1.3, on compact subsets of the parameter space which increase to the full space, taking appropriate limits of the constrained reference priors to define the reference prior on the full space. This is illustrated in Proposition 4.2, whose proof can be found in [5].

Proposition 4.2 (Two group reference prior under asymptotic normality.) The k replications result in model $\mathcal{M} \equiv \{p(\boldsymbol{x}^{(k)} \mid \theta_1, \theta^2), \boldsymbol{x}^{(k)} \in \boldsymbol{\mathcal{X}}^k, (\theta_1, \theta^2) \in \Theta_1 \times \Theta^2\}$. Let $I(\theta_1, \theta^2)$ denote the Fisher information for a single $\boldsymbol{x}$. Define $V(\theta_1, \theta^2) = I(\theta_1, \theta^2)^{-1}$ and, with $I_{11}(\theta_1, \theta^2)$ and $V_{11}(\theta_1, \theta^2)$ being scalar,

$$I(\theta_1,\theta^2) = \begin{pmatrix} I_{11}(\theta_1,\theta^2) & I_{12}(\theta_1,\theta^2) \\ I_{21}(\theta_1,\theta^2) & I_{22}(\theta_1,\theta^2) \end{pmatrix}, V(\theta_1,\theta^2) = \begin{pmatrix} V_{11}(\theta_1,\theta^2) & V_{12}(\theta_1,\theta^2) \\ V_{21}(\theta_1,\theta^2) & V_{22}(\theta_1,\theta^2) \end{pmatrix}.$$

Under asymptotic normality, the reference prior when θ_1 is the parameter of interest is defined as follows:

Case 1. If $c(\theta_1) = \int_{\Theta^2} \sqrt{I_{22}(\theta_1,\theta^2)}\, d\theta^2 < \infty$, let

$$\pi^R(\theta^2 \mid \theta_1) = \sqrt{I_{22}(\theta_1,\theta^2)}/c(\theta_1),\ \pi^R(\theta_1) \propto \exp\left\{ \int_{\Theta^2} \pi^R(\theta^2 \mid \theta_1) \log[V_{11}^{-1/2}(\theta_1,\theta^2)]\, d\theta^2 \right\}. \tag{4.28}$$

Then the reference prior is $\pi^R(\theta_1,\theta^2) = \pi^R(\theta^2 \mid \theta_1)\pi^R(\theta_1)$.

Case 2. If $c(\theta_1)$ is infinite, constrain θ^2 to compact subsets $\{\Theta_l^2,\ l = 1,2,\ldots\}$ of Θ^2, such that $\lim_{l\to\infty} \Theta_l^2 = \Theta^2$, and define

$$\begin{aligned} \pi_l(\theta^2 \mid \theta_1) &= \frac{1}{c_l(\theta_1)} \sqrt{I_{22}(\theta_1,\theta^2)} 1_{\Theta_l^2}(\theta^2),\ c_l(\theta_1) = \int_{\Theta_l^2} \sqrt{I_{22}(\theta_1,\theta^2)}\, d\theta^2, \\ \pi_l(\theta_1) &\propto \exp\left\{ \int_{\Theta_l^2} \pi_l(\theta^2 \mid \theta_1) \log[V_{11}^{-1/2}(\theta_1,\theta^2)]\, d\theta^2 \right\}, \\ \pi^R(\theta_1,\theta^2) &= \lim_{l\to\infty} \frac{\pi_l(\theta^2 \mid \theta_1)\pi_l(\theta_1)}{\pi_l(\theta_0^2 \mid \theta_{10})\pi_l(\theta_{10})}, \end{aligned}$$

for any interior point (θ_{10},θ_0^2). Then $\pi^R(\theta_1,\theta^2)$ is the reference prior.

Example 4.18 *Multinomial problem (continued).* For the multinomial problem, computations in [4] yield (this is a Case 1 scenario)

$$\pi^R(\theta_1,\theta^2) \propto \left(\prod_{i=1}^m \theta_i^{-1/2} \right) (1-\theta_1)^{-(m-1)/2} \left(1 - \sum_{i=1}^m \theta_i \right)^{-1/2}. \tag{4.29}$$

The marginal reference posterior for θ_1 from π^R can be shown to be

$$\pi^R(\theta_1 \mid \boldsymbol{x}) = \text{Be}(\theta_i \mid x_i + \tfrac{1}{2}, n - x_i + \tfrac{1}{2}). \tag{4.30}$$

There is a more general reference prior method called the *one-at-a-time reference prior algorithm* [4]. In this algorithm, one first finds the reference prior $\pi^R(\theta_m \mid \theta_1,\ldots,\theta_{m-1})$, forms the marginal model by integrating over θ_m, as was done over θ^2 in (4.28) (operating on compact sets if necessary), then finds the reference prior $\pi^R(\theta_{m-1} \mid \theta_1,\ldots,\theta_{m-2})$, repeating the process until finally obtaining $\pi^R(\theta_1)$, with the appropriate product of all of these being the reference prior. This is, indeed, our recommended reference prior algorithm, since it is based on one-dimensional information matrices, which all perspectives suggest are optimal.

This can be difficult to implement, however, and the approach discussed above – treating the parameter of interest separately, while grouping all the nuisance parameters together in a single step – has appeal as something much simpler to implement. It is, of interest, in this regard, that application of the one-at-a-time reference prior algorithm to the multinomial problem was shown in [4] to yield a reference prior that differed from (4.29), but resulted in the same marginal posterior (4.30) for the parameter of interest; as this is what is needed for inference about the parameter of interest, the cruder method still gave the optimal result.

4.4 Overall Objective Priors

In single parameter problems, the reference prior (usually the Jeffreys-rule prior) is uniquely defined but, in multiparameter models, we have seen that the reference prior often changes, depending on the quantity of interest. There are, however, situations where one is *simultaneously* interested in all the parameters of the model or, at least in several of them; in such situations having a single overall prior for use is appealing. In prediction or decision analysis for instance, all of the parameters of the model are typically relevant and often none are of major interest individually.

Another situation in which having an overall prior would be beneficial is when there is a non-standard quantity of interest (*e.g.*, a non-standard function of the model parameters), and formal derivation of the reference prior is not tenable. Computation can also be a consideration, since having to separately do Bayesian computations with different reference priors for each parameter can be challenging. Finally, when dealing with non-specialists, it may be best to just present them with one overall objective prior, rather than attempting to explain the reasons for preferring different priors for different quantities of interest.

4.4.1 Preliminaries

We have already encountered objective Bayes approaches that produce an overall prior, namely the use of the constant prior, the use of right-Haar priors in invariant situations, and the use of the Jeffreys-rule prior. We extensively discussed the limitations of the constant prior and the Jeffreys-rule prior, and having a suitably invariant model is rather special, so there is more to be done.

The approach taken in [8] to find an overall prior is to find a prior which yields marginal posteriors for each parameter that are close to the reference prior marginal posteriors for each parameter. To be more precise, assume that $\theta = \{\theta_1, \ldots, \theta_m\}$ is is the set of $m > 1$ parameters that are of interest. The goal is to find a prior, $\pi^o(\theta)$, whose corresponding marginal posteriors, $\{\pi^o(\theta_i \mid \boldsymbol{x}), i = 1, \ldots, m\}$, are close to the reference prior marginal posteriors $\pi^R_{\theta_i}(\theta_i \mid \boldsymbol{x})$ (the subscript θ_i indicating that this arose from the reference prior when θ_i is the parameter of interest).

Example 4.19 *Multinomial data.* Suppose $\boldsymbol{x} = (x_1, \ldots, x_m)$ is multinomial $\text{Mu}(\boldsymbol{x} \mid n; \theta_1, \ldots, \theta_m)$, with $\sum_{i=1}^m x_i = n$, and $\sum_{i=1}^m \theta_i = 1$. In Example 4.18, it was shown that the reference prior, when the parameter of interest is θ_i, is a different prior for each θ_i, and that the reference prior for θ_i results in a Beta reference marginal posterior $\pi^R_{\theta_i}(\theta_i \mid \boldsymbol{x}) = \text{Be}(\theta_i \mid x_i + \frac{1}{2}, n - x_i + \frac{1}{2})$.

It is natural here to consider the Dirichlet $\{\text{Di}(\theta \mid a, \ldots, a) : a \in (0, \infty)\}$ class of priors. The choice of a for which the resulting marginal posteriors are closest to the reference marginal posteriors is shown in [8] to be approximately $a = 1/m$. Thus the recommended overall objective prior would be $\pi^o(\theta) = \text{Di}(\theta \mid \frac{1}{m}, \ldots, \frac{1}{m})$.

Note that this overall prior overcomes the problem with the Jeffreys-rule prior that was discussed in Example 4.13; the Jeffreys-rule prior, corresponding to $a = 1/2$, was shown to yield unreasonable posterior means for the θ_i in the situation where $n = 3$, $m = 1000$, $x_{240} = 2$, $x_{876} = 1$, and the other x_i are zero. In particular, the posterior mean for θ_{240} was only 0.005 even though 2/3 of the observations were in that cell. The problem is that the Jeffreys-rule prior effectively adds 1/2 to each cell, so that the cells with $x_i = 0$ overwhelmed the cells with $x_i \neq 0$. In contrast, $\pi^o(\theta)$ only adds $1/m = 0.001$ to each cell, so that

$$\text{E}[\theta_i \mid \boldsymbol{x}] = \frac{x_i + 1/m}{\sum_{i=1}^m (x_i + 1/m)} = \frac{x_i + 1/m}{n+1} = \frac{x_i + 0.001}{4}.$$

Thus $\mathrm{E}[\theta_{240} \mid \boldsymbol{x}] \approx 0.5$, $\mathrm{E}[\theta_{876} \mid \boldsymbol{x}] \approx 0.25$, and $\mathrm{E}[\theta_i \mid \boldsymbol{x}] \approx \frac{1}{4000}$ otherwise, all sensible results.

We will discuss various possible approaches to the development of an overall prior in this section. Most of the approaches and examples are taken from either [8] or [12].

4.4.2 When the Reference Prior Is Common

If one is able to find a single joint prior $\pi^o(\theta)$ whose corresponding marginal posteriors are precisely equal to the reference posteriors for each of the θ_i's, so that, for all $\boldsymbol{x} \in \boldsymbol{\mathcal{X}}$, $\pi^o(\theta_i \mid \boldsymbol{x}) = \pi^R_{\theta_i}(\theta_i \mid \boldsymbol{x})$, $i = 1, \ldots, m$, then it is natural to argue that this should be an appropriate solution to the problem. The $\mathrm{N}(x_i \mid \mu, \sigma)$ model has already been mentioned as an example, where the reference priors when μ or σ are the parameters of interest are the same. (Their common reference prior is also excellent if μ and σ are jointly of interest, *e.g.*, if a joint credible set for them is sought.)

4.4.3 Hierarchical Reference Approach

The hierarchical reference prior approach consists of the following steps.

i. Choose a class of *proper* priors $\{\pi(\theta \mid \boldsymbol{a})\colon \boldsymbol{a} \in \mathcal{A}\}$ reflecting the desired structure of the problem.

ii. Form the marginal density for the hyperparmeter $\boldsymbol{a}$,

$$p(\boldsymbol{x} \mid \boldsymbol{a}) = \int p(\boldsymbol{x} \mid \boldsymbol{a})\, \pi(\theta \mid \boldsymbol{a})\; d\theta\,,$$

which is proper because the $\pi(\theta \mid \boldsymbol{a})$ are proper.

iii. Find the reference prior, $\pi^R(\boldsymbol{a})$, for $\boldsymbol{a}$ in this marginal model.

iv. Use, as the overall prior for θ,

$$\pi^o(\theta) = \int \pi(\theta \mid \boldsymbol{a})\, \pi^R(\boldsymbol{a})\, d\boldsymbol{a}\,. \tag{4.31}$$

Note that computation with this prior is often easiest if one keeps $\boldsymbol{a}$ as a random variable, rather than integrating it out and working directly with (4.31). This is especially true when $\{\pi(\theta \mid \boldsymbol{a})\colon \boldsymbol{a} \in \mathcal{A}\}$ is a conjugate family to the data distribution, so that Gibbs sampling can be employed.

4.4.3.1 Application to the Multinomial Distribution

The Dirichlet $\{\mathrm{Di}(\theta \mid a, \ldots, a)\colon a \in (0, \infty)\}$ class of hierarchical priors for θ is natural here, reflecting a desire to treat all the θ_i similarly. The marginal density is

$$\begin{aligned} p(\boldsymbol{x} \mid a) &= \int \binom{n}{x_1, \ldots, x_m} \left(\prod_{i=1}^{m} \theta_i^{x_i}\right) \frac{\Gamma(m\,a)}{\Gamma(a)^m} \prod_{i=1}^{m} \theta_i^{a-1} d\theta \\ &= \binom{n}{x_1, \ldots, x_m} \frac{\Gamma(m\,a)}{\Gamma(a)^m} \frac{\prod_{i=1}^{m} \Gamma(x_i + a)}{\Gamma(n + m\,a)}\,. \end{aligned} \tag{4.32}$$

This can be shown to be a regular, one-parameter model, so the reference prior, $\pi^R(a)$, for the hyper parameter a, is the Jeffreys prior for this marginal model. In [8], this was shown to be given by

$$\pi^R(a) \propto \left[\sum_{j=0}^{n-1} \left(\frac{Q(j \mid a, m, n)}{(a + j)^2} - \frac{m}{(m\,a + j)^2}\right)\right]^{1/2}, \tag{4.33}$$

where $Q(\cdot \mid a, m, n)$, for $j = 0, \ldots, n-1$, is given by

$$Q(j \mid a, m, n) = \sum_{l=j+1}^{n} \binom{n}{l} \frac{\Gamma(l+a)\,\Gamma(n-l+(m-1)a)\,\Gamma(m\,a)}{\Gamma(a)\,\Gamma((m-1)a)\,\Gamma(n+m\,a)} \,.$$

Finally, the overall prior for θ is

$$\pi^o(\theta) = \int \mathrm{Di}(\theta \mid a)\, \pi^R(a)\, da \,.$$

This prior turns out to be a proper distribution, which is rather unusual for an objective prior on an unbounded space. It turns out that the marginal likelihoood of a does not go to zero as a grows ([8]), so that the prior must be proper for the posterior to exist. It is a rather amazing property of reference/Jeffreys priors that they seem to always be proper when the likelihood does not go to zero.

4.4.4 Hierarchical Normal Models

Hierarchical normal models are the basis of much of applied Bayesian statistics, yet there is uncertainty as to which priors to use for hyperparameters (parameters at higher levels of the hierarchical model). The long history of efforts to develop objective hyperpriors is reviewed in [29] and [12], from which this section is primarily based.

Once the difficulties of using the Jefferys-rule prior in multi-parameter models and hierarchical models became recognized, it became common to use a constant prior for higher level variances or covariances, but the constant prior is much too diffuse, requiring twice as many observations to obtain posterior propriety as is logically needed (see [12], which also showed that the constant prior has markedly poor performance in risk simulations).

[10] and [11] approached the question of choice of hyperpriors in normal hierarchical models by looking at the frequentist notion of admissibility of resulting estimators; an estimator is admissible if it cannot be improved upon in terms of mean squared error and inadmissible if it can be improved. Hyperpriors that are too diffuse result in inadmissible estimators, while hyperpriors that are concentrated enough result in admissible estimators. Hyperpriors that are 'on the boundary of admissibility' thus seem to be sensible choices for objective priors, being as diffuse as possible without resulting in inadmissible procedures. This approach was used in [12] to develop the recommended hyperpriors presented here for any normal hierarchical model.

4.4.4.1 The Hierarchical Model Considered

Consider the following two-stage hierarchical model. Suppose that, independently and for $i = 1, \ldots, m$,

$$\boldsymbol{x}_i \mid \theta_i \sim \mathrm{N}_k(\cdot \mid \theta_i, \Sigma_i) \quad \text{and} \quad \theta_i \mid \beta, \boldsymbol{V} \sim \mathrm{N}_k(\cdot \mid \boldsymbol{Z}_i\beta, \boldsymbol{V})\,, \tag{4.34}$$

where the $\boldsymbol{x}_i$ are the $k \times 1$ observation vectors; the Σ_i are *known* $k \times k$ full rank covariance matrices; the θ_i are $k \times 1$ unknown mean vectors; β is an $l \times 1$ unknown "hyper-mean" vector; the $\boldsymbol{Z}_i$ are $k \times l$ known matrices of covariates of full rank; and $\boldsymbol{V}$ is an unknown $k \times k$ "hyper-covariance matrix." If the Σ_i are not known, they either need to be estimated from data and plugged in (resulting in a partly empirical Bayes approach) or incorporated into a larger hierarchical Bayesian analysis using, say, the reference prior for a first level covariance matrix, given in [35].

Example 4.20 (Following the path-breaking paper [21]:) Suppose $k = 1$ and, for $i = 1, \ldots, m$, x_i is the batting average, after one month, of baseball player i (after an arcsin transformation of the original binomial random variable to normality), with $\sigma_i^2 = 1/[4n_i]$ and n_i being the number of at-bats the player had in the first month. The goal is to estimate θ_i, the 'true' batting ability of the player, as would be determined (say) by batting average at the end of the year, using the $x_i \sim \text{N}(x_i \mid \theta_i, \sigma_i^2)$.

The (arcsin transformed) batting averages, b_i, of the players are known from the previous year and the θ_i are modeled by regression on these previous batting averages, leading to the model $\theta_i \sim \text{N}(\cdot \mid \beta_1 + b_i\beta_2, V)$. In the above notation, $\beta = (\beta_1, \beta_2)'$ (the unknown regression coefficients), $\boldsymbol{Z}_i = (1, b_i)$ are the covariates, and V is the unknown variance of the regression. To proceed, an objective prior distribution is needed for (β, V).

Example 4.21 At hospital i, the observations $\boldsymbol{x}_i = (x_{i1}, \ldots, x_{ik})'$ are the sample averages of the costs of k different medical treatments; θ_i is the corresponding unknown vector of true mean costs of the treatments at the hospital; and Σ_i is the covariance matrix associated with these sample averages.

Consider the jth coordinate, θ_{ij}, of each θ_i; this refers to the cost of the jth medical treatment at the ith hospital. It is natural to model this as depending on p hospital characteristics, such as the number of patients receiving the treatment, the average severity of the condition of the patients for the treatments, the average income of the patients, etc. Denoting these characteristics for the jth treatment at the ith hospital as $\boldsymbol{z}_{ij} = (z_{ij1}, \ldots, z_{ijp})$, a reasonable model would be a regression model

$$\theta_{ij} = \boldsymbol{z}_{ij}\alpha_j + \epsilon_{ij}, \quad \text{for } i = 1, \ldots, m; \quad j = 1, \ldots, k,$$

where α_j is a $p \times 1$ column vector of weights (specific to each treatment but assumed to be constant across hospitals) determining the effect of hospital characteristics on the cost of treatment j, and ϵ_{ij} is normal error. There would typically be a separate regression of this form for each treatment j, and stacking these regressions vertically leads to equation (4.34), where $l = kp$,

$$\boldsymbol{z}_i = \begin{pmatrix} \boldsymbol{z}_{i1} & \boldsymbol{0} & \ldots & \boldsymbol{0} \\ \boldsymbol{0} & \boldsymbol{z}_{i2} & \ldots & \boldsymbol{0} \\ \vdots & \vdots & \ddots & \vdots \\ \boldsymbol{0} & \boldsymbol{0} & \ldots & \boldsymbol{z}_{ik} \end{pmatrix}, \qquad \beta = \begin{pmatrix} \alpha_1 \\ \alpha_2 \\ \vdots \\ \alpha_k \end{pmatrix}. \tag{4.35}$$

If it were thought that each of the treatment cost means were independent, $\boldsymbol{V}$ could be taken as diagonal, but it is far more likely that there is considerable dependence, so that completely unknown $\boldsymbol{V}$ is reasonable.

In this example, there could be a third level of hierarchial modeling,

$$\alpha_i \sim \text{N}_p(\cdot \mid \xi, \Omega). \tag{4.36}$$

Whether or not this is appropriate depends on the precise context, but we will see that the analysis could still be done for this three-level hierarchical model with the methodology proposed herein.

Remark 4.1 Often the data vectors themselves arise from linear models $\boldsymbol{y}_i \sim N_r(\cdot \mid \boldsymbol{B}_i\theta_i, \Lambda_i)$, where the $\boldsymbol{B}_i$ are known design matrices of covariates. In this situation, one can transform to the least squares estimates for θ_i,

$$\boldsymbol{x}_i = (\boldsymbol{B}_i^t\Lambda_i^{-1}\boldsymbol{B}_i)^{-1}\boldsymbol{B}_i^t\Lambda_i^{-1}\boldsymbol{y}_i,$$

which will be distributed as in (4.34), with covariance matrix $\Sigma_i = (\boldsymbol{B}_i^t\Lambda_i^{-1}\boldsymbol{B}_i)^{-1}$.

4.4.4.2 The Recommended Prior

In [12], it is recommended to utilize independent priors for the unknown hyperparameters β and $\boldsymbol{V}$, i.e., $\pi(\beta, \boldsymbol{V}) = \pi(\beta)\pi(\boldsymbol{V})$, with the following choices:

$$\pi(\beta) \propto \frac{1}{[1+||\beta||^2]^{(l-1)/2}}, \quad \pi(\boldsymbol{V}) = \frac{1}{|\boldsymbol{V}|^{1-1/(2k)} \prod_{i<j}(d_i - d_j)}, \tag{4.37}$$

where $||\beta||^2 = \beta'\beta$ and $d_1 > d_2 \ldots > d_k$ are the ordered eigenvalues of $\boldsymbol{V}$. For $k = 1$, the second prior is defined as $\pi(V) = 1\sqrt{V}$. Extensive justification of these choices is given in [12].

One of the most important properties of this prior is that it can be used at any level of a hierarchical model for the mean parameters and the covariance matrix. This is not so for many other proposed priors such as 'marginal Jeffreys priors' (see [12]). That the resulting posterior distributions are guaranteed to be proper is shown in [12].

For the priors in (4.37) to be reasonable, one must deal with the fact that the covariates (defined by the columns of $\boldsymbol{Z}_i$) often use different units of measurement, units which can be arbitrary, and objective priors should be invariant to the choice of unit of measurement. The easiest way to achieve this is to first rescale the covariates so they are 'unitless' and then proceed with the use of the priors in (4.37). Here are two common methods of such re-scaling.

Method 1 - Standard Deviation Rescaling: Define c_j to be the sample standard deviation of all non-zero entries in the jth columns of all the $\boldsymbol{Z}_i$, and rescale to obtain new covariates $\boldsymbol{z}^*_{ij} = \boldsymbol{z}_{ij}/c_j$.

Method 2 - Max Coordinate Rescaling: Define c_j to be the maximum absolute value of all entries in the jth columns of all the $\boldsymbol{Z}_i$, and rescale as above. (This puts everything on the interval from -1 to 1.)

Remark 4.2 If covariate j is rescaled by c_j, for all j, the new hyper-mean becomes $\beta^* = (c_1\beta_1, \ldots, c_l\beta_l)'$. Then, if the rescaled covariates are used in Bayesian analysis, the analysis will yield the posterior for β^*, which must then be transformed back to the posterior for β, if the latter is desired. If only the posterior distributions of the θ_i are desired, the analysis carried out in the rescaled space is fine.

4.5 Conclusions

Objective Bayesian inference has been a central part of statistics for over 250 years and this has (necessarily) been a very brief introduction to the area. For a much more extensive review of objective Bayesian inference, see [9] and the over 1500 references cited therein. The study of the interface between the objective Bayesian approach and the frequentist approach is much more recent (necessarily, because the frequentist approach is only about 100 years old). For a more recent study of this interface, see [3] and the accompanying discussions.

Acknowledgments

James Berger was supported by the U.S. National Science Foundation grant DMS 1821289.

Note

1. The Haar measures, from which the Haar priors are derived, were introduced in the 1930's by the Hungarian mathematician Alfred Haar (cf. [32])

Bibliography

[1] Bayes, T., 1958. An essay towards solving a problem in the doctrine of chances. *Philosophical Transactions of the Royal Society A*, *53* and *54*, pp. 370–416; 370–416. Reprinted in *Biometrika*, *45*, pp. 293–315, with a biographical note by G. A. Barnard. Reproduced in Press (1989), 185–217, 1763.

[2] Berger, J. O., 1985. *Statistical Decision Theory and Bayesian Analysis* (2nd ed.). New York: Springer.

[3] Berger, J. O., 2022. Four types of frequentism and their interplay with Bayesianism (with discussion). *The New England Journal of Statistics in Data Science*, *1*, pp. 126–137. https://doi.org/10.51387/22-NEJSDS4.

[4] Berger, J. O. and Bernardo, J. M., 1992. On the development of reference priors. In *Bayesian Statistics 4*, eds J. M. Bernardo, J. O. Berger, A. P. Dawid and A. F. M. Smith, pp. 35–60. Oxford: Oxford University Press.

[5] Berger, J. O. and Bernardo, J. M., 1992. Ordered group reference priors, with applications to multinomial problems. *Biometrika*, *79*, pp. 25–37.

[6] Berger, J. O., Bernardo, J. M. and Mendoza, M., 1989. On priors that maximize expected information. In *Recent Developments in Statistics and Their Applications*, eds J. P. Klein and J. C. Lee, pp. 1–20. Seoul: Freedom Academy.

[7] Berger, J. O., Bernardo, J. M. and Sun, D., 2009. The formal definition of reference priors. *Annals of Statistics*, *37*, pp. 905–938.

[8] Berger, J. O., Bernardo, J. M. and Sun, D., 2015. Overall objective priors. *Bayesian Analysis*, *10*, pp. 189–221.

[9] Berger, J. O., Bernardo, J. M. and Sun, D., 2024. *Objective Bayesian Inference.* Singapore: World Scientific Publishing Company.

[10] Berger, J. O. and Strawderman, W. E., 1996. Choice of hierarchical priors: Admissibility in estimation of normal means. *Annals of Statistics*, *24*, pp. 931–951.

[11] Berger, J. O., Strawderman, W. E. and Tang, D., 2005. Posterior propriety and admissibility of hyperpriors in normal hierarchical models. *Annals of Statistics*, *33*, pp. 606–646.

[12] Berger, J. O., Sun, D. and Song, C., 2020. An objective prior for hyperparameters in normal hierarchical models. *Journal of Multivariate Analysis, 178*, p. 104606.

[13] Berger, J. O. and Wolpert, R. L., 1988. *The Likelihood Principle* (2nd ed.). Hayward, CA: IMS.

[14] Bernardo, J. M., 1979. Reference posterior distributions for Bayesian inference. *Journal of the Royal Statistical Society: Series B, 41*, pp. 113–147. Reprinted in *Bayesian Inference 1* (eds G. C. Tiao and N. G. Polson). Oxford: Edward Elgar, 1995, pp. 229–263.

[15] Chang, T. and Villegas, C., 1986. On a theorem of Stein relating Bayesian and classical inferences in group models. *Canadian Journal of Statistics, 14*, pp. 289–296.

[16] Cox, D. R. and Reid, N., 1987. Parameter orthogonality and approximate conditional inference (with discussion). *Journal of the Royal Statistical Society: Series B, 49*, pp. 1–39.

[17] Dale, A. I., 1999. *A History of Inverse Probability: From Thomas Bayes to Karl Pearson.* Berlin: Springer.

[18] Datta, G. S. and Ghosh, J. K., 1995. On priors providing frequentist validity for Bayesian inference. *Biometrika, 82*, pp. 37–45.

[19] Datta, G. S. and Mukerjee, R., 2004. *Probability Matching Priors: Higher Order Asymptotics.* New York: Springer.

[20] de Morgan, A., 1838. *An Essay on Probabilities.* London: Longman.

[21] Efron, B. and Morris, C. N., 1975. Data analysis using Stein's estimator and its generalizations. *Journal of the American Statistical Association, 70*, pp. 311–319.

[22] Fienberg, S. E., 2006. When did Bayesian inference become Bayesian? *Bayesian Analysis, 1*, pp. 1–40.

[23] Fraser, D. A. S., Reid, N., Marras, E. and Yi, G. Y., 2010. Default priors for Bayesian and frequentist inference. *Journal of the Royal Statistical Society: Series B, 72*, pp. 631–654.

[24] Hald, A., 1998. *A History of Mathematical Statistics from 1750 to 1930.* Hoboken, NJ: Wiley.

[25] Hunt, G. and Stein, C., 1946. Most stringent tests of statistical hypotheses. Technical Report, Stanford University.

[26] Jeffreys, H., 1946. An invariant form for the prior probability in estimation problems. *Proceedings of the Royal Society A, 186*, pp. 453–461.

[27] Laplace, P. S., 1812. *Théorie Analytique des Probabilités.* Paris: Courcier. Reprinted in *Oeuvres Complètes de Laplace 7*, 1878–1912. Paris: Gauthier-Villars.

[28] Lindqvist, B. H. and Taraldsen, G., 2018. On the proper treatment of improper distributions. *Journal of Statistical Planning and Inference, 195*, pp. 93–104.

[29] Morris, C. N. and Lysy, M., 2012. Shrinkage estimation in multilevel normal models. *Statistical Science, 27*, pp. 115–134.

[30] Peers, H. W., 1965. On confidence points and Bayesian probability points in the case of several parameters. *Journal of the Royal Statistical Society: Series B*, *27*, pp. 9–16.

[31] Stein, C., 1985. On the coverage probability of confidence sets based on a prior distribution. In *Sequential Methods in Statistics*, ed Ryszard Zielisnski, pp. 485–514. Varsaw: Polish Scientific Publications.

[32] Szökefalvi-Nagy, B., 1985. Alfred Haar (1885–1933). *Results in Mathematics*, *8*, pp. 194–196.

[33] Tibshirani, R., 1989. Noninformative priors for one parameter of many. *Biometrika*, *76*, pp. 604–608.

[34] Welch, B. L. and Peers, H. W., 1963. On formulae for confidence points based on intervals of weighted likelihoods. *Journal of the Royal Statistical Society: Series B*, *25*, pp. 318–329.

[35] Yang, R. and Berger, J. O., 1994. Estimation of a covariance matrix using the reference prior. *Annals of Statistics*, *22*, pp. 1195–1211.

5

Fiducial Inference, Then and Now

Philip Dawid
University of Cambridge

IN MEMORIAM
Mervyn Stone 1932–2020
Donald Fraser 1925–2020

5.1 Introduction

According to Zabell (1992), "the fiducial argument stands as Fisher's one great failure", a sentiment that has been echoed by others. Fisher never constructed a fully fledged theory of fiducial inference, but developed his ideas by means of examples and *ad hoc* responses to increasingly complex problems or challenges raised by others. Few other statisticians have taken the fiducial argument seriously, and after some sporadic activity (mostly critical) in the decades following Fisher's introduction of the idea in 1930, it almost completely disappeared from the scene. The trio of Encyclopedia articles Edwards (1983); Buehler (1983); Stone (1983) is a useful resource for the state of the enterprise up to 1982. In recent times, however, there has been a resurgence of interest in the fiducial programme, as evidenced by works such as Hannig et al. (2016) and Martin and Liu (2016), and the success of the series of annual *Bayesian, Fiducial & Frequentist* (BFF) conferences (http://bff-stat.org), since 2014.

In this chapter, I give a personal review of the main contributions, positive and negative, to fiducial inference, in both earlier and later periods. In Section 5.2, I describe the original argument, centered on inference for a correlation coefficient. Section 5.3 introduces an extension to more complex problems based on the idea of a pivotal function. Such a function arises naturally when the problem possesses the property of invariance under a group of transformations, as described in Section 5.4. A variant of this is Fraser's structural model described in Section 5.4.2, which starts by prescribing the relevant group, while a further extension is the functional model of Section 5.5. This forms the basis for the rest of this chapter.

In Section 5.6, we show that, in a functional model, two different, but equally intuitive, ways to marginalize a multivariate fiducial distribution to a component parameter give the same answer—so long as we can form an appropriate new, reduced, model, involving only that parameter. However, we soon encounter some problems with the fiducial argument (Section 5.7). In particular, the same statistical model may be representable in more than one way as a structural or functional model, and different representations can lead to different fiducial inferences, with no obvious way of choosing between them. Even within a single functional representation, marginalization is ill-defined when we can't form a

DOI: 10.1201/9780429341731-5

reduced model, and even when we can, if we wish to condition a multivariate fiducial distribution on the value of a component parameter, we again have two possible routes, both intuitively reasonable—but now they can yield different answers, and again it is not clear how one ought to choose between them.

To this point, we have only considered simple functional models—essentially, those where the dimensions of the parameter and the data are the same. As described in Section 5.8, in a non-simple model we need to condition on certain aspects of the data. But in order for such conditioning to be well-defined, we have to require an additional property, partitionability, of the model: this allows us to construct and condition on an appropriate ancillary statistic and leads to a unique fiducial distribution. In the absence of partitionability, however, there is simply no clearly defined fiducial distribution. Various ways of constructing something to take its place have recently been suggested, but a high degree of arbitrariness remains.

Section 5.9 considers cases in which the fiducial argument fails to yield a distribution for the parameter, but only a distribution for a set containing the parameter. This is linked to the Dempster-Shafer theory of belief functions and has been a focus of recent work. In this case too, a partitionability property is required for well-defined inference.

Section 5.10 considers the newer work on "generalized fiducial inference" as a pointer to directions that fiducial theory is likely to take in future. Some concluding thoughts are gathered in Section 5.11, while an Appendix organizes the various examples by model and by topic.

5.2 Fisher's Original Fiducial Argument

The fiducial argument was introduced by Fisher (1930) by means of the following example.

We have n observations from a bivariate normal distribution. Let random variable R be the sample correlation, and let parameter variable Φ be the population correlation. Then the sampling distribution of R depends only on (n and) the value ϕ of Φ. The form of this distribution (Fisher, 1915) is not expressible by means of simple functions[1]; however, the rest of Fisher's argument does not involve the specific form of this distribution. Indeed, letting $F(r;\phi) = \Pr(R \leq r \mid \Phi = \phi)$ be the cumulative distribution function of R when $\Phi = \phi$, the general argument applies to any problem satisfying the following sufficient (but not entirely necessary) regularity conditions:

i. The range of each of R and Φ is an open interval in the real line
ii. $F(r;\phi)$ is a continuous function of each of its arguments
iii. For fixed ϕ, $F(r;\phi)$ strictly increases as r increases, taking all values in $(0,1)$
iv. For fixed r, $F(r;\phi)$ strictly decreases as ϕ increases, taking all values in $(0,1)$

It follows from the probability integral transformation (Angus, 1994) that, for all ϕ, the distribution of $F(R;\phi)$, given $\Phi = \phi$, is the uniform distribution Unif$[0,1]$. That is, $E = F(R;\Phi) \sim \text{Unif}[0,1]$, independently of Φ: $E \perp\!\!\!\perp \Phi$.

Since, for any $\gamma \in [0,1]$, when $\Phi = \phi$

$$\Pr\{F(R;\phi) \leq \gamma\} = \gamma, \tag{5.1}$$

a level-γ confidence set for Φ is, for observed $R = r$, $I(r;\gamma) := \{\phi : F(r;\phi) \leq \gamma\}$. In fact, because $F(r;\phi)$ is a decreasing function of ϕ, this is an upper confidence interval: $I(r;\gamma) = [\phi(\gamma), \infty)$, where $F(r;\phi(\gamma)) = \gamma$.

Fisher now takes this argument further. He regards the uniform distribution for $E = F(R;\Phi)$, and so (5.1), as remaining valid, even after observing data with $R = r$. Equivalently, he takes $E \perp\!\!\!\perp R$ (compare the sampling property $E \perp\!\!\!\perp \Phi$). Thus he assumes, for any r,

$$F(r, \Phi) \sim \text{Unif}[0, 1]. \tag{5.2}$$

This argument has now assigned to the parameter Φ the status of a random variable. Its implied distribution is termed the *fiducial distribution* of Φ (based on data r): we denote it by Π_r. Then, after observing $R = r$, (5.2), in conjunction with (ii) and (iv), implies

$$\Pi_r(\Phi \leq \phi) = \Pi_r\{F(r;\Phi) \geq F(r;\phi)\} = 1 - F(r;\phi). \tag{5.3}$$

On account of (iv) this yields a full "fiducial distribution function" for Φ; under differentiability, the associated fiducial density of Φ is given by $\pi(\phi) = -\frac{\partial}{\partial\phi}F(r;\phi)$. In particular, $\Pi_r\{\Phi \in I(r;\gamma)\} = \Pi_r\{\Phi \geq \phi(\gamma)\} = F\{r, \phi(\gamma)\} = \gamma$, so transforming a confidence statement for Φ into a probability statement for Φ—an interpretation of a confidence interval that is typically castigated as a showing a gross misunderstanding of its nature.

Fisher, noting that (5.3) does not follow from standard probability arguments, termed it a "fiducial probability". However, he appeared to believe that (subject to some caveats—see Section 5.3.1 below) it can still be interpreted as a regular probability. Savage (1961) memorably described the fiducial argument as "a bold attempt to make the Bayesian omelet without breaking the Bayesian eggs". Lindley (1958) considered the general case (i)–(iv) of Fisher's construction, and showed that the fiducial distribution does not arise as a Bayesian posterior, for any prior, except in the special case of a (possibly transformed) location model with uniform prior on the location parameter. In particular, the above fiducial distribution of the population correlation cannot be obtained as a Bayesian posterior based on combining the likelihood based on the sample correlation with some prior distribution.

5.3 Pivotal Inference

A more general fiducial construction relies on the existence of a pivot (Barnard, 1980).

Let X have distribution governed by parameter Θ. A *pivot* E is a function of X and Θ with known distribution P_0, not depending on the value θ of Θ: $E \perp\!\!\!\perp \Theta$, with $E \sim P_0$.

Example 5.1. Fisherian Pivot
In Section 5.2, $E = F(R;\Phi)$ is a pivot, with distribution Unif$[0, 1]$. We have seen how this can be used to supply a fiducial distribution. □

Example 5.2. Normal Model
For a sample of size n from the normal distribution with unknown mean M and standard deviation Σ, having sample mean $\overline{X}$ and sample standard deviation S, $E_1 = (\overline{X} - M)/\Sigma$ is a pivot, having the normal distribution $N(0, 1/n)$; $E_2 = S/\Sigma$ is a pivot, having distribution $\sqrt{\chi^2_{n-1}/(n-1)}$; and $E_3 = \sqrt{n}E_1/E_2 = (\overline{X} - M)/(S/\sqrt{n})$ is a pivot, with the Student distribution t_{n-1}. □

Given a suitable pivot $E = f(X, \Theta)$, a fiducial distribution can be obtained by regarding the distribution P_0 of E as still relevant, even after observing the data. Equivalently, instead of $E \perp\!\!\!\perp \Theta$, we regard $E \perp\!\!\!\perp X$. Thus after observing $X = x$, we suppose $f(x, \Theta) \sim P_0$.

When Θ (or a desired function Ψ of Θ) can be expressed as a function of $f(x, \Theta)$—the case of *invertibility*—this delivers a fiducial distribution, Π_x, for Θ (or Ψ).

Let R be a set such that $P_0(R) = 1 - \alpha$, and consider the region $R_x = \{\theta : f(x, \theta) \in R\}$. For any x, this has fiducial probability $\Pi_x\{f(x, \Theta) \in R\} = P_0(R) = 1 - \alpha$. Also, it has sampling coverage $P_\theta\{f(X, \theta) \in R\} = P_0(R) = 1 - \alpha$, so is a $(1 - \alpha)$-level confidence region. That is, we can construct a family of $(1 - \alpha)$-level fiducial regions that has the confidence property.

In Example 5.2, using E_3 we have $(\overline{x} - M)/(s/\sqrt{n}) \sim t_{n-1}$, which can be solved as $M = \overline{x} - (s/\sqrt{n})E_3$, yielding fiducial distribution $M \sim \overline{x} + (s/\sqrt{n})t_{n-1}$. Let $R_x = [\overline{x} - cs/\sqrt{n}, \overline{x} + cs/\sqrt{n}]$, where $P(t_{n-1} \leq c) = 1 - \alpha$. Then the event $M \in R_X$ has, given $X = x$, fiducial probability $1 - \alpha$, as well as coverage probability $1 - \alpha$ in the sampling distribution with $\Theta = \theta$.

Similarly using E_2 we obtain a fiducial distribution for Σ: $\Sigma = s/E_2 \sim s/\sqrt{\chi^2_{n-1}/(n-1)}$. The pivot E_1, by itself, is not invertible, and does not yield a fiducial distribution. However, the bivariate pivot (E_1, E_2) is invertible and yields a joint fiducial distribution $\Pi_{\overline{x},s}$ for (M, Σ), represented by

$$M = \overline{x} - sE_1/E_2 \tag{5.4}$$

$$\Sigma = s/E_2 \tag{5.5}$$

where still $E_1 \sim N(0, 1/n)$, $E_2 \sim \sqrt{\chi^2_{n-1}/(n-1)}$ (these moreover retain their sampling distribution independence). In particular, the induced marginals for M and Σ agree with those above based directly on E_3 and E_2, as above. However, M and Σ are not independent. Learning $\Sigma = \sigma$ is equivalent to learning $E_2 = s/\sigma$. This does not change the $N(0, 1/n)$ distribution of E_1, and we now have $M = \overline{x} - \sigma E_1$. So we have conditional fiducial distribution

$$M \mid (\Sigma = \sigma) \sim N(\overline{x}, \sigma^2/n). \tag{5.6}$$

5.3.1 Validity

There will typically be many pivotal functions available. For instance, in Example 5.2 we could retain just the first $n/2$ (say) observations, and use the sample mean and variance computed from these. In his early writings, Fisher insisted that, to make use of all the available information, a fiducial distribution should be based on the minimal sufficient statistic.

In addition, Fisher indicated that a fiducial distribution should be regarded as yielding an appropriate inference only if the following vaguely stated conditions are satisfied:

i. there is no available prior information about the unknown parameter;

ii. (*"principle of irrelevance"*): the data are uninformative about the pivot.

While (i) obviously precludes having an informative Bayesian prior distribution, its intended scope is much wider (and much vaguer). As for (ii), Hacking (1965) (see also Harris and Harding, 1984) attempted to rigorize it as requiring that the likelihood function, expressed as function of the data and the pivot, be the same, up to proportionality, for any data. He then showed that, in the univariate case, this holds if and only if (possibly after transformation) we have a location model, in which case, as observed by Lindley (1958), the fiducial distribution agrees with the formal posterior based on an improper uniform prior.

5.4 Group-Structured Models

Let observable X take values in a space $\mathcal{X}$, and identifiable parameter Θ take values in $\mathcal{T}$. We suppose $\mathcal{T}$ can be identified with a group G of transformations acting on $\mathcal{X}$. The image of $x \in \mathcal{X}$ under $g \in G$ is denoted by $g \circ x$. We further suppose the group action is *exact*, so that, given $x_0, x_1 \in \mathcal{X}$ there is at most one $g \in G$ such that $g \circ x_1 = x_2$ (this condition can be relaxed: see Bondar, 1972).

We shall investigate cases in which the family $\mathcal{P} = \{P_\theta : \theta \in \mathcal{T}\}$ of distributions over $\mathcal{X}$ is *equivariant* under the action of G: that is, if $X \sim P_\theta$, then $g \circ X \sim P_{g\theta}$ (where $g\theta$ is the group product).

5.4.1 Simple Group Model

In the simplest case, G acts transitively (as well as exactly) on $\mathcal{X}$, so that, for any $x_1, x_2 \in \mathcal{X}$ there exists exactly one $g \in G$ such that $x_2 = g \circ x_1$. Fix some $x_0 \in \mathcal{X}$, and henceforth identify any $x \in \mathcal{X}$ with the unique $g \in G$ such that $x = g \circ x_0$. We can thus take $\mathcal{X} = G$, with $g \circ x$ becoming the group product gx. Let $P_0 = P_\iota$, with ι the identity element of G. Define $E := \Theta^{-1}X$. Then, by equivariance, conditional on $\Theta = \theta$, $E \sim P_0$. Hence E is a pivot, and could be used to construct a fiducial distribution Π_x: after observing $X = x$, take $\Theta^{-1}x \sim P_0$. It can be shown that this construction satisfies Hacking's version of the principle of irrelevance. Moreover (Fraser, 1961), assuming G is locally compact, the resulting fiducial distribution is identical to a Bayesian posterior distribution, based on the (typically improper) right-invariant distribution (right Haar measure) for Θ over G.

The region R_x described in Section 5.3, having both fiducial and coverage probability $1 - \alpha$, has the form $\{\theta : \theta^{-1}x \in R\} = xR^{-1}$, and is thus equivariant.

Example 5.3. Normal Model with Group Structure
In Example 5.2, we can consider both $(\overline{X}, S)$ and (M, Σ) as elements of the location-scale group, with multiplication $(a, b)(A, B) = (a + bA, bB)$, and identity $\iota = (0, 1)$. Under P_0, $\overline{X} \sim N(0, n^{-1})$ and $S \sim \sqrt{\chi^2_{n-1}/(n-1)}$, independently; then $(\mu, \sigma)(\overline{X}, S) = (\mu + \sigma\overline{X}, \sigma S) \sim (N(\mu, \sigma^2/n), \sigma\sqrt{\chi^2_{n-1}/(n-1)})$ (independently), which is $P_{(\mu,\sigma)}$. Thus, we have equivariance.

We have pivot

$$\begin{aligned} E = (E_1, E_2) &= (M, \Sigma)^{-1}(\overline{X}, S) \\ &= \left(\frac{\overline{X} - M}{\Sigma}, \frac{S}{\Sigma}\right). \end{aligned}$$

We recover the same joint fiducial distribution represented by (5.4) and (5.5). Moreover, this is the same as the posterior distribution based on the right-invariant prior, having density element $\mathrm{d}\mu\,\mathrm{d}\sigma/\sigma$. □

Example 5.4. Wishart Model with Lower Triangular Group
Let G be the group of lower triangular matrices with positive diagonal. An observable random 2×2 matrix S has the Wishart distribution $W(\nu; \Sigma)$, where $\nu \geq 2$ and Σ is positive definite. Then S is almost surely non-singular. We can alternatively represent S by the unique $L \in G$ such that $S = LL^{\mathrm{T}}$, and similarly Σ by $\Lambda \in G$ with $\Sigma = \Lambda\Lambda^{\mathrm{T}}$. We write the implied distribution of L, depending on Λ, as $L \sim \mathcal{L}(\nu; \Lambda)$. It is then easy to see that, for fixed $A \in G$, $AL \sim \mathcal{L}(\nu; A\Lambda)$, so that the problem is equivariant under G. It follows that a pivot is $E = \Lambda^{-1}L$, with distribution $\mathcal{L}(\nu; I)$—under which (Mauldon, 1955) the non-zero

entries of E are independent, with $E_{11} \sim \sqrt{\chi^2_\nu}$, $E_{22} \sim \sqrt{\chi^2_{\nu-1}}$, and $E_{21} \sim N(0,1)$. So the fiducial distribution of $\Sigma = \Lambda\Lambda^{\mathrm{T}}$, given data $S = s = ll^{\mathrm{T}}$, is that of $l(E^{\mathrm{T}}E)^{-1}l^{\mathrm{T}}$ when $E \sim \mathcal{L}(\nu; I)$. In this case, the right-invariant prior density element, under the action of G, can be expressed in terms of the entries of Σ as

$$(\mathrm{d}\sigma_{11}/\sigma_{11})\,\mathrm{d}\sigma_{12}\,\mathrm{d}\sigma_{22}, \tag{5.7}$$

and the fiducial distribution of Σ agrees with its posterior, based on (5.7) as prior. □

5.4.2 Structural Models

Fraser (1961; 1968) has a somewhat different take on group-structured models, which takes the group structure as part of the specification of the problem. He posits, as part of the very set-up, a nominated group G acting on $\mathcal{X}$, and an "error variable" E, with known distribution P_0 over $\mathcal{X}$. Both E and the parameter Θ, which takes values in G, are regarded as having independent existence. The observable X is then *defined* by $X = \Theta \circ E$. There is thus additional algebraic structure, over and above the implied parametric family of distributions for X given $\Theta = \theta$. This extended structure is termed a "structural model".

Since the implied distributional model is equivariant under G, we can now construct a fiducial distribution as in Section 5.4.1 (see also Section 5.8.2 below for the non-transitive case). Fraser terms this a "structural distribution".

Note that, as will be demonstrated in Section 5.7.1, distinct structural models can correspond to the same distributional model but yield different structural distributions for its parameter. Since a structural model is considered to comprise more than just its induced distributional model—including, in particular, specification of the group G as a key ingredient—this is not regarded as an inconsistency.

5.5 Functional Models

Dawid and Stone (1982) propose the *functional model*, a generalization of the structural model. We have arbitrary sample space $\mathcal{X}$ and parameter space Θ. We again consider an "error variable" E, taking values in a space $\mathcal{E}$ that now may be different from $\mathcal{X}$. The observable X is defined, algebraically, as $X = f(\Theta, E)$, where $f : \mathcal{T} \times \mathcal{E} \to \mathcal{X}$ is a specified function, and E has a known distribution P_0 over $\mathcal{E}$, independently of the value of Θ. For simplicity, we notate the functional relationship as $X = \Theta \circ E$. Then the distribution P_θ of X given $\Theta = \theta$ is that of $\theta \circ E$ where $E \sim P_0$.

The structural approach, which is a special case of the functional approach, identifies $\theta \in \mathcal{T} = G$ with the function (an element of G) $e \mapsto \theta \circ e$ on $\mathcal{X}$. In the functional approach, by contrast, it is more helpful to consider $e \in \mathcal{E}$ as the function $\theta \mapsto \theta \circ e$, mapping $\mathcal{T}$ into $\mathcal{X}$. (Note that, as a function, e is written to the right of its argument θ.)

5.5.1 Simple Functional Model (SFM)

In the simplest case, for any $x \in \mathcal{X}$, $e \in \mathcal{E}$, there exists exactly one θ such that $x = \theta \circ e$: we write $\theta = x \circ e^{-1}$, since this determines the inverse function $e^{-1} : \mathcal{X} \to \mathcal{T}$. In this case, the fiducial distribution, for data x, is obtained from $\Theta = x \circ E^{-1}$, with $E \sim P_0$.

In the special case that E can be expressed as a function of (X, Θ), it serves as a pivot. The model is then termed *pivotal*, and the fiducial distribution agrees with that constructed as in Section 5.3.

5.5.1.1 Monotonic Functional Model

When $\mathcal{X} = \mathcal{T} = \mathbb{R}$ and each e acts as a strictly monotonic function of θ, the fiducial distribution is fully determined by the distributional model for X given Θ: in particular, the finer details of the functional model do not enter. Thus when e is a strictly increasing function, the fiducial probability $\Pi_x(\Theta \leq \theta) = P_0(x \circ E^{-1} \leq \theta) = P_0(x \leq \theta \circ E) = \Pr_\theta(X \geq x)$; similarly, when e is strictly decreasing, $\Pi_x(\Theta \leq \theta) = \Pr_\theta(X \leq x)$. One can show that, under regularity conditions parallel to those in Section 5.2, and by a similar argument, a 1-sided fiducial interval is also a confidence interval.

Example 5.5. Monotonic Model
Let $\mathcal{X} = \mathcal{T} = \mathbb{R}$, $\mathcal{E} = \mathbb{R}^+ \times \mathbb{R}^+ \times \mathbb{R}$. The function $x = \theta \circ e$ is given by $x = (\theta e_1 + e_3)/e_2$, which is strictly increasing in θ. This model is not pivotal, but we can solve for θ: $\theta = x \circ e^{-1} = (xe_2 - e_3)/e_1$. So the fiducial distribution of Θ, for data $X = x$, is that of $(xE_2 - E_3)/E_1$, with E having its initially assigned distribution P_0.

As a special case, suppose that, under P_0, $E_1 \sim \sqrt{\chi^2_{n-1}}$, $E_2 \sim \sqrt{\chi^2_{n-2}}$, and $E_3 \sim N(0, 1)$, all independently. Define $R = X/\sqrt{1 + X^2}$, $\Phi := \Theta/\sqrt{1 + \Theta^2}$. It then turns out (see Example 5.7 below) that (compare Section 5.2):

i. The sampling distribution of R is that of a sample correlation coefficient, based on n independent observations from a bivariate normal distribution with population correlation coefficient Φ.

ii. The fiducial distribution of Φ agrees with Fisher's fiducial distribution, based on R. This follows from (i) as a consequence of the monotonic structure of this model.

□

5.6 Marginalization Consistency

In a SFM $X = \Theta \circ E$, with $E \sim P_0$, let $D = d(X)$ be a function of X. Suppose that $d = d(\theta \circ e)$ can be expressed as a function of δ and e, where $\delta = \delta(\theta)$ is some function of θ: we write $d = \delta * e$. With $\Delta := \delta(\Theta)$ we thus have a new model $D = \Delta * E$: in particular, the sampling distribution of D depends only on the value of Δ. We require that this model itself be a SFM, so that, given (d, e), we can solve $d = \delta * e$ for δ, which solution we write as $\delta = d * e^{-1}$.[2] We term the model $D = \Delta * E$ a *reduction* of $X = \Theta \circ E$.[3]

Given data $X = x$, we have two different routes to computing the fiducial distribution of $\Delta = \delta(\Theta)$:

i. Obtain the fiducial distribution of Θ based on data $X = x$, using the full SFM $X = \Theta \circ E$; then marginalize this to get the implied distribution of $\Delta = \delta(\Theta)$.

ii. Start from the reduced SFM $D = \Delta * E$ and obtain the associated fiducial distribution of Δ, based on the reduced data $D = d(x)$.

To see that these give the same result, we argue as follows. Route (i) represents $\Theta = x \circ E^{-1}$, and so produces the distribution of $\delta(x \circ E^{-1})$. Route (ii) represents $\Delta = d(x) * E^{-1}$. In both cases $E \sim P_0$. Now if $x = \theta \circ e$ then $\theta = x \circ e^{-1}$. Also $d(x) = \delta(\theta) * e$, so $\delta(\theta) = d(x) * e^{-1}$. That is, $\delta(x \circ e^{-1}) = d(x) * e^{-1}$, whence $\delta(x \circ E^{-1}) = d(x) * E^{-1}$, showing that both routes yield the same representation, and hence the same fiducial distribution, for Δ (in particular, the marginal fiducial distribution of Δ in the route (i) analysis must depend on the data x only through $d = d(x)$).

Example 5.6. Normal Coefficient of Variation
Example 5.3 can be regarded as a SFM (in fact a structural model): $(\overline{X}, S) = (M, \Sigma) \circ (E_1, E_2)$, where $\circ$ is group product in the location-scale group. That is,

$$\overline{X} = M + \Sigma E_1 \tag{5.8}$$
$$S = \Sigma E_2 \tag{5.9}$$

with $E_1 \sim N(0, 1/n)$, $E_2 \sim \sqrt{\chi^2_{n-1}/(n-1)}$, independently.

Define $D = \overline{X}/S$, $\Delta = M/\Sigma$ (the reciprocals, respectively, of the sample and population coefficients of variation). Then

$$D = \frac{\Delta + E_1}{E_2}. \tag{5.10}$$

This is itself a SFM $D = \Delta * E$ (though not structural), so is a reduction of the original SFM.[4]

Dempster (1963) noted that (as indeed follows from the above) the sampling distribution of D depends only on Δ, so that a fiducial distribution for Δ can be constructed from these univariate sampling distributions using Fisher's approach of inverting the distribution function. By monotonicity, the route (ii) analysis of the reduced model $D = \Delta * E$ will also deliver this fiducial distribution. Dempster (1963) further showed that this agrees with the distribution of $\Delta = M/\Sigma$ obtained by marginalizing the joint fiducial distribution of (M, Σ) represented by (5.4) and (5.5)—as also arises from the route (i) analysis of the initial SFM (5.8)–(5.9). Here we see marginal consistency in action. □

Example 5.7. Reduction of Wishart Model
Example 5.4 can be regarded as a (structural) SFM $L = \Lambda E$ (all lower triangular matrices), with $E_{11} \sim \sqrt{\chi^2_\nu}$, $E_{22} \sim \sqrt{\chi^2_{\nu-1}}$, and $E_{21} \sim N(0,1)$, independently. Thus

$$\begin{aligned} L_{11} &= \Lambda_{11} E_{11} \\ L_{12} &= \Lambda_{12} E_{11} + \Lambda_{22} E_{12} \\ L_{22} &= \Lambda_{22} E_{22}. \end{aligned}$$

Defining $X = L_{12}/L_{22}$, $\Theta = \Lambda_{12}/\Lambda_{22}$, we obtain a reduction[5] $X = \Theta * E$, given by the SFM $X = (\Theta E_{11} + E_{12})/E_{22}$. Note that this is identical with the special case considered in Example 5.5 (on identifying (E_1, E_2, E_3) there with (E_{11}, E_{22}, E_{12}) here).

In Example 5.4 we have $S = LL^{\mathrm{T}}$, *i.e.*,

$$\begin{pmatrix} S_{11} & S_{12} \\ S_{12} & S_{22} \end{pmatrix} = \begin{pmatrix} L_{11}^2 & L_{11}L_{12} \\ L_{11}L_{12} & L_{12}^2 + L_{22}^2 \end{pmatrix},$$

with a similar expression for Σ in terms of Λ. The sample correlation based on S is

$$R := \frac{S_{12}}{\sqrt{S_{11}S_{22}}} = \frac{X}{\sqrt{1+X^2}},$$

and similarly the population correlation is $\Phi := \Theta/\sqrt{1+\Theta^2}$. The former identity explains the distribution of R asserted in (i) of Example 5.5. There we deduced that the fiducial distribution of Φ, based on R, agrees with that derived by Fisher. By marginalization consistency, this must thus be true for the marginal distribution of $\Phi = \Sigma_{12}/\sqrt{\Sigma_{11}\Sigma_{22}}$ formed from the full fiducial distribution for Σ given S, as described in Example 5.4, based on the lower-triangular structural model. □

5.6.1 Marginalization Paradox

Although the marginalization consistency property seems to speak in favor of fiducial inference, at least in some problems, it becomes a problem for Bayesian inference with improper priors. We know that the full fiducial distribution, used in the route (i) analysis, is also the Bayesian posterior, based on the right-invariant prior distribution. Thus the output of the route (i) analysis is the marginal distribution of Δ in this Bayesian posterior. By marginalization consistency, this is the same as the output of route (ii) analysis—which necessarily depends on the data X only through D. It therefore seems reasonable to expect that the marginal Bayesian posterior distribution of Δ from the full model, depending as this does only on D, could arise as a Bayesian posterior based on the likelihood in the reduced model for D—which only depends on Δ. But by the result of Lindley (1958) (see Section 5.2), if—as in both the above examples—the reduced model has univariate D and Δ but is not equivalent to a location model, this cannot be the case. We then have an example of a *marginalization paradox* (Dawid et al., 1973) in improper Bayesian inference.

5.7 Some Difficulties

5.7.1 Choice of Group

Mauldon (1955) pointed out a problem with Example 5.4: if we simply interchange the order in which we consider the variables (equivalent to now using equivariance under the upper triangular, rather than lower triangular, group), the analysis proceeds essentially as before, but we obtain a different fiducial distribution. This can most easily be seen by noting that the right-invariant prior (5.7) is altered on interchanging the suffices 1 and 2, leading to a different posterior, hence fiducial, distribution for Σ.

One possible escape from this bind is not to allow the use of just any group G under which the statistical model happens to be equivariant, but to specify an appropriate group as part of the very structure of the problem—this thus requiring an additional ingredient in the model, over and above its purely distributional properties. This tallies with the position adopted in Fraser's structural modeling—see Section 5.4.2.

Although use of the upper triangular group produces a different fiducial distribution for Σ than that based on the lower triangular group, nevertheless, by an argument parallel to that of Example 5.7, the implied distribution for the population correlation coefficient Φ again agrees with Fisher's fiducial distribution, and thus is the same in both cases.

5.7.2 Marginalization Inconsistency

Example 5.8. Many Normal Means
Consider the n-variate SFM $\mathbf{X} = \mathbf{\Theta} \circ \mathbf{E}$ given by $X_i = \Theta_i + E_i$ $(i = 1, \ldots, n)$, with $E_i \sim N(0, 1)$, all independently. On observing $\mathbf{X} = \mathbf{x}$, the fiducial distribution $\Pi_{\mathbf{x}}$ has $\Theta_i \sim N(x_i, 1)$, independently.

Let $D = \sum_{i=1}^{n} X_i^2$, $\Delta = \sum_{i=1}^{n} \Theta_i^2$. The marginal fiducial distribution of Δ, given data $\mathbf{x}$, depends only on $d = \sum_{i=1}^{n} x_i^2$; it is non-central χ^2 with n degrees of freedom and non-centrality parameter d: $\Delta \sim \chi_n^2(d)$. Also, the sampling distribution of D, when $\boldsymbol{\Theta} = \boldsymbol{\theta}$, depends only on $\delta = \sum_{i=1}^{n} \theta_i^2$; it is $\chi_n^2(\delta)$. Nevertheless, $D = \sum_{i=1}^{n}(\Theta_i + E_i)^2$ cannot be expressed as a function of Δ and E, so we do not have a reduction of the initial model. We note that $D - \Delta$ has sampling expectation n, but fiducial expectation $-n$, which suggests a serious inadequacy in the marginalized fiducial distribution.

We can attempt to derive a "route (ii)"-type fiducial distribution of Δ, by Fisherian inversion of the distribution function of D given Δ. In this case, condition (iii) of Section 5.2 does not hold, and we obtain only an incomplete distribution—which obviously cannot agree with the (complete) marginal fiducial distribution obtained from route (i) analysis, so we do not have marginalization consistency. In fact (Wilkinson, 1977), for $n = 50$ and $d = 100$ we get 95% central fiducial interval $(109, 196)$ by marginalizing the full fiducial distribution to Δ, compared to $(21, 89)$ based on the distribution of D given Δ. □

This example indicates that sensible marginalization of a joint fiducial distribution may not be possible when it is not based on a reduction of a functional model. Wilkinson (1977) embraces inconsistencies such as in this example by his *noncoherence principle*, which allows the overall joint fiducial to coexist with the "marginal" based on the reduced data—which is not the actual marginal. But then fiducial distributions do not satisfy the axioms of probability theory.

Other examples of marginalization inconsistency, evidenced by incompatibilities between fiducial and confidence statements, are the Behrens-Fisher problem, looking at the difference between the means of two normal distributions with different, unknown, variances, and the Fieller-Creasy problem, looking at the ratio of two normal means, with known variances (Wallace, 1980).

5.7.3 Conditional Consistency?

Example 5.9. Normal Model Conditioned on its Standard Deviation
Consider again Example 5.2, and suppose we want to construct a conditional fiducial distribution for M, given $\Sigma = \sigma$. Again we have two possible routes to do this:

i. Condition the joint fiducial distribution on $\Sigma = \sigma$, leading to (5.6).
ii. Note that, when $\Sigma = \sigma$ is fixed, the sampling model becomes

$$\overline{X} = M + \sigma E_1 \tag{5.11}$$
$$S = \sigma E_2. \tag{5.12}$$

On observing $(\overline{x}, s)$ we learn $E_2 = s/\sigma$, so should condition on this. The same reasoning that led to (5.6) again applies, so yielding the same answer. We have "conditional consistency".

□

Example 5.10. Normal Model Conditioned on its Coefficient of Variation
Introduce D and Δ as in Example 5.6, related by the reduced SFM (5.10). We have seen that the marginal fiducial distribution for Δ is the same under the two routes of computation. What about the conditional fiducial distribution of Σ, given $\Delta = \delta$?

We can reexpress the full model as

$$D = \frac{\Delta + E_1}{E_2} \tag{5.13}$$
$$S = \Sigma E_2. \tag{5.14}$$

Again we can identify two routes to construct a conditional distribution for Σ, given $\Delta = \delta$.

i. Form the joint fiducial distribution of (Δ, Σ), and condition this on $\Delta = \delta$.
Given data (x, s) (with $\overline{x}/s = d$) the joint fiducial distribution is represented by:

$$\begin{aligned} \Delta &= dE_2 - E_1 \\ \Sigma &= s/E_2. \end{aligned}$$

So conditioning on $\Delta = \delta$ is equivalent to conditioning on

$$dE_2 - E_1 = \delta. \tag{5.15}$$

We should therefore condition E_2 on this, and then invert (5.14), so obtaining $\Sigma = s/E_2$, where E_2 has its distribution conditioned on (5.15).

ii. Alternatively we can argue as follows, using (5.13). We have observed $D = d$. Since we are assuming $\Delta = \delta$, we have thus learned

$$\frac{\delta + E_1}{E_2} = d. \tag{5.16}$$

The conditioning of E_2 should therefore be on (5.16).

Dempster (1963) showed that we get different answers, depending on whether we condition on (5.15) or on (5.16). This is an example of the "Borel-Kolmogorov paradox" when conditioning on an event of probability 0. So here we have conditioning inconsistency—and it is not clear how we should resolve it. □

In both Example 5.9 and Example 5.10, we wished to condition on a parameter function that itself figures in a reduced functional model. However, only in Example 5.9 is this model pivotal. When this is the case, but not more generally, we will obtain the same result by following each of the two routes, (i) and (ii).

5.7.4 Conditioning and Partitioning

The logical information expressed by (5.15) and (5.16) is the same in both cases. How then can it matter which we condition on?

We can interpret (5.15) as conditioning on the "fattened" positive-probability event $|dE_2 - E_1 - \delta| \leq \epsilon$, for small $\epsilon \downarrow 0$; and likewise (5.16) as conditioning on $|\{(\delta + E_1)/E_2\} - d| \leq \epsilon$. This latter is equivalent to $|dE_2 - E_1 - \delta| \leq \epsilon E_2$. The additional term E_2 on the right-hand side here gives a different fattened event, and so a different conditional distribution.

Otherwise put, when we condition on some function $f(E)$ taking some value ϕ, we are embedding that information in the family of possible alternative informations $f(E) = \phi'$, for varying ϕ' (with local variation of ϕ' around ϕ determining the appropriate fattened event $|f(E) - \phi| \leq \epsilon$). At a more abstract level, the family of events $f(E) = \phi'$, as ϕ' varies, constitutes a partition of the whole space; the conditioning is on the event in the partition containing E; and the partition determines how the conditioning is to be effected.

Thus, when we condition on information, we must be clear what question the received information answers. In Example 5.10, for both cases (i) and (ii) the logical information is $dE_2 - E_1 - \delta = 0$, but in (i) the question is "What is the value δ' of $dE_2 - E_1$?", whereas in (ii) the question is "What is the value d' of $(\delta + E_1)/E_2$?" Only when $d' = d$ and $\delta' = \delta$ do the answers even contain equivalent logical information. And even then, since the partitions differ, so do the conditional distributions.

It may be thought that the above difficulties depend on the conditioning event having probability 0. But essentially the same paradox can arise even when we have positive probabilities. A parable may help explain this.

Example 5.11. How Many Boys?
Suppose Mr Smith tells you: "I have two children, who are not twins." At this point you regard each of them as equally likely to be a boy (B) or a girl (G), independently. He then says: "One of them is a boy". Given this information, what is the probability he has two boys?

Argument 1 Initially, you assessed four equally likely cases: BB, BG, GB, and GG. The new information rules out GG, leaving three cases, just one of which is BB. The conditional probability is thus 1/3.

Argument 2 You might consider that, if he had two boys, he would have said "They are both boys". The fact that he did not then implies a conditional probability of 0.

Moral: Again here, when conditioning on information, we must take account of what other information might have been obtained. Otherwise put, we must specify the question, explicit or implicit, that the received information answers (and whose different possible answers define the relevant partition). Was it the question "Do you have a boy?", or the question "How many boys do you have?" ? □

5.8 Non-simple Models

5.8.1 Ancillary Information

In many cases, there is no simple sufficient statistic. Fisher (1956), towards the end of the book, suggested—as usual by means of examples—an alternative approach. Suppose we can identify a statistic S that is ancillary, *i.e.*, has the same distribution under any P_θ; and a further statistic T such that, together, (S, T) are equivalent to the full data X (or, more generally, are jointly sufficient). Given data $(S, T) = (s, t)$, we can first restrict attention to the conditional distribution of T, given $S = s$; and then try to identify a pivotal function of (T, Θ) in this conditional distribution. Finally we invert this pivot to obtain a fiducial distribution.

In the case that there exists a complete sufficient statistic, we can take this as T. By a result of Basu (1955), T will then be independent of any ancillary statistic, so the above conditioning has no effect, and we can restrict attention to T and its marginal distribution, thus recovering a simple model. However, this is a very special case, and conditioning will usually be required.

Example 5.12. Location Model
Let $\mathbf{X} = (X_1, \ldots, X_n)$ arise as a random sample from a general location model, with sampling density of the form

$$f(x \mid \theta) = g(x - \theta).$$

Typically there is no simple sufficient statistic. However, it seems natural to base inference on a location statistic, such as the sample mean $\overline{X}$, or (for n odd) the sample median, $\widetilde{X}$. But since the sampling distributions of the pivots $E_1 = \overline{X} - \Theta$ and $E_2 = \widetilde{X} - \Theta$ are typically very different, we seem to have a problem of choice.

This can be resolved as follows. Let $T = t(\mathbf{X}) := \overline{X}$, and $\mathbf{S} = \mathbf{s}(\mathbf{X}) := (X_i - \overline{X} : i = 1, \ldots, n)$. Then $(\mathbf{S}, T)$ are together equivalent to $\mathbf{X}$, $\mathbf{S}$ is ancillary, and $E = T - \Theta$ is a pivot. Since the joint distribution of $(\mathbf{S}, E)$ does not depend on Θ, E is still pivotal after conditioning on $\mathbf{S}$. Letting $P^{\mathbf{s}}$ denote the distribution of E given $\mathbf{S} = \mathbf{s}$ (the same for all θ), a fiducial distribution, for data $\mathbf{X} = \mathbf{x}$, can be obtained by regarding $t - \Theta$ as having distribution $P^{\mathbf{s}}$, with $t = t(x)$, $\mathbf{s} = \mathbf{s}(\mathbf{x})$. It is easy to show that, if we had instead used $T = \widetilde{X}$ (or any other location statistic), the identical fiducial distribution would have been obtained. □

The above seemingly well-specified procedure becomes less so when we take into account the results of Basu (1959) that there is, typically, a plethora of incommensurate choices of an ancillary to condition on.

5.8.2 Non-transitive Group Models

We now consider the general case of a group-structured model, as introduced in Section 5.4, where we continue to assume exactness, but do not assume transitivity: given $x_1, x_2 \in \mathcal{X}$, there may be no $g \in G$ such that $x_2 = g \circ x_1$. When there is such a g (which by exactness will be unique) we write $x_1 \sim_G x_2$. It is easily checked that $\sim_G$ is an equivalence relation on $\mathcal{X}$, and the equivalence classes are termed the *orbits* of G in $\mathcal{X}$. Let $\mathcal{S}$ be a set labelling the orbits, and let $s(x) \in \mathcal{S}$ be the label of the orbit containing x. Then $S = s(X)$ is an ancillary statistic: this is the *group ancillary*, and, unlike general ancillaries, is essentially unique. Now for each $s \in S$ choose, arbitrarily, some representative point ξ_s in the orbit labelled by s. Then for any $x \in \mathcal{X}$ there is a unique $g \in G$ such that $x = g \circ \xi_{s(x)}$. Let $t(x)$ denote this g, and define $T = t(X)$, a statistic taking values in G. Then $\Theta^{-1}T$ is a pivot, even conditional on S. We can thus apply the construction of Section 5.8.1 to obtain a fiducial distribution. Again, this will coincide with the Bayesian posterior distribution based on the right-invariant prior—which may be easier to compute.

Example 5.13. Location Model, Continued
In Example 5.12, the problem is equivariant under the location group, and S and T satisfy the above requirements. Since the right-invariant prior has density element $\mathrm{d}\theta$, the fiducial distribution, which could be daunting to compute directly, must have density element proportional to $\prod_{i=1}^{n} g(x_i - \theta)\,\mathrm{d}\theta$. □

As noted in Section 5.7.1, a given problem may be equivariant under more than one group, and these may induce different fiducial distributions. This problem is defined away when we start with a structural model, which includes specification of the relevant group G. Then the above recipe yields a unique fiducial distribution.

5.8.3 Non-simple Functional Models

In a general functional model $X = \Theta \circ E$, we term $x \in \mathcal{X}$ and $e \in \mathcal{E}$ *compatible* when $x = \theta \circ e$ for some $\theta \in \mathcal{T}$. We first assume *invertibility*: that such θ is unique (compare the group-theoretic concept of exactness), and write $\theta = x \circ e^{-1}$, noting that in this case e^{-1} is a partial function, operating only on x's that are compatible with e.

Let $\mathcal{E}_x = \{e : x \text{ and } e \text{ are compatible}\}$. On observing $X = x$, we learn the logical information $E \in \mathcal{E}_x$, but no other logical information about E.

We should thus aim to adjust the distribution of E to account for this new information, yielding a revised distribution $E \sim P_x$, say—where P_x is confined to $\mathcal{E}_x$. Then $\Theta = x \circ E^{-1}$ is well-defined, and a fiducial distribution can be formed by assigning to E the distribution P_x over $\mathcal{E}_x$.

But how might we compute P_x? As seen in Example 5.10, conditioning on the logical information $E \in \mathcal{E}_x$ is only well-defined when $\mathcal{E}_x$ is embedded in a suitable partition. We have to consider what other information we might have obtained, in other circumstances. Such information would be of the form $E \in \mathcal{E}_y$, as y varies in $\mathcal{X}$. Conditioning would thus be justified when the $\{\mathcal{E}_y : y \in \mathcal{X}\}$ form a partition, which will be the case when, for $x, y \in \mathcal{X}$, $\mathcal{E}_x$ and $\mathcal{E}_y$ are either identical or disjoint—in which case we term the FM *partitionable*. Equivalently, there exist, essentially unique, functions $a(\cdot)$ on $\mathcal{X}$, $u(\cdot)$ on $\mathcal{E}$, such that $x \in \mathcal{X}$ and $e \in \mathcal{E}$ are compatible just when $a(x) = u(e)$. Since the observable $X = \Theta \circ E$ is necessarily compatible with the error variable E, $a(X) = u(E)$, and so is ancillary—the *functional ancillary*. The fiducial distribution of Θ, for data $X = x$, is now that of $x \circ E^{-1}$, with $E \sim P$ conditioned on $u(E) = a(x)$.

However, a non-partitionable FM may not support unambiguous fiducial inference.

5.8.4 Examples

We do not have a general necessary and sufficient condition for a functional model to be partitionable. This will, however, hold when the model is structural, or a reduction of a structural model.

Example 5.14. Location-Scale Model
Let $\mathcal{E} = \mathcal{X} = \mathbb{R}^n$, $\Theta = (M, \Sigma) \in \mathcal{T} = \mathbb{R} \times \mathbb{R}^+$. The structural model $\mathbf{X} = \Theta \circ \mathbf{E}$ is given by $X_i = M + \Sigma E_i$, $i = 1, \ldots, n$. The functional ancillary can be taken as $a(\mathbf{x}) = ((x_i - \overline{x})/s_x : i = 1, \ldots, n)$, where $s_x^2 = \sum_{i=1}^n (x_i - \overline{x})^2/(n-1)$; and $u(\mathbf{e}) = a(\mathbf{e})$. The fiducial distribution is represented by $\Theta = (\overline{x} - s_x \overline{E}/s_E, s_x/s_E)$, where the initial distribution of $(\overline{E}, s_E)$ is conditioned on $(E_i - \overline{E})/s_E = (x_i - \overline{x})/s_x$, $i = 1, \ldots, n$. It can alternatively, and more readily, be derived as the Bayesian posterior distribution based on the right-invariant prior, having density element $\mathrm{d}\mu\, \mathrm{d}\sigma/\sigma$. □

Example 5.15. Reduced Structural Model
We have observable $\mathbf{D} \in \mathcal{D} = \{\mathbf{d} \in \mathbb{R}^n : s_d = 1\}$, parameter $\Delta \in \mathbb{R}$, error variable $\mathbf{E} \in \mathcal{E} = \mathbb{R}^n$. The functional model $\mathbf{D} = \Delta \circ \mathbf{E}$ is given by $D_i = (\Delta + E_i)/s_E$, $i = 1, \ldots, n$. This is a reduction of the structural model of Example 5.14, induced by $\mathbf{D} = \mathbf{X}/s_X$, $\Delta = M/\Sigma$. It is partitionable, with $u(\mathbf{e}) = ((e_i - \overline{e})/s_e : i = 1, \ldots, n)$, $a(\mathbf{d}) = ((d_i - \overline{d}) : i = 1, \ldots, n)$. The fiducial distribution is represented by $\Delta = \overline{d} s_E - \overline{E}$, with the distribution of $(\overline{E}, s_E)$ conditioned on $(E_i - \overline{E})/s_E = d_i - \overline{d}$, $i = 1, \ldots, n$. It is not a Bayesian posterior based on the likelihood in the reduced model, though it does agree with the marginal for Δ in the full fiducial distribution of Example 5.14 (which is a Bayesian posterior). □

Example 5.16. Truncated Model
Let $\mathcal{X} = \mathcal{E} = \mathbb{R}$, $\mathcal{T} = [0, \infty)$. Consider the functional model $X = \Theta + E$. Then $\mathcal{E}_x = (-\infty, x]$. Conditioning on $E \in \mathcal{E}_x$ appears, *prima facie*, straightforward: just truncate the initial distribution of E to $(-\infty, x]$ (that is, take the new distribution of E to be restricted to $(-\infty, x]$, with the original density renormalized to integrate to 1 over that range.) Equivalently, we initially ignore the constraint $\Theta \geq 0$, form the unrestricted fiducial distribution, and then truncate that to satisfy the constraint. However, as the $\{\mathcal{E}_x : x \in \mathbb{R}\}$ do not form a partition, it is arguable whether this is appropriate. Indeed, from the knowledge "$E \in \mathcal{E}_x$" we can deduce the exact value x of X, and so, it might be argued, we should condition on $X = x$. But while this is possible in a Bayesian framework, it is not possible in this functional model, since it does not correspond to a constraint on E.

As pointed out by Taraldsen and Lindqvist (2018), Fisher (1956) favored a different fiducial distribution in this problem, essentially that obtained by (attempted) inversion of the sampling distribution function (and so exhibiting some confidence properties.) This retains the initial fiducial density, unaltered, for $\theta > 0$, and assigns all the remaining probability to the point $\theta = 0$. □

Example 5.17. What to Condition on?
Consider the following functional model (a variation on Example 4 of Hannig (2009)):

$$
\begin{aligned}
X_1 &= \Theta_1/E_1 && (5.17)\\
X_2 &= (\Theta_1+\Theta_2)/E_2 && (5.18)\\
X_3 &= (\Theta_1+2\Theta_2)/E_3. && (5.19)
\end{aligned}
$$

Since we have three equations for the two unknowns (Θ_1, Θ_2), there will not be any solution unless an appropriate compatibility condition is satisfied.

i. Fixing data $X = x$ and solving (5.17) and (5.18), we obtain

$$
\begin{aligned}
\Theta_1 &= x_1E_1 && (5.20)\\
\Theta_2 &= x_2E_2 - x_1E_1, && (5.21)
\end{aligned}
$$

and then inserting these in (5.19) we get the compatibility condition, $E \in \mathcal{E}_x$, expressed as

$$\frac{2x_2E_2 - x_1E_1}{E_3} = x_3. \tag{5.22}$$

ii. A similar analysis that starts by solving (5.18) and (5.19) yields compatibility condition

$$\frac{2x_2E_2 - x_3E_3}{E_1} = x_1, \tag{5.23}$$

Both (5.22) and (5.23) are (necessarily) equivalent to each other, and to

$$x_1E_1 - 2x_2E_2 + x_3E_3 = 0. \tag{5.24}$$

The partitions generated by the variables on the left-hand sides of (5.22), (5.23) and (5.24) are all different, so that conditioning (5.20) and (5.21) on them will give different answers. And indeed, since the model is non-partitionable, there is no correct answer as to how (or whether) we should condition. To see non-partitionability directly, note that, as expressed by (5.24), each $\mathcal{E}_x$ is a plane through the origin in $\mathbb{R}^3$. When not identical, any two such planes must intersect in a line, so cannot be disjoint. □

5.8.5 Recent Developments

For a continuous non-partitionable model, such as in Example 5.17, with a specific choice having been made of what compatibility function to condition on, Hannig (2009) supplies a formula for the implied "generalized fiducial density": it is proportional to the product of the likelihood function and a possibly data-dependent term. In the case of Example 5.17, the additional term is proportional to $1/\theta_1(\theta_1+\theta_2)$ if we condition on $(2x_2E_2-x_1E_1)/E_3$, as in (i), while it is proportional to $1/(\theta_1+\theta_2)(\theta_1+2\theta_2)$ if we condition on $(2x_2E_2-x_3E_3)/E_1$, as in (ii). This illustrates how, in the absence of partitionability, where there is no unique way of conditioning on the logical information $E \in \mathcal{E}_x$, there is no well-defined fiducial distribution.

The above approach involves choosing a subset of the generating equations that can be solved exactly for Θ as a function of (x, E), and inserting these in the remaining equations to develop a compatibility criterion, which is then conditioned on. A more recent variation of this (Hannig et al., 2016) defines $\theta^*(x, e)$ as the minimizer over θ of some measure of the discrepancy between x and $\theta \circ e$, and constructs a generalized fiducial distribution as the distribution of $\theta^*(x, E)$, conditional on the discrepancy being less than ϵ (similar to the "fattening" construction of Section 5.7.4)—finally taking a limit as $\epsilon \downarrow 0$. Again, under some regularity conditions, this procedure results in applying a possibly data-dependent adjustment factor to the likelihood function. And again, that term is not uniquely defined, depending in particular on the discrepancy function used.

5.9 Non-invertible Models

Consider a functional model $X = \Theta \circ E$, $E \sim P_0$. Dropping the invertibility requirement, $\tau_{x,e} := \{\theta : x = \theta \circ e\}$ may be a set with more than one element.

5.9.1 Simple Non-invertible Functional Model

We first suppose the model *simple*, so that any $x \in \mathcal{X}$ and $e \in \mathcal{E}$ are compatible: equivalently, $\tau_{x,e}$ is never empty.

On observing $X = x$, no new logical information is obtained about E. The usual fiducial argument now implies that we can still regard $E \sim P_0$. But even were we to know the realized value e of E, we could only infer $\Theta \in \tau_{x,e}$. In the absence of knowledge of e, the fiducial argument represents our knowledge of Θ by $\Theta \in T$, where $T := \tau_{x,E}$, with $E \sim P_0$, is a random subset of $\mathcal{T}$.

5.9.1.1 Belief Functions

The above kind of partial probabilistic knowledge, based on random sets, lies at the heart of the Dempster-Shafer theory of inference (Dempster, 2008). Using it, we can go on to define the *belief* and *plausibility* functions for Θ, after observing $X = x$:

$$\begin{aligned} \text{Bel}_x(\Theta \in A) &= P_0(T \subseteq A) \\ \text{Pl}_x(\Theta \in A) &= P_0(T \cap A \neq \emptyset). \end{aligned}$$

5.9.1.2 Monotonic Model

Majumder and Hannig (2016, Theorem 2.1) consider a variation of the model $X = \Theta \circ E$ of Section 5.5.1, where each e is a non-decreasing function of θ, and $\tau_{x,e}$ is never empty. It is easy to see that $F_\theta(x) = \Pr_\theta(X \leq x) = \Pr(\theta \circ E \leq x)$ is a non-increasing function of θ; that is, the sampling distributions of X are stochastically non-decreasing in θ.

Suppose $\theta_1 \leq \theta \leq \theta_2$, where θ_1 and θ_2 are both in $\tau_{x,e} = \{\theta : x = \theta \circ e\}$. Since e is non-decreasing,

$$x = \theta_1 \circ e \leq \theta \circ e \leq \theta_2 \circ e = x.$$

So $\theta \in \tau_{x,e}$. It follows that $\tau_{x,e}$ is an interval in $\mathbb{R}$. Let its lower bound be $l(x, e)$ and upper bound be $u(x, e)$, and define $L = l(x, E)$, $U = u(x, E)$. Thus T is the random interval (L, U) (where the end-points may or may not be included), and the fiducial inference is $\Theta \in T$. The end-points L and U have a joint fiducial distribution Π_x.

We have $\theta > u(x, e) \Rightarrow \theta \circ e > x \Rightarrow \theta \geq u(x, e)$ whence $\Pi_x(U < \theta) \leq \Pr_\theta(X > x) \leq \Pi_x(U \leq \theta)$. Noting that $\Pi_x(U \leq \theta) = \lim_{\phi \downarrow \theta} \Pi_x(U \leq \phi) = \lim_{\phi \downarrow \theta} \Pi_x(U < \phi)$, we deduce $\Pi_x(U \leq \theta) = 1 - \lim_{\phi \downarrow \theta} F_\phi(x)$. By a similar argument, $\Pi_x(L \leq \theta) = 1 - \lim_{\phi \downarrow \theta} F_\phi(x-)$. Since these results depend only on the sampling distributions, they will hold for every monotonic model consistent with them.

If every $F_\theta(x)$ is continuous in x, then T reduces to a single point Θ, and we have $\Pi_x(\Theta \leq \theta) = 1 - \lim_{\phi \downarrow \theta} F_\phi(x)$, which is a slight generalization of the argument of Section 5.2, allowing discontinuities of $F_\theta(x)$ as a function of θ. But the more general argument also allows discontinuities of $F_\theta(x)$ as a function of x, as occurs in discrete distributions.

Example 5.18. Binomial Distribution
Let $\mathcal{X} = [0, n]$, $\mathcal{T} = \mathcal{E} = (0, 1)$, and consider the functional model $X = \Theta \circ E$, where $E \sim \text{Unif}(0, 1)$, and $\theta \circ e = \inf\{y : F_\theta(y) \geq e\}$, F_θ being the distribution function of the binomial distribution $\text{Bin}(n; \theta)$ $(0 < \theta < 1)$. Then the sampling distribution of X, when $\Theta = \theta$, is $\text{Bin}(n; \theta)$ (Angus, 1994).

We note that $F_\theta(x)$ is a discontinuous function of x, with a jump at each integer between 0 and n inclusive, being flat in between. We further note the identity: $\Pr(\text{Bin}(n; \theta) \leq x) = \Pr(\text{Beta}(\lfloor x \rfloor + 1, n - \lfloor x \rfloor) \geq \theta)$, where $\lfloor x \rfloor$ denotes the greatest integer less than or equal to x (and, for $x = n$, $\text{Beta}(n + 1, 0)$ is taken as the one-point distribution at 1). Hence, for fixed $x < n$, $F_\theta(x)$ is a strictly decreasing continuous function of θ, taking all values in $(1, 0)$ as θ goes through $(0, 1)$. That is, the distributions F_θ are strictly stochastically increasing in θ. This implies that $\theta \circ e$ is a non-decreasing function of θ, so the model is monotonic. Furthermore, for all $e \in \mathcal{E}$, $x \in \mathcal{X}$ there exists $\theta \in \mathcal{T}$ with $x = \theta \circ e$, so the model is simple—though not invertible.

On applying the general results above, we find that, for x an integer in $[0, n]$, $\Pi_x(U \leq \theta) = 1 - F_\theta(x)$, and $\Pi_x(L \leq \theta) = 1 - F_\theta(x - 1)$. It then follows that $U \sim \text{Beta}(x + 1, n - x)$, while $L \sim \text{Beta}(x, n - x + 1)$. □

5.9.2 Variations

To finish with a probability distribution for Θ itself, rather than for an interval containing it, Hannig (2009) adds a further step, in which, given the compatible set T, a single value in T is selected at random, from some specified conditional distribution. There is of course sensitivity to this specification, and there does not seem to be any principled way to resolve this. The theory of inferential models (Martin and Liu, 2016) uses a different auxiliary construction, which effectively replaces Bel_x by a new belief function Bel_x^*, bounded above by Bel_x. Again there is a choice of the extra specification. In both approaches, some guidance on this may be found by aiming towards compliance with frequentist (*e.g.*, confidence) properties. In the case of the binomial distribution of Example 5.18, Hannig (2009) compared a number of alternatives, based on simulations of their sampling properties, and suggested that taking Θ as uniformly distributed in T is a reasonable choice.

5.9.3 General Non-invertible Functional Model

We now generalize by allowing $\tau_{x,e} = \emptyset$, equivalent to x and e being incompatible.

On observing $X = x$, we obtain new logical information about E, namely

$$E \in \mathcal{E}_x := \{e : \tau_{x,e} \neq \emptyset\}.$$

Only when $e \in \mathcal{E}_x$ could we have made the observation $X = x$ (for some $\theta \in \mathcal{T}$). To support fiducial inference, the initial distribution P_0 of E must be adjusted, somehow, to a new distribution, P^x, supported on $\mathcal{E}_x$. But how?

Again, things are reasonably straightforward if the model is *partitionable*, *i.e.*, , for all $x, x' \in \mathcal{X}$, $\mathcal{E}_x$ and $\mathcal{E}_{x'}$ are either identical or disjoint. This will hold if and only if there exist functions $a(\cdot)$ on $\mathcal{X}$ and $u(\cdot)$ on $\mathcal{E}$, such that $e \in \mathcal{E}_x$ exactly when $u(e) = a(x)$. Then learning $X = x$ is equivalent to learning $u(E) = a(x)$, and conditioning on this information is unproblematic: letting (P_a) be the family of conditional distribution of E given $u(E) = a$ (well-defined under partitionability), we take $E \sim P_{a(x)}$—which is a distribution supported on $\mathcal{E}_x$. We can finally use this to define the distribution of the random set T (and so Bel_x, Pl_x). However, when the model is not partitionable it is again not clear how (or, indeed, whether) to construct P^x.

Example 5.19. Bernoulli Model
Let $\mathcal{X} = \mathcal{E} = [0,1]^n$, $\mathcal{T} = [0,1]$. Under P_0, $(E_i : i = 1,\ldots,n)$ are independently uniform over $[0,1]$. The functional model $\mathbf{X} = \Theta \circ \mathbf{E}$ is given by $X_i = \mathbb{1}(E_i \leq \Theta)$, $i = 1,\ldots,n$. Then, when $\Theta = \theta$, the X_i are n independent Bernoulli(θ) variables. This functional model is the basis of Example 6 of Hannig (2009).

We see that $\mathbf{e}$ and $\mathbf{x}$ are compatible ($\tau_{\mathbf{x},\mathbf{e}} \neq \emptyset$) when $x_i = 1, x_j = 0$ if and only if $e_i < e_j$. Thus on observing $\mathbf{X} = \mathbf{x}$, we learn

$$\mathbf{E} \in \mathcal{E}_{\mathbf{x}} := \{\mathbf{e} : e_i < e_j \text{ just when } x_i = 1, x_j = 0\}, \tag{5.25}$$

and then Θ lies in the random interval between two order statistics:

$$T := [\,E_{(r)}, E_{(r+1)}\,) \quad (r = \sum x_i).$$

We have to confine the distribution of E to $\mathcal{E}_{\mathbf{x}}$, but how? Noting that $P_0(\mathcal{E}_{\mathbf{x}}) > 0$, an obvious approach is simply to truncate P_0 to the set $\mathcal{E}_{\mathbf{x}}$. This is the approach of Hannig (2009). We may then note that (5.25) is a condition on the way in which the (E_i) are ordered, which is independent of their order-statistic—which is what determines T. So the distribution of the random interval T will be unaffected by the truncation to (5.25)—allowing us to use its unconditional distribution, based on the order statistics of a random sample from the uniform distribution on $[0,1]$. We note that then $E_{(r)} \sim \mathrm{Beta}(r, n-r+1)$, $E_{(r+1)} \sim \mathrm{Beta}(r+1, n-r)$, in agreement with the conclusions of Example 5.18. This gives some reassurance that the above procedure is reasonable.

Nevertheless, this model is not partitionable, as may be seen by noting that, when $\mathbf{x} = \mathbf{1}$ (the vector with all $x_i = 1$), we have $\mathcal{E}_{\mathbf{1}} = \mathcal{E}$—which is not disjoint from or identical with any other $\mathcal{E}_{\mathbf{x}}$. So, in the light of examples such as Example 5.11, and in the absence of a clear question that is answered by the information $E \in \mathcal{E}_{\mathbf{x}}$, it remains debatable whether the above argument is fully justifiable. □

5.10 Where Next?

The more recent contributions to fiducial theory, such as Hannig et al. (2016), recognize, explicitly or implicitly, the various amibiguities in defining a fiducial distribution. Rather than throwing in the towel, they propose ways of choosing between different functional model representations, and between ways of "conditioning" in the case of non-partitionability. This approach echoes Wilkinson's non-coherence principle, in regarding inconsistencies such as in Example 5.8 as less a problem than an opportunity.

From the very beginning, connexions (sometimes strong, sometimes weak) have been noted between fiducial theory and confidence intervals. One focus of the modern works is to develop methods leading to good frequentist coverage of a fiducial interval in a wide range of problems. Sonderegger and Hannig (2014) show that, under some regularity conditions, a generalized fiducial distribution has the same asymptotically normal form as a Bayesian posterior—which leads to asymptotically correct frequency coverage for a one-sided fiducial interval. Hannig et al. (2016, Theorem 3) give a more general result on asymptotic coverage, which can apply in some non-regular problems where we do not have asymptotic normality. They also conduct simulations to demonstrate good finite-sample coverage. Majumder and Hannig (2016) develop conditions under which a $(1-\alpha)$-level one-sided fiducial interval will have frequency coverage $1-\alpha+o(n^{-s/2})$, for $s=1$ or 2. In suitable problems, reverse engineering these conditions can inform the choice between different ways of constructing a generalized fiducial distribution.

5.11 Concluding Comments

While the fiducial argument has considerable *prima facie* appeal, all attempts to formulate a fully coherent theory of fiducial inference have fallen foul of inconsistency and counter-examples. The investigations of Dawid and Stone (1982), based on functional models, were an attempt to see just how far the theory could be taken before crashing onto the rocks—but crash it eventually did. Many of the difficulties are associated with the need to specify, unambiguously, a relevant partition for performing probabilistic conditioning. In some cases, as in Example 5.10, two equally natural routes to take account of new or assumed information lead to different partitions and hence conflicting fiducial distributions—a parallel inconsistency in predictive inference was exhibited by Dawid and Wang (1993). In other cases, there is no natural embedding of the information obtained within any partition whatsoever, rendering fiducial inference undefined.

However, even though methods motivated by fiducial theory may have limited validity from a fully principled theoretical standpoint, they can prove useful for other purposes—for example, for constructing exact or approximate confidence regions. This emphasis of the more recent works can be seen as moving the middle F in BFF away from the initial B and towards the final F.

Appendix: Examples Roadmaps

The examples in this chapter relate to each other, and to broad topics, in various ways. To assist navigation, Table 5.1 organizes the examples by statistical model or parameter and Table 5.2 by topic.

TABLE 5.1
Roadmap of Models

Model	Examples
Normal location-scale	5.2, 5.3, 5.6, 5.8, 5.9, 5.10
Normal correlation	5.1, 5.5, 5.7
Wishart	5.4, 5.7
Location	5.12, 5.13
Location-scale	5.14, 5.15
Binomial	5.18, 5.19

TABLE 5.2
Roadmap of Topics

Topic	Examples	Notes
Monotonic model	5.5	Inference depends only on sampling distributions
Inversion of cdf	5.1, 5.18	Confidence property
Pivotal model	5.1, 5.2	Confidence property
Group-structured model	5.3, 5.4, 5.12, 5.13, 5.14	Inference depends on group as well as sampling model
Reduced model	5.6, 5.7, 5.10, 5.15	5.6 and 5.10 from 5.3, 5.7 from 5.4, 5.15 from 5.14
Marginalization consistency?		
Yes	5.6, 5.7	Using a reduced model
No	5.8	No reduction available
Conditionalization consistency?		
Yes	5.9	Pivotal reduction, unique partition
No	5.10	Non-pivotal reduction, two distinct partitions
Partitionable?		
Yes	5.14, 5.15	(Reduction of) structural model
No	5.11, 5.16, 5.17, 5.19	Conditioning not well-defined

Acknowledgments

I am grateful to Xiao-Li Meng and three anonymous referees for helpful comments.

Notes

1. See Taraldsen (2021) for representations of the sampling distribution of R and the fiducial distribution of Φ.
2. Note that the notation e^{-1} here, signifying a function acting on d, has a different meaning than in $\theta = x \circ e^{-1}$, where it signified a function acting on x.
3. Appendix A2 of Dawid and Stone (1982) characterizes such a reduction in terms of group actions. In particular, if the initial SFM is structural, with $\mathcal{T}$ a group G of transformations of $\mathcal{X} = \mathcal{E}$, a reduction is obtained by taking D and Δ as maximal invariants under a subgroup K of G, acting on $\mathcal{X}$ and $\mathcal{T}$ respectively.
4. This is an example of endnote 3, with K the scale subgroup.
5. Again an example of endnote 3, with K the subgroup of diagonal matrices.

Bibliography

Angus, J. E., 1994. The probability integral transform and related results. *SIAM Review, 36*, pp. 652–654.

Barnard, G. A., 1980. Pivotal inference and the Bayesian controversy. *Trabajos de Estadistica y de Investigacion Operativa, 31*, pp. 295–318.

Basu, D., 1955. On statistics independent of a complete sufficient statistic. *Sankhyā, 15*, pp. 377–380.

Basu, D., 1959. The family of ancillary statistics. *Sankhyā, 21*, pp. 247–256.

Bondar, J. V., 1972. Structural distributions without exact transitivity. *Annals of Mathematical Statistics, 43*, pp. 326–339.

Buehler, R. J., 1983. Fiducial inference. In *Encyclopedia of Statistical Sciences*, volume 3, eds S. Kotz, N. L. Johnson and C. B. Read, pp. 76–81. New York: Wiley-Interscience. https://doi.org/10.1002/9781118445112.stat01529.

Dawid, A. P. and Stone, M., 1982. The functional-model basis of fiducial inference (with discussion). *Annals of Statistics, 10*, pp. 1054–1074.

Dawid, A. P., Stone, M. and Zidek, J. V., 1973. Marginalization paradoxes in Bayesian and structural inference (with discussion). *Journal of the Royal Statistical Society: Series B (Methodological), 35*, pp. 189–233.

Dawid, A. P. and Wang, J., 1993. Fiducial prediction and semi-Bayesian inference. *Annals of Statistics, 21*, pp. 1119–1138.

Dempster, A. P., 1963. Further examples of inconsistencies in the fiducial argument. *Annals of Mathematical Statistics, 34*, pp. 884–891.

Dempster, A. P., 2008. The Dempster-Shafer calculus for statisticians. *International Journal of Approximate Reasoning, 48*, pp. 365–377.

Edwards, A. W. F., 1983. Fiducial distributions. In *Encyclopedia of Statistical Sciences*, volume 3, eds S. Kotz, N. L. Johnson and C. B. Read, pp. 70–76. New York: Wiley-Interscience. https://doi.org/10.1002/0471667196.ess0775.pub2.

Fisher, R. A., 1915. Frequency distribution of the values of the correlation coefficient in samples from an indefinitely large population. *Biometrika, 10*, pp. 507–521.

Fisher, R. A., 1930. Inverse probability. *Mathematical Proceedings of the Cambridge Philosophical Society, 26*, pp. 528–535.

Fisher, R. A., 1956. *Statistical Methods and Scientific Inference* (3rd ed.). Oliver and Boyd, Edinburgh. New York: Hafner, 1973.

Fraser, D. A. S., 1961. On fiducial inference. *Annals of Mathematical Statistics, 32*, pp. 661–676.

Fraser, D. A. S., 1968. *The Structure of Inference.* New York: Wiley.

Hacking, I., 1965. *Logic of Statistical Inference.* Cambridge: Cambridge University Press.

Hannig, J., 2009. On generalized fiducial inference. *Statistica Sinica, 19*, pp. 491–544.

Hannig, J., Iyer, H. K., Lai, R. C. S. and Lee, T. C. M., 2016. Generalized fiducial inference: A review and new results. *Journal of the American Statistical Association, 111*, pp. 1346–1361.

Harris, R. R. and Harding, E. F., 1984. The fiducial argument and Hacking's principle of irrelevance. *Journal of Applied Statistics, 11*, pp. 170–181.

Lindley, D. V., 1958. Fiducial distributions and Bayes' theorem. *Journal of the Royal Statistical Society: Series B, 20*, pp. 102–107.

Majumder, A. P. and Hannig, J., 2016. Higher order asymptotics of generalized fiducial distribution. *arXiv e-prints*, arXiv:1608.07186v1.

Martin, R. and Liu, C., 2016. *Inferential Models.* New York: Chapman and Hall/CRC.

Mauldon, J. G., 1955. Pivotal quantities for Wishart's and related distributions, and a paradox in fiducial theory. *Journal of the Royal Statistical Society: Series B, 17*, pp. 79–85.

Savage, L. J., 1961. The foundations of statistics reconsidered. In *Proceedings of the Fourth Berkeley Symposium on Mathematical Statistics and Probability*, volume 1, pp. 575–586. University of California Press.

Sonderegger, D. and Hannig, J., 2014. Fiducial theory for free-knot splines. In *Contemporary Developments in Statistical Theory, a Festschrift in Honor of Professor Hira L. Koul*, eds S. Lahiri, A. Schick, A. SenGupta and T. N. Sriram, pp. 155–189. New York: Springer.

Stone, M., 1983. Fiducial probability. In *Encyclopedia of Statistical Sciences*, volume 3, eds S. Kotz, N. L. Johnson and C. B. Read, pp. 81–86. New York: Wiley-Interscience. https://doi.org/10.1002/9781118445112.stat01530.

Taraldsen, G., 2021. The confidence density for correlation. *Sankhya A: The Indian Journal of Statistics, 85*(1), pp. 600–616. https://doi.org/10.1007/s13171-021-00267-y.

Taraldsen, G. and Lindqvist, B., 2018. Conditional fiducial models. *Journal of Statistical Planning and Inference*, *195*, pp. 141–152.

Wallace, D. L., 1980. The Behrens-Fisher and Fieller-Creasy problems. In *R. A. Fisher: An Appreciation*, eds S. E. Fienberg and D. V. Hinkley, pp. 119–147. New York: Springer.

Wilkinson, G. N., 1977. On resolving the controversy in statistical inference (with discussion). *Journal of the Royal Statistical Society: Series B*, *39*, pp. 119–171.

Zabell, S. L., 1992. R. A. Fisher and the fiducial argument. *Statistical Science*, *7*, pp. 369–387.

6

Bridging Bayesian, Frequentist and Fiducial Inferences Using Confidence Distributions

Suzanne Thornton
Swarthmore College

Min-ge Xie
Rutgers University

6.1 Introduction

Bayesian, frequentist, and fiducial (BFF) inferences are much more congruous than have been perceived historically in the scientific community (e.g., Reid and Cox (2015); Kass (2011); Efron (1998)). Most practitioners are probably more familiar with the two dominant statistical inferential paradigms, Bayesian inference and frequentist inference. The third, lesser known fiducial inference paradigm was pioneered by R.A. Fisher in an attempt to define an inversion procedure for inference as an alternative to Bayes' theorem. Although each paradigm has its own strengths and limitations subject to their different philosophical underpinnings, this chapter intends to bridge these different inferential methodologies by calling upon *confidence distribution* theory and Monte-Carlo simulation procedures, thereby increasing the range of possible techniques available to both statistical theorists and practitioners across all fields.

One motivation behind the development of confidence distributions is to address a simple question: "Can frequentist inference rely on a distribution function, or a *distribution estimator*, in a manner similar to a Bayesian posterior?" (e.g., Xie and Singh, 2013). The notion of a confidence distribution answers this question in the affirmative. Many statisticians have considered confidence distributions to be a frequentist analogue to a Bayesian posterior (Schweder and Hjort, 2003, Chapter 2). Although by definition, a confidence distribution is defined within a frequentist framework, this special type of estimator can potentially unite Bayesian, frequentist, and fiducial inferential methods. In the subsequent sections, we provide a brief review of the concept of a confidence distribution and illustrate the flexibility of inferential framework this provides. We further highlight two deeply rooted connections involving the estimation and uncertainty quantifications among confidence distributions, bootstrap distributions, fiducial distributions, and Bayesian posterior distributions. Here, we use the term "uncertainty" to specifically refer to the variability associated with the random sample under an assumed model, called "empirical variability" in Reid and Cox (2015). We present a unified platform from which one may compare and combine inference across different paradigms and derive new simulation-based inference approaches. We conclude by presenting a novel assertion that, across all inferential paradigms, a parameter has both a "fixed" and a "random" version.

 DOI: 10.1201/9780429341731-6

6.2 Confidence Distribution (CD): A Distribution Estimator

6.2.1 Definitions

Traditionally, statistical analysis uses a single point or perhaps an interval to estimate an unknown target parameter. However, another valid option is to consider a sample-dependent function, specifically, a sample-dependent function on the parameter space, to estimate the parameter of interest. This is precisely the purpose of a confidence distribution. A confidence distribution is a sample-dependent, distribution (or density) function whose main purpose is estimation, including an assessment of uncertainty, for an unknown target parameter, similar to any other statistical estimator. A confidence distribution can be defined using a simple, prescriptive definition. In the following definition and throughout the remainder of this chapter, we use $\mathcal{Y}$ and Θ to denote the sample space and parameter space, respectively. We also use the conventional notation to clarify the difference between random variables, $\mathbf{Y}$, and their observations, $\mathbf{y}$.

Definition 6.1 *A sample-dependent function on the parameter space, i.e. a function on* $\mathcal{Y} \times \Theta$, $H_n(\cdot) = H_n(\mathbf{Y}, \cdot)$, *is called a* confidence distribution (CD) *for* $\theta \in \Theta$ *if*
[R1] For each given sample $\mathbf{y} \in \mathcal{Y}$, *the function* $H_n(\cdot) = H_n(\mathbf{y}, \cdot)$ *is a distribution function on the parameter space* Θ; *and*
[R2] The function can provide confidence intervals (or regions) for the parameter θ *at all confidence levels.*

If [R2] only holds asymptotically (or approximately), then $H_n(\cdot)$ *is called an* asymptotic (or approximate) CD (aCD).

In plain language, a CD is a function of both the parameter and the random sample with two requirements: *[R1]* that the estimator is a sample-dependent (distribution) function on the parameter space and *[R2]* that a particular performance property holds. This type of definition is analogous to the way consistent (or unbiased) estimators are defined in point estimation. Specifically, consistent (or unbiased) point estimators are determined as such if first (*[R1]*), the estimator is a sample-dependent point in the parameter space and second (*[R2]*), the estimator satisfies a particular performance-based criteria. For a consistent estimator, the latter (*[R2]*) requirement is that the estimator approaches the true parameter value as the sample size increases; for unbiased estimators, this requirement is instead that the expectation of the estimator is equal to the true parameter value. For the distributional estimator CD, the requirement *[R2]* instead states that this estimator must provide valid confidence intervals (or regions) for any significance level.

When the parameter θ is scalar, requirement *[R2]* in the CD definition can be restated as the requirement that, at the true parameter value $\theta = \theta_0$, $H_n(\theta_0) \equiv H_n(\mathbf{Y}, \theta_0)$ as a function of the random data follows a standard uniform distribution (e.g., Xie and Singh (2013); Schweder and Hjort (2016, Chapter 3)). The classical definition of CD in Cox (1958) and Efron (1993) is obtained by inverting the upper endpoints of a set of nested one-sided confidence intervals of all levels. This classical definition is covered by the descriptive definition of a CD in Definition 1 (e.g., Xie and Singh (2013, Sections 2.1 and 2.2)). When θ is a vector, there may not be a simple or neat mathematical expression for *[R2]*. In some cases, simultaneous CDs of a certain form for multiple parameters may not even exist (e.g., Xie and Singh (2013)). Fortunately, *[R2]* in Definition 1 only requires one set of confidence regions of all levels. In this manner, there are several options for defining multivariate CDs, e.g., Singh et al. (2007); Xie and Singh (2013); Schweder and Hjort (2016, Chapters 9 and 15) and Liu et al. (2021).

The idea of using a sample-dependent distribution function as an estimator of the target parameter is not new or unique to CD theory. A bootstrap distribution is sample-dependent and it is a distribution estimator for the unknown target parameter (Efron, 1998). In Bayesian inference, the posterior distribution is also a distribution estimator for the unknown target parameter, and it too depends upon the observed sample. The benefit of using an entire function as an estimator, rather than an interval or a point, is that these distribution functions carry a greater wealth of information about the parameter of interest. A CD estimator, like a Bayesian posterior distribution, enables the statistician to draw inferential conclusion about an unknown parameter. For example, if provided a CD for some parameter of interest, one can readily derive a point estimate, an interval estimate, and a p-value, among other features. This is illustrated in Figure 6.1. More details can be found in Singh et al. (2007); Xie and Singh (2013) and Schweder and Hjort (2016, Chapter 3).

The simple, illustrative example presented next will be revisited throughout the chapter.

Example 6.1 *Suppose we observe an independent sample,* $\mathbf{y} = (y_1, \ldots, y_n)$*, from a* $N(\theta, 1)$ *distribution. To estimate the unknown* θ *we may consider 1) a point estimate, e.g.* $\bar{y} = (1/n)\sum_{i=1}^{n} y_i$*; 2) an interval estimate, e.g.* $(\bar{y} - 1.96/\sqrt{n}, \bar{y} + 1.96/\sqrt{n})$ *or 3) a distribution estimate, e.g.* $H_n(\theta) = \Phi(\sqrt{n}(\theta - \bar{y}))$*, were* $\Phi(\cdot)$ *is the cumulative distribution function of the standard normal distribution. This distribution estimate is a CD for* θ*.*

By itself, $H_n(\theta)$ *provides information about* θ *including a point estimate, e.g. the center of the CD:* $H_n^{-1}(1/2) = \bar{y}$*; an interval estimate, e.g., the interval formed by the upper and lower level-*$(\alpha/2)$ *CD quantiles:* $\left(H_n^{-1}(\alpha/2), H_n(1-\alpha/2)\right) = \left(\bar{y}+\Phi^{-1}(\alpha/2)/\sqrt{n}, \bar{y}+\Phi^{-1}(1-\alpha/2)/\sqrt{n}\right)$*; and a p-value for the test* $K_0 : \theta \geq b$ *versus* $K_1 : \theta < b$*, e.g. the tail mass of the CD:* $H_n([b, \infty)) = 1 - H_n(b) = 1 - \Phi(\sqrt{n}(b - \bar{y}))$*.*

Another important concept related to CD-based inference is that of a *CD-random variable* (henceforth CD-r.v.). For a given sample, a CD is a distribution function defined on the parameter space. We can construct a random variable (or vector), θ^*_{CD}, on the parameter

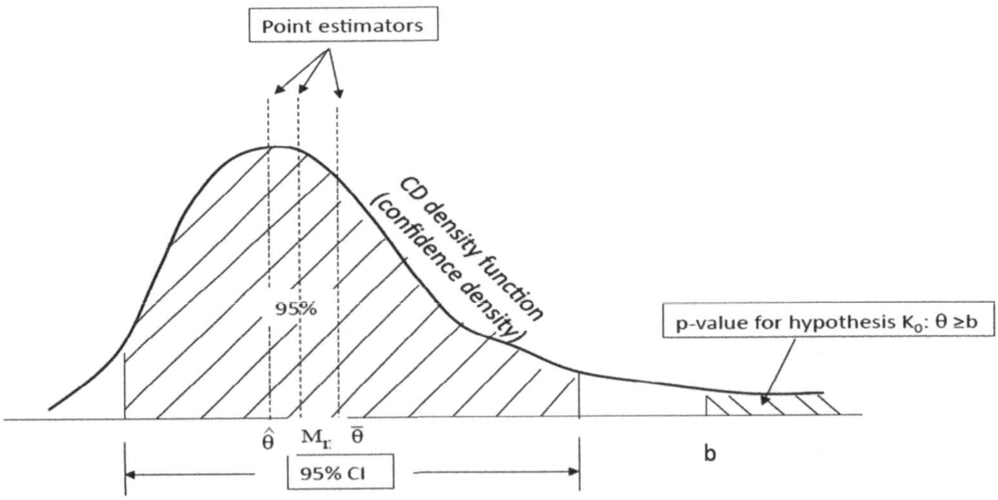

FIGURE 6.1
A graphical illustration of the inferential information contained in a CD as found in Xie and Singh (2013). Examples pictured include point estimators (mode $\hat{\theta}$, median M_n, and mean $\bar{\theta}$), a 95% CI, and a one-sided p-value.

space such that, conditional upon the observed sample of data, θ^*_{CD} follows the CD. In the context of Example 6.1, we can simulate a CD-r.v. θ^*_{CD} by generating $\theta^*_{CD} \mid \bar{Y} = \bar{y} \sim N(\bar{y}, 1/n)$. This θ^*_{CD} can be understood as a "CD-randomized" estimator of θ (cf., Xie and Singh, 2013; Zhang et al., 2020).

Definition 6.2 *Let* $\mathbf{y} = (y_1, \dots, y_n)$ *be an observed sample of data drawn from a distribution involving parameter* θ *and let* $H_n(\theta)$ *be a CD for* θ. *Then,* $\theta^*_{CD}|\mathbf{y} \sim H_n(\cdot)$, *is referred to as a* CD-random variable.

Defining the CD-r.v. by conditioning upon the data is similar to the conditioning required in bootstrap and Bayesian inferences. Indeed, a CD-r.v. has many similarities to a bootstrap estimator or a random θ generated from a posterior distribution. We will explore these connections further in Sections 6.3 and 6.4.

CD-based inference is incredibly flexible and can be presented in three forms: as a density function, a cumulative distribution function, or a *confidence curve*. If the target parameter θ is a scalar, a confidence curve is simply a mathematical representation of all two-sided confidence intervals for θ along every level of α on the vertical axis, i.e., $CV(\theta) = 2\min\{H_n(\theta), 1 - H_n(\theta)\}$; e.g., Xie and Singh (2013); Schweder and Hjort (2016, Chapter 1). Also, see Liu et al. (2021) for discussions of confidence curves when the target parameter θ is a vector. In Example 6.1, a confidence density is $h_n(\theta) = (1/\sqrt{n})\phi(\sqrt{n}(\theta - \bar{Y}))$, where $\phi(\cdot)$ is the density function of the standard normal distribution. Alternatively, a CD in the cumulative distribution function form is $H_n(\theta) = \Phi(\sqrt{n}(\theta - \bar{Y}))$ and a confidence curve is the function $CV(\theta) = 2\min\{\Phi(\sqrt{n}(\theta - \bar{Y}), 1 - \Phi(\sqrt{n}(\theta - \bar{Y})\}$. Figure 6.2 (a) shows a CD density, CD function, and confidence curve for the unknown location parameter using a random sample of size $n = 12$ drawn from a $N(\theta_0 = 10, \sigma^2 = 1)$ distribution.

In settings with nuisance parameters, a CD can provide an informative estimator for the parameter of interest (or the focus parameter). The following Neyman-Scott example is a slight modification of an example in Schweder and Hjort (2016, Section 4.4).

Example 6.2 *In the classical Neyman-Scott problem, we observe* $2n$ *independent pairs of data such that*

$$Y_{i,1}, Y_{i,2} \sim N(\mu_i, \sigma^2), \text{ for } i = 1, \dots, n;$$

i.e. each data pair is drawn from a distribution with the same variance σ^2 *but with different mean* μ_i. *The focus parameter is the common* σ^2. *Even with a large number of nuisance parameters, we still can construct a CD for* σ^2 *once we realize that*

$$\frac{(1/n)\sum_{i=1}^n [(Y_{i,1} - \bar{Y}_i)^2 + (Y_{i,2} - \bar{Y}_i)^2]}{\sigma^2} \sim \frac{\chi^2_{(n)}}{n}$$

is free of the nuisance parameters μ_i*'s. Here,* $\bar{Y}_i = (Y_{i,1} + Y_{i,2})/2$. *Specifically, let* C_ν *represent the cumulative distribution function for a* χ^2 *random variable with* ν *degrees of freedom. A CD for* σ^2 *is then* $H_n(\sigma^2) = 1 - C_n(2n\hat{\sigma}^2_{obs}/\sigma^2)$, *where* $\hat{\sigma}^2_{obs} = (1/2n)\sum_{i=1}^n [(y_{i,1} - \bar{y}_i)^2 + (y_{i,2} - \bar{y}_i)^2]$. *The corresponding confidence density and confidence curve are* $h_n(\sigma^2) = [2n\hat{\sigma}^2_{obs}/\sigma^4] c_n(2n\hat{\sigma}^2_{obs}/\sigma^2)$ *and* $CV(\sigma^2) = 2\min\{1 - C_n(2n\hat{\sigma}^2_{obs}/\sigma^2), C_n(2n\hat{\sigma}^2_{obs}/\sigma^2)\}$, *respectively, where* $c_n(t) = C_n'(t)$ *is the density function of the* χ^2_n *distribution. Figure 6.2 (b) shows a CD density, CD function, and confidence curve for the unknown common variance from a sample of* $n = 100$ *independent data pairs. In this particular example, the data was drawn from a* $N(\mu_j, \sigma_0^2 = 1)$, *for* $j = 1, \dots, n$ *and the* μ_j*'s are drawn from a* $U(0, 10)$ *distribution.*

When the sample data are drawn from a discrete distribution, most CD-related work in the literature considers the case where the sample size is large and one may use an asymptotic CD, e.g., Singh et al. (2007); Xie and Singh (2013); Hannig and Xie (2012); Schweder and Hjort (2016, Chapter 3). However, the concept of a CD is also useful in the case where the sample size is finite and the data are drawn from a discrete distribution. In these situations, one may examine the difference between the discrete distribution estimator and the standard uniform distribution to get an idea of the under/over coverage of the confidence intervals derived from the CD. The definitions of lower and upper CDs are provided below for a scalar parameter $\theta \in \Theta$. Upper and lower CDs thus provide inferential statements in applications to discrete distributions with finite sample sizes.

Definition 6.3 *A function* $H_n^+(\cdot) = H_n^+(\mathbf{Y}, \cdot)$ *on* $\mathcal{Y} \times \Theta \to [0,1]$ *is said to be an* upper CD *for* $\theta \in \Theta$ *if*
[R1] For each given $\mathbf{y} \in \mathcal{Y}$, $H_n^+(\mathbf{y}, \cdot)$ *is a monotonic increasing function on* Θ *with values ranging within* $(0,1)$; *and*
[R2] At the true parameter value $\theta = \theta_0$, $H_n^+(\theta_0) = H_n^+(\mathbf{Y}, \theta_0)$, *as a function of the sample* $\mathbf{Y}$, *is stochastically less than or equal to a uniformly distributed random variable* $U \sim U(0,1)$, *i.e.,*

$$\mathbb{P}(H_n^+(\mathbf{Y}, \theta_0) \leq t) \geq t, \text{ for all } t \in (0,1). \tag{6.1}$$

A lower CD $H_n^-(\cdot) = H_n(\mathbf{Y}, \cdot)$ *for parameter* θ *can be defined similarly, but with (6.1) replaced by* $\mathbb{P}(H_n^-(Y, \theta_0) \leq t) \leq t$ *for all* $t \in (0,1)$.

Sometimes, even if condition *[R1]* in Definition 6.3 does not hold, $H_n^+(\cdot)$ and $H_n^-(\cdot)$ are still referred to as upper and lower CDs, respectively. This is because regardless of whether or not the monotonicity condition holds, the stochastic dominance inequalities in the definition of upper and lower CDs ensure that for any $\alpha \in (0,1)$

$$\mathbb{P}\left(\theta_0 \in \{\theta : H_n^+(\mathbf{Y}, \theta) \leq \alpha\}\right) \geq \alpha \quad \text{and} \quad \mathbb{P}\left(\theta_0 \in \{H_n^-(\mathbf{Y}, \theta) \leq \alpha\}\right) \leq \alpha.$$

Thus, a level $(1-\alpha)$ confidence interval (or region) $\{\theta : H_n^+(\mathbf{y}, \theta) \leq 1-\alpha\}$ or $\{\theta : H_n^-(\mathbf{y}, \theta) \geq \alpha\}$ has guaranteed coverage rate of at least $(1-\alpha)$. Without condition *[R1]* however, $H_n^+(\cdot)$ and $H_n^-(\cdot)$ may not be distribution functions and confidence intervals (or sets) may no longer follow the nesting condition, e.g., Blaker (2000); Xie and Singh (2013). This means that a level $(1-\alpha)$ confidence set, $C_{1-\alpha}$, may not necessarily be contained in a corresponding level $(1-\alpha')$ confidence set, $C_{1-\alpha'}$, for some $\alpha' < \alpha$. This is a departure from the classical definition of a CD found in Cox (1958), where a CD is obtained by inverting the bounds of nested confidence intervals at every α level.

The following is an example of an upper and lower CD for the parameter in a discrete distribution. For more details on this example, see Hannig and Xie (2012). A more complex example can be found in Luo et al. (2021), which links Fisher's sharp null test in causal inference with the concepts of lower and upper CDs and their corresponding (extended) confidence curves.

Example 6.3 *Suppose we observe* y *from a Binomial*(n, p_0) *distribution. Let* $H_n(p, y) = Pr(Y > y) = \sum_{y<k\leq n_i} \binom{n}{k} p^k (1-p)^{n-k}$. *It can be shown that* $Pr(H_n(p_0, Y) \leq t) \geq t$ *and* $P(H_n(p_0, Y-1) \leq t) \leq t$. *Thus,* $H_n^+(p, y) = H_n(p, y)$, *and* $H_n^-(p, y) = H_n(p, y-1)$ *are upper and lower CDs for the true success rate* p_0. *The half-corrected CD used in the literature (e.g., Schweder and Hjort 2002, Section 3.8),*

$$\frac{H_n^-(p,y) + H_n^+(p,y)}{2} = \sum_{y<k\leq n} \binom{n}{k} p^k (1-p)^{n-k} + \frac{1}{2}\binom{n}{k} p^y (1-p)^{n-y},$$

is an aCD. In the finite sample setting, we do not use the half-corrected CD, and a level $(1-\alpha)$ *confidence interval that guarantees the coverage rate is* $\{\theta : H_n^+(\mathbf{y},\theta) \leq 1-\frac{\alpha}{2}, H_n^-(\mathbf{y},\theta) \geq \frac{\alpha}{2}\}$.

6.2.2 Construction with Different Inferential Methods

The concept of a CD includes a broad class of common inferential methods, e.g., Xie and Singh (2013). Among this class of methods are bootstrap distributions, (normalized) likelihood functions, (normalized) empirical likelihood functions, *p*-value functions, fiducial distributions, Bayesian posterior distributions, and more. As summarized in Cox (2013), the usefulness of inference from a CD perspective lies in its ability to "provide simple and interpretable summaries of what can reasonably be learned from data (and an assumed model)". Regardless of whether the setting is parametric or nonparametric, normal or non-normal, exact or asymptotic, CD inference is possible as long as one can create confidence intervals (or regions) of all levels for the parameter of interest. This fact illustrates the appeal of CD-based inference as uniting different inferential paradigms. Before moving on to the next section, we present a CD framework based on likelihood inference, and on hypothesis testing, with a few relevant examples.

Likelihood-based approaches and their extensions are arguably the most commonly used methods to answer questions of statistical inference. Asymptotic, large sample arguments are frequently used to quantify the uncertainty in our point estimators from likelihood-based approaches. The very same consequences of the central limit theorem also provide a connection between the likelihood function and a CD, provided $\int L(\boldsymbol{\theta} \mid data)d\boldsymbol{\theta} < \infty$, where $L(\boldsymbol{\theta} \mid data)$ is either the likelihood function or a generalized likelihood function. For instance, under some regularity conditions, Fraser and McDunnough (1984) proved that a normalized likelihood function is an aCD and Singh et al. (2007) showed that a normalized profile likelihood function is also an aCD. The inferential conclusions based on these CDs asymptotically match the first-order conclusions based on the likelihood or profile likelihood functions. Higher-order asymptotic developments relating likelihood inference to CDs include, for instance, Reid and Fraser (2010); Reid and Cox (2015); Pierce and Bellio (2017), among others. These higher-order asymptotic developments can also be used to derive more accurate CDs, e.g., Xie and Singh (2013).

Example 6.1 (continued) *The likelihood function is* $L(\theta \mid \mathbf{y}) \propto \prod_i f(y_i \mid \theta) \propto e^{-\frac{n}{2}(\bar{y}-\theta)^2}$. *Normalizing* $L(\theta \mid \mathbf{y})$ *with respect to* θ, *we get* $L(\theta \mid \mathbf{y})/\int L(\theta \mid \mathbf{y})d\theta = \sqrt{n}\phi(\sqrt{n}(\theta-\bar{y}))$ *which is exactly a CD density function. Furthermore, since we are only dividing by a constant when normalizing (standardizing) the likelihood function, the mode of the CD is the same as that of the likelihood function. Thus, inferential conclusions drawn from the likelihood function* $L(\theta \mid \mathbf{y})$ *match the conclusions drawn from the CD* $H_n(\theta) = \Phi(\sqrt{n}(\theta - \bar{Y}))$.

There may be instances in which we are interested in a scalar parameter in a model with a vector of parameters. In this situation, it is common to use a profile likelihood for inference about the focus parameter, and the profile likelihood can also be used to derive a CD for the focus parameter; e.g., Singh et al. (2007). Example 6.4 outlines one such case where a CD for the shape parameter of a Gamma distribution is derived from the profile likelihood function.

Example 6.4 *Suppose we observe a random sample of size* n, *say* $\mathbf{y} = y_1, \ldots, y_n$, *from a* $Gamma(\theta,\beta)$ *distribution, where the shape parameter,* $\theta > 0$, *is the focus parameter and the rate parameter,* $\beta > 0$, *is a nuisance parameter. The likelihood function is* $L(\theta,\beta \mid \mathbf{y}) = \prod_i [\beta^\theta/\Gamma(\theta)] y_i^{(\theta-1)} e^{-\beta y_i}$, *where* $\Gamma(x) = \int_0^\infty z^{(x-1)} e^{-z} dz$. *The profile likelihood is found by substituting* $\hat{\beta}(\theta) = \theta/\bar{y}$ *for* β *in* $L(\theta,\beta \mid \mathbf{y})$. *This can be normalized with respect to the* *parameter of interest thereby deriving a CD density for* θ

$$h_n(\theta) \propto \frac{1}{\Gamma(\theta)^n}\left(\frac{\theta}{\bar{y}}\right)^{n\theta}\left(\prod_i y_i\right)^{(\theta-1)} e^{-n\theta}.$$

The corresponding CD, $H_n(\theta) = \frac{\int_0^\theta h(s)ds}{\int_0^\infty h(s)ds}$, *can be obtained numerically.*

Likelihood-based inference represents only one particular generative method for obtaining a CD since CDs can be obtained in many other cases without likelihood functions. In the example below, we consider an hypothesis testing approach, another common inferential method intimately linked to CDs. As with likelihood-based inference, hypothesis tests represent only one of many other generative methods by which a CD may be derived. For simplicity, we consider only the case that the focus parameter θ is a scalar; In the case when θ is a vector, an example connecting CD with a testing problem can be found in Liu et al. (2021).

Consider a one-sided hypothesis testing problem $K_0 : \theta \leq \theta_0$ versus $K_1 : \theta > \theta_0$ or a two-sided test $K_0 : \theta = \theta_0$ versus $K_1 : \theta \neq \theta_0$. Denote by $p_n = p_n(\theta_0)$ the p-value from a testing method; this p-value depends on both the observed sample and the tested value of the parameter, θ_0. As θ_0 varies in the parameter space Θ, $p_n = p_n(\theta_0)$ is a function on the parameter space called a *p-value function* or a *significance function*; see, e.g., Fraser (1991). Often, $p_n(\theta_0)$ (as a function of the random sample) follows a standard uniform distribution under the the null hypothesis. As function of θ_0, $p_n(\theta_0)$ is typically monotonically increasing in the one-sided hypothesis test setting and first increases and then decreases in the case of a two-sided hypothesis test. Singh et al. (2007) and Xie and Singh (2013) showed that the p-value function from the one-sided test corresponds to a regular CD function and the p-value function from a two-sided test corresponds to a confidence curve.

Example 6.1 (continued) *For a one-sided hypothesis test of* θ, $K_0 : \theta \leq \theta_0$ *versus* $K_1 : \theta > \theta_0$, *the p-value is calculated by* $q_n(\theta_0) = \Phi(\sqrt{n}(\theta_0 - \bar{y}))$. *As* θ_0 *varies in* Θ, $q_n(\theta_0)$ *is the cumulative distribution function of a* $N(\bar{y}, 1/n)$ *distribution and is therefore a CD for* θ. *For the two-sided hypothesis test* $K_0 : \theta = \theta_0$ *versus* $K_1 : \theta \neq \theta_0$, *the p-value is calculated by* $2\min\{q_n(\theta_0), 1 - q_n(\theta_0)\} = 2\min\{\Phi(\sqrt{n}(\theta_0 - \bar{y})), \Phi(\sqrt{n}(\bar{y} - \theta_0))\}$, *which is a confidence curve for* θ. *See Figure 6.2 (a).*

The next example considers the Mann–Whitney test for determining whether two independent samples are drawn from the same population. For a careful discussion on setting up the hypotheses for a Mann–Whitney test, we refer the reader to Divine et al. (2018).

Example 6.5 *Consider two independent samples* $\mathbf{x} = (x_1, \ldots, x_{n_1})$ *and* $\mathbf{y} = (y_1, \ldots, y_{n_2})$ *each drawn from an unknown continuous distribution (say,* $F(\cdot)$ *and* $G(\cdot)$*, respectively) that differ in location only. Without loss of generality, suppose* $n_1 > n_2$ *and let* $n_1 + n_2 = n$. *In this case, the Mann-Whitney test for a difference in the population distributions is equivalent to testing the null that* $K_0 : F(t) = G(t)$, *for all* t *versus the alternative* $K_1 : F(t) = G(t - \theta)$, *for every* t *and for some* $\theta \neq 0$. *The Mann-Whitney test statistic is* $U_\theta = n_1 n_2 + \frac{n_2(n_2+1)}{2} - R_\theta$, *where* R_θ *is the sum of the ranks of* $\mathbf{X}$ *among the set* $\{X_1, \ldots, X_{n_1}, Y_1 + \theta, \ldots, Y_{n_2} + \theta\}$. *Let* $A_{\theta_0}(t) = P(U_\theta \leq t \mid \theta = \theta_0)$ *be the distribution of* U_θ *under the null hypothesis* K_0. *Then, the p-value for this test is calculated by*

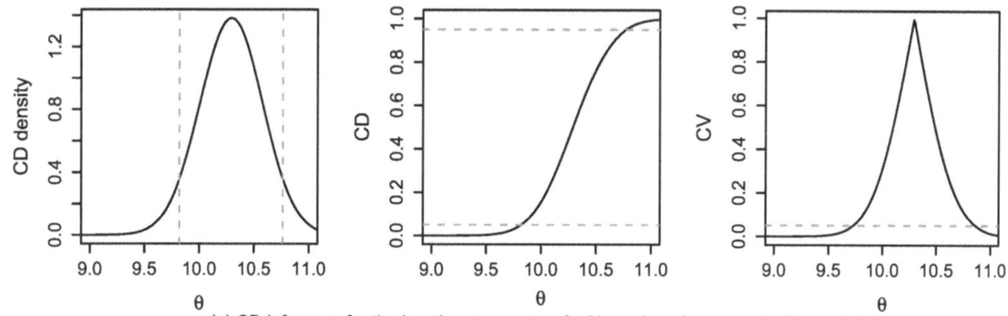

(a) CD-inference for the location parameter of a Normal random sample, Example 6.1.

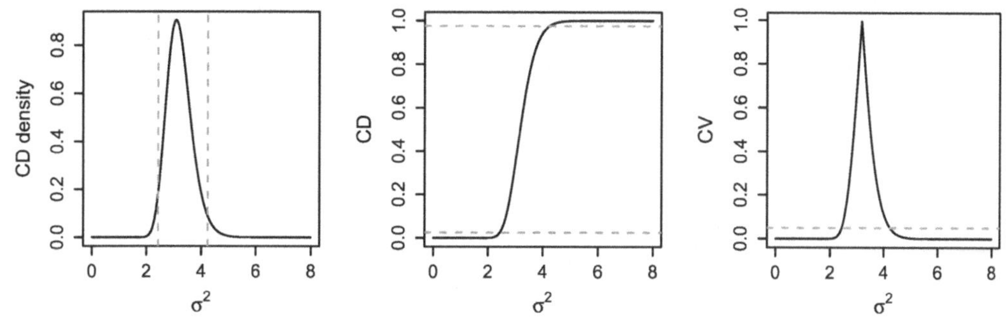

(b) CD-inference for the common variance in the Neyman-Scott setting, Example 6.2.

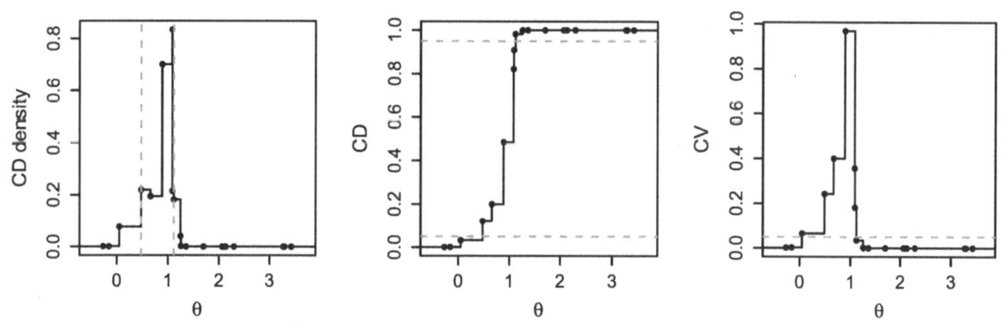

(c) CD-inference for the location shift from the Mann-Whitney setting, Example 6.5.

FIGURE 6.2
Each row displays a CD density, a CD curve, and a confidence curve (CV) for various examples in this chapter. Plot (a) is derived from Example 6.1 with a random sample of size $n = 12$. Plot (b) is derived from Example 6.2 for $n = 100$ observations of independent, paired data points. Plot (c) is derived from Example 6.5 with two independent samples of size $n_1 = 10$ and $n_2 = 9$. In each plot, the values of the horizontal axis at intersections between the gray dashed line(s) and the solid (CD) curves are the lower and upper bounds of the corresponding 95% confidence interval.

$2\min\{A_\theta(u_\theta), 1-A_\theta(u_\theta)\}$, *where* u_θ *is the value of* U_θ *evaluated at* $\mathbf{x}$ *and* $\mathbf{y}$. *This generates a confidence curve for* θ. *Also, we can show that* $H_n(\theta) = A_\theta(u_\theta)$ *is a CD function, from which a CD density can be readily derived. Figure 6.2 (c) shows an image of a CD density, CD function, and confidence curve for the unknown location shift where one sample* ($n_1 = 10$) *is drawn from a t-distribution with* $\nu = 5$ *degrees of freedom, and the other* ($n_2 = 9$) *is independently drawn from a t-distribution also with* $\nu = 5$ *degrees of freedom but also with a non-centrality parameter* $\theta_0 = 1$.

6.3 Using the CD Framework to Unify and Combine Inference across the BFF Paradigms

A distinguishing feature of CDs is that this type of estimation provides a framework under which results across different inferential paradigms can be directly compared or even combined, as will be demonstrated in Example 6.6. One may better understand how this is possible by recognizing that CD theory extends several practical considerations behind point estimation to the class of distribution estimators. To make this clear, let us first consider the context of statistical point estimation.

Point estimation as a concept itself is inherently arbitrary. An estimator must be a sample dependent function that produces a point in the parameter space. The rich mathematical literature on particular methods of point estimation and their comparison somewhat paradoxically thrives through setting restrictions. For instance, one may be interested in estimators that behave according to certain impartiality conditions such as unbiasedness or equivariance. Still more desirable properties may include consistency or efficiency. These traits are ubiquitously recognized as favorable among statisticians working within any of the Bayesian, fiducial, or frequentist paradigms, but are nothing more than performance-based descriptors of a potential estimator. That is, the pursuit of statistical estimators from this perspective reflects a behavioral reasoning process. A point estimator defined by behavioral reasoning, however, does not necessarily include instructions for obtaining the estimate. For example, the definition of the consistent estimator does not provide any direction on how the consistent estimate can be obtained. Instead, a consistent estimate is often obtained in collaboration with a procedural-based approach.

Examples of procedurally defined estimators include the maximum likelihood estimator, moment estimator, Bayes estimator, and others. Most of these procedurally defined estimators are rooted in a certain type of inductive reasoning often supported by decision theory, whereby a risk function is minimized. These procedural methods are frequently applied in the search for desirable behaviorally-defined estimators, but they generally do not provide universal guarantees about the performance of the resulting estimator. Instead, the starting point afforded by these various procedures provides some orientation in a search that may otherwise be quite arbitrary. The reasons for choosing a particular procedure-based approach are varied and highly problem-dependent but what is universally appealing is that the resulting estimators behave according to some performance-based quality.

Now, we may broaden our perspective to also consider distribution estimators. A CD is defined according to two performance-based descriptors; recall *[R1]* and *[R2]* from Definition 6.2.1. The requirement *[R2]* describes the performance of a distribution estimator that covers the parameter of interest with at least a specified frequency in repeated, independent experiments. This so-called relative frequency principle has become a (nearly) universally appealing standard for useful statistical inference regardless of the inferential framework. Thus one may consider different inferential methods as a starting point in the search for a well-behaved distribution estimator. For example, the Bayesian method derives a posterior distribution which can be thought of as a distribution estimator for the unknown parameter. As mentioned earlier, within a frequentist framework one may derive a p-value function or a normalized likelihood function to obtain a distribution estimator. The fiducial distribution is another distribution estimator that can be derived in many settings through the method of generalized fiducial inference (Hannig et al., 2016). There are innumerable ways in which one may derive a distribution estimator, but a CD inferential framework suggests restricting this endeavour according to a tried-and-true (frequency) property that can be easily explained and communicated to the broader scientific community.

TABLE 6.1
A Table Depicting a Range of Ways in Which a Statistical Estimator May be Defined

Estimator	**Definition**	
Type	Behavior (Desired performance)	Procedure (Inductive reasoning and/or operational decision)
Point	Consistent estimator	Maximum likelihood estimator, M-estimation, Moment estimator, Bayes estimator
Distribution	Confidence distribution	Bootstrap distribution, Fiducial distribution, (normalized) likelihood, *p*-value function, Bayesian posterior

Table 6.1 compares these different motivating (or restrictive) qualities and methods across the broad landscape of statistical estimation. For instance, a consistent estimator is defined by requiring a certain behavior, and it is often obtained via a procedural-based estimator, such as an MLE approach or some other method. On its own, the procedural-based MLE is obtained through steps devised in likelihood inference. This estimator is often consistent; however, this is not guaranteed and the statistician must verify its performance in different contexts. Similarly, a CD is a behaviorally defined estimator, and this definition does not provide any direction on how it may be obtained. Practically, a CD can often be obtained via a procedural-based approach, for example, from a p-value function via hypothesis testing, or a bootstrap distribution via resampling, or a fiducial distribution via inversion, or a posterior distribution via Bayes formula, and so forth. A fiducial distribution and Bayesian posterior, on the other hand, are procedurally defined estimators supported by inductive reasoning. A fiducial distribution and Bayesian posterior are often, but not always, a CD or aCD, and one must verify the requirements hold in different contexts.

By approaching the problem of statistical inference from a more general perspective targeting well-behaved distribution estimators, CD theory provides a common platform to compare and even combine inferential outcomes across different paradigms in ways that would otherwise not be possible. The following example, inspired by Cui and Xie (2021), presents a CD-based meta-analysis approach that combines the results from four independent studies across three different inferential paradigms. The unique feature of this particular example is that, in each of the four independent studies, one particular inferential method, either frequentist, Bayesian, or fiducial, was adopted to initially analyze the data and the results of these initial analyses provide only summary statistics for use in a CD-based meta-analysis that combines all results. We refer the interested reader to Xie et al. (2011) and Singh et al. (2005) for more detailed developments on CD-based meta-analysis and CD combining methodologies.

Example 6.6 *Suppose we have $k = 4$ independent studies in a typical meta-analysis setup where each study collects n independent observations of some count data before and after a new treatment, $\mathbf{y}^{(k)} = (\mathbf{y}_1^{(k)}, \mathbf{y}_2^{(k)})^T$. For each individual i in the kth study, the paired observations are assumed to follow a bivariate normal distribution*

$$\mathbf{y}_i^{(k)} = \begin{pmatrix} y_{i1}^{(k)} \\ y_{i2}^{(k)} \end{pmatrix} \sim N\left(\begin{pmatrix} \mu_1 \\ \mu_2 \end{pmatrix}, \begin{pmatrix} \sigma_1^2 & \rho\sigma_1\sigma_2 \\ \rho\sigma_1\sigma_2 & \sigma_2^2 \end{pmatrix} \right), \quad \text{for } i = 1, \dots, n \text{ and } k = 1, 2, 3, 4. \tag{6.2}$$

Consider the correlation coefficient, ρ, as the parameter of interest and $(\mu_1, \mu_2, \sigma_1^2, \sigma_2^2)$ as nuisance parameters.

Say study $k = 1$ uses Fisher's Z method (Fisher, 1915) and, by inverting the distribution of an asymptotic pivot statistic to obtain a distribution estimator (Singh et al., 2007), we *treat it as a fiducial approach. Study $k = 2$ uses a bias-corrected and accelerated (BCa) bootstrap method (DiCiccio and Efron, 1996; Davison and Hinkley, 1997; Angelo and Ripley, 2020) and study $k = 3$ uses a profile likelihood approach. Hence, each of these two studies can produce a CD and draw inferential conclusions within a frequentist framework. Study $k = 4$ produces a Bayesian posterior distribution based on a uniform prior. The combined CD-approach draws from each of these $k = 4$ studies by treating all four distribution estimators as CDs, and produces a fifth distribution estimator. We study the performance of the fifth distribution estimator and compare it with the other four distribution estimators.*

In particular, $H_1(\rho) = H_1(\rho, \mathbf{y}^{(1)}) = 1 - \Phi\left[\frac{\sqrt{n-3}}{2}(\log\frac{1+r}{1-r} - \log\frac{1+\rho}{1-\rho})\right]$ is the distribution estimator derived from the Fisher's Z test statistic, where r is the sample correlation coefficient; (cf, e.g., Singh et al., 2007). The corresponding CD density is $h_1(\rho) = \frac{\sqrt{n-3}}{2(1-\rho^2)}\phi\left[\frac{\sqrt{n-3}}{2}(\log\frac{1+r}{1-r} - \log\frac{1+\rho}{1-\rho})\right]$. Let $h_2(\rho) = h_2(\rho, \mathbf{y}^{(2)})$ be the bootstrap (density) distribution using the bootstrap BC method. Similarly, let $h_3(\rho) = h_3(\rho, \mathbf{y}^{(3)})$ be the normalized profile likelihood function and $p_4(\rho) = p_4(\rho|\mathbf{y}^{(4)})$ be the posterior distribution. The key to the combined CD approach is realizing that each of $h_i(\rho)$, $i = 1, \ldots, 4$ are individual CD densities that summarize the inferential information available for ρ from each individual study. Thus, a combined CD that incorporates the information from all four studies can be expressed explicitly as

$$H^{(c)}(\rho) = G_c\left(g_c\left(\int_{-1}^{\rho} h_1(s)ds, \int_{-1}^{\rho} h_2(s)ds, \int_{-1}^{\rho} h_3(s)ds, \int_{-1}^{\rho} p_4(s)ds\right)\right),$$

where $g_c(u_1, u_2, u_3, u_4)$ is a given monotonic mapping function from the quartic cube to the real line and $G_c(t) = P(g_c(U_1, U_2, U_3, U_4) \le t)$ with $(U_1, \ldots, U_4)$ being IID $U(0,1)$ random variables. In the numerical study below, we have used $g_c(u_1, \ldots, u_4) = DE^{-1}(u_1) + \ldots + DE^{-1}(u_4)$, where $DE^{-1}(\cdot)$ is the inverse cumulative distribution function of the standard Laplace distribution. With this choice of g_c, the combining approach is Bahadur efficient (Singh et al., 2005, 2007).

Figure 6.3 illustrates the similarities of the individual inferential functions $h_1(\rho)$, $h_2(\rho)$, $h_3(\rho)$, and $p_4(\rho)$ and the combined CD density, $h^{(c)}(\rho) = \frac{\partial}{\partial\rho}H^{(c)}(\rho)$ as computed by the gmeta function in the R package of the same name (Yang et al., 2017). This figure is based on four independent samples drawn from a bivariate normal distribution with $n = 200$ and true parameter values $\mu_1 = 3.288$, $\mu_2 = 4.093$, $\sigma_1^2 = 0.657$, $\sigma_2^2 = 1.346$, and $\rho_0 = 0.723$. Table 6.2 summarizes the relative performance of each of these five different approaches based on a simulation study with 1000 repeated samples from the same bivariate normal distribution. We see that the combined approach still maintains the nominal coverage level though the interval width of this approach is roughly half that of the other four individual approaches. This result is unsurprising since theoretically each study provides a $\frac{1}{\sqrt{n}}$-confidence interval and the combined method provides a $\frac{1}{2\sqrt{n}}$-confidence interval, noting that the total sample size of the combined data is $4n$. This example demonstrates how inference across different paradigms can be combined through a CD approach.

6.4 BFF Union: Randomness Matching and Parameter Duality

Statistical inference is primarily concerned with the quantification of uncertainty inherent in sampling data from a population. Since we observe only a single copy of the data, assessing this uncertainty typically requires an assumed statistical model under which the observed

Study Method	Statistical Paradigm	95% Confidence/Credible Interval Estimate	Distribution Estimate
Fisher's Z Method	Fiducial	[0.61, 0.75]	
Bootstrap BCa	Frequentist	[0.67, 0.80]	
Profile Likelihood	Frequentist	[0.67, 0.79]	
Posterior (flat prior)	Bayesian	[0.62, 0.77]	
Combined CD	-	[0.66, 0.75]	

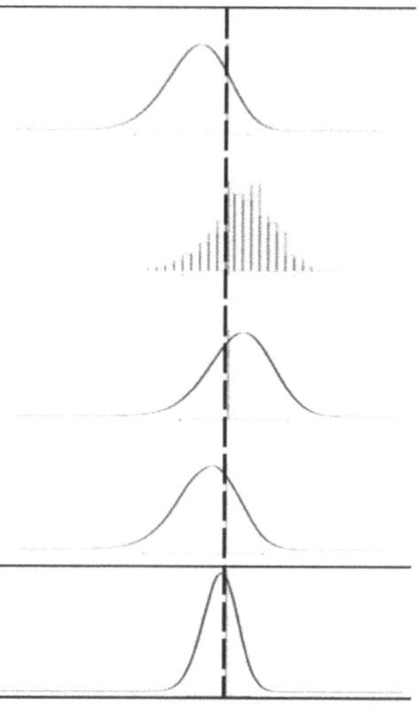

FIGURE 6.3
A visual comparison of one instance of the meta-analysis problem from Example 6.6. The four individual inferential functions for ρ are derived from four independent studies across three inferential paradigms and the combined CD approach (bottom row) incorporates information from all four studies. The target value of $\rho_0 = 0.723$ is indicated by the dashed vertical line.

TABLE 6.2
Coverage Results from 1000 Simulated Data Sets for Each of the Five Different Inferential Methods Mentioned in Example 6.6. The Last Row Represents a CD Approach That Combines Individual-Study Inference from the Four Independent Studies

Method	**95% CI**	
	Coverage	**Mean Length (sd)**
Fisher's Z method	0.939	0.132 (0.014)
Bootstrap BC_a	0.931	0.132 (0.018)
Profile likelihood	0.952	0.130 (0.013)
Bayes (uniform prior)	0.951	0.139 (0.015)
Combined CD	0.944	0.064 (0.006)

data are considered to be a particular realization of a random sample set. Sometimes an inference problem is tractable mathematically and assessments of empirical variability are possible through an explicit mathematical formulation. In more complicated situations, incorporating additional simulation-driven methods is necessary to assess the empirical variability. This section focuses on the role simulation-driven methods play in each of the three BFF paradigms.

We specifically highlight a common trait that matches two types of randomness: the artificial randomness induced by a simulation mechanism and the empirical randomness inherited from the sampling procedure under the assumed model. Following the exploration of this matching condition, we revisit the discussion as to whether or not a parameter should be viewed as random or fixed or if perhaps both perspectives are applicable. We try to avoid a philosophical discussion of *aleatory* versus *epistemic* probability and instead focus on matching the probability measures induced by artificial simulation (i.e., the one that governs the artificial randomness) and inherited from the assumed model (i.e., the one that describes the empirical randomness). A brief comment on how these probability measures may relate to aleatory and epistemic definitions of probability is provided in Section 6.5.

For ease of presentation, this section uses the notation $\stackrel{m}{\sim}$ to relate two real-valued random quantities of the same dimension, say, W_n and V_n of dimension p, that are defined on two different probability spaces, one associated with the artificial Monte-Carlo sample space and the other from the standard sample space. Specifically, the expression $W_n \stackrel{m}{\sim} V_n$ refers to either an *exact* matching such that

$$\mathbb{P}_A(W_n \in C|data) = \mathbb{P}(V_n \in C), \tag{6.3}$$

in probability, or an *approximate* matching (often reliant upon $n \to \infty$) such that

$$\left|\mathbb{P}_A(W_n \in C|data) - \mathbb{P}(V_n \in C)\right| \to 0, \tag{6.4}$$

in probability, for any Borel set, C, in the Borel σ-field on $\mathbf{R}^p$. Here, the probability statement $\mathbb{P}_A(\cdot|data)$ is defined according to the artificial randomness governed by the Monte-Carlo simulation procedure, conditioned upon an observed data set, and $\mathbb{P}(\cdot)$ is defined according to the usual empirical randomness as determined by the sampling procedure under the assumed model. The uncertainty described by $\mathbb{P}(\cdot)$ is typically the target of interest for frequentist inference. As we will see next, most literature exploring such matching conditions rely on asymptotic arguments; however, our consideration includes the exact matching of (6.3) as well.

6.4.1 Monte-Carlo Simulation and a Matching Condition for Statistical Inference

Monte-Carlo simulation methods have proven to be incredibly useful and effective tools for statistical inference for which there exists a rich literature, e.g., Efron and Tibshirani (1993); Fishman (1996); Glasserman (2003), and references therein. One popular approach is to create many copies of artificial (or "fake") data using a Monte-Carlo procedure and then to use these fake data to assess statistical uncertainty. This type of schema appears throughout the BFF paradigms including, as will be explored in this section, the bootstrap sampling method, generalized fiducial inference (GFI), and approximate Bayesian computing (ABC) (cf., e.g., Efron and Tibshirani, 1993; Ernst, 2004; Robert, 2016; Hannig et al., 2016). Viewing these simulation methods within the context of CD development reveals a deep connection among different inferential paradigms.

6.4.1.1 Frequentist: Bootstrap Estimators and CD Random Variables

The bootstrap method simulates artificial data sets through a re-sampling process, which is often referred to as *bootstrapping* a sample. The mechanism underlying the success of this simulation approach is the so-called bootstrap central limit theorem (Singh, 1981; Freedman and Bickel, 1981). This theorem states general conditions under which an estimator, $\hat{\theta}$, for some parameter θ (with true value θ_0) satisfies

$$\theta^*_{BT} - \hat{\theta} \mid data \stackrel{m}{\sim} \hat{\theta} - \theta \mid \theta = \theta_0, \tag{6.5}$$

as the sample size $n \to \infty$. In (6.5), θ^*_{BT} represents the bootstrap estimator, that is, the estimator evaluated over a random set of bootstrapped samples. The matching statement in equation (6.5) is a particular example of the approximate probability statement of equation (6.4). In the bootstrapping method for inference, $\mathbb{P}_A(\cdot|data)$ is defined by the multinomial-distributed randomness of the bootstrap sampling. Although the multinomial-distributed randomness is typically not the same as the empirical randomness inherited by the statistical model, each side of equation (6.5) converges to the same normal distribution by application of the central limit theorem. Therefore, (6.5) can be exploited for the purpose of valid frequentist inference about θ.

Consider again the setting of Example 6.1 where $\sqrt{n}(\bar{Y} - \theta_0) \sim N(0,1)$ and also, by the central limit theorem, $\sqrt{n}(\bar{Y}^*_{BT} - \bar{Y}) \mid \bar{Y} = \bar{y} \to N(0,1)$. Here the matching expression of (6.5) may be written as: $(\bar{Y}^*_{BT} - \bar{Y}) \mid \bar{Y} = \bar{y} \overset{m}{\sim} (\bar{Y} - \theta)|\theta = \theta_0$, since both sides are (exactly or asymptotically) $N(0, 1/n)$. Because the simulated variability of the bootstrap estimator, conditioned upon the observed sample, matches the sampling variability of the estimator as a function of a random sample from the study population, the bootstrap method can provide valid frequentist inferential conclusions.

When the bootstrap procedure is applicable, the bootstrap estimator is closely connected to the concept of a CD-r.v. In our earlier discussion of Example 6.1, the CD-r.v. $\theta^*_{CD}|\bar{Y} = \bar{y} \sim N(\bar{Y}, 1/n)$ and thus $(\theta^*_{CD} - \bar{Y})\big|\bar{Y} = \bar{y} \overset{m}{\sim} (\bar{Y} - \theta)\big|\theta = \theta_0$. Here, however, the matching is exact as in (6.3) and $\mathbb{P}_A(\cdot|data)$ is defined with respect to the Monte-Carlo randomness of simulating from a CD, rather than a resampling procedure.

In general, just as the matching in (6.5) ensures the validity of bootstrap inference, a similar argument underscores the inferential validity of CD-based inference. That is

$$(\theta^*_{CD} - \hat{\theta})\big|data \overset{m}{\sim} (\hat{\theta} - \theta)\big|\theta = \theta_0, \tag{6.6}$$

where $\hat{\theta}$ is a point estimator (often the MLE) for θ (e.g., Xie and Singh, 2013). Again, the matchingness of equation (6.6) can be exact or approximate (holding as $n \to \infty$), depending on whether the CD is itself exact or asymptotic. From this, one concludes that θ^*_{CD} essentially performs in the same manner as the bootstrap estimator, θ^*_{BT}. Their common trait is that the simulated artificial variability of the random estimator, θ^*_{CD} or θ^*_{BT}, given an observed sample, matches the sampling variability of the estimator, $\hat{\theta}$, as a function of a random sample from the population.

In contrast to the bootstrap procedure, a CD-r.v. can be obtained through many different simulation schemes besides the multinomial re-sampling mechanism. As discussed earlier in this chapter, a CD-r.v. can be simulated from a CD that itself is obtained through the normalized likelihood method or a p-value function. A CD may also be obtained from a pivotal statistic or through Bayesian or fiducial procedures, as discussed next. Therefore, although the bootstrap estimator performs similarly to a CD-r.v., a CD-r.v. is a more broadly defined concept.

6.4.1.2 Fiducial: Generalized Fiducial Inference and Fiducial Sampling

In modern statistical practice, R.A. Fisher's fiducial method is understood as *an inversion method* that solves a structural model (or algorithmic model) for some parameter θ (Brenner et al., 1983; Fraser, 1966, 1968; Hannig et al., 2016).

Assume the data can be generated by the following algorithmic model

$$\mathbf{Y} = G(\theta, \mathbf{U}). \tag{6.7}$$

Here, the form of $G(\cdot,\cdot)$ is known, θ is the model parameter with true value θ_0 , and $\mathbf{U} \sim D(\cdot)$, where $D(\cdot)$ is also known, but $\mathbf{U}$ is an unobserved random vector of noise. In Example 6.1, the algorithmic model is $Y_i = \theta + U_i$, for $i = 1, \dots, n$, where $\mathbf{Y} = (Y_1, \dots, Y_n)'$ and $\mathbf{U} = (U_1, \dots, U_n)$ follows a standard multivariate normal distribution of dimension n.

In the classical fiducial context, model (6.7) is called a structural model, especially in the case where $\mathbf{Y}$ is replaced with a summary statistic (Fraser, 1966, 1968). In Example 6.1, for instance, a structural model may be $\bar{Y} = \theta + \bar{U}$, where $\bar{Y} = \sum Y_i/n$ and $\bar{U} = \sum U_i/n$. Fisher suggested a fiducial inversion process that rearranges this structural model so that $\theta = \bar{Y} - \bar{U}$ and claimed that this equality holds for any observed $\bar{Y} = \bar{y}$ (i.e., the equation $\theta = \bar{y} - \bar{U}$ holds). Since $\bar{U} \sim N(0, 1/n)$, the *fiducial distribution* of θ is $N(\bar{y}, 1/n)$. This fiducial distribution is identical to the CD derived in Example 6.1 and in this instance, it is an effective distribution estimator for inferential conclusions regarding the unknown θ. However, an infamous paradox of "hidden subjectivity" underlies this classical fiducial argument: the method continuously treats $\mathbf{U}$ as random while conditioning upon an observed sample, $\mathbf{Y} = \mathbf{y}$ (Dempster, 1963). As evident in equation (6.7), the data, $\mathbf{Y}$, and the random noise, $\mathbf{U}$, are completely dependent. If one is given, the other must be as well (even if it may not be observed). Therefore, if $\mathbf{Y} = \mathbf{y}$ is observed, the correct structural equation for Example 6.1 is $\theta = \bar{y} - \bar{u}$, rather than $\theta = \bar{y} - \bar{U}$ where $\bar{u}$ is an unobserved realization of $\bar{U}$ corresponding to an observed realization $\bar{y}$ from the sampling distribution of $\bar{Y}$. We refer interested readers to (Dawid, 2022) for a comprehensive review of fiducial inference in the classical fiducial inference context.

With a slight adjustment in perspective, namely by treating the fiducial method as an stochastic inversion algorithm that solves for a random estimator of θ, modern fiducial arguments avoid the paradoxical reasoning of the classical argument and treat the fiducial distribution in a manner consistent with frequentist inference.

This perspective maintains that once the data is observed, i.e. $\mathbf{Y} = \mathbf{y}$, the corresponding value of $\mathbf{U} = \mathbf{u}$ is also realized. Model (6.7) is thus replaced by

$$\mathbf{y} = G(\theta, \mathbf{u}), \tag{6.8}$$

where $\mathbf{u}$ is an unobserved realization from a known distribution, $D(\cdot)$. One may simulate an artificial copy of $\mathbf{U}$, say $\mathbf{u}^* \sim D(\cdot)$, and use this to estimate the unobserved realization, $\mathbf{u}$. A random estimator, θ^* (also known as a *fiducial sample*), for the parameter θ is derived by solving $\mathbf{y} = G(\theta^*, \mathbf{u}^*)$. For repeated draws of $\mathbf{u}^*$, we have multiple copies of θ^*. The distribution of these θ^* is called a *fiducical distribution.* In Example 6.1, $\theta^* = \bar{y} + \bar{u}^*$ with $\bar{u}^* \sim N(0, 1/n)$ and thus $\theta^*|\mathbf{y} \sim N(\bar{y}, 1/n)$; therefore, $N(\bar{y}, 1/n)$ is the fiducial distribution for θ. As it turns out, once again $(\theta^* - \bar{y})|\bar{Y} = \bar{y} \overset{m}{\sim} (\bar{Y} - \theta)|\theta = \theta_0$ and therefore the fiducial sample, θ^*, plays the same role as CD-r.v., θ^*_{CD}, and as the bootstrap estimator, $\bar{Y}^*_{BT}$. In the fiducial context, $\mathbb{P}_A(\cdot|data)$ from equation (6.3) is defined with respect to the artificial randomness of generating $\bar{u}^* \sim N(0, 1/n)$.

Although θ_0 will always solve equation (6.8) for the realized pair $(\mathbf{y}, \mathbf{u})$, replacing $\mathbf{u}$ with $\mathbf{u}^*$ introduces the possibility that a solution to $\mathbf{y} = G(\theta, \mathbf{u}^*)$ may not exist. The method of *generalized fiducial inference* (GFI) addresses this issue by introducing an optimization procedure to solve for

$$\theta^*_{FD,\epsilon} = \text{argmin}_{\theta \in \{\theta:\ ||\mathbf{y} - G(\theta, \mathbf{u}^*)||^2 \leq \epsilon\}} ||\mathbf{y} - G(\theta, \mathbf{u}^*)||^2, \tag{6.9}$$

where $\epsilon \to 0$ at a fast rate (Hannig, 2009; Hannig et al., 2016). Hannig (2009) proved a *Bernstein von Mises theorem* for the fiducial distribution that justifies the validity of the GFI approach, which states: under some general regularity conditions,

$$(\theta^*_{FD,\epsilon} - \hat{\theta}) \mid data \overset{m}{\sim} (\hat{\theta} - \theta) \mid \theta = \theta_0, \tag{6.10}$$

as $n \to \infty$ and $\epsilon \to 0$ at a fast enough rate. Here, $\hat{\theta}$ is often the MLE. The proof of this statement involves derivations showing that both sides of equation (6.10) converge to the normal distribution centered at zero with variance equal to the inverse of the expected Fisher information. The matching in (6.10) is approximate as defined in (6.4), where $\mathbb{P}_A(\cdot|data)$ is defined with respect to $\theta^*_{FD,\epsilon}$, given observed data $\mathbf{Y} = \mathbf{y}$, and hence with respect to the artificial variability in the simulation of $\mathbf{u}^* \sim D(\cdot)$.

As with bootstrap estimators and CD-r.v.s, the validity of generalized fiducial inference hinges upon a match between the simulated variability conditioned upon the observed sample and the sampling variability under the assumed statistical model. The fiducial matching equation (6.10) thus establishes an *equivalence* among the fiducial sample (θ^*_{FD}), the bootstrap estimator (θ^*_{BT}), and a CD-r.v. (θ^*_{CD}) in that each estimator plays the same role in justifying the validity of their respective inferential approaches.

6.4.1.3 Approximate Bayesian Computing and Bayesian Inference

The *approximate Bayesian computing (ABC)* method is a Bayesian algorithm that attempts to obtain a Bayesian posterior without a direct use of Bayes' formula. It can be viewed as a Bayesian stochastic inversion algorithm to solve the same model equation (6.8), where $\mathbf{y}$ is the observed data generated from the model (6.7) and θ and $\mathbf{u}$ are unknown realizations from known prior and error distributions, $\pi(\theta)$ and $D(\mathbf{u})$, respectively. The premise for establishing inferential conclusions for parameter θ in ABC is the following rejection algorithm.

Approximate Bayesian Computing (ABC) Algorithm

Step 1 Simulate $\theta^* \sim \pi(\cdot)$ and $\mathbf{u}^* \sim D(\cdot)$ to compute $\mathbf{y}^* = G(\theta^*, \mathbf{u}^*)$.

Step 2 If $\mathbf{y}^*$ is equal or approximately equal to $\mathbf{y}_{obs}$ (i.e., $\mathbf{y}^* \approx \mathbf{y}_{obs}$), retain θ^*; otherwise repeat Step 1.

Effectively, the set of retained θ^* resulting from this algorithm satisfy the equation $\mathbf{y} \approx G(\theta^*, \mathbf{u}^*)$, so this stochastic algorithm solves equation (6.8) for θ^*. This inversion bears a striking resemblance to the fiducial stochastic inversion described above, except that the Bayesian inversion method assumes a prior distribution and is limited to finding those θ^* that correspond to simulations from this prior. In contrast to the θ^*_{FD} obtained from the fiducial inversion algorithm, the θ^* in the ABC algorithm contain information about both $\pi(\cdot)$ and $D(\cdot)$.

In applications of the ABC algorithm, the approximate equality in Step 2 is difficult to achieve and is instead replaced by an approximate equality of some summary statistics while permitting a small degree of mismatch, $\epsilon > 0$. Thus, in practice, Step 2 of the ABC algorithm is replaced with Step 2' below, where $d(\cdot, \cdot)$ is a distance metric (often Euclidean distance), $t(\cdot)$ is a summary statistic:

Step 2' If $d(t(\mathbf{y}^*), t(\mathbf{y}_{obs})) < \epsilon$, retain θ^*; otherwise repeat Step 1.

Given the observed data, the underlying distribution of the output θ^* from the ABC algorithm is called the *ABC posterior*, which we denote by $f_a(\theta \mid \mathbf{y})$. If the summary statistic $t(\cdot)$ is sufficient, then the ABC-posterior (approximately) matches the target posterior distribution, $f(\theta \mid \mathbf{y})$, derived from Bayes formula, when $\epsilon \to 0$; cf, e.g., Frazier et al. (2018); Thornton (2019). In such circumstances, the ABC method is a computational approach to obtain the posterior distribution without the use of Bayes' formula or a direct evaluation of the likelihood function. Note that, when given the observed data $\mathbf{y}$, the randomness in θ^* is induced by artificial Monte-Carlo simulations of $\theta^* \sim \pi(\cdot)$ and $\mathbf{u}^* \sim D(\cdot)$.

The *Bernstein von Mises theorem* for the posterior distribution is a well-known bridge relating Bayesian and frequentist inference for large sample sizes, e.g., van der Vaart (1998). Denote by θ_0 the realized value of θ from the prior and let θ^*_{BY} be an outcome from an ABC algorithm, or, more generally, a Monte-Carlo sample draw from a posterior distribution, given the observed data $\mathbf{y}$. We can restate the Bernstein von Mises theorem in this context as follows:

Under some general regularity conditions and as $n \to \infty$,

$$(\theta^*_{BY} - \hat{\theta}) \mid data \overset{m}{\sim} (\hat{\theta} - \theta) \mid \theta = \theta_0, \tag{6.11}$$

where the above follow a normal distribution centered at zero with variance equal to the inverse of the expected Fisher information, and $\hat{\theta}$ is (often) the MLE.
The underlying probability measure, $\mathbb{P}_A(\cdot|data)$, on the left-hand side of equation (6.11) is defined with respect to the posterior probability. If θ^*_{BY} is instead a Monte-Carlo draw from the posterior distribution exactly, then $\mathbb{P}_A(\cdot|data)$ is defined with respect to the artificial randomness of the Monte-Carlo simulation only. In the ABC algorithm presented above, the randomness of $\mathbb{P}_A(\cdot|data)$ is derived from the randomness of the Monte-Carlo sampling of θ^* and $\mathbf{u}^*$ in Step 1.

ABC methods are valued as a Bayesian approach that does not require the direct specification of the likelihood function (or direct use of Bayes' formula). As such, these methods are most useful in applications where one has little to no knowledge about the sufficiency (or near sufficiency) of the summary statistic. If the summary statistic, $t(\cdot)$, is not a sufficient statistic, then the ABC posterior distribution can be quite different from the targeted posterior distribution as demonstrated in the Cauchy example discussed next. Thus, considerations for the inferential validity of conclusions based on $f_a(\theta \mid \mathbf{y})$ for insufficient $t(\cdot)$ are critical to consider for any practical use of the ABC algorithm. To this end, Thornton (2019); Thornton et al. (2023) derived some regularity conditions under which the matching equation (6.11) still holds either exactly (as in (6.3)) or asymptotically (as in (6.4)), and thus an ABC posterior, $f_a(\theta|\mathbf{y})$, is arguably a valid CD, regardless of whether or not $t(\mathbf{y})$ is sufficient. Example 6.7 explores such a case.

Example 6.7 *Suppose we observe $\mathbf{y} = (y_1, y_2, \ldots, y_n)$, an independent sample from a Cauchy$(\theta, 1)$ distribution. We would like to use the ABC algorithm to obtain ABC posteriors $f_a(\theta \mid \mathbf{y})$ using a flat prior. We consider two popular summary statistics, the sample mean, $t_1(\mathbf{y}) = \bar{y}$, and the sample median, $t_2(\mathbf{y}) = y_{(m)}$. Since, the distribution of the Cauchy sample mean follows a Cauchy$(\theta, 1)$ distribution, one can show that $f_a(\theta \mid \mathbf{y})$ is Cauchy$(\bar{y}, 1)$, as $\epsilon \to 0$ (for any given finite n) (e.g., Thornton, 2019). Thus for the ABC method based on the summary statistic $\bar{y}$, the matching equation $(\theta^*_{BY} - \bar{Y})|\mathbf{Y} = \mathbf{y} \overset{m}{\sim} (\bar{Y} - \theta)|\theta = \theta_0$ holds for finite n and as $\epsilon \to 0$. Similarly, since the distribution of the Cauchy sample median follows a $N(\theta, \pi^2/4n)$ distribution as $n \to \infty$, one can show that $f_a(\theta \mid \mathbf{y})$ is asymptotically $N(y_{(m)}, \pi^2/4n)$ as well. Once again, for the ABC method based on the summary statistic $y_{(m)}$, the matching equation $\theta^*_{BY} - Y_{(m)}|\mathbf{Y} = \mathbf{y} \overset{m}{\sim} (Y_{(m)} - \theta)|\theta = \theta_0$, holds as $n \to \infty$ and $\epsilon \to 0$. (The rate at which $\epsilon \to \infty$ is dependent on the sample size and must be faster than $O(n^{-1/2})$ (Li and Fearnhead, 2018).) In both cases, $\mathbb{P}_A(\cdot|data)$ on the left-hand sides of the matching equations represents the simulation randomness of the Monte-Carlo draws in the ABC algorithm. A particularly noteworthy feature of this example, illustrated in Figure 6.4, is that neither summary statistic, t_1 nor t_2, is sufficient and both corresponding ABC posteriors fail to converge to the actual posterior of the target parameter θ. However, because both summary statistic choices yield a distribution such that equation (6.11) holds, Thornton (2019) uses the CD definition to argue that the ABC posteriors are both CDs. Therefore, although the ABC algorithm does not lead us to the traditional Bayesian inference for either t_1 or t_2, the output of the ABC algorithm is still useful for valid (frequentist) inference.*

Despite the dramatic difference in efficiency of these two example CDs, the inferential validity of statistical conclusions from either method remains intact according to the

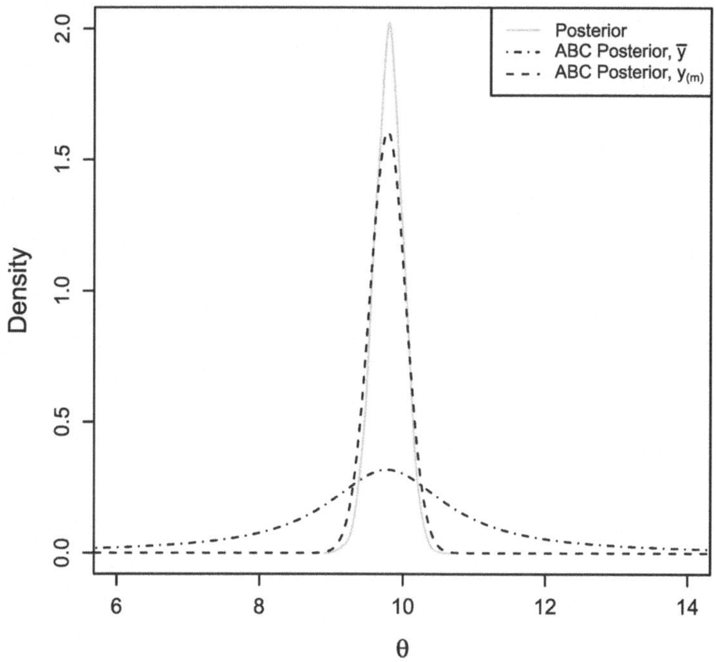

FIGURE 6.4
Given a sample of size $n = 40$ from a Cauchy$(\theta, 1)$ distribution, these are the densities for both ABC posterior distributions (in black) and for the target posterior distribution (in gray). Here we set $\theta_0 = 10$ and select a flat prior, $\pi(\theta) \propto 1$. Note that the much tighter density curve for the ABC posterior based on the sample median indicates that this method is far more efficient than the ABC posterior based on the sample mean.

TABLE 6.3
Coverage Results Over 800 Random Samples for the Two ABC Posterior Distributions and the Target Posterior Distribution in the Same Settings as Figure 6.4. The Width of the Confidence Intervals from the ABC Posterior Based on the Sample Median is Much Smaller Indicating This Method is (Far) More Efficient Than the Other Method Based on the Sample Mean

	95% Confidence or Credible Intervals	
Method	**Coverage**	**Width**
ABC posterior with $t(\mathbf{y}) = \bar{y}$	0.934	25.412
ABC posterior with $t(\mathbf{y}) = y_{(m)}$	0.944	0.974
Bayesian posterior	0.964	0.897

definition of a CD. This conclusion is demonstrated for a sample of size $n = 40$ *in Table 6.3. See Thornton (2019); Thornton et al. (2023) for more discussion on the role of equation (6.11) in establishing a CD.*

As indicated in the above Cauchy sample mean case, (6.11) can also hold without relying on the (Bayesian) Bernstein von Mises theorem. The structure of (6.11) is analogous to (6.5), (6.6) and (6.10); the only distinction is that Bayesian inference imposes an additional model assumption (the prior) and this information factors into the inferential conclusions along with the model uncertainty of $\mathbf{U}$. The common conclusion from all four of these analogous statements is that the random estimators, $\hat{\theta}^*_{BT}$, θ^*_{CD}, θ^*_{FD} and θ^*_{BY}, behave according to the same standard of frequentist inference. These random estimators are useful precisely because of this common trait: the artificial Monte-Carlo randomness incorporated into each of these estimators matches the sampling variability of $\hat{\theta}$. Thus, we have the final piece of a puzzle that connects Bayesian, frequentist and fiducial inference through the construction of random estimators (θ^*_{BY}, θ^*_{BT}, and θ^*_{FD}) that can all be described as CD-random variables, θ^*_{CD}.

Finally, to conclude this subsection, we remark that the matching equations (6.5), (6.6), (6.10), and (6.11) rely on the first-order large sample theory. The large sample Bernstein von Mises theorem for the posterior distribution mitigates the impact of the prior on inference in favor of growing data to make this alignment possible. These matching equations can be extended to higher-order results (e.g., higher-order results for the boostrap method in Hall (1992), for likelihood-based methods in Reid and Fraser (2010), and for objective Bayesian methods in Berger et al. (2015)). Nevertheless, these matching equations can also be exact as in (6.3), leading to exact inference results that are valid under a finite sample size. Beside the non-normal matching example of the Cauchy sample using $t_1 = \bar{y}$ in Example 6.7, there are some other recent works that utilize the exact matching (6.3) to make finite-sample inferential statements where neither the central limit theorem nor a high-order asymptotic development is required; e.g., Liu et al. (2018).

6.4.2 Duality of Parameters in BFF Inference

In statistics, we have traditionally distinguished inferential perspectives by asserting that the model parameter is a fixed, unknown constant in frequentist inference but is random in Bayesian inference. From this perspective, these inferential paradigms seem to conflict. Can we reconcile these different understandings of a parameter to create a meaningful bridge between the information gained from either perspective? Our answer to this question is an emphatic "yes" as we argue in the remainder of this subsection that there are two reconcilable versions of the parameter, random and fixed, in each of the Bayesian, frequentist, and fiducial inferential paradigms. We will use our understanding of the common trait of random estimators and the matching equations from the previous subsection to support this argument.

Let's start our discussion by considering *Bayes' Billiard Table Experiment*, from whence the field of statistics was developed:

> *Assume that billiard balls are rolled on a line of length one, with a uniform probability. Ball W is rolled and stops at $\theta_0 \in (0,1)$. Then a different ball O is rolled n times under the same assumptions. Let y be the number of times that ball O stops to the left of ball W. Bayes proposed the motivating question: "Given y, what inference can we make concerning θ_0?"(Bayes, 1764)*

In modern statistical terms, Bayes' Billiard Table experiment assumes a $U(0,1)$ prior distribution and collects a binomial sample of data. In this context, θ_0 (where ball W landed) is a realization from the prior distribution, $\theta \sim U(0,1)$, and the sampling scheme follows a binomial model $Y|\theta = \theta_0 \sim Bin(n, \theta_0)$. For the sake of discussion and without loss of generality, suppose ball W lands at the location $\theta_0 = 0.3896$ and suppose out of the $n = 14$ times ball O was rolled, we observe $y = 5$ instances where O landed to the left of ball W. This $y = 5$ is a realization from $Y|\theta = 0.3896 \sim Bin(n = 14, \theta = 0.3896)$. We emphasize that the target quantity we are interested is the fixed, unknown value $\theta_0 = 0.3896$, the location where ball W landed, producing the observation $y = 5$. By Bayes' formula, the posterior distribution is $\theta|y = 5 \sim Beta(6, 10)$. In this experiment, the target of interest is not a random quantity that follows either a $U(0, 1)$ or a $Beta(6, 10)$ distribution, rather it is a fixed value for parameter θ that resulted in the observation of $y = 5$.

Bayes' Billiard Table experiment clearly displays two versions of the parameter at play, which we call the *duality of a parameter*. The scientific, application-oriented point of view, is interested in the fixed version of the parameter that generated the observed data, i.e. θ_0. There is uncertainty however in exactly how to assess θ_0; to address this uncertainty, a Bayesian inferential approach elects to work with a random version of θ through the prior and posterior distributions. As stated in Berger et al. (2015), in a Bayesian perspective, "parameters must have a distribution describing the available information about their values ... this is not a description of their variability (since they are fixed unknown quantities), but a description of the uncertainty about their true values."

The duality of a parameter is actually present in other inferential paradigms as well. The model parameter has two versions: (a) a fixed, unknown true (or realized) value which generated the observed data, and (b) a random version that is used to describe the uncertainty (but not the variability) in estimating the fixed, unknown parameter value. The duality of parameters was first recognized in the fiducial argument, although the philosophical underpinnings of the fiducial interpretation have probably generated more confusion and controversy than clarity (see, e.g. Rao, 1973). From a frequentist perspective, modern developments regarding CDs, as summarized in the previous subsection, support lucid arguments for considering a random version of the model parameter in tandem with a fixed, data-generating version. In particular, a CD-r.v. θ^*_{CD} can be viewed a random version of the parameter. As demonstrated in the previous subsection, θ^*_{CD} plays the same role as the Bayesian random parameter, θ^*_{BY}, and a fiducial sample, θ^*_{FD}, and the variability in θ^*_{CD} matches the empirical uncertainty inherited from the assumed population model.

Altogether, these two versions of a parameter apply across each inferential paradigm. The fixed version of the parameter is the target of interest from which the sample of data is assumed to have been generated. It has the same standard interpretation in frequentist and fiducial inference and it refers to the realized parameter value, θ_0, from the prior under the Bayesian framework. The random version is used to describe the uncertainty in our inference about the fixed version. In Bayesian inference, this has a natural interpretation, but we argue that we can also adopt the same interpretation for the corresponding CD-r.v. (or bootstrap estimator in special cases) and fiducial sample in the frequentist and fiducial frameworks.

Whereas much previous research has highlighted differences among Bayesian, frequentist and fiducial inference, this unifying perspective can deepen our understanding of the foundational principles of statistical inference and provides a philosophical framework whereby potentially any inferential method can benefit from each of the three dominant paradigms. For instance, recognizing and establishing the random version of parameters in frequentist and fiducial inference suggests that the powerful statistical computational tools that have been so successful in many Bayesian applications can also be applied to frequentist and fiducial inference. Alternatively, introducing the interpretation of a fixed target parameter in the context of Bayesian inference can help directly and unambiguously connect the interpretation of parameters to the physical meanings of parameters in physics and other applied scientific domains. Furthermore, the CD-based connection among bootstrap sampling and approximate Bayesian computing reinforce the usefulness of developing new sampling methods to address more difficult inference problems that lie beyond the reach of likelihood-based inference and extensions of the central limit theorem (e.g., Liu et al., 2018; Luo et al., 2021).

6.5 Further Discussion and Concluding Remarks

A CD is a sample-dependent distribution function on the parameter space that can represent confidence intervals (or regions) of all levels for some parameter of interest. The CD approach summarizes the information available from the observed data in a distribution form. An emerging theme from the development of CD estimators is that any approach, regardless of whether it is Bayesian, fiducial, or frequentist, can potentially be unified under the concept of CDs as long as the method can be used to build confidence intervals (regions) of all levels, exactly or asymptotically. This union provides a platform to compare and also combine a variety of distribution estimators that may have been derived through various procedural methods. Furthermore, we have also demonstrated that each of the frequentist, fiducial, and Bayesian inferential frameworks share two important features: an ability to describe parameter uncertainty with a random version and a view that there is some fixed, unknown value as the target of interest. The random version of the parameter is associated with distribution estimators, namely, a posterior distribution, fiducial distribution or CD, across all inferential paradigms. When the model is given, the common theory critical to the success of each of these inferential frameworks aligns the variability of the random estimator (conditional upon the observed sample) with the empirical uncertainty about the fixed version of the parameter. These key similarities hint at the broad range of inference problems that may be solvable by simulating Monte-Carlo samples that mimic the observed data.

In addition to supporting inferential unity among different statistical paradigms, this CD-based perspective promises many new methodological developments that can provide novel inferential solutions in cases where traditional solutions were previously impossible. This potential for new developments includes prediction (e.g., Shen et al., 2018; Vovk et al., 2019; Xie and Zheng, 2021; Tian et al., 2021), simulation inspired novel testing and inference methods (e.g., Liu et al., 2021; Luo et al., 2021; Xie and Wang, 2022; Wang et al., 2022), and combining information from various inferential sources (e.g., Xie et al., 2011; Hannig and Xie, 2012; Clagget et al., 2014; Liu et al., 2015; Shen et al., 2020; Cunen and Hjort, 2021; Cai et al., 2021).

We remark that the two versions of randomness explored in this chapter do not have to mirror the classical discussions of aleatory and epistemic probabilities. Rather, one may connect the uncertainty in the probability statement $\mathbb{P}(\cdot)$ on the right hand side of (6.3) and (6.4), with the uncertainty of randomly sampling from an assumed population. The latter represents empirical variability, and apparently can also be explained as an aleatory probability. On the other hand, the Monte-Carlo induced randomness, described by $\mathbb{P}_A(\cdot|data)$, does not necessarily have a straightforward interpretation as an epistemic probability. Although we may argue that the Monte-Carlo randomness is intended to describe the uncertainty in knowledge, it is derived from a concrete Monte-Carlo procedure that mimics the observed sampling mechanism and does not have an obvious interpretation as "a measure of the 'degree of belief' of the individual assessing the uncertainty of a particular situation" (Ramsey, 1931, Chapter 7). The so-called "personal belief" is not under consideration in this chapter, since, when a prior is assumed, the desired matching that we promote either requires a matching prior or that the information in the prior can be "washed out" by a large sample size through the Bernstein von Mises theorem or its generalizations. When the prior is determined from historical data or knowledge, one may still attempt to model the uncertainty both in the prior and assumed model through CDs, e.g., Xie et al. (2013). In this case, the epistemic probability discussed in a typical Bayesian analysis is replaced by an empirical (aleatory) probability inherited from the prior model (historical experiments) and the observed sample. Finally, there are different views on how to interpret the probability underscored by the CD. For instance, Schweder and Hjort (2016, Chapter 15) take the position that the probability described by a CD is epistemic, which differs from our perspective. Our perspective is more aligned with that offered in Reid and Cox (2015, Section 2.2): "We may avoid the need for a different version of probability by appeal to a notion of calibration, measured by the behaviour of a procedure under hypothetical repetition." More discussions on this aspect and other principles of statistical inference can be found in Reid and Cox (2015).

Although this chapter emphasizes the unity among the Bayesian, frequentist and fiducial inferences, these procedures are distinct in several respects. A Bayesian method includes an additional model assumption through the prior distribution. Depending on the situation, this additional assumption may result in inferential conclusions that do not match analogous frequentist conclusions, especially in the case of a finite sample size or higher order asymptotic developments, e.g., Fraser (2011); Reid and Cox (2015). Furthermore, the conclusions reached through a fiducial inversion procedure or through a Bayesian approach do not automatically (even if they do typically) lead to the correct frequentist inference with respect to a CD. This issue is similar to the fact that an MLE is usually but not automatically a consistent estimator and also that a consistent estimator does not need to be an MLE. (See Table 6.1.) Nevertheless, the differences among Bayesian, frequentist and fiducial procedures do not negate underlying congruencies among these methods namely, a practical duality in the interpretation of model parameters and an inferential connection through confidence distributions.

Bibliography

Angelo, C. and Ripley, B. D., 2020. *boot: Bootstrap R (S-Plus) Functions.* R package version 1.3-25.

Bayes, T., 1764. An essay towards solving a problem in the doctrine of chances. *Philosophical Transactions, Royal Society, London*, *53*, pp. 370–418 .

Berger, J. O., Bernardo, J. M. and Sun, D. 2015. Overall objective priors. *Bayesian Analysis*, *10*, pp. 189–221.

Blaker, H., 2000. Confidence curves and improved exact confidence intervals for discrete distributions. *Canadian Journal of Statistics*, *28*, pp. 783–798.

Brenner, D., Fraser, D. and Monette, G., 1983. On models and theories of inference; structural or pivotal analysis. *Statistische Hefte*, *24*, pp. 7–19.

Cai, C., Chen, R. and Xie, M., 2021. Individualized group learning. *Journal of the American Statistical Association*, *118*(541), pp. 622–638. https://doi.org/10.1080/01621459.2021.1947306.

Clagget, B., Xie, M. and Tian, L., 2014. Meta analysis with fixed, unknown, study-specific parameters. *Journal of the American Statistical Association*, *109*, pp. 1667–1671.

Cox, D. R., 1958. Some problems connected with statistical inference. *Annals of Mathematical Statistics*, *29*, pp. 357–372.

Cox, D. R., 2013. Discussion of confidence distribution, the frequentist distribution estimator of a parameter: A review. *International Statistical Review*, *81*, pp. 40–41.

Cui, Y. and Xie, M., 2021. Confidence distribution and distribution estimation for modern statistical inference. In *Springer Handbook of Engineering Statistics* (2nd ed.), ed H. Pham, pp. 575–592. London: Springer-Verlag. https://doi.org/10.1007/978-1-4471-7503-2_29.

Cunen, C. and Hjort, N. L., 2021. Combining information across diverse sources: The II-CC-FF paradigm. *Scandinavian Journal of Statistics; Danish Society for Theoretical Statistics; Finnish Statistical Society; Norwegian Statistical Association; Swedish Statistical Association*, *49*(2), pp. 625–656.

Davison, A. C. and Hinkley, D. V., 1997. *Bootstrap Methods and Their Application.* New York: Cambridge University Press.

Dawid, P., 2022. Fiducial inference then and now. In *Handbook on Bayesian, Fiducial and Frequentist (BFF) Inference*, eds J. Berger, X.-L. Meng, N. Reid and M. Xie, ch 5. New York: Chapman & Hall.

Dempster, A. P., 1963. Further examples of inconsistencies in the fiducial argument. *The Annals of Mathematical Statistics*, *34*, pp. 884–891.

DiCiccio, T. J. and Efron, B., 1996. Bootstrap confidence intervals. *Statistical Science*, *11*, pp. 189–228.

Divine, G. W., Norton, H. J., Baron, A. E. and Juarez-Colunga, E. 2018. The Wilcoxon-Mann-Whitney procedure fails as a test of medians. *The American Statistican*, *72*, pp. 278–286.

Efron, B., 1993. Bayes and likelihood calculations from confidence intervals. *Biometrika*, *80*, pp. 3–26.

Efron, B., 1998. R. A. Fisher in the 21st century. *Statistical Science*, *13*, pp. 95–122.

Efron, B. and Tibshirani, R. J., 1993. *An Introduction to the Bootstrap.* Monographs on Statistics and Applied Probability. Boca Raton: Chapman & Hall/CRC.

Ernst, M. D., 2004. Permutation methods: A basis for exact inference. *Statistical Science*, *19*, pp. 676–685.

Fisher, R., 1915. Frequency distribution of the values of the correlation coefficient in samples of an indefinitely large population. *Biometrika*, *10*, pp. 507–521.

Fishman, G., 1996. *Monte Carlo: Concepts, Algorithms, and Applications.* New York: Springer Science & Business Media.

Fraser, D., 1991. Statistical inference: Likelihood to significance. *Journal of the American Statistical Association*, *86*, pp. 258–265.

Fraser, D. A. S., 1966. Structural probability and a generalization. *Biometrika*, *53*, pp. 1–9.

Fraser, D. A. S., 1968. *The Structure of Inference.* New York: Wiley.

Fraser, D. A. S., 2011. Is bayes posterior just quick and dirty confidence? *Statistical Science*, *26*, pp. 299–316.

Fraser, D. A. S. and McDunnough, P., 1984. Further remarks on asymptotic normality of likelihood and conditional analyses. *The Canadian Journal of Statistics*, *12*, pp. 183–190.

Frazier, D. T., Martin, G. M., Robert, C. P. and Rousseau, J., 2018. Asymptotic properties of approximate Bayesian computation. *Biometrika*, *105*, pp. 593–607.

Freedman, D. A. and Bickel, P. J., 1981. Some asymptotic theory for the bootstrap. *Annals of Statistics*, *9*, pp. 1196–1217.

Glasserman, P., 2003. *Monte Carlo Methods in Financial Engineering*, Volume 53. New York: Springer Science & Business Media.

Hall, P., 1992. On the removal of skewness by transformation. *Journal of the Royal Statistical Society: Series B (Methodological)*, *54*, pp. 221–228.

Hannig, J., 2009. On generalized fiducial inference. *Statistica Sinica*, *19*, pp. 491–544.

Hannig, J., Iyer, H., Lai, R. C. S. and Lee, T. C. M., 2016. Generalized fiducial inference: A review and new results. *Journal of American Statistical Association*, *111*, pp. 1346–1361.

Hannig, J. and Xie, M., 2012. On Dempster-Shafer recombinations of confidence distributions. *Electrical Journal of Statistics*, *6*, pp. 1943–1966.

Kass, R. E., 2011. Statistical inference: The big picture. *Statistical Science*, *26*, pp. 1–9.

Li, W. and Fearnhead, P., 2018. Convergence of regression-adjusted approximate bayesian computation. *Biometrika*, *105*, pp. 301–318.

Liu, D., Liu, R. and Xie, M., 2015. Multivariate meta-analysis of heterogeneous studies using only summary statistics: Efficiency and robustness. *Journal of the American Statistical Association*, *110*, pp. 326–340.

Liu, D., Liu, R. Y. and Xie, M., 2021. Nonparametric fusion learning for multiparameters: Synthesize inferences from diverse sources using data depth and confidence distribution. *Journal of the American Statistical Association*, *117*(540), pp. 2086–2104. https://doi.org/10.1080/01621459.2021.1902817.

Liu, S., Tian, L., Lee, S. and Xie, M., 2018. Exact inference on meta-analysis with generalized fixed-effects and random-effects models. *Biostatistics & Epidemiology*, *2*, pp. 1–22.

Luo, X., Dasgupta, T., Xie, M. and Liu, R., 2021. Leveraging the fisher randomization test using confidence distributions: Inference, combination and fusion learning. *Journal of the Royal Statistical Society: Series B (Methodological)*, *83*(4), pp. 777–797. https://doi.org/10.1111/rssb.12429.

Pierce, D. A. and Bellio, R., 2017. Modern likelihood-frequentist inference. *International Statistical Review*, *85*, pp. 519–541.

Ramsey, F. P. (1931). *The Foundations of Mathematics and Other Logical Essays (reprint)*. London: Routledge and Kegan Paul Ltd.

Rao, C. (1973). *Linear Statistical Inference and Its Applications*. Probability and Mathematical Statistics. New York: Wiley.

Reid, N. and Cox, D. R., 2015. On some principles of statistical inference. *International Statistical Review* *83*, 293–308.

Reid, N. and Fraser, D. A. S., 2010. Mean log likelihood and higher-order approximations. *Biometrika* *97*, 159–170.

Robert, C. P., 2016. Approximate Bayesian computation: A survey on recent results. In *Monte Carlo and Quasi-Monte Carlo Methods*, Volume 163, eds R. Cools and D. Nuyens, pp. 185–205. Switzerland: Springer Proceedings in Mathematics and Statistics.

Schweder, T. and Hjort, N., 2016. *Confidence, Likelihood and Probability*. Cambridge: Cambridge University Press.

Schweder, T. and Hjort, N. L., 2002. Confidence and likelihood. *Scandinavian Journal of Statistics*, *29*, pp. 309–332.

Schweder, T. and Hjort, N. L., 2003. Frequentist analogues of priors and posteriors. In *Econometrics and the Philosophy of Economics: Theory-Data Confrontations in Economics*, ed Bernt P Stigum, pp. 285–317. Princeton: Princeton University Press.

Shen, J., Liu, R. Y. and Xie, M., 2018. Prediction with confidence, a general framework for predictive inference. *Journal of Statistical Planning and Inference*, *195*, pp. 126–140.

Shen, J., Liu, R. Y. and Xie, M., 2020. ifusion: Individualized fusion learning. *Journal of the American Statistical Association*, *115*, pp. 1251–1267.

Singh, K., 1981. On the asymptotic accuracy of Efron's bootstrap. *The Annals of Statistics*, *9*, pp. 1187–1195.

Singh, K., Xie, M. and Strawderman, W. E., 2005. Combining information from independent sources through confidence distributions. *The Annals of Statistics*, *33*, pp. 159–183.

Singh, K., Xie, M. and Strawderman, W. E., 2007. Confidence distribution (cd) - distribution estimator of a parameter. In *IMS Lecture Notes-Monograph Series*, Volume 54, eds Regina Liu, William Strawderman and Cun-Hui Zhang, pp. 132–150. Beachwood, OH: Institute of Mathematical Statistics.

Thornton, S., 2019. *Advanced Computing Methods for Statistical Inference*. PhD thesis, Rutgers, The State University of New Jersey.

Thornton, S., Li, W. and Xie, M., 2023. Approximate confidence distribution computing: An effective likelihood-free method with statistical guarantees. *The New England Journal of Statistics in Data Science*, 1, pp. 270–282.

Tian, Q., Meng, F., Nordman, D. and Meeker, W., 2021. Predicting the number of future events. *Journal of the American Statistical Association*, *117*(539), pp. 1296–1310. https://doi.org/10.1080/01621459.2020.1850461.

van der Vaart, A., 1998. *Asymptotic Statistics*. Statistical and Probabilistic Mathematics. Cambridge: Cambridge University Press.

Vovk, V., Shen, J., Manokhin, V. and Xie, M., 2019. Nonparametric predictive distributions by conformal prediction. *Machine Learning*, *108*, pp. 445–474.

Wang, P., Xie, M. and Zhang, L., 2022. Finite and large sample inference for model and coefficients in high dimensional linear regression with repro samples. *arXiv e-prints*, arXiv:2209.4500515.

Xie, M., Liu, R. Y., Damaraju, C., Olson, W. H. et al., 2013. Incorporating external information in analyses of clinical trials with binary outcomes. *The Annals of Applied Statistics*, *7*, pp. 342–368.

Xie, M. and Singh, K., 2013. Confidence distribution, the frequentist distribution estimator of a parameter (with discussions). *International Statistical Review*, *81*, pp. 3–39.

Xie, M., Singh, K. and Strawderman, W. E., 2011. Confidence distributions and a unifying framework for meta-analysis. *Journal of the American Statistical Association*, *106*, pp. 320–333.

Xie, M. and Wang, P., 2022, June. Repro samples method for finite- and large-sample inferences. *arXiv e-prints*, arXiv:2206.06421.

Xie, M. and Zheng, Z., 2021. Homeostasis phenomenon in conformal prediction and predictive distribution functions. *International Journal of Approximate Reasoning*, *141*, pp. 131–145. https://doi.org/10.1016/j.ijar.2021.09.001.

Yang, G., Cheng, J., Xie, M. and Qian, W., 2017. *gmeta: Meta-Analysis via a Unified Framework of Confidence Distribution.* https://CRAN.R-project.org/package=gmeta. R package version 2.3-0.

Zhang, Z., Dai, J. and Yang, Z., 2020. Randomized statistical inference: A unified statistical inference frame of frequentist, fiducial, and Bayesian inference. *Science China Mathematics*, *63*, pp. 1007–1028.

7

Objective Bayesian Testing and Model Uncertainty

James Berger
Duke University

Gonzalo García-Donato
University of Castilla-La Mancha

Elias Moreno
University of Granada

Luis Pericchi
University of Puerto Rico, Rio Piedras

7.1 Introduction

This chapter gives an introduction to Bayesian hypothesis testing and model uncertainty. The focus is on methodology that is, in principle, general and which can be automatically applied. Therefore, this falls within the domain of objective Bayesian analysis.

In Section 7.2, the key concepts of Bayesian hypothesis testing are introduced and illustrated, including the discussion of common mistakes in choosing prior distributions. A specific prior distribution (the intrinsic prior) is recommended for testing any point null hypothesis; it is a relatively simple function of the null and alternative models. Finally, testing, when there are nuisance parameters, is discussed.

Section 7.3 introduces the basics of dealing with model uncertainty. The choice of prior probabilities of models is discussed, with emphasis on the default choice that adjusts for multiple testing. Topics such as model averaging, prediction, and the median probability model are also discussed.

The key difficulty in model uncertainty is that of choosing prior distributions for model parameters, especially when there are many models involved. A brief review of the use of mixtures of g-priors and fractional priors is given in Section 7.4.

Section 7.5 focuses on the development of intrinsic priors and expected posterior priors. These are our recommended approaches to the development of automatic priors because they are extremely general and have excellent properties.

Other issues arising in model uncertainty are considered in Section 7.6, in particular, computation and search and approximation and asymptotics. While not discussed in this chapter, note that there are a number of software solutions, in the form of R packages, that

 DOI: 10.1201/9780429341731-7

make the application of Bayesian tools for testing and model uncertainty extremely easy (see [39]). In this chapter, we have used BayesVarSel ([46]), for many of the computations, and this package has been updated to include the recommendations and new methods that are provided in this chapter.

Bayesian testing and model uncertainty is a huge methodological area in statistics, and many important techniques and approaches will be mentioned herein only briefly, if at all. Other useful reviews, with many more references, include [64], [12], [30], [72], [4].

7.2 Testing a Point Null Hypothesis

Most of this section only considers the simple problem of testing a point null hypothesis, to introduce notation and key concepts. Section 7.2.4 introduces composite testing. General testing in the linear model is considered in Section 7.5.6.

7.2.1 Basics and Notation

Suppose data $\boldsymbol{x}$ is observed, arising from density $f(\boldsymbol{x} \mid \theta)$, with $\theta \in \Theta$ unknown. It is desired to test

$$H_0 : \theta = \theta_0 \quad vs \quad H_1 : \theta \neq \theta_0.$$

Bayesian analysis proceeds by choosing a prior distribution for the unknowns. Here, this consists of two parts, the prior probabilities $\Pr(H_0)$ and $\Pr(H_1)$ of the hypotheses, and a proper prior density $\pi(\theta)$ on Θ. Recommended default choices are $\Pr(H_0) = \Pr(H_1) = 0.5$, for the prior probabilities, and the *intrinsic prior*, defined in Section 7.2.3.2, for $\pi(\theta)$.

A key quantity in the Bayesian approach is the *marginal likelihood* under H_1, given by

$$m(\boldsymbol{x} \mid H_1) = \int_\Theta f(\boldsymbol{x} \mid \theta)\pi(\theta)\, d\theta.$$

(Strictly, this is an integral over $\{\theta \neq \theta_0\}$, but we will only be dealing with continuous spaces Θ, so it makes no difference.) The *Bayes factor* of H_0 to H_1 is then defined to be

$$B_{01} = \frac{f(\boldsymbol{x} \mid \theta_0)}{m(\boldsymbol{x} \mid H_1)}.$$

This is commonly interpreted as the odds of H_0 to H_1; a Bayes factor of 10 implies that H_0 is 10 times more likely to be true than H_1, while a Bayes factor of $1/100$ implies that H_1 is 100 times more likely to be true than H_0. Bayes factors have a long history, being introduced during the first decades of the past century in a series of papers by Harold Jeffreys and Dorothy Wrinch [37]. These culminated in the book [65], which is nowadays the standard original citation for Bayes factors.

The posterior distribution of the unknowns has two parts, the posterior probabilities of the hypotheses (with $\Pr(H_1 \mid \boldsymbol{x}) = 1 - \Pr(H_0 \mid \boldsymbol{x})$)

$$\Pr(H_0 \mid \boldsymbol{x}) = \frac{\Pr(H_0) f(\boldsymbol{x} \mid \theta_0)}{\Pr(H_0) f(\boldsymbol{x} \mid \theta_0) + \Pr(H_1) m(\boldsymbol{x} \mid H_1)} = \left[1 + \frac{\Pr(H_1)}{\Pr(H_0)} \cdot \frac{1}{B_{01}}\right]^{-1}.$$

and the posterior density on $\theta \neq \theta_0$, $\pi(\theta \mid \boldsymbol{x}) = f(\boldsymbol{x} \mid \theta)\pi(\theta)/m(\boldsymbol{x} \mid H_1)$. A useful relation is

$$\underset{\text{(posterior odds)}}{\frac{\Pr(H_0|\boldsymbol{x})}{\Pr(H_1|\boldsymbol{x})}} \quad = \quad \underset{\text{(prior odds)}}{\frac{\Pr(H_0)}{\Pr(H_1)}} \quad \times \quad \underset{\text{(Bayes factor)},}{B_{01}.}$$

since this clarifies that the prior odds of the hypotheses and the Bayes factor have separate influences on the posterior odds.

Example 7.1 *Normal testing:* Suppose $\{x_i, i = 1, 2, \ldots, n\}$ are i.i.d. $N(x_i \mid \theta, \sigma^2)$, i.e. are normally distributed with unknown mean θ and known variance σ^2, and it is desired to test $H_0 : \theta = \theta_0$ versus $H_1 : \theta \neq \theta_0$. One can reduce to the sufficient statistic $\bar{x} \sim N(\bar{x} \mid \theta,\, \sigma^2/n)$.

The intrinsic prior on H_1 will later be seen to be $\pi(\theta) = N(\theta \mid \theta_0, 2\sigma^2)$ leading to the marginal likelihood $m(\bar{x} \mid H_1)) = N(\bar{x} \mid \theta_0, 2\sigma^2 + \sigma^2/n) = N(\bar{x} \mid \theta_0, \sigma^2(2 + 1/n)$ (for a convolution of two normal distributions, the variances just add) and posterior probability (for the default choice $\Pr(H_0) = \Pr(H_1) = 0.5$)

$$\begin{aligned} \Pr(H_0 \mid \bar{x}) &= \left[1 + \frac{\frac{1}{\sqrt{2\pi}\sigma(2+1/n)^{1/2}} \exp\left\{-\frac{1}{2}\frac{1}{\sigma^2(2+1/n)}(\bar{x}-\theta_0)^2\right\}}{\frac{1}{\sqrt{2\pi}\sigma/\sqrt{n}} \exp\left\{-\frac{1}{2}\frac{1}{\sigma^2/n}(\bar{x}-\theta_0)^2\right\}}\right]^{-1} \\ &= \left[1 + \frac{\exp\{\frac{1}{2}\, z^2\, [1 + \frac{1}{2n}]^{-1}\}}{\sqrt{1+2n}}\right]^{-1}, \end{aligned}$$

where $z = \sqrt{n}(\bar{x} - \theta_0)/\sigma$ is the usual (frequentist) test statistic for this problem. For various values of z and n, this posterior probability is given in Table 7.1. Also given is the p-value corresponding to z (which does not depend on n). Because the p-value is often misinterpreted as the error probability in rejecting H_0, its vast discrepancy with the objective posterior probability is a primary motivation for considering the Bayesian approach.

There are interesting issues regarding the interface of the Bayesian and frequentist approaches to testing; see [13] for a discussion and review. Note that Bayesian tests are essentially always admissible (cannot be beaten) from a frequentist perspective, and one can examine their frequentist properties, if desired. For instance, Figure 7.1 (left panel) presents, for the normal testing example, the probability of rejecting H_0 as a function of the true value of θ (i.e., the *power* function) for several values of n, assuming that H_0 is rejected when $Pr(H_0 \mid \bar{x}) < 0.5$. Many comparisons of Bayesian and frequentist methods, in terms of overall performance, have also been made; see [42] for a recent illustration in which the Bayesian approach is shown to have excellent comparative performance.

TABLE 7.1
For Point Null Testing of a Normal Mean (Known Variance), the Posterior Probabilities of H_0 and the p-values are Given, as a Function of the Test Statistic z and for Different Values of n

		$Pr(H_0 \mid z)$		
z	p-value	$n = 5$	$n = 20$	$n = 100$
1.645	0.1	0.49	0.63	0.79
1.960	0.05	0.37	0.50	0.68
2.576	0.01	0.14	0.20	0.35

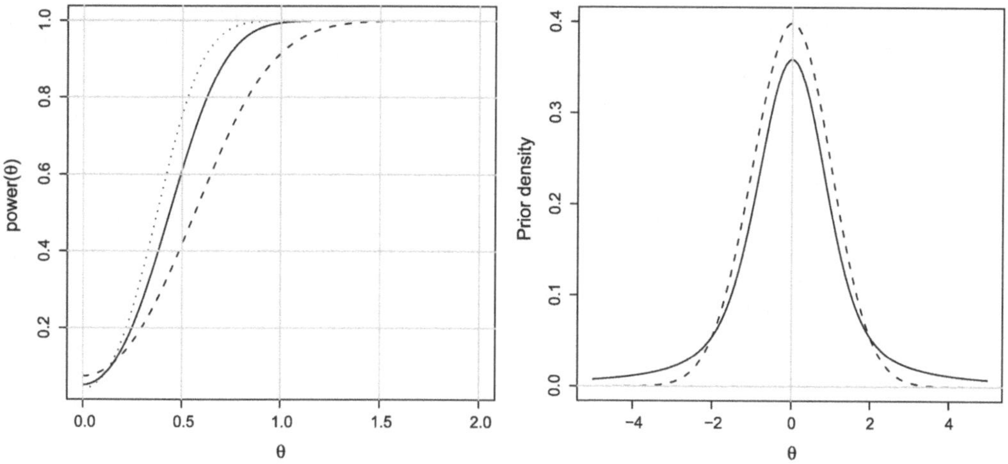

FIGURE 7.1
Testing a normal mean $H_0 : \theta = 0$. Left (known σ): power function for $n = 10$ (dashed), $n = 20$ (solid) and $n = 30$ (points). Right (unknown σ): graphs of the conditional prior $\pi(\theta \mid \sigma = 1)$ for the intrinsic prior (solid line) and conjugate normal prior (dashed line).

7.2.2 Formulating a Problem as a Precise Hypothesis Test

It is only sensible to conduct a hypothesis test if both hypotheses are 'plausible,' i.e., they must have reasonable prior probability of being true. To better understand this, consider the example of comparing the effects of two treatments A and B. Letting θ denote the difference between the mean treatment effects, one commonly sees tests of $H_0 : \theta = 0$ versus $H_1 : \theta \neq 0$. The key question is then whether H_0 is a plausible hypothesis. Consider, for instance, the following two possible scenarios involving treatment for cancer:

Scenario 1: Treatment A = chemotherapy; Treatment B = chemotherapy + steroids.

Scenario 2: Treatment A = chemotherapy; Treatment B = a new radiation therapy.

In Scenario 1, $H_0 : \theta = 0$ is plausible, since the effect of steroids could well be essentially negligible. In Scenario 2, however, $H_0 : \theta = 0$ is *not* plausible; it would be a bizarre coincidence if the new radiation therapy would have essentially the same effect as standard chemotherapy (assuming we can rule out both being completely ineffective or completely effective). It would thus be much more appropriate to test $H_0 : \theta < 0$ versus $H_1 : \theta > 0$.

Note that this is not an issue of one-sided versus two-sided testing. In Scenario 1, for instance, one could test $H_0 : \theta = 0$ versus $H_1 : \theta < 0$ versus $H_2 : \theta > 0$, which involves one-sided testing. The key issue is, rather, whether or not $\theta = 0$ is a point that has positive probability of being true. If so $H_0 : \theta = 0$ must be part of the formulation of the problem, it is not simply a modeling 'choice.'

Of course, a precise null hypothesis, such as $H_0 : \theta = \theta_0$, is typically never true *exactly*; rather, the point null is used as a surrogate for an approximate precise hypothesis

$$H_0^{\epsilon} : |\theta - \theta_0| < \epsilon, \quad \epsilon \text{ small.}$$

But it is typically the case that one obtains essentially the same answer whether using H_0 or H_0^{ϵ}. Indeed, in [6], it was shown that $Pr(H_0^{\epsilon}|\boldsymbol{x}) \approx Pr(H_0|\boldsymbol{x})$ if $\epsilon < 1/[4\sigma_{\hat{\theta}}]$, where $\sigma_{\hat{\theta}}$ is the standard error of the estimate of θ. Thus, one simply needs to assess that ϵ satisfies this inequality to utilize the precise null. Note that, often, $\sigma_{\hat{\theta}} \approx c/\sqrt{n}$ so, for very large n, the aforementioned condition can be violated and using a precise null may not be appropriate. For additional insights into the correspondence between H_0 and H_0^ϵ, see [73].

7.2.3 Choice of the Prior $\pi(\theta)$

We first discuss two commonly used choices of $\pi(\theta)$ that are suboptimal. Then we present our recommended prior, the intrinsic prior.

7.2.3.1 Erroneous Use of Improper or Vague Proper Priors

When testing a precise hypothesis, one cannot generally employ usual objective Bayes estimation priors.

Example 7.2 *Normal testing continued:* In estimation of the mean θ, it is perfectly reasonable to use the improper objective estimation prior $\pi(\theta) = c$, where c is any constant, since the posterior density of θ is then $\pi(\theta \mid x) = f(x \mid \theta)c/\int f(x \mid \theta)cd\theta = f(x \mid \theta)$ (since $\int f(x \mid \theta)d\theta = 1$) and this does not depend on the choice of the constant c. If, however, $\pi(\theta) = c$ were used for testing $H_0 : \theta = \theta_0$ versus $H_1 : \theta \neq \theta_0$, the Bayes factor would be

$$B_{01}(c) = \frac{f(x \mid \theta_0)}{\int_{-\infty}^{\infty} f(x \mid \theta)(c)d\theta} = \frac{f(x \mid \theta_0)}{c},$$

depending on the arbitrary choice of c.

Even worse here would be use of *vague proper priors*, such as a $Uniform(-K, K)$ prior for θ, with K chosen very large; many Bayesians erroneously view use of vague proper priors to be superior to the use of improper priors. But, for this prior in the normal example, the Bayes factor is

$$B_{01}(K) = \frac{f(x \mid \theta_0)}{\int_{-K}^{K} f(x \mid \theta)(2K)^{-1}d\theta} \approx \frac{2K\, f(x \mid 0)}{\int_{-\infty}^{\infty} f(x \mid \theta)d\theta} = 2K\, f(x \mid \theta_0),$$

which depends dramatically and inappropriately on the arbitrary choice of K and results in a terrible answer if K is large. This issue with vague proper priors is often referred to as Bartlett's Paradox (from [3])

7.2.3.2 The Intrinsic Prior for Testing a Precise Hypothesis

The prior we recommend for precise hypothesis testing is the intrinsic prior, formally derived in Section 7.5.2. Here we simply present the algorithm to compute the prior. The parameter could be multi-dimensional, so consider the test of $H_0 : \boldsymbol{\theta} = \boldsymbol{\theta}_0$ versus $H_1 : \boldsymbol{\theta} \neq \boldsymbol{\theta}_0$, where $\boldsymbol{\theta}$ is a k-dimensional unknown parameter.

Step 1. Choose a good estimation objective prior $\pi^O(\boldsymbol{\theta})$ (using $\pi^O(\boldsymbol{\theta}) = 1$ will almost always work fine) and a *minimal training sample* $\boldsymbol{x}^*$, which is the smallest sized sample for which

$$0 < m^O(\boldsymbol{x}^* \mid H_1) = \int f(\boldsymbol{x}^* \mid \theta)\pi^O(\theta)\, d\theta < \infty.$$

Usually, the minimal training sample size is k, the dimension of $\boldsymbol{\theta}$.

Step 2. Noting that the posterior distribution is $\pi^O(\boldsymbol{\theta} \mid \boldsymbol{x}^*) = f(\boldsymbol{x}^* \mid \boldsymbol{\theta})\pi^O(\boldsymbol{\theta})/m^O(\boldsymbol{x}^* \mid H_1)$, the intrinsic prior (which will be proper) is given by

$$\pi^I(\boldsymbol{\theta}) = \int \pi^O(\boldsymbol{\theta} \mid \boldsymbol{x}^*) f(\boldsymbol{x}^* \mid \boldsymbol{\theta}_0)\, d\boldsymbol{x}^*.$$

$\pi^I(\boldsymbol{\theta})$ is often available in closed form but, even if not, computation of the Bayes factor is often a straightforward numerical exercise.

Example 7.3 *Normal testing continued:* Here θ is one dimensional, so a single observation is a minimal training sample using the objective estimation prior $\pi^O(\theta) = 1$, since $m^O(\theta \mid x) = \int N(x \mid \theta, \sigma^2)(1)\, d\theta = 1 < \infty$. Also, $\pi^O(\theta \mid x) = N(\theta \mid x, \sigma^2)$, so that

$$\pi^I(\theta) = \int \pi^O(\theta \mid x) f(x \mid \theta_0)\, dx = \int N(\theta \mid x, \sigma^2) N(x \mid \theta_0, \sigma^2)\, dx = N(\theta \mid \theta_0, 2\sigma^2).$$

7.2.4 Precise Hypothesis Testing with Common Nuisance Parameters

Consider the normal testing example, but now assume that σ^2 is unknown. One is now testing $H_0 : \{\theta = \theta_0,\ \sigma^2 > 0\}$ versus $H_1 : \{\theta \neq \theta_0,\ \sigma^2 > 0\}$ and σ^2 is called a common nuisance parameter for the two hypotheses. This is still considered a precise hypothesis test because H_0 has a lower dimensional parameter space than H_1.

7.2.4.1 Use of Objective Priors for the Common Nuisance Parameters

Luckily, a common nuisance parameter does not greatly complicate the analysis because improper objective priors can be used for the common parameter. In the normal testing example with unknown variance, the typical objective prior for the variance is $1/\sigma^2$, so one can restrict attention to priors, under H_0 and H_1, respectively, of the form

$$\pi_0(\sigma^2) = \frac{1}{\sigma^2}, \qquad \pi_1(\theta, \sigma^2) = \pi_1(\theta \mid \sigma^2)\frac{1}{\sigma^2}.$$

Attempting to justify this by claiming that σ^2 is the same parameter in each problem is incorrect; strictly we should be writing σ_0^2 and σ_1^2 for the different variances in each model. A rather crude justification is to note that if we had, instead, used the common prior c/σ^2 in each model, then the constants c would cancel in the Bayes factor, as opposed to what happened in the normal problem with known variance. The best justification arises from invariance theory and notions of predictive matching; see [11] and [4] for discussion.

7.2.4.2 Choice of $\pi_1(\theta \mid \sigma^2)$

When available, it is common to choose $\pi_1(\theta \mid \sigma^2)$ to be a 'conjugate' prior to simplify computation. In the normal example with unknown variance for instance the usual conjugate prior choice is $\pi_1(\theta \mid \sigma^2) = N(\theta \mid \theta_0, \sigma^2)$. The Bayes factor of H_1 to H_0 for this choice (and $1/\sigma^2$ as the priors for the variances) can be computed to be

$$B_{10} = \sqrt{n+1}\left(\frac{1 + t^2/(n+1)}{1+t^2}\right)^{n/2},$$

where $t = \sqrt{n}(\bar{x} - \theta_0)/s$ is the usual t-statistic. Notice that B_{10} has the following strange behavior (called *information inconsistency* and studied in [12], [72], [4], and [79]): when $|t|$ is extremely large, $B_{10} \simeq (n+1)^{-(n-1)/2} > 0$. For instance,

$$\text{if } n = 3, \quad B_{10} \simeq 1/4; \qquad \text{if } n = 5, \quad B_{10} \simeq 1/36.$$

This is counter to common intuition that a huge value of t is compelling evidence against H_0, and is enough of a concern that we do not recommend conjugate priors.

Instead, we recommend using the intrinsic prior here. Testing a normal mean with unknown variance is a particular case of model selection in normal linear models, the problem considered in Section 7.5.2.2. Indeed, the intrinsic prior for the testing problem is a special case of Equation 7.5[1], in that section, and is

$$\pi_1^I(\theta,\sigma^2) = \pi_1^I(\theta \mid \sigma^2)\ \frac{1}{\sigma^2} = \frac{1}{\pi}\int_1^\infty N(\theta \mid 0, \frac{g}{2}\sigma^2)\frac{1}{g\sqrt{g-1}}\ dg\ \frac{1}{\sigma^2}.$$

The graph of $\pi_1^I(\theta \mid \sigma^2)$ and the conjugate $N(\theta \mid \theta_0, \sigma^2)$ prior are given in Figure 7.1 (right panel); the intrinsic prior has a much flatter tail; indeed it has no moments, as recommended by [65]. It can be shown that this results in an information-consistent Bayes factor, whereas the conjugate prior does not. An expression for this Bayes factor, following from (7.6) in Section 7.5.2.2, is

$$B_{10} = \frac{1}{\pi}\int_1^\infty \left(1+\frac{gn^2}{2(n+t^2)}\right)^{-n/2}\left(1+\frac{gn}{2}\right)^{(n-1)/2}\frac{1}{g\sqrt{g-1}}\ dg,$$

which is the Bayesian counterpart of the t-test.

The first illustration of a t-test was in [102]; Student considered the problem of testing whether a certain hypnotic treatment (labelled *Laevo*) produces an effect on hours of sleep. The sample consists of the difference in sleep hours in 10 patients (with and without treatment) resulting in $t = 3.68$ ($\bar{x} = 2.33$, $s = 1.9$). The posterior probability of a treatment effect (equal prior probabilities of hypotheses) can be computed to be 0.93 ($B_{10} = 12.8$), a quite large probability of a treatment effect. (The p-value here was 0.005, which is often erroneously viewed as being overwhelming evidence of a treatment effect.)

7.3 Bayesian Model Uncertainty

The Bayesian logic in dealing with model uncertainty is the same as the logic in dealing with hypothesis testing. However, there can be so many models – in variable selection with k possible variables for, instance, there are 2^k models – that the methodology and interpretation necessarily become more involved.

7.3.1 Notation for Model Uncertainty

Assume that data, $\boldsymbol{x}$, can arise from one of q possible models $M_i, i = 1, ..., q$. Under M_i,

$$M_i\text{: } \boldsymbol{x} \text{ has density } f_i(\boldsymbol{x} \mid \boldsymbol{\theta}_i),$$

where $\boldsymbol{\theta}_i$ is the k_i-vector of unknown parameters in M_i. The prior distribution now consists of the model probabilities, $\Pr(M_i)$, as well as the prior densities of each of the parameters $\boldsymbol{\theta}_i$, to be denoted $\pi_i(\boldsymbol{\theta}_i)$. Because of the potentially large number of models involved, the specification of these quantities must be automated. Choice of the model prior probabilities is discussed in the next section. Discussion of choice of the $\pi_i(\boldsymbol{\theta}_i)$ is delayed until Sections 7.4 and 7.5.

Once the prior distributions have been chosen, the analysis proceeds by determining the marginal density of $\boldsymbol{x}$ under each model $M_i, i = 1, ..., q$, namely

$$m_i(\boldsymbol{x}) = \int f_i(\boldsymbol{x} \mid \boldsymbol{\theta}_i)\pi_i(\boldsymbol{\theta}_i)\, d\boldsymbol{\theta}_i.$$

The posterior probability of each of the models is

$$\Pr(M_i \mid \boldsymbol{x}) = \frac{\Pr(M_i)\, m_i(\boldsymbol{x})}{\sum_{j=1}^q \Pr(M_j)\, m_j(\boldsymbol{x})}.$$

The Bayes factor of model M_j to model M_i is

$$B_{ji} = \frac{m_j(\boldsymbol{x})}{m_i(\boldsymbol{x})}\,;$$

this is typically interpreted as the odds provided by the data for the correctness of model M_j to that of M_i.

It is often convenient to choose a base model M_1 (usually the null model, if defined), and then report all the Bayes factors B_{j1}. It is then possible to reconstruct all the posterior model probabilities as

$$\Pr(M_i \mid \boldsymbol{x}) = \frac{\Pr(M_i)B_{i1}}{\sum_{j=1}^{q}\Pr(M_j)B_{j1}}.$$

Prediction also utilizes the posterior densities of the $\boldsymbol{\theta}_i$ under model M_i, given by

$$\pi_i(\boldsymbol{\theta}_i \mid \boldsymbol{x}) = \frac{f_i(\boldsymbol{x}|\boldsymbol{\theta}_i)\pi_i(\boldsymbol{\theta}_i)}{m_i(\boldsymbol{x})}.$$

7.3.2 Assigning Model Prior Probabilities

A common choice of the model prior probabilities is equal probabilities, $\Pr(M_i) = 1/q$, but this is often inappropriate, from several perspectives. Consider variable selection, for instance, in which there are variables (e.g. regression coefficients) $\theta_1, \ldots, \theta_k$ that could be included in the model, or could be excluded. Giving equal prior probability to all models can be shown to be equivalent to saying that each variable is in or out of the model with probability 0.5, independently of the others. This produces a pernicious effect, which is particularly severe if k is large. For instance, suppose one is testing 1,000,000 genes to detect gene expression related to some disease. The prior assignment $\Pr(M_i) = 1/q$ can be shown to imply that there is an extremely large prior probability that the number of influential genes is around 500,000 (e.g., the models, for which the number of positive genes is in the interval [499,000; 501,000], have probability of about 0.95).

The common solution to this problem is to state that each variable will *conditionally* independently be in the model with some unknown probability p – called the *prior inclusion probability* – and then assign p a prior distribution. If p is given the uniform distribution, the prior probability of a model M_i that has k_i variables (with k total number of variables) is

$$\Pr(M_i) = \int_0^1 p^{k_i}(1-p)^{k-k_i}dp = \text{Beta}(1+k_i, 1+k-k_i) = \frac{k_i!\,(k-k_i)!}{(k+1)!}\,,$$

which has the attractive feature of being the probability assignment recommended by [65]. The argument therein was to give each potential model size an equal probability of $1/(k+1)$, and then divide this probability up among all models of a given size, the result equaling the expression above.

García-Donato and Steel [45] (a study of gene association involving $k = 4088$ genes) illustrated the dramatic difference in the use of the two priors. With the Jeffreys choice, only one gene is clearly positive (posterior inclusion probability of 0.97), while the great majority are clearly ruled out (97% have inclusion probability below 0.005). With the constant prior, the posterior distribution tends to highly mimic the prior, suggesting that about half of the genes may be important (with most of the inclusion probabilities hovering around 0.05). For other discussions of the choice of prior probabilities of models, see [52, 51], [26], [99], [89], [94, 95], and [71].

7.3.3 Model Averaging

Selecting a single model and using only it do draw statistical conclusions ignores model uncertainty; this can result in biased inferences and considerable overestimation of accuracy. The Bayesian approach to model uncertainty is *model averaging*, viewing the uncertain model as another unknown with the uncertainty represented by the model posterior probabilities.

7.3.3.1 Basics of Model Averaging

Assume inference is desired concerning ξ, a generic quantity such as prediction of a future observation, the overall mean response, or the overall probability that a variable should be in models. In model averaging, this would be based on

$$\pi(\xi \mid \boldsymbol{x}) = \sum_{i=1}^{q} \Pr(M_i \mid \boldsymbol{x})\, \pi(\xi \mid \boldsymbol{x}\,, M_i)\,,$$

where $\pi(\xi \mid \boldsymbol{x}\,, M_i)$ is the posterior distribution of ξ under model M_i alone; the reason for the name 'model averaging' is apparent.

When there are thousands or millions of models (as is common in, say, variable selection), one must study the overall features of the posterior distribution, and model averaging is, again, the vehicle for doing so. For instance, a key quantity in variable selection is the probability with which a parameter, θ_i, is in or out of the models. In regression, for example, θ_i would be the coefficient of a covariate, and its presence or absence from models reflects whether or not the covariate affects the response variable. In graphical models, θ_i could be related to a link in the graph, its presence or absence indicating whether that link is present in the graph. In such situations, key model-averaged quantities are the *posterior inclusion probabilities*

$$q_i = \Pr(\theta_i \in \text{model} \mid \boldsymbol{x}) = \sum_{j:\theta_i \in M_j} \Pr(M_j \mid \boldsymbol{x})\,, \tag{7.1}$$

i.e., the posterior probability that the ith feature of interest is in the model.

Example 7.4 *Hald regression example.* This is a classical data set often used for illustration (see [18] for references). The sample size of the data $\boldsymbol{y} = (y_1, y_2, \ldots y_{13})$ is $n = 13$ and there are four possible regressors: X_1, X_2, X_3, X_4. The full model for an observation is

$$y = \theta_0 + \theta_1 X_1 + \theta_2 X_2 + \theta_3 X_3 + \theta_4 X_4 + \epsilon, \quad \epsilon \sim N(\epsilon \mid 0, \sigma^2)\,,$$

with σ^2 unknown. Under consideration are all of the models that can be defined with subsets of regressors, with the intercept being present in all models. For example,

$$\{1, 3, 4\} \quad \text{denotes the model} \quad Y = \theta_0 + \theta_1 X_1 + \theta_3 X_3 + \theta_4 X_4 + \epsilon\,.$$

Table 7.2 from [2] (to which the reader is referred for the details of the analysis) reports the model posterior probabilities (and risks, $\Delta R(M_i)$, that will be discussed later.)

The posterior inclusion probabilities are

$$\begin{aligned} q_1 &= \textstyle\sum_{j:\theta_1 \in M_j} \Pr(M_j \mid \boldsymbol{y}) = 0.95, \quad q_2 = \sum_{j:\theta_2 \in M_j} \Pr(M_j \mid \boldsymbol{y}) = 0.73\,, \\ q_3 &= \textstyle\sum_{j:\theta_3 \in M_j} \Pr(M_j \mid \boldsymbol{y}) = 0.43, \quad q_4 = \sum_{j:\theta_4 \in M_j} \Pr(M_j \mid \boldsymbol{y}) = 0.55. \end{aligned} \tag{7.2}$$

Thus X_1 is the most important regressor, with lesser evidence of the importance of the other X_i as regressors.

Among the many references to model averaging are [48], [35], [89], and [25]. An excellent review of model averaging techniques in the context of economics is [101].

TABLE 7.2
Posterior Model Probabilities and Corresponding Excess Predictive Risks for the Hald Regression Example

Model	$\Pr(M_i \mid y)$	$\Delta R(M_i)$	Model	$\Pr(M_i \mid y)$	$\Delta R(M_i)$
null	0.000003	2652.44	{2,3}	0.000229	353.72
{1}	0.000012	1207.04	{2,4}	0.000018	821.15
{2}	0.000026	854.85	{3,4}	0.003785	118.59
{3}	0.000002	1864.41	{1,2,3}	0.170990	1.21
{4}	0.000058	838.20	{1,2,4}	0.190720	0.18
{1,2}	0.275484	8.19	{1,3,4}	0.159959	1.71
{1,3}	0.000006	1174.14	{2,3,4}	0.041323	20.42
{1,4}	0.107798	29.73	{1,2,3,4}	0.049587	0.47

7.3.3.2 Prediction

Another common role for model averaging is in prediction. Suppose, for instance, that the goal is to predict a future observation x, given the past data $\boldsymbol{x}$. The *posterior predictive distribution* of x (its posterior distribution given $\boldsymbol{x}$) is

$$m(x \mid \boldsymbol{x}) = \sum_{i=1}^{q} \Pr(M_i \mid \boldsymbol{x})\, m_i(x \mid \boldsymbol{x}, M_i)\,,$$

where

$$m_i(x \mid \boldsymbol{x}, M_i) = \int f(x \mid \theta_i)\, \pi_i(\theta_i \mid \boldsymbol{x})\, d\theta_i$$

is the posterior predictive distribution when M_i is true. Thus the usual predictor of x would be the posterior predictive mean, which would be $\hat{x} = \sum_{i=1}^{q} \Pr(M_i \mid \boldsymbol{x})\, \hat{x}_i$, where $\hat{x}_i$ is the posterior predictive mean arising from model M_i. This is the optimal predictor of x under squared error loss (and assuming the Bayesian setup is correct).

7.3.3.3 The Median Probability Model

Computational or other considerations sometimes require the choice of a single model to be used for prediction. Standard practice is to choose the maximum posterior probability model (resulting in 'MAP' estimates), but this is not generally optimal. For instance, in variable selection in linear models, the optimal single predictive model is often the *median probability model*, defined as the model consisting of those parameters whose posterior inclusion probability (7.1) is at least 1/2.

In the Hald regression example for instance, it follows from (7.2) that the median probability model is $\{1, 2, 4\}$, since the corresponding parameters have probability greater than 1/2 of being in the model. Note from Table 7.2 that this is not the maximum posterior probability model, which is $\{1, 2\}$. (It should be mentioned, however, that a later re-analysis of this example using the intrinsic prior found that the median probability model coincided with the maximum posterior probability model, both being $\{1, 2\}$).

Conditions under which the median probability model is guaranteed to yield better predictive performance than the maximum probability model (unless they coincide) are given in [2]. Even when the conditions are not satisfied, however, the median probability model is often optimal. The Hald regression example is not a situation which satisfies the conditions in [2], yet the predictive risk of $\{1, 2, 4\}$ is smaller than the predictive risk of any other model. Indeed, Table 7.2 gives $\Delta R(M_i) = E[(y - \hat{y}_i)^2] - E[(y - \hat{y})^2]$, the difference between the expected squared error loss of the single model (best) predictor, $\hat{y}_i$, and the expected squared error loss of the optimal overall predictor, which is the model-averaged predictor. Note that $\{1, 2, 4\}$ does almost as well in this example as the optimal model-averaged predictor.

7.4 Background in Choice of the $\pi(\boldsymbol{\theta}_i)$

There is a vast literature on choosing the model parameter priors; some reviews can be found in [12], [54, 55], [72], and [4]. Many of the same considerations that were discussed in the section on hypothesis testing apply here. For instance, improper priors cannot be used, unless they are for nuisance parameters that are common to *all* models. As an example, in choice among linear models, it is common to include the intercept term in all models (i.e., to not consider the null model of zero mean effect). Also, conjugate priors can quite generally suffer from information inconsistency ([79]).

As mentioned in the introduction, the primary purpose of this chapter is simply to present, in complete detail, the prior choices we recommend. It is useful, however, to mention a more general class of priors for linear models, since the priors we recommend for linear models are in this class.

7.4.1 Mixture g-priors in the Linear Model

Most analyses of model uncertainty in the linear model utilize g-priors ([107]) or Zellner-Siow priors ([108])). These and more modern alternatives can all be derived as mixtures of g-priors, the subject of this section. In Section 7.5.2.2 we will present the prior we recommend, the intrinsic prior for linear models, and will use its g-prior mixture representation.

The full linear model for data $\boldsymbol{y} = (y_1, \ldots, y_n)'$ is given by

$$\boldsymbol{y} = \boldsymbol{X}\boldsymbol{\theta} + \boldsymbol{\varepsilon}, \tag{7.3}$$

where $\boldsymbol{X}$ $(n \times k)$ is the matrix of covariates (full rank), $\boldsymbol{\theta} = (\theta_1, \ldots, \theta_k)'$ is an unknown vector of regression coefficients, and $\boldsymbol{\varepsilon}$ is $N(\boldsymbol{\varepsilon} \mid \mathbf{0}, \sigma^2 \boldsymbol{I})$. Typically, $\boldsymbol{X} = (\mathbf{1}\ \boldsymbol{X}^*)$, where $\mathbf{1}$ is the column vector of ones and θ_1 is the corresponding unknown mean level (intercept). For simplicity in this section, we make the common assumption that $\mathbf{1}^t \boldsymbol{X} = \mathbf{0}$, which results from 'centering' the columns of $\boldsymbol{X}$ by their means.

The models under consideration are the submodels from (7.3) that arise by deleting components of $\boldsymbol{\theta}^* = (\theta_2, \ldots, \theta_k)$ and corresponding columns in $\boldsymbol{X}^*$. We assume that all models have the intercept. A convenient way to label submodels is to define $\boldsymbol{\gamma} = (\gamma_2, \ldots, \gamma_k)$, where $\gamma_i = 1$ if variable X_i enters the model (that is, $\theta_i \neq 0$), and $\gamma_i = 0$ otherwise ($\theta_i = 0$). $M_{\boldsymbol{\gamma}}$ will denote the corresponding linear model, with parameters $(\theta_1, \boldsymbol{\theta}_{\boldsymbol{\gamma}})'$ and covariates $(\mathbf{1}\ \boldsymbol{X}^*_{\boldsymbol{\gamma}})$. The generalization of this variable selection context to the situation where the set of potential predictors contains factors (qualitative variables) has been given full consideration in [44].

Note first that the parameters θ_1 and σ^2 are common parameters in all models. Thus, as mentioned in Section 7.2.4.1, they can be assigned the common objective prior $\pi^O(\theta_1, \sigma^2) = 1/\sigma^2$ in all models. See [84] and [4] for further discussion.

Since the $\boldsymbol{\theta}_{\boldsymbol{\gamma}}$ are not common to all models, recall that they must be proper. An interesting class of priors for each $\boldsymbol{\theta}_{\boldsymbol{\gamma}}$, given σ^2, is

$$\pi_{\boldsymbol{\gamma}}(\boldsymbol{\theta}_{\boldsymbol{\gamma}} \mid \theta_1, \sigma^2) = \int_0^\infty N_{k_{\boldsymbol{\gamma}}}(\boldsymbol{\theta}_{\boldsymbol{\gamma}} \mid \mathbf{0}, g\, \sigma^2\, ((\boldsymbol{X}^*_{\boldsymbol{\gamma}})^t \boldsymbol{X}^*_{\boldsymbol{\gamma}})^{-1}) \pi_{\boldsymbol{\gamma}}(g) dg\,, \tag{7.4}$$

TABLE 7.3

Specific Proposals for the Hyperparameter g in the Literature. $a = 3$ is the Recommended Choice in (8). (1):[70] (2):[107], (3):[40], (4):[38], (5):[65], (6):[108], (7):[109], (8):[72], (9):[4], (10):[56].

Proposal	Reference	Name
Fixed g		
$g = n$	(1,2)	Unit Information prior
$g = n/k_\gamma$	(10)	Little informative prior
$g = (1 + k_\gamma)^2$	(3)	Risk inflation criterion prior
$g = \max\{n, (1 + k_\gamma)^2\}$	(4)	Benchmark prior
$g = \log(n)$	(4)	Hannan-Quinn
Random g		
$g \sim IGa(1/2, n/2)$	(5, 6, 7)	Cauchy prior
$g \mid a \sim p(g) \propto (1 + g)^{-a/2}$, $g > 0$	(8)	hyper-g
$g \mid a \sim p(g) \propto (1 + g/n)^{-a/2}$, $g > 0$	(8)	hyper-g/n
$g \sim p(g) \propto (1 + g)^{-3/2}$, $g > \frac{1+n}{k_\gamma+1} - 1$	(9)	Robust prior

where k_γ is the dimension of $\boldsymbol{\theta}_\gamma$ and $\pi_\gamma(g)$ is a proper density on $g > 0$ for all models. Centering the prior at $\mathbf{0}$ is a natural objective choice. Choosing the covariance matrix to be proportional to $\sigma^2\ ((\boldsymbol{X}^*_\gamma)^t \boldsymbol{X}^*_\gamma)^{-1}$, which is the covariance matrix of the least squares estimate of $\boldsymbol{\theta}_\gamma$, greatly simplifies the ensuing computations. It can also be justified from invariance and 'predictive matching' perspectives; see [4] for discussion.

Priors such as this are called *mixtures of g-priors*, and are extensively studied in [72] and [4]. Various choices that have appeared in the literature are given in Table 7.3; several proposals have been for a fixed g, while others have been for random g. By far the most commonly used are the original g-prior ([107]), with recommended choice $g = n$, and the Cauchy prior (also called the Zellner-Siow prior, [108]). The original g-prior is information inconsistent ([79]) and has sharp tails, whereas Jeffreys had argued for priors with no moments. For this reason, the Cauchy prior is typically preferred over the original g-prior. The Obesity example, in Section 7.5.4, demonstrates use of the $g = n$-prior in an applied setting.

The robust prior was introduced in [4] as a prior that satisfied a set of desiderata that were proposed for model uncertainty analysis. It also resulted in a closed form Bayes factor of M_γ to the intercept only model M_0:

$$B_{\gamma 0} = \left[\frac{n+1}{k_\gamma + 1}\right]^{-\frac{k_\gamma}{2}} \frac{Q_{\gamma 0}^{-\frac{n-1}{2}}}{k_\gamma + 1}\, {}_2F_1\left[\frac{k_\gamma + 1}{2}; \frac{n-1}{2}; \frac{k_\gamma + 3}{2}; \frac{(1 - Q_{\gamma 0}^{-1})(k_\gamma + 1)}{(1+n)}\right],$$

where ${}_2F_1$ is the standard hypergeometric function (see [1]) and $Q_{\gamma 0} = S^2_\gamma / S^2_0$ is the ratio of the sum of squared errors of M_γ and M_0.

There is a huge literature on priors for linear and generalized linear models, with many other ideas and suggestions. A few of the references are [27], [50], [23], [66, 67], [31], [104], [74], [77], [32], [61], [75], [76] and [17].

7.4.2 Fractional Bayes Factors

This was introduced in [81, 82]. The idea is to use a fraction of the likelihood to develop priors for model parameters. The algorithm is as follows:

Step 1. For model M_i, choose a "fraction" $0 < b < 1$. Typically recommended is is $b = k_i/n$, where n is the sample size and k_i is the dimension of the parameter in M_i. It has been observed, however, that different fractions may sometimes be needed for different parts of the likelihood ([33]).

Step 2. For model M_j, use the prior $\pi_j^*(\boldsymbol{\theta}_j) \propto [f_j(\boldsymbol{x}|\boldsymbol{\theta}_j)]^b \cdot \pi_j^O(\boldsymbol{\theta}_j)$, where $\pi_j^O(\boldsymbol{\theta}_j)$ is an objective estimation prior; often the constant prior is fine.

Step 3. Compute Bayes factors using these priors and the unused fractions of the likelihoods, $[f_j(\boldsymbol{x}|\boldsymbol{\theta}_j)]^{(1-b)}$, resulting in

$$B_{ji}^F = \frac{\int [f_j(\boldsymbol{x}|\boldsymbol{\theta}_j)]^{(1-b)}\, \pi_j^*(\boldsymbol{\theta}_j) d\boldsymbol{\theta}_j}{\int [f_i(\boldsymbol{x}|\boldsymbol{\theta}_i)]^{(1-b)}\, \pi_i^*(\boldsymbol{\theta}_i) d\boldsymbol{\theta}_i} = \frac{\int f_j(\boldsymbol{x}|\boldsymbol{\theta}_j)\pi_j^O(\boldsymbol{\theta}_j) d\boldsymbol{\theta}_j}{\int f_i(\boldsymbol{x}|\boldsymbol{\theta}_i)\pi_i^O(\boldsymbol{\theta}_i) d\boldsymbol{\theta}_i} \cdot \frac{\int [f_i(\boldsymbol{x}|\boldsymbol{\theta}_i)]^b\, \pi_i^O(\boldsymbol{\theta}_i) d\boldsymbol{\theta}_i}{\int [f_j(\boldsymbol{x}|\boldsymbol{\theta}_j)]^b\, \pi_j^O(\boldsymbol{\theta}_j) d\boldsymbol{\theta}_j}.$$

The posterior probabilities are then computed from these Bayes factors.

This approach is particularly attractive when the Bayes factors for objective estimation priors are available in closed form. It is also broadly applicable, except for serious concerns in irregular problems, as in Example 7.5.

7.5 Expected Posterior Priors

Efforts to utilize the data itself or imaginary data to help determine the prior have a long history, starting with [58], [65], and [98]. There have been very bad versions of this idea, such as repeated efforts to utilize the likelihood itself as the prior. The successful versions of this idea utilize only a small fraction of the data – or a small amount of imaginary data – to help determine the prior; the data used in this way is typically called *training samples.* To avoid confusion with the data $\boldsymbol{x}$, we will denote the training samples by $\boldsymbol{x}^*$; sometimes this will be imaginary data independent of $\boldsymbol{x}$, but sometimes it will be subsets of $\boldsymbol{x}$.

7.5.1 General Definition

Following [86, 87], and [80], the expected posterior prior algorithm is as follows:

Step 1. Starting with objective priors $\pi_i^O(\boldsymbol{\theta}_i)$ for each model M_i, a key notion is that of a *minimal training sample* for model M_i, which is defined to be the smallest sample $\boldsymbol{x}_i^*$ for which $m_i(\boldsymbol{x}^*) = \int f_i(\boldsymbol{x}^* \mid \boldsymbol{\theta}_i)\pi_i^O(\boldsymbol{\theta}_i)\, d\boldsymbol{\theta}_i < \infty$; define m_i to be the corresponding size of a minimal training sample. Note that then the posteriors $\pi_i^O(\boldsymbol{\theta}_i \mid \boldsymbol{x}_i^*) = f_i(\boldsymbol{x}_i^* \mid \boldsymbol{\theta}_i)\pi_i^O(\boldsymbol{\theta}_i)/m_i(\boldsymbol{x}_i^*)$ exist.

Step 2. Choose a density $m^*(\boldsymbol{x}^*)$ for training samples $\boldsymbol{x}^*$ of the largest size $m^u = \max\{m_i\}$. The *expected posterior prior (EPPrior for short)* for $\boldsymbol{\theta}_i$ is then given by

$$\pi_i^{\mathrm{EP}}(\boldsymbol{\theta}_i) = \int \pi_i^O(\boldsymbol{\theta}_i \mid \boldsymbol{x}_i^*) m^*(\boldsymbol{x}_i^*)\, d\boldsymbol{x}_i^*\,,$$

where $m^*(\boldsymbol{x}_i^*)$ is the marginal density of $m^*(\boldsymbol{x}^*)$ for samples of size m_i.

The name 'expected posterior prior' reflects the fact that the prior is computed as an expectation of posteriors.

This is not the only version of the EPPrior. Another version has the expected posterior prior defined as $\int \pi_i^O(\boldsymbol{\theta}_i \mid \boldsymbol{x}^*)\, m^*(\boldsymbol{x}^*)\, d\boldsymbol{x}^*$, i.e., the largest training sample size is used for every model. The rationales for using the minimal training sample size for each model are given in [7]. (See [91] for evidence supporting the alternative.)

The expected posterior priors, $\pi_i^{\mathrm{EP}}(\boldsymbol{\theta}_i)$, will not be proper unless m^* itself is proper, but they are always properly 'calibrated' across models, because they are trained by the same $m^*(\boldsymbol{x}^*)$. The key issue is, of course, choice of m^*, and we will be proposing two different choices, one for nested models in the next section and the other, for non-nested models, in Section 7.5.3.

While we are focusing on the objective Bayes approach in this chapter, it is worth mentioning that EPPriors are also compatible with the subjective approach. It is often argued by subjectivists (starting with deFinetti) that one should not elicit prior distributions directly, but should elicit the distribution of future observables, e.g. $m^*(\boldsymbol{x}^*)$. Sometimes this is also available from past studies. For instance, suppose a study of a new drug is to be performed, and there exists a previous study for a drug in the same class. One would typically not want to use the posterior distribution from the old study as the prior distribution for the new study, but using the previous posterior to define $m^*(\boldsymbol{x}^*)$ would often be sensible and appropriate.

7.5.2 Expected Posterior Priors (Intrinsic Priors) for Nested Models

The next section introduces the expected posterior prior for general nested models. The subsequent section specializes to the normal linear model.

7.5.2.1 Introduction

Suppose that there is a model that is nested in all the other models; call this M_0, having data density $f_0(\boldsymbol{x} \mid \boldsymbol{\theta}_0)$; this will be called the *base model.* Then a natural choice of m^* is

$$m^*(\boldsymbol{x}^*) = \int f_0(\boldsymbol{x}^* \mid \boldsymbol{\theta}_0)\pi_0^O(\boldsymbol{\theta}_0)\, d\boldsymbol{\theta}_0,$$

with associated marginal densities $m^*(\boldsymbol{x}_i^*) = \int f_0(\boldsymbol{x}_i^* \mid \boldsymbol{\theta}_0)\pi_0^O(\boldsymbol{\theta}_0)\, d\boldsymbol{\theta}_0$. Although these will typically be improper, the idea is that one is generating imaginary samples under the nested model, and using these samples to train all the priors for the other models. Thus one is centering the other model priors at the nested model.

In the problem of testing a point null hypothesis, $\boldsymbol{\theta}_0$ was given, so $m^*(\boldsymbol{x}^*)$ was $f(\boldsymbol{x}^* \mid \boldsymbol{\theta}_0)$. Thus the training samples were generated directly from H_0.

The ensuing EPPrior for model M_i is

$$\pi_i^{\mathrm{EP}}(\boldsymbol{\theta}_i) = \int \pi_i^O(\boldsymbol{\theta}_i \mid \boldsymbol{x}_i^*)\, m^*(\boldsymbol{x}_i^*)\, d\boldsymbol{x}_i^* = \int f_i(\boldsymbol{x}_i^* \mid \boldsymbol{\theta}_i)\pi_i^O(\boldsymbol{\theta}_i)\frac{m^*(\boldsymbol{x}_i^*)}{m_i(\boldsymbol{x}_i^*)}\, d\boldsymbol{x}_i^*.$$

For the base model M_0, $m^*(\boldsymbol{x}_0^*) = m_0(\boldsymbol{x}_0^*)$, so that $\pi_0^{\mathrm{EP}}(\boldsymbol{\theta}_0) = \pi_0^O(\boldsymbol{\theta}_0)$, the original objective prior. Note that $\boldsymbol{\theta}_0$ appears in all models (because of the nesting); the EPPriors for the $\boldsymbol{\theta}_0$ will thus be $\pi_0^O(\boldsymbol{\theta}_0)$ in every model.

The ensuing marginal likelihood for M_i and the actual data $\boldsymbol{x}$ is

$$m_i(\boldsymbol{x}) = \int f_i(\boldsymbol{x} \mid \boldsymbol{\theta}_i)\pi_i^{\mathrm{EP}}(\boldsymbol{\theta}_i)\, d\boldsymbol{\theta}_i = \int\int f_i(\boldsymbol{x} \mid \boldsymbol{\theta}_i) f_i(\boldsymbol{x}_i^* \mid \boldsymbol{\theta}_i)\pi_i^O(\boldsymbol{\theta}_i)\frac{m^*(\boldsymbol{x}_i^*)}{m_i(\boldsymbol{x}_i^*)}\, d\boldsymbol{x}_i^* d\boldsymbol{\theta}_i.$$

This looks formidable, but typically most of these integrals can be evaluated in closed form, switching orders of integration as needed. This will be seen to be the case in the linear model, for which all the above integrals reduce to a one-dimensional numerical integral.

It can be shown that EPPriors are equal to the intrinsic priors defined in [8] and [20] in the nested model scenario.

Example 7.5 *An irregular example:* Suppose x arises from the exponential location density $f(x \mid \theta) = \exp(-(x-\theta))$, $x > \theta$. The three models (more commonly called hypotheses), are $M_0 : \theta = 0$, $M_1 : \theta < 0$ and $M_2 : \theta > 0$. Note that, if $x < 0$, then M_1 must be true, since a negative x could not arise from M_0 or M_2. Hence we only consider positive x.

Using the objective prior $\pi^O(\theta) = 1$, a single observation yields proper posteriors

$$\pi_1^O(\theta \mid x) = \frac{\exp(-(x-\theta))1_{(-\infty,0)}(\theta)}{\int_{-\infty}^0 \exp(-(x-\theta))\, d\theta} = \exp(\theta)1_{(-\infty,0)}(\theta),$$

$$\pi_2^O(\theta \mid x) = \frac{\exp(-(x-\theta))1_{(0,x)}(\theta)}{\int_0^x \exp(-(x-\theta))\, d\theta} = \frac{\exp(\theta)1_{(0,x)}(\theta)}{\exp(x)-1}.$$

Choosing M_0 to be the base model means that $m^*(x^*) = f(x^* \mid 0) = \exp(-x^*)$, $x^* > 0$. Clearly $\pi_1^{\text{EPP}}(\theta) = \exp(\theta)1_{(-\infty,0)}(\theta)$ (the posterior is not affected by x^*, provided $x^* > 0$), while

$$\pi_2^{EPP}(\theta) = \int_0^\infty \frac{\exp(\theta)1_{(0,x^*)}(\theta)}{\exp(x^*)-1} \exp(-x^*)\, dx^* = -[1 + \exp(\theta)\log(1-\exp(-\theta))],$$

(a proper prior). The density of $\boldsymbol{x} = (x_1, \ldots, x_n)$ is, if all of the x_i are positive (if any are negative, M_1 is the true model), $f(\boldsymbol{x} \mid \theta) = \exp(-[\sum x_i - n\theta])1_{\{\theta < \min\{x_i\}\}}$, so that the intrinsic prior Bayes factor of M_0 to M_1 is

$$B_{01} = \frac{f(\boldsymbol{x} \mid 0)}{\int f(\boldsymbol{x} \mid \theta)\pi_1^{\text{EPP}}(\theta)\, d\theta} = \frac{\exp(-\sum x_i)}{\int_{-\infty}^0 \exp(-(\sum x_i - n\theta)\exp(\theta)\, d\theta} = n+1.$$

Thus postive $\boldsymbol{x}$ always favors M_0 over M_1, and by an increasing amount as the number of positive observations increases. B_{02} is not available in closed form, but it is an easy numerical computation. It behaves as one would expect, favoring M_0 over M_2 when $\min\{x_i\}$ is small, and favoring M_2 when $\min\{x_i\}$ is large. A variant of this example was presented in [12], in which the Fractional Bayes Factor was also computed; B_{02}^F was always less than one, for any value of $\min\{x_i\}$, and thus always favors M_2, an inappropriate result. The comparison of three hypotheses, zero vs negative vs positive is important in practice and is tractable from an intrinsic-EPP Bayesian perspective, even in irregular settings.

7.5.2.2 EP (Intrinsic) Priors for the Linear Model

We resume consideration of the linear model, defined in Section 7.4.1. We utilize the name 'intrinsic prior', rather than EPPrior, in this section, because the name intrinsic prior considerably pre-dated the EPPrior label. Nevertheless, we adopt the EPPrior representation, because it is much easier to work with and allows much more general implementations than the previous derivations of intrinsic priors.

We introduce two generalizations to the situation considered in Section 7.4.1. First, we allow any model to be the base model, as long as it is nested in all the other models under consideration. Again denote this base model by M_0 and suppose it has k_0 of the covariates; to simplify notation, rearrange the covariates (if necessary) so that they appear as the first k_0 coordinates of $\boldsymbol{\theta}$. Thus the parameter for M_0 is $\boldsymbol{\theta}_0 = (\theta_1, \ldots, \theta_{k_0})$. Write $\boldsymbol{X} = (\boldsymbol{X}_0\ \boldsymbol{X}^*)$, with $\boldsymbol{X}_0$ being the covariate matrix corresponding to M_0; the second generalization made herein is that we do not require $\boldsymbol{X}_0$ and $\boldsymbol{X}^*$ to be orthogonal. For generalizations to generalized linear models, see [41].

As in Section 7.4.1, let M_γ denote another model, which has parameter $(\boldsymbol{\theta}_0, \boldsymbol{\theta}_\gamma)$, where now $\boldsymbol{\gamma}$ is a $k - k_0$ vector of ones and zeroes, indicating which additional variables are in the model; let k_γ denote the number of additional variables in the model (the dimension of $\boldsymbol{\theta}_\gamma$). The corresponding covariate matrix will be $(\boldsymbol{X}_0\ \boldsymbol{X}^*_\gamma)$. Since $\boldsymbol{\theta}_0$ and σ^2 are common to all models, the intrinsic prior will assign them the usual objective prior $\pi^O(\boldsymbol{\theta}_0, \sigma^2) = 1/\sigma^2$.

The intrinsic prior and resulting Bayes factor: When M_0 is the base model, the *intrinsic hyper-g prior* (IHG prior, for short) for model M_γ, based on minimal training sample size $m_\gamma = k_0 + k_\gamma + 1$, is a mixture of g-priors given by

$$\pi^{IHG}_\gamma(\boldsymbol{\theta}_\gamma \mid \boldsymbol{\theta}_0, \sigma^2) = \int_1^\infty N_{k_\gamma}\left(\boldsymbol{\theta}_\gamma \mid \boldsymbol{0}, g\,\frac{(n-k_0)}{(k_\gamma+1)}\sigma^2\ \boldsymbol{\Sigma}_\gamma\right) \frac{1}{\pi\ g\sqrt{g-1}}\ dg\,, \tag{7.5}$$

where $\boldsymbol{\Sigma}_\gamma^{-1} = (\boldsymbol{X}^*_\gamma)'(\boldsymbol{I} - \boldsymbol{X}_0(\boldsymbol{X}_0'\boldsymbol{X}_0)^{-1}\boldsymbol{X}_0')\boldsymbol{X}^*_\gamma$. The resulting Bayes factor of M_γ to M_0 is

$$B_{\gamma 0} = \int_1^\infty \left(1 + g\frac{(n-k_0)}{(k_\gamma+1)}Q_{\gamma 0}\right)^{-(n-k_0)/2} \left(1 + \frac{(n-k_0)}{(k_\gamma+1)}g\right)^{(n-k_0-k_\gamma)/2} \frac{1}{\pi\ g\sqrt{g-1}}\ dg\,, \tag{7.6}$$

recalling that $Q_{\gamma 0} = S^2_\gamma / S^2_0$.

Properties of the intrinsic prior

1. The presence of n in the exponent of the intrinsic prior is because, typically, $n\boldsymbol{\Sigma}_\gamma \to \boldsymbol{C}$ as n grows, where $\boldsymbol{C}$ is a constant matrix. (In ANOVA, for example $\boldsymbol{C} = \boldsymbol{I}$.) Choices of g that are not of order n will lead to covariance matrices that either go to zero or infinity, as n grows, neither of which is acceptable from a Bayesian perspective. Note that the original g-prior, the 'little informative prior', the Zellner-Siow (Cauchy) prior, the hyper-g/n prior, and the robust prior all also have this behavior.
2. The intrinsic prior has a further scale factor of k_γ in the denominator. That this was important was indicated in [4] (for the robust prior) and [56] (for the $g = n/k$-prior).
3. The conditional intrinsic prior had no moments, an important property first recognized by [65], and will not suffer from information inconsistency.

Comments on the derivation of the intrinsic prior: The proof of the above result is rather lengthy and can be found in [7]. It is worth mentioning, however, how training samples are handled, since each observation y_i is associated with a vector of covariates $\boldsymbol{x}_i$. If the $\boldsymbol{x}_i$ arise randomly from some distribution, then one could consider the pairs $(y_i, \boldsymbol{x}_i)$ as the training samples and derive m^* accordingly, incorporating also the distribution of the covariates. More common, however, is treating the covariates as fixed; this is the situation we considered in deriving the above result.

Training samples of size $k+1$ (k was the dimension of the full model) are needed. A training sample from the actual data $\boldsymbol{y} = (y_1, \ldots, y_n)'$ would consist of some subset, $\boldsymbol{y}(l) = (y_{i_1}, \ldots, y_{i_{k+1}})'$, of the observations in $\boldsymbol{y}$. Associated with this subset is the $(k+1) \times k$ matrix of covariates $\boldsymbol{X}(l)$, consisting of the corresponding rows of $\boldsymbol{X}$. Since M_0 is the base model, which has the objective prior density $\pi^O(\boldsymbol{\theta}_0, \sigma^2) = 1/\sigma^2$, this suggests defining the density m^*, for imaginary training samples $\boldsymbol{y}^*$ of size $k+1$, as

$$m^*(\boldsymbol{y}^* \mid \boldsymbol{X}(l)) = \int f_0(\boldsymbol{y}^* \mid \boldsymbol{X}(l), \boldsymbol{\theta}_0, \sigma^2)\frac{1}{\sigma^2}\ d\boldsymbol{\theta}_0\ d\sigma^2\,,$$

where $f_0(\boldsymbol{y}^* \mid \boldsymbol{X}(l), \boldsymbol{\theta}_0, \sigma^2)$ is the normal density arising from the base model.

Of course, each choice of a subsample leads to a (possibly) different $\boldsymbol{X}(l)$, so that the $m^*(\boldsymbol{y}^* \mid \boldsymbol{X}(l))$ need to be averaged in some way. Arithmetic averaging is most natural, but leads to a rather unwieldy intrinsic prior. Hence, a (somewhat adhoc) geometric averaging is employed in [7], to reach the intrinsic prior defined above (for the full model).

For any other model M_γ, one needs a marginal distribution of size $k_0+k_\gamma+1$ from $m^*(\boldsymbol{y}^* \mid \boldsymbol{X}(l))$. Randomly subsampling $(\boldsymbol{y}^*, \boldsymbol{X}(l))$ and then randomly subsampling $\boldsymbol{y}^*_{k_0+k_\gamma+1}$ and corresponding covariates from $(\boldsymbol{y}^*, \boldsymbol{X}(l))$ is the same as randomly sampling $\boldsymbol{y}^*_{k_0+k_\gamma+1}$ and corresponding covariates from $(\boldsymbol{y}, \boldsymbol{X})$, so the same argument can be used to derive the intrinsic priors for each model.

7.5.3 The Empirical Expected Posterior Prior for Non-nested Models

In the previous section, a nested model was used to generate training samples for all other models. If there are non-nested models, this does not work. For instance, even if there were just two non-nested models, one would typically get different answers by generating training samples from one, versus the other.

An interesting choice for m^* that overcomes this difficulty (see [86, 87]) is the *empirical distribution*; i.e, given observations $\boldsymbol{x} = \{x_1, \ldots, x_n\}$ (we forgo the $\boldsymbol{x}^*$ notation, as we will be subsampling from the actual data)

$$m^*(\boldsymbol{x}) = \frac{1}{L} \sum_l I_{\{\boldsymbol{x}(l)\}}(\boldsymbol{x}),$$

where $\boldsymbol{x}(l) = (x_{l_1}, \ldots, x_{l_{m^u}})$ is a subsample of the observations of minimal size, such that all the $m^*(\boldsymbol{x}_i)$ are finite; thus $L = \binom{n}{m^u}$ and m^* is to be interpreted as assigning equal probabilities of $1/L$ to each of the subsamples. Note that the marginal distribution $m^*(\boldsymbol{x}_i)$ is just

$$m^*(\boldsymbol{x}_i) = \frac{1}{L_i} \sum_l I_{\{\boldsymbol{x}_i(l)\}}(\boldsymbol{x}_i),$$

where $\boldsymbol{x}_i(l) = (x_{l_1}, \ldots, x_{l_{m_i}})$ is a subsample of size m_i and $L_i = \binom{n}{m_i}$. The appeal here is that one is not generating training samples from a possibly true model; one is generating training samples from the *true* model.

The ensuing empirical EPPrior (EEPPrior) for model M_i is

$$\pi_i^{\text{EEP}}(\boldsymbol{\theta}_i) = \int \pi_i^O(\boldsymbol{\theta}_i \mid \boldsymbol{x}_i)\, m^*(\boldsymbol{x}_i)\, d\boldsymbol{x}_i = \frac{1}{L_i} \sum_l \pi_i^O(\boldsymbol{\theta}_i \mid \boldsymbol{x}_i(l)).$$

This is obviously proper and can be easy to work with. In particular, the resulting marginal likelihood for the model uncertainly analysis is

$$m_i(\boldsymbol{x}) = \int f_i(\boldsymbol{x} \mid \boldsymbol{\theta}_i) \pi_i^{\text{EEP}}(\boldsymbol{\theta}_i)\, d\boldsymbol{\theta}_i = \frac{1}{L_i} \sum_l \int f_i(\boldsymbol{x} \mid \boldsymbol{\theta}_i) \pi_i^O(\boldsymbol{\theta}_i \mid \boldsymbol{x}_i(l))\, d\boldsymbol{\theta}_i, \tag{7.7}$$

and often the latter integrals are available in closed form. Note also that, if L_i is very large, one can simply take a subsample of the empirical distribution to perform the averaging.

Example 7.6 *A non-nested example.* The data is $\boldsymbol{x} = (x_1, \ldots, x_n)$, where the x_i are non-nonnegative integers arising independently from either the Poisson (M_1) or the Negative Binomial (M_2) densities:

$$M_1 : x \sim f_1(x \mid \lambda) = \exp(-\lambda)\, \frac{\lambda^x}{x!}, \;\; \lambda > 0; \quad M_2 : x \sim f_2(x \mid \theta) = \theta\, (1-\theta)^x, \;\; 0 < \theta < 1.$$

The standard objective priors for these models are the Jeffreys priors, given, respectively, by $\pi_1^O(\lambda) = 1/\lambda^{1/2}$ and $\pi_2^O(\theta) = 1/[\theta\,(1-\theta)^{1/2}]$. The minimal training sample size is $m = 1$ (since the objective posteriors are proper for one observation). It is immediate that the EEPpriors for M_1 and M_2 are, respectively,

$$\pi_1^{\text{EEP}}(\lambda) = \frac{1}{n}\sum_{l=1}^{n} \frac{\exp(-\lambda)\lambda^{x_l-1/2}}{\Gamma(x_l+1/2)}, \quad \pi_2^{\text{EEP}}(\theta) = \frac{1}{n}\sum_{l=1}^{n} \frac{\Gamma(x_l+3/2)}{\Gamma(x_l+1/2)}(1-\theta)^{(x_l-1/2)}. \tag{7.8}$$

The resulting Bayes factor can be computed in closed-form and, with $x_+ = \sum x_i$, is given by

$$B_{12}^{\text{EEP}} = \frac{\int f_1(\boldsymbol{x}\mid\lambda)\;\pi_1^{\text{EEP}}(\lambda)d\lambda}{\int f_2(\boldsymbol{x}\mid\theta)\;\pi_2^{\text{EPP}}(\theta)d\theta} = \frac{\frac{1}{n\cdot\prod x_i!}\sum_{l=1}^{n}\frac{\Gamma(x_++x_l+1/2)}{\Gamma(x_l+1/2)\cdot(n+1)^{(x_++x_l+1/2)}}}{\frac{\Gamma(n+1)}{n}\sum_{l=1}^{n}\frac{\Gamma(x_++x_l+1/2)\cdot\Gamma(x_l+3/2))}{\Gamma(n+x_++x_l+3/2)\cdot\Gamma(x_l+1/2)}}. \tag{7.9}$$

For instance, if $n = 2$ and $x_1 = x_2 = 0$, $B_{12}^{\text{EEP}} = 0.924$ while, if $x_1 = x_2 = 2$, $B_{12}^{\text{EEP}} = 0.33$. This same example was considered in Bernardo and Smith (1994, page 393), but with varying conjugate priors. The conjugate priors gave Bayes factors ranging from 0.27 to 5.5 in the first case, but only from 0.29 to 0.38 in the second case. Some light can be shed on this by looking at the likelihood ratio, which is undetermined in the first case and is 0.3 in the second case. So there is extreme sensitivity to conjugate priors in the first case, but not in the second. The B_{12}^{EEP} were quite reasonable in both cases. This example was also studied in [78] from the intrinsic prior perspective.

An issue of some concern is revealed in (7.7); there is a clear double use of the data, since the $\boldsymbol{x}_i(l)$ are part of $\boldsymbol{x}$. The effect of this slight double use of data is, of course, small when the sample size n is much larger than the training sample size m^u.

Sometimes it is possible to avoid the use of the EEPPrior for non-nested models, particularly when the models have common parameters with an invariance structure. For instance, consider testing $M_1 : x_i \sim N(x_i \mid \mu, \sigma^2)$ versus $M_2 : x_i \sim \text{Cauchy}(x_i \mid \mu, \sigma)$. These are non-nested models, but (μ, σ) is a common-location scale parameter in each model and so one can utilize the usual objective prior $\pi(\mu, \sigma) = 1/\sigma$ in both models to compute the Bayes factor. This would be preferable to using the EEPPrior.

7.5.4 Choice of the Base Model

For linear models, it might seem natural to choose the base model to be the null model having mean 0 and use that to derive the intrinsic priors for all other models. It is much more common, however to choose the base model to be the simplest model with a nonzero mean; typically this model is that with a constant mean θ_1 (the intercept term in regression). Of course, if mean 0 is deemed to be plausible, using it as the base model would be natural. But usually, in linear model analysis, a zero mean is deemed highly implausible, with at least an intercept term being needed.

The folklore here is that one should choose the base model for intrinsic priors to be the smallest *plausible* model. There are three motivations for this:

- In the EPPrior formulation of the intrinsic prior, the imaginary training samples arise from the base model marginal distribution. If the base model is extremely implausible, then the intrinsic prior is being trained by training samples that are not at all like the real data.

- One can encounter the *conditional Lindley paradox*, discovered in [100], when an implausible base model is used.

- Simulations in [20] show that using the null model as the base model can give inferior results.

Even the intercept model is implausible if there are one or more other covariates that are virtually certain to be in the model. Including any such variables in the base model is then natural. Indeed, we recommend catering to this possibility by proceeding in two stages.

Stage 1. Perform the intrinsic prior analysis using the intercept model (or possibly the null model) as the base model and compute the posterior inclusion probabilities of all the variables.

Stage 2. If any variables have extremely high posterior inclusion probability (e.g., 0.99), include them in the base model with the intercept, yielding a new base model M_0^*, and re-derive the intrinsic priors with respect to this new base model.

If one has progressed to Stage 2 above, there are two choices concerning what to do with the simpler models that were also considered in Stage 1. First, one could simply ignore them, since they will have extremely small posterior probabilities. The second possibility is to also utilize the Stage 1 Bayes factors (of the new base model M_0^* to the simpler models) in the analysis. There is a slight incoherency here, since different parameter priors are being used for M^* when comparing to smaller versus larger models, but it should be negligible. Indeed, here is a result established in [7].

Lemma 7.1 Suppose the variable corresponding to θ_i has posterior inclusion probability p_i (the posterior probability that the variable occurs in models). Let $\mathcal{M}^*$ denote the set of models that do not contain all of $\{\theta_1, \ldots, \theta_{k_0}\}$ (the variables in the base model M_0; also the common variables in all Stage 2 models). Then the posterior probability of $\mathcal{M}^*$ is less than $\sum_{i=1}^{k_0}(1 - p_i)$.

A final consideration in the analysis is what to choose for the model prior probabilities at Stage 2. One does not want to alter the prior odds of any of the models in Stage 2, so the simple answer is just to take the model prior probabilities assigned to these models in Stage 1, and renormalize them to sum to one.

Example 7.7 *Children obesity dataset (OBICE study).* This dataset, available in the current distribution of the R package BayesVarSel [43], has $n = 996$ and $k = 16$ possible covariates, listed in the first column of Table 7.4. The intercept is included in all models. Two different prior distributions are considered: the hyper-g intrinsic prior (IHG) and the standard $g = n$-prior.

The table entries are the posterior inclusion probabilities of the variables. The Stage 1 inclusion probabilities were computed with the base model, M_0, being the intercept-only model. The variables for which this inclusion probability exceeded 0.99, were then included (with the intercept) in the updated base model; these variables are labeled as IN. The model uncertainty analysis is then rerun with this new base model, yielding the final Step 2 posterior inclusion probabilities.

There are modest differences between the inclusion probabilities at the two stages. Note, however, that the median probability model (which contains all variables with posterior inclusion probability greater than 1/2) remains the same under both priors (although 'sex' is almost eliminated from the median probability model under the IHG prior at Stage 2). In all cases, in the table, the posterior inclusion probabilities at Stage 2 either stay the same or decrease, but this is not always true; in particular, if the Stage 1 base model has poor SSE, the inclusion probabilities are likely to increase at Stage 2.

TABLE 7.4
Posterior Inclusion Probabilities of Variables at Stage 1 and Stage 2, under the Hyper-g Intrinsic Prior (IHG) and the $g = n$-prior

	IHGprior		**$g = n$-prior**	
Variable	Step 1	Step 2	Step 1	Step 2
Age	1.000	IN	1.00	IN
Weight at birth	0.993	IN	0.849	0.845
Height at birth	0.987	0.987	0.817	0.812
Sex	0.567	0.503	0.173	0.172
The father is obese	1.000	IN	1.000	IN
The mother is obese	1.000	IN	1.000	IN
...has 5 daily meals	0.998	IN	0.991	IN
...eats vegetables	0.388	0.326	0.064	0.064
...eats fruit	0.348	0.289	0.053	0.053
...afternoon snacks.	0.841	0.793	0.431	0.423
...was breastfed	0.442	0.376	0.082	0.082
...practices sports	0.885	0.860	0.676	0.672
...watches TV	1.000	IN	1.000	IN
...plays electronic	0.409	0.343	0.066	0.065
...sleeps	0.445	0.380	0.084	0.084
Daily candy consumption	0.987	0.982	0.930	0.928

The Stage 2 analysis restricts the model space to always include the intercept and the IN variables. Thus the model space goes from 2^{16} models at Stage 1, to 2^{10} models for the IHG prior and 2^{11} for the $g = n$-prior at Stage 2. Thus the vast majority of models are being ignored. Lemma 7.1, however, gives upper bounds of $0 + 0.007 + 0 + 0 + 0.002 + 0 = 0.009$ and $0 + 0 + 0 + 0.009 + 0 = 0.009$ on the sum of the posterior probability of these ignored models for the two priors. So Stage 2 results in a considerable simplification of the model space by eliminating models of negligible posterior probability.

Finally, note that the posterior inclusion probabilities for the $g = n$-prior are usually considerably smaller than those for the IHG priors, so that the $g = n$-prior favors small models much more strongly than the IHG prior. This is likely due to the very sharp tails of the $g = n$-prior; since it is centered at $\mathbf{0}$, the prior gives very little mass to the more complex models whose likelihoods are not close to zero. The much longer tails of the IHG prior can reach these complex model likelihoods.

7.5.5 Studying Robustness by Varying the Training Sample Size

Using the minimal training sample size is natural, in that the resulting intrinsic priors are somewhat diffuse, which seems desirable from an objective perspective. There is no compelling reason to use the minimal training sample size, however, and there are situations in which consideration of larger training sample sizes is reasonable.

The typical situation in which this issue arises is when the sample size n is large, and the parameter $\boldsymbol{\theta}_\gamma$ in the more complex model is small. The likelihood for $\boldsymbol{\theta}_\gamma$ is then concentrated near zero, and the intrinsic prior will typically be comparatively diffuse and give little weight to this likelihood. As one increases the training sample size, however, the intrinsic prior will become more concentrated near zero and can end up giving much more weight to the likelihood.

For this reason, we recommend doing a robustness study with respect to training sample size as follows.

- Add a constant q to all the minimal training sample sizes, and then graph the results (Bayes factors, posterior model probabilities, posterior inclusion probabilities, ...) as q varies between 0 and n. (The Bayesian expressions are continuous in q, so there is no need to restrict q to be an integer.)

- If the answers change markedly with q, this should be reported. It may be that essentially the same overall conclusion is reached (e.g., the drug is shown to be effective over the entire range of answers). But, if the overall conclusion can change, subjective input might be needed to resolve the situation.

7.5.6 General Testing

Suppose

$$\boldsymbol{y} = \boldsymbol{X}\boldsymbol{\theta} + \sigma\boldsymbol{\varepsilon}, \ \ \boldsymbol{\varepsilon} \sim N_n(0, I),$$

where $\boldsymbol{y}$ is $n \times 1$ and $\boldsymbol{X}$ is $n \times k$. Suppose it is desired to test $H_1 : \Omega\boldsymbol{\theta} = \mathbf{0}$ versus $H_2 : \Omega\boldsymbol{\theta} \neq \mathbf{0}$, where Ω $(k_2 \times k)$ is of full rank. For instance, if one has two normal means and wants to test $\theta_1 = \theta_2$, set $\Omega = (1, -1)$.

The following lemma, whose proof can be found in [7], gives the intrinsic prior Bayes factor.

Lemma 7.2 The Bayes factor corresponding to the hyper-g intrinsic prior is

$$B_{21} = \int_0^\infty \left(1 + g\frac{\text{SSE}}{\text{SSE}_\text{R}}\right)^{-(n-k+k_2)/2} (1+g)^{(n-k)/2}\, p^I(g)\, dg,$$

where SSE_R is the sum of squared errors for the reduced model (restricted to $\Omega\boldsymbol{\theta} = \mathbf{0}$) and SSE is the SSE for the unrestricted model.

7.6 Other Issues with Bayesian Model Uncertainty

This section briefly addresses computational issues in Bayesian model uncertainty analysis.

7.6.1 Computation and Moving in Model Space

There are two main computational difficulties in dealing with Bayesian model uncertainty. The first is that the marginal likelihoods $m_i(\boldsymbol{x}) = \int f_i(\boldsymbol{x} \mid \boldsymbol{\theta}_i)\pi_i(\boldsymbol{\theta}_i)\, d\boldsymbol{\theta}_i$ can be difficult to calculate, since they can involve high dimensional numerical integration. There is a huge literature addressing this issue, including [49], [19], [59], [68], [103], [89], [25], [22], [34], [57], [62], [10], [21], [97], [96], [90], [92], [24].

The second computational issue involves the potential size of the model space. If one has, say, 100 variables in a variable selection problem, the number of possible models is 2^{100}, which is far too large for enumeration; indeed, only a very small fraction of these models can be visited in any scheme for moving around in model space. This is not an issue if the variables are orthogonal (see, e.g. [29]), but otherwise can cause problems. For approaches to dealing with this problem, see [10], [63], [16], [105], [28], [47] and [60]

7.6.2 Approximations and Asymptotics

An important approximation to a Bayes factor is the Laplace approximation, given by

$$B_{21}^L = \frac{\int f_2(\boldsymbol{x} \mid \boldsymbol{\theta}_2)\pi_2(\theta_2)d\theta_2}{\int f_1(\boldsymbol{x} \mid \boldsymbol{\theta}_1)\pi_1(\theta_1)d\theta_1} \approx \frac{f_2(\boldsymbol{x} \mid \widehat{\boldsymbol{\theta}}_2) \mid \widehat{\mathbf{I}}_2|^{-1/2}}{f_1(\boldsymbol{x} \mid \widehat{\boldsymbol{\theta}}_1) \mid \widehat{\mathbf{I}}_1|^{-1/2}} \cdot \frac{(2\pi)^{k_2/2}\pi_2(\widehat{\boldsymbol{\theta}}_2)}{(2\pi)^{k_1/2}\pi_1(\widehat{\boldsymbol{\theta}}_1)}, \tag{7.10}$$

where $\widehat{\boldsymbol{\theta}}_1$ and $\widehat{\boldsymbol{\theta}}_2$ are the m.l.e.'s for $\boldsymbol{\theta}_1$ and $\boldsymbol{\theta}_2$ – which have dimensions k_1 and k_2 – and $\widehat{\mathbf{I}}_1$ and $\widehat{\mathbf{I}}_2$ are observed information matrices. It would be somewhat more accurate to base the $\widehat{\boldsymbol{\theta}}_i$ and $\widehat{\mathbf{I}}_i$ on the products of likelihoods and priors in (7.10), rather than only the likelihoods, but the observed information matrices are readily available in statistical software.

This approximation converges to the actual marginal likelihood as the sample size grows and, often, is surprisingly accurate for even small sample sizes. Indeed, the Laplace approximation has become one of the fundamental tools in the popular Bayesian analysis package INLA (see www.r-inla.org/ where software, references and examples are provided).

Finally, the Laplace approximation is useful in developing simple model selection tools such as BIC (the Bayes Information Criterion, from [93]). Following [88], one can obtain BIC from (7.10) by choosing the $\pi_j(\boldsymbol{\theta}_j)$ to be $N(\boldsymbol{\theta}_j \mid \widehat{\boldsymbol{\theta}}_j, n\widehat{\mathbf{I}}_j^{-1})$, where n is the sample size. With this choice, (7.10) becomes

$$B_{21} \approx \frac{f_2(\boldsymbol{x} \mid \widehat{\boldsymbol{\theta}}_2)}{f_1(\boldsymbol{x} \mid \widehat{\boldsymbol{\theta}}_1)} \cdot n^{\frac{1}{2}(k_2-k_1)},$$

which is the Bayes factor form of BIC. Notice, however, that the 'BIC priors' are centered at the m.l.e.'s, which is a clear (and potentially problematical) double use of the data. Further discussion of these issues can be found in [70], [36], [69], [83], [12], [54], [9], [53], [106], and [85].

In [5], BIC is generalized (the generalization is named PBIC) to include a term arising from the robust prior, and to reflect the fact that there are often different effective sample sizes for different parameters. A method for determining effective sample sizes for parameters in linear models is given in [14].

Acknowledgments

James Berger was supported by the U.S. National Science Foundation grant DMS 1821289. G. García-Donato was supported by grant PID2019-104790GB-I00 from the Spanish Ministerio de Ciencia e Innovación. Elias Moreno was supported by Junta de Andalucía Grant A462 FQM-456-UGR18. Luis Pericchi was supported by UPR/MDACC CA096297/CA063000 INBRE P20 GM103475 (NIGMS).

Note

1. Use $y_i = x_i - \theta_0$, $k_1 = 0$, $X_\gamma^* = 1$, $k_\gamma = 1$ and hence $\Sigma_\gamma^{-1} = n$.

Bibliography

[1] Abramowitz, M. and Stegun, I. A., 1964. *Handbook of Mathematical Functions with Formulas, Graphs, and Mathematical Tables.* New York: Dover.

[2] Barbieri, M. M. and Berger, J. O., 2004. Optimal predictive model selection. *Annals of Statistics, 32*, pp. 870–897.

[3] Bartlett, M., 1957. A comment on D. Lindley's statistical paradox. *Biometrika, 44*, pp. 533–534.

[4] Bayarri, M. J., Berger, J. O., Forte, A. and García-Donato, G., 2012. Criteria for Bayesian model choice with application to variable selection. *The Annals of Statistics, 40*, pp. 1550–1577.

[5] Bayarri, M. J., Berger, J. O., Jang, W., Ray, S., Pericchi, L. R. and Visser, I., 2019. Prior-based Bayesian information criterion. *Statistical Theory and Related Fields, 3*(1), pp. 2–13.

[6] Berger, J. and Delampady, M., 1987. Testing precise hypotheses (with discussion). *Statistical Science, 3*, pp. 317–352.

[7] Berger, J., García-Donato, G., Moreno, E. and Pericchi, L., 2022. The intrinsic hyper-g prior for normal linear models. *in preparation.*

[8] Berger, J. and Pericchi, L., 1996. The intrinsic Bayes factor for model selection and prediction. *Journal of the American Statistical Association, 91*, pp. 109–122.

[9] Berger, J. O., Ghosh, J. K. and Mukhopadhyay, N., 2003. Approximations and consistency of Bayes factors as model dimension grows. *Journal of Statistical Planning and Inference, 112*, pp. 241–258.

[10] Berger, J. O. and Molina, G., 2003. Discussion of "recognition of faces versus greebles: A case study in model selection" by viele et al. In *Case Studies in Bayesian Statistics, Vol. VI*, eds C. Gatsonis, A. Carriquiry, D. Higdon, R. E. Kass, D. Pauler and I. Verdinelli, pp. 111–124. New York: Springer-Verlag.

[11] Berger, J. O., Pericchi, L. R. and Varshavsky, J. A., 1998. Bayes factors and marginal distributions in invariant situations. *Sankhya: The Indian Journal of Statistics, Series A, 60*, pp. 307–321.

[12] Berger, J. O. and Pericchi, R. L., 2001. Objective Bayesian methods for model selection: Introduction and comparison (with discussion). In *Model Selection*, ed P. Lahiri, pp. 135–207. Institute of Mathematical Statistics Lecture Notes - Monograph Series, volume 38, Hayward, CA.

[13] Berger, J., 2022. Four types of frequentism and their interplay with Bayesianism. *The New England Journal of Statistics in Data Science,1*, pp. 126–137.

[14] Berger, J., Bayarri, M. J. and Pericchi, L. R., 2014. The effective sample size. *Econometric Reviews, 33*(1–4), pp. 197–217.

[15] Bernardo, J. M. and Smith, A. F. M., 1994. *Bayesian Theory.* New York: Wiley, John & Sons.

[16] Bottolo, L. and Richardson, S., 2010. Evolutionary stochastic search for Bayesian model exploration. *Bayesian Analysis, 5*, pp. 583–618.

[17] Sabanés Bové, D. and Held, L., 2011. Hyper-g priors for generalized linear models. *Bayesian Analysis, 6*, pp. 1–24.

[18] Burnham, K. P. and Anderson, D. R., 1998. *Model Selection and Inference - A Practical Information-Theoretic Approach.* New York: Springer-Verlag.

[19] Carlin, B. P. and Chib, S., 1995. Bayesian model choice via Markov chain Monte Carlo methods. *Journal of the Royal Statistical Society: Series B (Methodological), 57*(3), pp. 473–484.

[20] Casella, G. and Moreno, E., 2006. Objective Bayesian variable selection. *Journal of the American Statistical Association, 101*(473), pp. 157–167.

[21] Chen, M. H., 2005. Bayesian computations, from posterior densities to Bayes factors, marginal likelihoods and posterior model probabilities. In *Handbook of Statistics, Volume 25: Bayesian Thinking, Modeling and Computation*, eds D. K. Dey and C. R. Rao, pp. 437–458. Amsterdam: North Holland.

[22] Chib, S. and Jeliazkov, I., 2001. Marginal likelihood from the Metropolis-Hastings output. *Journal of the American Statistical Association, 96*, pp. 270–281.

[23] Chipman, H., George, E. I. and McCulloch, R., 2001. The practical implementation of Bayesian model selection. In *Model Selection - IMS Lecture Notes Volume 38*, ed P. Lahiri, pp. 67–116. Hayward, CA: Institute of Mathematical Statistics.

[24] Chopin, N. and Robert, C. P., 2010. Properties of nested sampling. *Biometrika, 97*, pp. 741–755.

[25] Clyde, M., 1999. Bayesian model averaging and model search strategies (with discussion). In *Bayesian Statistics 6*, eds J. M. Bernardo, J. O. Berger, A. P. Dawid and A. F. M. Smith, pp. 157–185. Oxford: Oxford University Press.

[26] Clyde, M., DeSimone, H. and Parmigiani, G., 1996. Prediction via orthogonalized model mixing. *Journal of the American Statistical Association, 91*, pp. 1197–1208.

[27] Clyde, M. and George, E. I., 1999. Empirical Bayes estimation in wavelet nonparametric regression. In *Bayesian Inference in Wavelet-Based Models*, eds P. Muller and B. Vidakovic, pp. 309–322. New York: Springer-Verlag.

[28] Clyde, M., Ghosh, J. and Littman, M., 2011. Bayesian adaptive sampling for variable selection and model averaging. *Journal of Computational and Graphical Statistics, 20*, pp. 80–101.

[29] Clyde. M. and Parmigiani, G., 1996. Orthogonalizations and prior distributions for orthogonalized model mixing. In *Modelling and Prediction: Honoring Seymour Geisser*, eds J. C. Lee, W. O. Johnson and A. Zellner, pp. 206–227. New York: Springer-Verlag.

[30] Clyde, M. A. and George, E., 2004. Model uncertainty. *Stastical Science, 19*, 81–94.

[31] Cripps, E., Carter, C. and Kohn, R., 2005. Variable selection and covariance selection in multivariate regression models. In *Handbook of Statistics, Volume 25: Bayesian Thinking, Modeling and Computation*, eds D. K. Dey and C. R. Rao, pp. 519–552. Amsterdam: Elsevier.

[32] Cui, W. and George, E. I., 2008. Empirical Bayes vs. fully Bayes variable selection. *Journal of Statistical Planning and Inference, 138*, pp. 888–900.

[33] de Santis, F. and Spezzaferri, F., 2001. Consistent fractional Bayes factor for linear models. *Journal of Statistical Planning and Inference, 97*, pp. 305–321.

[34] Dellaportas, P., Forster, J. J. and Ntzoufras, I., 2002. On Bayesian model and variable selection using MCMC. *Statistics and Computing, 12*, pp. 27–36.

[35] Draper, D., 1995. Assessment and propagation of model uncertainty. *Journal of the Royal Statistical Society: Series B, 57*, pp. 45–98.

[36] Dudley, R. M. and Haughton, D., 1997. Information criteria for multiple data sets and restricted parameters. *Statistica Sinica, 7*, pp. 265–284.

[37] Etz, A. and Wagenmakers, E.-J., 2017. J.B.S. Haldane's contribution to the Bayes factor hypothesis test. *Statistical Science, 32*(2), pp. 313–329.

[38] Fernández, C., Ley, E. and Steel, M. F., 2001. Benchmark priors for Bayesian model averaging. *Journal of Econometrics, 100*, pp. 381–427.

[39] Forte, A., García-Donato, G. and Steel, M., 2018. Methods and tools for Bayesian variable selection and model averaging in normal linear regression. *International Statistical Review, 86*(2), pp. 237–258.

[40] Foster, D. P. and George, E. I., 1994. The risk inflation criterion for multiple regression. *The Annals of Statistics, 22*, pp. 381–427.

[41] Fouskakis, D., Ntzoufras, I. and Perrakis, K., 2018. Power-expected-posterior priors for generalized linear models. *Bayesian Analysis, 13*(3), pp. 721–748.

[42] García-Donato, G., Castellanos, M. E. and Quirós, A., 2021. Bayesian variable selection with applications in health sciences. *Mathematics, 9*, p. 218.

[43] García-Donato, G. and Forte, A., 2018. Bayesian testing, variable selection and model averaging in linear models using R with BayesVarSel. *The R Journal, 10*(1), pp. 155–174.

[44] Garcia-Donato, G. and Paulo, R., 2022. Variable selection in the presence of factors: A model selection perspective. *Journal of American Statistical Association*, 117, pp. 1847–1857.

[45] García-Donato, G. and Steel, M. F. J., 2021. Bayes factors based on g-priors for variable selection. In *Handbook of Bayesian Variable Selection*, pages 367–391, Chapter 15, eds M. G. Tadesse and M. Vanucci. New York: Chapman and Hall/CRC.

[46] Garcia-Donato, G. and Forte, A., 2015. *BayesVarSel: Bayes Factors, Model Choice and Variable Selection in Linear Models.* R package version 1.6.1. R Foundation.

[47] Garcia-Donato, G. and Martinez-Beneito, M. A., 2013. On Sampling strategies in Bayesian variable selection problems with large model spaces. *Journal of the American Statistical Association, 108*(501), pp. 340–352.

[48] Geisser, S., 1993. *Predictive Inference: An Introduction.* New York: Chapman & Hall.

[49] Gelfand, A. E. and Smith, A. F. M., 1990. Sampling-based approaches to calculating marginal densities. *Journal of the American Statistical Association, 85*, pp. 398–409.

[50] George, E. I. and Foster, D. P., 2000. Calibration and empirical bayes selection. *Biometrika, 87*, pp. 731–748.

[51] George, E. I. and McCulloch, R., 1997. Approaches for Bayesian variable selection. *Statistica Sinica, 7*, pp. 339–374.

[52] George, E. I. and McCulloch, R. E., 1993. Variable selection via Gibbs sampling. *Journal of the American Statistical Society, 88*, pp. 881–889.

[53] Ghosh, J. K. and Ramamoorthi, R. V., 2003. *Bayesian Nonparametrics.* New York: Springer.

[54] Ghosh, J. K. and Samanta, T., 2001. Model selection – An overview. *Current Science, 80*, pp. 1135–1144.

[55] Ghosh, J. K. and Samanta, T., 2002. Nonsubjective Bayes testing–an overview. *Journal of Statistical Planning and Inference, 103*, pp, 205–223.

[56] Javier Girón, F., 2021. *Bayesian Testing of Statistical Hypotheses.* Malaga: Editorial Arguval.

[57] Godsill, S. J., 2001. On the relationship between Markov chain Monte Carlo methods for model uncertainty. *Journal of Computational Graphics and Statistics, 10*, pp. 230–248.

[58] Good, I. J., 1950. *Probability and the Weighing of Evidence.* London: Charles Griffin.

[59] Green, P., 1995. Reversible jump Markov chain Monte Carlo computation and Bayesian model determination. *Biometrika, 82*, pp. 711–732.

[60] Griffin, J. E. and Steel, M. F. J., 2021. Adaptive computational methods for Bayesian variable selection. In *Handbook of Bayesian Variable Selection*, Chapter 5, eds M. G. Tadesse and M. Vanucci. New York: Chapman and Hall/CRC.

[61] Gupta, M. and Ibrahim, J., 2009. An information matrix prior for Bayesian analysis in generalized linear models with high dimensional data. *Statistica Sinica, 19*, pp. 1641–1663.

[62] Han, C. and Carlin, B. P., 2001. Markov chain Monte Carlo methods for computing Bayes factors: A comparative review. *Journal of the American Statistical Association, 96*, pp. 1122–1132.

[63] Hans, C., Dobra, A. and West, M., 2007. Shotgun stochastic search for "large p" regression. *Journal of the American Statistical Association, 102*, pp. 507–516.

[64] Hoeting, J. A., Madigan, D., Raftery, A. E. and Volinsky. C. T., 1999. Bayesian model averaging: A tutorial. *Statistical Science, 14*(4), pp. 382–417.

[65] Jeffreys, H., 1961. *Theory of Probability.* London: Oxford University Press.

[66] Johnstone, I. M. and Silverman, B. W., 2004. Needles and hay in haystacks: Empirical Bayes estimates of possibly sparse sequences. *Annals of Statistics, 32*, pp. 1594–1649.

[67] Johnstone, I. M. and Silverman, B. W., 2005. Empirical Bayes selection of wavelet thresholds. *Annals of Statistics, 33*, pp. 1700–1752.

[68] Kass, R. E. and Raftery, A., 1995. Bayes factors. *Journal of the American Statistical Association, 90*, pp. 773–795.

[69] Kass, R. E. and Vaidyanathan, S. K., 1992. Approximate Bayes factors and orthogonal parameters, with application to testing equality of two binomial proportions. *Journal of the Royal Statistical Society, 54*, pp. 129–144.

[70] Kass, R. E. and Wasserman, L., 1995. A reference Bayesian test for nested hypotheses and its relationship to the Schwarz criterion. *Journal of the American Statistical Association, 90*, pp. 928–934.

[71] Ley, E. and Steel, M. F. J., 2009. On the effect of prior assumptions in Bayesian model averaging with applications to growth regression. *Journal of Applied Econometrics (Wiley InterScience, Online), 24*(1002), pp. 651–674.

[72] Liang, F., Paulo, R., Molina, G., Clyde, M. A. and Berger, J. O., 2008. Mixtures of g-priors for Bayesian variable selection. *Journal of the American Statistical Association, 103*(481), pp. 410–423.

[73] Ly, A. and Wagenmakers, E.-J., 2021. Bayes factors for peri-null hypotheses. *arXiv*, arxiv:2102.07162.

[74] Marin, J. M. and Robert, C. P., 2007. *Bayesian Core: A Practical Approach to Computational Bayesian Statistics.* New York: Springer.

[75] Maruyama, Y. and George, E. I., 2011. A fully Bayes factor with a generalized g-prior. *Annals of Statistics, 5*, pp. 2740–2765.

[76] Maruyama, Y. and Strawderman, W. E., 2011. Robust Bayesian variable selection with sub-harmonic priors. Technical Report, Center for Spatial Information Science, University of Tokyo. arXiv:1009.1926v2 [stat.ME].

[77] Moreno, E. and Girón, F. J., 2008. Comparison of Bayesian objective procedures for variable selection in linear regression. *Test, 17*(3), pp. 472–490.

[78] Moreno, E., Martínez, C. and Vázquez-Polo, F. J., 2021. Objective Bayesian model choice for non-nested families: The case of the poisson and the negative binomial. *Test, 30*(1), pp. 255–273.

[79] Mulder, J., Berger, J. O., Peña, V. and Bayarri, M. J., 2021. On the prevalence of information inconsistency in normal linear models. *Test, 30*(1), pp. 103–132.

[80] Neal, R., 2001. Transferring prior information between models using imaginary data. Technical Report, Department Statistics, University of Toronto.

[81] O'Hagan, A., 1995. Fractional Bayes factors for model comparisons. *Journal of the Royal Statistical Society: Series B, 57*, pp. 99–138.

[82] O'Hagan, A., 1997. Properties of intrinsic and fractional Bayes factors. *Test, 6*, pp. 101–118.

[83] Pauler, D., 1998. The Schwarz criterion and related methods for normal linear models. *Biometrika, 85*, pp. 13–27.

[84] Paulo, R., Clyde M., Berger, J., Ibrahim, J., Mallick, B.,Huber, M., Liang, F., Zhao, H., 2002. Notes on model selection for the normal multiple regression model. Model selection group of the 2002 stochastic computation SAMSI Program. SAMSI.

[85] Peña, V. and Berger, J. O., 2020. Restricted type II maximum likelihood priors on regression coefficients. *Bayesian Analysis, 15*(4), pp. 1281–1297.

[86] Pérez, J. M. and Berger, J., 2001. Analysis of mixture models using expected posterior priors, with application to classification of gamma ray bursts. In *Bayesian Methods, with Applications to Science, Policy and Official Statistics*, eds E. George and P. Nanopoulos, pp. 401–410. Luxembourg: Official Publications of the European Communities.

[87] Pérez, J. M. and Berger, J., 2002. Expected posterior prior distributions for model selection. *Biometrika, 89*, pp. 491–512.

[88] Raftery, A. E., 1999. Bayes factors and BIC - comment on 'a critique of the Bayesian information criterion for model selection'. *Sociological Methods and Research, 27*, pp. 411–427.

[89] Raftery, A. E., Madigan, D. and Hoeting, J. A., 1997. Bayesian model averaging for regression models. *Journal of the American Statistical Association, 92*, pp. 179–191.

[90] Raftery, A. E., Newton, M. A., Satagopan, J. and Krivitsky, P., 2007. Estimating the integrated likelihood via posterior simulation using the harmonic mean identity (with discussion). In *Bayesian Statistics 8*, eds J. M. Bernardo, M. J. Bayarri, J. O. Berger, A. P. Dawid, D. Hechkerman, A. F. M. Smith and M. West, pp. 1–45. Oxford: Oxford University Press.

[91] Rodríguez, A. and Pericchi, L. R., 2001. Intrinsic Bayes factors for dynamic linear models. In *ISBA 2000 Proceedings*, pp. 459–468. Washington, DC.

[92] Rue, H., Martino, S. and Chopin, N., 2009. Approximate Bayesian inference for latent Gaussian models by using integrated nested Laplace approximations. *Journal of the Royal Statistical Society: Series B, 71*, pp. 319–392.

[93] Schwarz, G., 1978. Estimating the dimension of a model. *Annals of Statistics, 6*, pp. 461–464.

[94] Scott, J. and Berger, J., 2005. An exploration of aspects of Bayesian multiple testing. *Journal of Statistical Planning and Inference, 136*, pp. 2144–2162.

[95] Scott, J. and Berger, J., 2010. Bayes and empirical-Bayes multiplicity adjustment in the variable-selection problem. *Annals of Statistics, 38*, pp. 2587–2619.

[96] Sivia, D. S. and Skilling, J., 2006. *Data Analysis: A Bayesian Tutorial* (2nd ed.). Oxford: Oxford University Press.

[97] Skilling, J., 2006. Nested sampling for general Bayesian computation. *Bayesian Analysis, 1*, pp. 833–860.

[98] Smith, A. F. M. and Spiegelhalter, D. J., 1980. Bayes factors and choice criteria for linear models. *Journal of the Royal Statistical Society: Series B, 42*, pp. 213–220.

[99] Smith, M. and Kohn, R., 1996. Nonparametric regression using Bayesian variable selection. *Journal of Econometrics, 75*, pp. 317–344.

[100] Som, A., Hans, C. M. and MacEachern, S. N., 2016. A conditional Lindley paradox in Bayesian linear models. *Biometrika, 103*(4), pp. 993–999.

[101] Steel, M. F., 2020. Model averaging and its use in economics. *Journal of Economic Literature, 58*(3), pp. 644–719.

[102] Student, 1908. The probable error of a mean. *Biometrika, 6*(1), 1–25.

[103] Verdinelli, I. and Wasserman, L., 1996. Bayes factors, nuisance parameters, and imprecise tests. In *Bayesian Statistics 5*, eds J. M. Bernardo, J. O. Berger, A. P. Dawid and A. F. M. Smith, pp. 765–771. London: Oxford University Press.

[104] Wang, X. and George, E. I., 2007. Adaptive Bayesian criteria in variable selection for generalized linear models. *Statistica Sinica, 17*, 667–690.

[105] Wilson, M. A., Iversen, E. S., Clyde, M. A., Schmidler, S. C. and Schildkraut, J. M., 2010. Bayesian model search and multilevel inference for SNP association studies. *Annals of Applied Statistics, 4*, 1342–1364.

[106] Zak-Szatkowska, M. and Bogdan, M., 2011. Modified versions of Bayesian information criterion for sparse generalized linear models. *Computational Statistics and Data Analysis, 55*, pp. 2908–2924.

[107] Zellner, A., 1986. On assessing prior distributions and Bayesian regression analysis with g-prior distributions. In *Bayesian Inference and Decision techniques: Essays in Honor of Bruno de Finetti*, ed A. Zellner, pp. 389–399. Cheltenham: Edward Elgar Publishing Limited.

[108] Zellner, A. and Siow, A., 1980. Posterior odds for selected regression hypotheses. In *Bayesian Statistics*, eds J. M. Bernardo, M. H. DeGroot, D. V. Lindley and A. F. M. Smith, pp. 585–603. Valencia: Valencia University Press.

[109] Zellner, A. and Siow, A., 1984. *Basic Issues in Econometrics.* Chicago: University of Chicago Press.

8

A BFFer's Exploration with Nuisance Constructs: Bayesian p-value, H-likelihood, and Cauchyanity

Xiao-Li Meng
Harvard University

8.1 Becoming an Adolescent BFFer

8.1.1 A Journey of Managing Nuisance Constructs

Scientific inquiries seek evidence to establish, support, or challenge theories and arguments. Therefore, hypothesis testing and statistical estimation, including prediction and clustering, form the pivotal subjects in all four branches of the statistical discipline: foundations, methods, computation, and applications. However, variations in data create both information and noise, concepts that themselves defy static treatments. As a result, statistical evidence and its attainment processes come with many grains of salt, and many flavours, not all of which are palatable to everyone or even anyone. The ongoing controversy around p-values and the contrast between frequentist hypothesis testing and Bayesian hypothesis assessment, despite a shared goal, serve as examples.

A silver lining of such a state of affairs is the opportunity for learning and growth. This chapter samples an idiosyncratic learning journey about managing nuisance constructs, undertaken by a pragmatic Bayesian (B), a fiducialist (F) sympathizer, and an evaluative frequentist (F). The term *nuisance constructs* pertains to any necessary but unknown elements in a learning process that are not the primary target. The most recognized nuisance constructs are nuisance parameters, such as the variance parameter σ^2 in the Normal model $N(\mu, \sigma^2)$ when our interest lies in μ. The value of σ^2 is crucial because it quantifies the difficulty of the learning problem: the higher the σ^2, the greater the uncertainty in estimating μ. The adjective 'nuisance' therefore is unfortunate, as it signifies our annoyance of not knowing the value of σ^2, not its vital role in our learning problem. However, dealing with unknowns is the basis for statistics as a discipline, making the phenomenon of nuisance-but-essential not exclusive to model parameters.

For example, in the monograph on the likelihood principle by Berger and Wolpert (1988)—a must-read for any BFFer wannabe—the authors classify unobservable variables into variables of interest and 'nuisance variables,' which include measurement errors, missing data, unobserved confounders, etc. While they are nuisances in the sense that we pray for their absences, they can have a significant impact on the estimation or testing of quantities of interest. Therefore, properly managing nuisance constructs is essential for accurate and reliable statistical inference and learning.

DOI: 10.1201/9780429341731-8

Furthermore, specifying a prior distribution is a paramount step for any Bayesian inference, but the prior itself is not a learning target by its very ideation. Hence, any unknown component of a prior distribution is a nuisance construct. A grand challenge for the Bayesian paradigm is that the entire shape of the prior distribution is a nuisance construct since it is virtually impossible to pinpoint knowledgeably the relative frequencies of all *a priori* plausible states. The challenge is not addressable by invoking subjective prescriptions because for such prescriptive actions to be internally coherent, they must include one for prescribing complete ignorance. However, complete ignorance cannot be logically captured by any probabilistic distribution by its very requirement of specifying relative frequencies, all of which represent knowledge.

Empirical Bayes or data-driven priors are sometimes used in practice. However, they are not part of the Bayesian prescription, but rather a pragmatic remedy for managing the headache created by the missing ingredients required by the prescription. The term 'construct' is especially apt for describing prior specifications, as the phrase typically carries a flavor of subjectivity[1]. All in all, managing nuisance constructs constitutes much of the statistical enterprise, and activities recounted in this chapter take place ubiquitously. It also distinguishes statistics from pure pattern recognition activities, but it can significantly benefit the latter; see a recent demonstration by Lo et al. (2019) on predicting drug approval, where constructing statistical imputation before machine learning outperformed significantly the execution of the latter by itself.

8.1.2 A Trio of BFF-Learning (and Practicing) Stories

Risking a cliché that challenges create opportunities, my experiential learning about statistical schools of thought has exemplified the wisdom of this adage. This chapter samples three learning and researching experiences in dealing with nuisance constructs, after tracing back to my earliest encounter with nuisance parameter, a hard lesson for a fearless but clueless college student.

The first story recounts my experiment with injecting Bayesian ingredients into a frequentist recipe to formalize the notion of posterior predictive p-value for managing nuisance parameters in hypothesis testing. This would have been a risky adventure for a tenure-seeking junior researcher, if not for the intellectual support and rigor of my statistical colleagues at the University of Chicago. This recounting also provides an opportunity to clarify some historical confusion about what posterior predictive p-value can or cannot do, issues that I was too inexperienced then to anticipate.

The second story reveals a riskier adventure, but for a different reason. It took place during the busiest year of my department chairship, dealing with the aftermath of the 2008 financial crisis, which was a true nuisance for any time-pressed researcher. Perhaps because my Harvard colleague Robert Langdon was too occupied with *Angels & Demons* (2009), I was called in to decode the h-likelihood mystery. What on earth is h-likelihood and what could 'h' possibly mean? In case a reader is as curious as I was then, be warned that curiosity was exactly how I got myself into a risky, or in the words of Edward George (in his discussion of Meng (2011)), a "heroic" voyage into the world of nuisance constructs, trying to traverse the entire BFF labyrinth with not much more time than was available to Professor Langdon for preventing an antimatter explosion. No explosives in this story, but there are a few traps as detailed in Meng (2009, 2011), and further discussed in this chapter, in managing nuisance variables and nuisance distributions.

The third story comes with a pleasant surprise. It started with a dinner conversation in 2014 regarding a conjecture of Drton and Xiao (2016), which states that a convex linear combination of certain type of *dependent* Cauchy variables is still Cauchy. The conjecture was proved in Pillai and Meng (2016), which confirms that there are uncountably many previously unknown pivotal quantities for estimating a multivariate Normal mean that is invariant to the Normal covariance matrix as a nuisance object, regardless of the dimension of the covariance matrix. The story reflects on the result and proof, which is built on a little known property about the Cauchy distribution given in Pitman and Williams (1967) and Williams (1969). This property has the potential to turn Cauchyanity from a theoretical curiosity into a surprisingly efficient—at least computationally—way of dealing with nuisance covariance matrices, regardless of their dimensions.

This chapter also interlaces research explorations with pedagogical reflections, a hybrid style blending a self-serving purpose with (hopefully) an altruistic one. I volunteered to contribute this chapter to document some milestones and millstones in my learning journey, as stepping stones for students to shorten theirs, even just by a stone's throw. The included stories enlivened my scholarly life at their own leisure, and the hybrid format allows me to regale readers chronologically as they unfold, rather than being curated around a particular theme. An orchestrated theme may indeed provide a sharper message, but this is an apt occasion to invoke a Pascalian apology: "I'm sorry that I don't have time to write a shorter one." All excuses aside, the chapter aims to evince the inevitability and benefit of multi-perspective thinking, as well as its traps to watch over, as we seek and learn about the essence of statistics, not unlike when we contemplate the meaning of life.

With that as the prologue, I invite readers who can indulge my meandering through a 35-year memory lane to join me for an expedition still in progress. I do not know where it is heading, but I do hope that it is a BFF bonding journey. To increase the rewards and joy of the expedition, let us warm-up our appreciation of each other by first cherishing the diversity of our world views through understanding and respecting different paths to approach a shared destination.

8.2 A Warm-up Exercise: Appreciating the BFF Trinity

8.2.1 A Common Imaginary for BFF

The concept of *imaginary representation* or *imaginary* in short, as a mind construct, appears more often in philosophical writings (e.g., Leonelli, 2021) than in statistical literature. But it is a rather useful construct for comparing and contrasting the different world views of Bayesian, frequentist, and fiducialist (Craiu et al., 2022). The construct starts with an imaginary triplet $\mathcal{T} = \{\mathbf{X}, \theta, U\}$, where $\mathbf{X}$ denotes the observable (e.g., data), which may not be fully observed. The notation θ collects all our learning objects (e.g., model parameters), which may contain auxiliary objects (e.g., nuisance parameters). The quantity U is a nuisance construct, representing all unobservables responsible for the indeterminacy of θ by $\mathbf{X}$ alone, such as measurement errors or latent traits.

Imagine a location-scale representation for n observations X_i's in the form of a linear model

$$X_i = \mu + \sigma U_i, \quad i = 1, \dots, n. \tag{8.1}$$

Hence $\mathbf{X} = \{X_1, \dots, X_n\}$, $\theta = \{\mu, \sigma\}$, and $U = \{U_1, \dots, U_n\}$, which is endowed with a (joint) probability distribution, $f(U)$, that is free of θ (but f itself may not be known otherwise).

Equations such as (8.1) have been given several names, from the most descriptive—data generating equation (Hannig et al., 2016), to more mathematical—stochastic representation

(Ding and Blitzstein, 2018), and to the most inferential—structural equation (Fraser and Haq, 1969). Regardless of its label, they postulate a set of algebraic relationships among $\{\mathbf{X}, \theta, U\}$, which in turn transfer the variations in U (and possibly in θ) into that for $\mathbf{X}$. Learning about θ then is an intellectual game of separating the variations in $\mathbf{X}$ into *revelatory variations* that update our knowledge about θ, and *obfuscatory variations* that contribute to uncertainty in our learning. The three BFF perspectives correspond to choosing respectively the three components in $\mathcal{T}$, namely, θ for Bayesians, $\mathbf{X}$ for frequentists, and U for fiducialists, as their primary players of the game.

8.2.2 Contrasting Frequentist and Bayesian Imaginaries

Most of us started with frequentist imaginaries because textbooks introduce them prior to other schools of thoughts, and because they are more easily relatable from the data themselves. When we observe $\mathbf{X} = \{X_i, i = 1, \ldots, n\}$, where i indexes subjects in a population, it does not require much mental gymnastic to imagine some variations of it from the same population, as hinted by the fact that $\mathbf{X}$ contains n elements. For example, a slightly more creative mind may contemplate creating n data sets each with $n-1$ subjects in them, by removing each X_i from $\mathbf{X}$ in turn, call it $\mathbf{X}_{-i}$. Clearly $\mathbf{X}_{-i}$ differs from $\mathbf{X}$ mathematically, but is the former qualitatively different from $\mathbf{X}$ in revealing the data variations expressible by the latter? Indeed, historically, such thought experiments created the jackknife, which ultimately led to the all-mighty or perhaps all-too-mighty bootstrap (Efron, 1992).

Conceptually speaking, in a frequentist imaginary, all pairs $\{(\mathbf{X}'_j, \theta), j = 1, \ldots\}$ share the same learning target θ, as depicted in the left-most diagram in Figure 8.1, adopted from Craiu et al. (2022). (For simplicity of illustration, we assume that there are only countably many pairs.) Here $\mathbf{X}'_j$s are imagined to be generated (typically) independently from some sampling distribution $p(\mathbf{X}'|\theta)$, creating the pairs $\{(\mathbf{X}'_j, \theta), j = 1, \ldots, \}$, for which the actual realized pair $(\mathbf{X}, \theta)$ is one of them. Any process that leads to a learning outcome for θ, denoted by $\hat{\theta}(\mathbf{X})$, can then be repeated on $\{(\mathbf{X}'_j, \theta), j = 1, \ldots\}$, analytically or by simulation. The replications are needed to study the sampling properties of $\hat{\theta}(\mathbf{X})$, such as the frequency for $\theta \in \hat{\theta}(\mathbf{X}')$, if $\hat{\theta}(\mathbf{X})$ is a confidence set for θ.

The conceptual simplicity of the frequentist imaginary comes at the expense of its relevance for our learning objectives, that is, it violates the *relevancy principle* articulated by Dempster (1963a). We want to learn reliably about θ with the data $\mathbf{X}$ at hand. It takes a

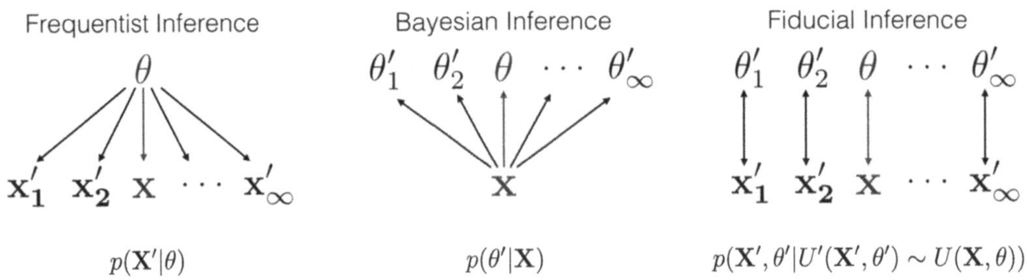

FIGURE 8.1
Reproduced from Craiu et al. (2022) (but see the text for a correction on a notation in conditioning) to depict the imaginary replications of $\mathcal{T}$ for frequentist, Bayesian, and fiducial inferential paradigms. Here $\mathbf{X}$ represents observable, θ the model parameters, U the unobservable, and p the respective conditional distribution.

leap of faith to trust that the reliability of $\hat{\theta}(\mathbf{X}')$ over many hypothetical $\mathbf{X}'$s is transferable to its reliability for a particular setting (Craiu et al., 2022), and indeed such a transference is a philosophical construct contemplated since the time of Galen in the Roman empire (Li and Meng, 2021). The Bayesian imaginary replaces this leap of faith with a different but no less troublesome one, namely, that we can reliably construct a distribution to describe our *a priori* knowledge about θ. The varying states of θ provided by this prior distribution then form the Bayesian imaginary, i.e., all θ_j''s that could have generated the data at hand, $\mathbf{X}$. Thus the Bayesian formulation turns the frequentist imaginary upside down, as seen in the middle diagram of Figure 8.1, because all pairs of $(\mathbf{X}, \theta_j')$ share the same data $\mathbf{X}$.

Whether one has difficulty to imagine the varying state for a single θ or to appreciate the use of probabilistic distribution for describing knowledge, the middle diagram of Figure 8.1 makes it crystal clear that a set of imagined replications for θ, that is, a prior specification, is paramount for Bayesian paradigm. Without these replications, there will be only one pair $\{\mathbf{X}, \theta\}$ left when we do not allow ourselves to consider hypothetical replications for $\mathbf{X}$. The single pair $\{\mathbf{X}, \theta\}$ construct is perfectly relevant in the sense of Dempster (1963b) and of Liu and Meng (2016), but it is utterly useless for Bayesian inference or any *probabilistic* inference framework, which requires counterfactual states for operation, regardless of the origin or the designation of our probabilistic journey. (One can certainly conduct deterministic inferences without imaginary replications, such as the direct likelihood approach, but that opens a different can of worms in terms of assessing their reliability or even meaning; see Liu and Meng (2016) for a comparative discussion.)

8.2.3 Imagine the Fiducial Imaginary

Since the frequentist's replications for $\mathbf{X}$ and Bayesian's replications for θ both require a leap of faith, it is natural to explore the only remaining component, U, in the hope that it can lead to something less problematic. At first sight, this thought experiment seems ill-formed, because U apparently is the least important element in $\mathcal{T}$, being neither the data nor the learning target, that is, U is inherently a nuisance object. However, since the trio $\{\mathbf{X}, \theta, U\}$ are connected via a data generating equation such as (8.1), we can express θ in terms of $(\mathbf{X}, U)$. Hence once $\mathbf{X}$ is observed, any assessment about U can be translated into that for θ. For example, from (8.1), we have

$$\frac{1}{\sigma^2}\sum_{i=1}^{n}(X_i - \bar{X}_n)^2 = \sum_{i=1}^{n}(U_i - \bar{U}_n)^2, \tag{8.2}$$

where $\bar{X}_n$ and $\bar{U}_n$ are the sample averages of $\{X_1, \ldots, X_n\}$ and $\{U_1, \ldots, U_n\}$ respectively. Consequently, for any replication $\{\mathbf{X}', \theta'\}$ such that its corresponding U' is consistent with the realized U (i.e., the U corresponds to the data $\mathbf{X}$), $\sum_i(X_i' - \bar{X}_n')^2/(\sigma')^2$ will follow the same distribution that generates the right-hand side of (8.2), under the replications as depicted in the right-most diagram in Figure 8.1. Note the conditioning in the fiducial setting $U'(\mathbf{X}', \theta') \sim U(\mathbf{X}, \theta)$ corrects the notation $U'(\mathbf{X}', \theta') = U(\mathbf{X}, \theta)$ in the original plot of Craiu et al. (2022). The notation $U' = U$ would suggest that they are the same algebraically, not that the U' and $(\mathbf{X}', \theta')$ are probabilistically compatible with the realized U corresponding to $\mathbf{X}$. Indeed the use of the "$\sim$" relationship itself is correct only when $\mathbf{X}$ does not restrict the state space of U. This complication signals the incoherence of the fiducial operation, because it uses the knowledge of $\mathbf{X}$ to restrict the state space of U, but not its distribution otherwise; see Dawid and Stone (1982) and Dawid (2023).

Nevertheless, the fiducial approach provides a means to construct confidence intervals for σ, just as with the frequentist approach, as well as mechanisms to consider replications for σ, as a Bayesian would. That is, operating on the U component seems to build a bridge between the frequentist approach and the Bayesian goal, but without actually needing to construct a prior distribution for θ. Historically when Fisher (1935) proposed the fiducial argument, it was indeed an attempt to "make the Bayesian omelet without breaking the Bayesian egg", as Savage (1961) quipped, that is, to conduct a Bayesian-esque inference but without specifying a prior.

To see how this might be done, let us assume that $U_1, \ldots, U_n \overset{\text{i.i.d}}{\sim} N(0,1)$. It follows then that $\sum_i (U_i - \bar{U}_n)^2 \sim \chi^2_{n-1}$, a fully known distribution because it is a pivotal quantity by design. There is nothing new or unusual here, and indeed it is equivalent to the corresponding frequentist setting with $\sum_i (X_i - \bar{X}_n)^2 \sim \sigma^2 \chi^2_{n-1}$. However, the fiducial approach rewrites (8.2) to declare that

$$\sigma^2 = \frac{\sum_i (X_i - \bar{X}_n)^2}{\sum_i (U_i - \bar{U}_n)^2} \sim \frac{\sum_i (X_i - \bar{X}_n)^2}{\chi^2_{n-1}}. \tag{8.3}$$

This declaration mixes a logical derivation with a leap of faith of the third kind. The first equality follows directly the identity (8.3), and hence it is logical. The second equality "$\sim$", in terms of the distribution, is where the trouble begins. When we assert that $\sum_i (U_i - \bar{U}_n)^2 \sim \chi^2_{n-1}$, it is with respect to the distribution of $\{U_1, \ldots, U_n\}$, irrespective of any potential values of $\{X_1, \ldots, X_n\}$ generated by them. However, in (8.3), since the value of $\sum_i (X_i - \bar{X}_n)^2$ is kept the same when we move from the second expression to the third expression, it effectively assumes that the χ^2_{n-1} distribution is applicable to $\sum_i (U_i - \bar{U}_n)^2$ even among all the pairs $\{(U_i, X_i), i = 1, \ldots, n\}$ for which the corresponding value of $\sum_i (X_i - \bar{X}_n)^2$ never changes. That is, this assertion amounts to assume $\sum_i (U_i - \bar{U}_n)^2$ are independent of $\sum_i (X_i - \bar{X}_n)^2$ on the joint space for the pairs $\{(X_i, U_i), i = 1, \ldots, n\}$.

However, relationship (8.1) tell us that X and U are perfectly correlated conditioning on σ. This should remind us that any *probabilistic* imaginary we can come up with for the fiducial argument must involve replications for σ if we want to fix $\mathbf{X}$ because otherwise the aforementioned independence is impossible. Therefore, even if the fiducial argument is designed to operationally avoid constructing a prior, conceptually it must endow the learning target—here being σ—a probabilistic structure *a priori*, just as under the Bayesian framework. A key question then is in order for the right-hand side of (8.3) to be a legitimate posterior distribution for σ^2, what should be its corresponding prior? Does such prior even exist? Without the Bayesian collar, can we unleash our imaginations to create an imaginary under which the last assertion in (8.3) can be reified, for example by a simulation experiment without any internal inconsistency? My learning about how to answer such questions has been a long one, as I shall report after reflecting on even earlier learning experiences of the frequentist and Bayesian frameworks.

8.3 A Frequentist Lesson: A Student's Taste of t

8.3.1 Why Is the Null Hypothesis Typically a Dull Hypothesis?

In a classical hypothesis testing setup, we typically start with a hypothesis of substantive interest, such as whether a COVID vaccination is effective. However, the substantive hypothesis usually is not directly testable since it is often not well specified; for example, how effective is *effective*? The standard statistical hypothesis testing therefore sets the null hypothesis, denoted by H_0, to be the opposite of the substantive hypothesis of interest in such cases. That is, the null usually is the hypothesis that we are not excited by or interested in, a 'dull hypothesis'. For example, the hypothesis that the vaccination is not effective may be formulated mathematically as an equality of probabilities of suffering from COVID between those who receive the vaccination and those who do not, an equality amenable to falsification by data. By collecting evidence to cast sufficient doubt on a 'dull hypothesis', we can be excited to the according degree about its antithesis.

Applying this negation logic in statistical inference is not by choice but by necessity, because probabilistic inferences sharpen our knowledge by eliminating states that are probabilistically incompatible with our data by some declared criterion, not by logically pinpointing a particular knowledge state. The latter is infeasible by the very nature that we adopt a probabilistic treatment of data as imperfect representations of objects associated with our learning target, not as logical constraints to reduce our solution space, as with algebraic relationships, for instance.

By this necessity, once H_0 is explicated, our task is to choose a test statistic T that can detect some probabilistic differences when data are incompatible with H_0. The phrase 'choose' signifies plurality and complexity. For example, whether we allow T to capture the direction of the difference would reflect whether we *a priori* contemplate the harmfulness of vaccination on COVID suffering itself. However, the term *statistic* has an unambiguous requirement, which I learned the hard way, during my first course in statistics as a pure math major at Fudan University (1978–1982).

8.3.2 An Unforgettable Introduction to Pivotal Quantity

During my years at Fudan, being a teenager without distractions of electronic gadgets or social media (but with plenty of socialist media), I had all the curiosity, restlessness, and unfilled brain space to do well on exams: nearly all my exam scores exceeded 95 points out of 100. This was an ego-stroking achievement, since Fudan's math faculty had this tradition of ending an exam with a 5-point special exam problem that they believed to possess the magic power of testing whether a student could become one of them, figuratively and literally (back then, it is a common practice in Chinese universities to hire their own top students as faculty). It was *nearly all* because on the statistics final exam, the first problem asked us to perform a hypothesis testing, but without any data, real or fake. I completed it briskly since it was an undemanding Z-test problem, and I needed to save as much time—and brain power—for the 5-point problem. The time-saving strategy apparently worked, for I had sufficient time to nail the special problem. I was therefore rather sanguine that my ego would be stroked for the $(n+1)st$ time, that is, until I got my score: 90.

What? How could that be? Losing a couple of points would be expected since we were not tested merely on being correct, but also on being rigorous and succinct—but not laconic—in our proofs. It is not always easy to know when to invoke the most beloved and hated mathematical assertion "It is easy to show that ...", for which I had another hard lesson[2]. However, losing 10 points could only be justified by a wrong answer unless a student was extraordinarily sloppy, a trait I was never permitted to explore by my mother because I was born in a year of the tiger.

Discombobulated, I visited my instructor, who revealed unhesitatingly that I bombed the Z-test problem. More dumbfounded, I contented that my derivation there was airtight. The instructor agreed, but with a gotcha smile: "You solved a wrong problem. It's t-test!" Indeed, the problem didn't assume that the variance σ^2 was known. But without any data to alert me that the Z-test statistic couldn't be evaluated, my airtight derivation provided only evidence of my statistical airhead, instead of (potentially) usable evidence for hypothesis testing. This incident permanently imprinted my encephalon with a requirement of a *statistic*: a function of the data only.

However, it would take me almost another decade to realize the real essence of going from Z to t: the avoidance of infinite regress, or in a more statistical term, the attainment of pivotality. That is, under the null, not only the test statistic but also its distribution is free of any unknown quantity. Hence, whatever statistical evidence we obtain will itself be free of uncertainty (under the null), avoiding the need to estimate *that* uncertainty, an estimate may come with its own uncertainty, and hence need yet another estimate, a process that may never end. It took me this long because while I was a mathematically competent student, statistically I was neither trained properly nor self-taught adequately for a very simple yet fatal reason: no problem I had worked on involved any data or substantive inferential contemplation until my Ph.D. training at Harvard (1986-1990).

8.4 A Bayesian Motivator: A Teacher's Taste of Two-Sample t

8.4.1 A Nuanced (or Nuisance) Question

If my college experience is a case of a student's sour taste of t, then an encounter during my University of Chicago years as a faculty (1991–2001) suggested that a teacher's taste of t could be even bitter (without any Guinness involved), when hypothesis tests are treated merely as mathematical procedures. This time I was on the side of "grading" for hiring a lecturer to teach introductory level statistics. The interview involved a mock teaching session by the candidates, and we suggested the topic of two-sample t tests. One candidate provided a fairly clear presentation, going through the usual order from two-sample tests with equal variances to cases without the equal-variance assumption. Chicago's department of statistics had a reputation for being one of the most rigorous programs in scholarship, which interweaves research and education to their full extent. As a beneficiary of that environment, I was compelled to continue the tradition of raising thought-provoking skepticism for speakers and insight-revealing questions for candidates. In this case, after the candidate talked about how to conduct the approximate t-test for the unequal variance case, I asked that given the modern computational power, could we replace the theoretical approximation by Monte-Carlo simulation from the exact null distribution?

Readers who have some knowledge about the Behrens-Fisher problem (Wallace, 1980) may immediately recognize that this is a nuanced question, designed to test if the candidate had sufficient statistical understanding and insights about the nature of the t approximation. A main reason for using an approximate t-distribution is that the exact null distribution depends on a nuisance parameter, that is, the ratio of the two variances, which is unknown. Hence it is not possible to simulate the null distribution without imposing assumptions that are not a part of the H_0 being tested. For one-sample tests, the problem is avoided by the celebrated 'student' t-statistic, by William Sealy Gosset; see Fienberg and Lazar (2001). But we have no such luck for two samples.

8.4.2 A Seriously Missed Educational Opportunity

Sadly, current introductory courses almost never highlight sufficiently the intellectual treasure unearthed by Student's t, seriously missing an opportunity for broadly conveying statistical intelligence, to borrow a trendy phrase. However, the question I raised belongs to the kind of questions that astute undergraduate students likely would pose.[3] My worry about the candidate's inability to address such questions unfortunately was confirmed, as the response was that exact simulation can be implemented but maybe it is too complicated for the students in the introductory courses.

My concern stems from the observation that despite possessing considerable teaching credentials, the candidate's understanding seemed limited, missing opportunities to encourage students to think beyond the mechanics of procedures or approximations. Over time, witnessing the poor state of statistical competence has become less shocking but more concerning. This concern is amplified by the procedure-oriented culture in various parts of the current data science. The inattention to statistical insights and sagacity, in research and in education, can negatively impact the long-term development of both research and researchers. As a statistics educator, I am compelled to push myself and fellow statisticians to engage in deep teaching, i.e., engaging students in deep statistical contemplation in every course we teach. As the rest of this chapter shows, statistical theories and methods are innately open to challenge. Hence, opportunities to engage and thought-provoke students can always be found or created.

In particular, when pivotal quantities are elusive, it creates an opportunity to introduce essential statistical thinking and methods for managing nuisance objects. These methods typically fall into the three major paradigms discussed in this handbook: Bayesian, frequentist, and fiducial, excluding hybrid methods. Bayesian methods are perhaps the most straightforward to motivate and describe, as they adhere to the standard Bayesian recipe. The principal challenge with Bayesian methods lies in specifying a reasonable prior distribution for all unobserved quantities, regardless of whether they are nuisance or not. Once the prior is established and the object of interest identified, the remaining challenge is largely computational. This can be prohibitively expensive in practice, necessitating approximations such as those offered by Markov chain Monte Carlo (Brooks et al., 2011). (Bayesian methods don't exclusively hold interpretational challenges once the prior has been rationalized.)

In the realm of hypothesis testing, however, there are two types of 'objects of interest': the hypotheses under consideration and the various metrics used to test them. When these metrics depend on nuisance parameters, they become unknown quantities for which we can seek a Bayesian estimate. This line of thought guided me when I was addressing a contemporary iteration of the Behrens-Fisher problem within hypothesis testing using multiply imputed data. It led me to explore the concept of the posterior predictive p-value (Meng, 1994), an adventure with several lessons of its own, as I shall report next.

This venture also introduced me to fiducial thinking, because my treatment of the Behrens-Fisher problem closely resembled Fisher's fiducial solution, as my Chicago colleague David Wallace pointed out. This sparked my curiosity about a school of thought that I never learned in school, and ultimately, despite its logical challenges, led me to sympathize with the fiducial perspective. This sympathy, akin to relating to those striving to exceed their limits, was borne out of its difficulties. So read on, but be prepared for some unease as you witness a coherent Bayesian mind infiltrated by incoherent fiducial thoughts.

8.5 A Pragmatic Bayesian Experiment: A Taste of Mixing Two Perspectives

8.5.1 Posterior Predictive p-values

In a frequentist setting, when H_0 fully specifies the null distribution of a test statistic $T(X)$, then the p-value, typically in the form of a tail-area probability, $p = \Pr\{T(X) \geq T(x) \mid H_0\}$, is computable, analytically, numerically, or by simulation. When H_0 fails to do so, this p

value in general depends on some nuisance constructs ν, unless $T(X)$ happens to be a pivotal quantity. Here ν may represent parameters, such as the variance ratio for Berhren-Fisher problem, or distributions, such as in testing the independence of two predictors with unknown (marginal) distributions. We then write $p = p(\nu)$, where ν is often replaced by an estimate $\hat{\nu}$ or maximized out to produce the worst p-value: $\sup_\nu p(\nu)$; see Meng (1994) for a discussion.

However, if we allow ourselves to inject some Bayesian flavor, we can treat $p(\nu)$ as an unknown object and integrate out ν by imposing a prior on it. Consider a simple case with parameter $\psi = \{\theta, \nu\}$, ν being the nuisance parameter, and $\theta = \theta_0$ the null hypothesis H_0. If we adopt a joint prior density $\pi(\theta, \nu)$, then under H_0, the only logical prior for ν is its implied conditional prior $\pi(\nu \mid \theta = \theta_0)$. This H_0–induced prior will lead to the H_0-induced posterior for ν:

$$\pi_0(\nu \mid x) = \frac{f(x \mid \theta_0, \nu)\pi(\nu|\theta_0)}{\int_A f(x \mid \theta_0, \nu)\pi(\nu|\theta_0)d\nu}, \quad \nu \in A, \tag{8.4}$$

where $f(x|\psi)$ is the sampling model for linking the observation x to the parameter ψ. We can then compute the expected $p(\nu)$ with respect to $\pi_0(\nu \mid x)$, which yields

$$p_B = \Pr\{T(\mathbf{X}^{\text{rep}}) \geq T(x) \mid x, H_0\}. \tag{8.5}$$

This tail-area probability was termed *posterior predictive p-value (ppp)* because the replication X^{rep} here is generated by the posterior predictive distribution under H_0, that is,

$$f(\mathbf{X}^{\text{rep}} \mid x, H_0) = \int f(\mathbf{X}^{\text{rep}} \mid \theta_0, \nu)\, \pi_0(d\nu \mid x). \tag{8.6}$$

Consequently, p_B can be viewed as applying the frequentist construct of a tail-area probability to the posterior predictive distribution of a test statistic. The notion of a Bayesian p-value (Gelman, 2005) is often regarded as oxymoronic due to its use of unobserved replications, $\mathbf{X}^{\text{rep}}$, for evaluating statistical evidence. This appears to directly contradict the Bayesian emphasis on conditioning on available data. However, this emphasis on relevance is itself conditioned on the reliability of the Bayesian constructs used for evaluation, including the validity of the sampling model and the prior specification. Utilizing hypothetical replications to assess this validity can enhance the reliability and credibility of the Bayesian approach (Rubin, 1984). Moreover, the posterior predictive prediction at the heart of (8.5) attempts to simulate replications under H_0 with the same value of ν that produced the observed data x. This reflects the Bayesian's objective for relevance, and averaging over $\pi(\nu \mid x)$ is a coherent way to achieve this objective while recognizing the unknown nature of ν.

Adopting the Bayesian formulation also relaxes the restriction that test statistic must be free of unknown quantities. This relaxation leads to the introduction of *discrepancy measure*, $D(x, \psi)$, a function of both data and parameter (Meng, 1994) for assessing a metrizable discrepancy between the data and our model, of which H_0 is only a part. The corresponding *ppp* then is given by

$$p_B = \Pr\{D(\mathbf{X}^{\text{rep}}, \psi) \geq D(\mathbf{X}, \psi) \mid \mathbf{X}, H_0\}, \tag{8.7}$$

where the probability is taken over the joint posterior distribution of $(\mathbf{X}, \psi)$ given H_0. Permitting parameter-dependent discrepancy can be computationally beneficial by avoiding for example the computation of the maximum likelihood estimator for ψ under H_0, $\hat{\psi}_0$, in order to form a parameter-free test statistic $T(\mathbf{X}) = D(\mathbf{X}^{\text{rep}}, \hat{\psi}_0)$. The statistical implications of this introduction, however, are more nuanced, which I certainly did not fully appreciate when I decided to explore this idea. Retrospectively, I wondered if my train of thought then was still in danger of being derailed by a desire for vindication for losing some pivotal points at Fudan.

8.5.2 A Bayesian History of Posterior Predictive p-value

Whereas the impetus for my work on *ppp* (Meng, 1994; Gelman et al., 1996) was the desire to eliminate the nuisance parameter for general hypothesis testing, the idea of using the posterior predictive distribution for testing model fitness traced back to when I was still an infant. In a pioneering but now largely overlooked article, Guttman (1967) put forward the idea of using "a future observation" for checking the fitness of a Bayesian model. Here the notion *future observation* came from an earlier work by Guttman and Tiao (1964), which termed the posterior predictive distribution as the marginal distribution for a future observation. The temporal reference of "future" suggests an imaginary where the current study is repeated. The key underpinning of the proposal by Guttman (1967) is that if our Bayesian model is any good for the current data, then any imaginary created by the very model should not be significantly different from the current data.

I choose the phrase *significantly* for it signifies Guttman's approach to construct a goodness-of-fit test to quantifying the significance via a p-value. The specific test Guttman proposed was the usual Pearson's χ^2 test for goodness-of-fit, by first sorting the data into k bins, and then comparing the observed bin counts to the expected ones derived from the posterior predictive distribution. The degrees of freedom of the χ^2 test are set at $k-r-1$, where r is the dimension of ψ, the parameter of the model. This null distribution is asymptotically exact when the maximum likelihood estimates for ψ are the same with or without binning, but otherwise it is conservative because Pearson's statistic is bounded between a χ^2_{k-1} variable and a χ^2_{k-r-1} variable, as noted by Guttman (1967).

Guttman's idea of using posterior predictive distribution for model checking was also central to Rubin (1984), which circumvents some issues with prior predictive checking advocated by Box (1980) (e.g., the non-existence of prior predictive distribution with improper prior). I learned a great deal from Rubin (1984), which led to Meng (1994) and Gelman et al. (1996). Both articles are highly cited among my publications, but a sizable number of citations are criticisms of the conservative nature of *ppp*, especially using the broader discrepancy variables, instead of the traditional test statistics (e.g., Bayarri and Berger, 2000; Robins et al., 2000; Dahl, 2006; Hjort et al., 2006). This case hence demonstrates the complicated nature of measuring scholarly accomplishments via citation accounts. Negative citations, in terms of the reason for citation, are still positive citations, in terms of the total number of citations. However, negative citations should not be entirely dismissed, since criticisms often prompt subsequent advances, as in the current context.

In fact, my thought journey to appreciate the criticism of *ppp* is a *nuanced meandering*. Putting on a frequentist hat, I could see myself criticizing anything that is given a label of p-value but without requiring it to have (approximately) a uniform distribution under the null, i.e., a "U-value", as termed by Gelman (2005). Or in the words of Robins et al. (2000), without knowing the frequentist property of a p-value as a benchmark, knowing merely the value of a p-value has the same value as being told that the temperature is 30°, but without knowing if it is in centigrade or Fahrenheit, which can make the difference between going swimming or skiing (for most people).

However, realizing that *ppp* is an authentic tail area *probability*, as established in (8.5) or (8.7), I questioned myself why I would want a benchmark for a *probability*? Is the purpose of calculating probability to form a quantitative benchmark for evaluating and comparing evidences under uncertainty? Why do I need a benchmark for a benchmark?

8.5.3 A Bayesian Imaginary for Posterior Predictive p-value: $H_{B,0}$ verses H_0

To answer these questions for myself, I contemplated three more questions. First, what is the essence of a tail-area probability? The term 'tail-area' typically is endowed with an imaginary of 'being more extreme' than a threshold, which generates the criticism that p-values violate the likelihood principle (Berger and Wolpert, 1988) because they involve hypothetical unobserved values. However, this popular narrative for p-values is ungainly and unnecessary. A p-value merely pins the observed value of the test statistic, T_o, to a specific location on the reference distribution. A p-value equal to 0.1 simply situates T_o at the 90% percentile of the null distribution. Once the reference distribution is settled, whether its 90% percentile is an extreme position or not is a decision question separate from how that position is identified. Recognizing a tail-area probability as a mere position locator should remind us that p-values are not exclusive to frequentists. Therefore, a *ppp* is no less or more legitimate as a p-value than a classical p-value.

Secondly, What is the imaginary of a *ppp*? In the absence of the nuisance parameter ν, the replications defined by (8.6) are the same as the classical replications: all possible probabilistic samples that can be generated from the null distribution. In the presence of ν, the imaginary becomes all possible probabilistic samples generated under the null and with the corresponding values of nuisance parameter that could have generated the data at hand if the null holds. That is, the possible values of ν are from its posterior distribution given the observed data and given the null hypothesis. This imagination for ν allows us to deal with the issue of nuisance parameter, but it comes with the expense of positing a prior distribution for ν (under the null).

This in turn leads to the third question: What hypothesis is being tested by *ppp*? Since a *ppp* is calculated based on all assumptions made, the actual tested hypothesis, which we denote by $H_{B,0}$, has more ingredients than the original H_0. In particular, the prior we assume, whether for the whole parameter ψ, or just for the nuisance parameter under the null, can make a big difference. In other words, testing $H_{B,0}$ based on a *ppp* follows the same reasoning as testing H_0 based on a frequentist p-value (if it exists). However, whether the test results can be transformed from the former to the latter will depend critically on the relationship between $H_{0,B}$ and H_0.

For example, when we use a noninformative prior—in the sense of Kass and Wasserman (1996)—for the nuisance parameter ν, $H_{0,B}$ tends to be a less restrictive hypothesis than H_0. Therefore, it requires stronger evidence from data to reject the former than the latter. This implies that *ppp* for $H_{0,B}$ should be used as a quick-and-dirty screening test for H_0 in the sense that an extreme *ppp* is a good indication of the falsehood of H_0. A non-significant *ppp*, however, should be interpreted as the failure of the screening test itself, because of its low power against H_0, not the usual 'failure to reject H_0' cautious conclusion we statisticians advocate. The situation is not unlike taking a COVID home-test, where being positive says a lot more than being negative; indeed we often follow up with a PCR test when symptoms persist, but the home-test stays being negative.

On the other hand, when we inject a strong prior on the nuisance parameter, a rejection of $H_{B,0}$ by a *ppp* could either be a sign of discordance of the information provided by the data and the prior, in the sense of Reimherr et al. (2021), or a rejection of H_0. This could make the matter even more complicated because whether *ppp* can be used as a screening test will depend on the (approximate) validity of the prior we use for the nuisance parameter.

Consider the case of testing $H_0 : \mu = \mu_0$, from data $\mathbf{X} = \{X_1, \ldots, X_n\} \overset{\text{i.i.d}}{\sim} N(\mu, \sigma^2)$, with σ^2 unknown. An obvious discrepancy measure is $n(\bar{X}_n - \mu)^2/\sigma^2$, or equivalently

$D(\mathbf{X}, \psi) = (\bar{X}_n - \mu)^2$. To deal with the nuisance parameter σ^2, let us assume *a priori* $\sigma^2 \sim$ Scaled-inverse-$\chi^2(m, \tau^2)$, where the degrees of freedom m can be imagined as the size of the previous study for estimating σ^2, and τ^2 was the estimate from that study. This setup implies that the marginal distribution of X_i, also known as the prior predictive distribution (Box, 1980), is $t_m(\mu, \tau)$, the distribution of a t random variable with m degrees of freedom, with location and scale parameters μ and τ respectively.

Therefore, even assuming the validity of both the Normal likelihood and the inverse-χ^2 prior distribution, we have effectively changed the problem of testing a Normal mean μ with unknown variance (i.e., H_0) based on i.i.d data $\mathbf{X} = \{X_1, \ldots, X_n\}$, to test a t-location parameter with known scale parameter τ and the degrees of freedom m (i.e., $H_{B,0}$) based on $\mathbf{X}$ being a single sample from the prior predictive distribution, i.e., a multivariate t distribution, $t_p(\mu\vec{\mathbf{1}}_n, I_n)$, where $\vec{\mathbf{1}}_n$ is a vector of 1's, and I_n is the $n \times n$ identity matrix. We say *effectively* because the imaginary for *ppp* is aimed to be more relevant to the data at hand than the prior predictive distribution, and hence the *ppp* test is more involved. Nevertheless, formulating $H_{B,0}$ as the effective null for *ppp* helps to highlight the critical role of the hyperparameters on the behavior of *ppp*, as we shall see below.

8.5.4 Prior Influence on Posterior Predictive p-value

Given the conjugacy of the inverse-χ^2 distribution, the posterior for σ^2 under the null $\mu = \mu_0$ is $\sigma^2|\mathbf{X}, H_0 \sim$ Scaled-inverse-$\chi^2\left(m+n, \tau^2_{m,n}(\mathbf{X})\right)$, where

$$\tau^2_{m,n}(\mathbf{X}) = \frac{m\tau^2 + nS_0^2}{m+n}, \quad \text{with} \quad S_0^2 = \frac{1}{n}\sum_{i=1}^{n}(X_i - \mu_0)^2. \tag{8.8}$$

Note that S_0^2 is the MLE for σ^2 under H_0. Under this setup, the posterior predictive distribution of the discrepancy measure $D(\mathbf{X}^{\text{rep}}, \psi) = (\bar{X}_n^{\text{rep}} - \mu)^2$ under H_0 has the stochastic representation

$$D(\mathbf{X}^{\text{rep}}, \psi_0) = (\bar{X}_n^{\text{rep}} - \mu_0)^2 = \frac{\sigma^2}{n}Z^2 = \frac{\tau^2_{m,n}(\mathbf{X})}{n}\frac{Z^2}{\chi^2_{m+n}/(m+n)} = \frac{\tau^2_{m,n}(\mathbf{X})}{n}F_{1,m+n}, \tag{8.9}$$

where $Z \sim N(0,1)$, independent of σ^2, and hence $F_{1,m+n} \sim F(1, n+m)$, i.e., an F random variable with degrees of freedom 1 and $m+n$. Consequently, the *ppp*, as a function of $\eta \equiv \{m, \tau^2, \mu_0\}$ and data $\mathbf{X}$ (which includes its size n), is given by

$$p(\mathbf{X}; \eta) = \Pr\left(F_{1,m+n} \geq D^*(\mathbf{X}; \tau_{m,n})\right), \quad \text{with } D^*(\mathbf{X}; \tau) = \frac{n(\bar{X}_n - \mu_0)^2}{\tau^2}. \tag{8.10}$$

To compare $p(\mathbf{X}; \eta)$ with its classical predecessor based on the t-test, we re-express its implied test statistic $D^*(\mathbf{X}; \tau_{m,n})$ in terms of the t statistic $T(\mathbf{X}) = \sqrt{n}(\bar{X}_n - \mu_0)/S_n$, where S_n^2 is the usual sample variance. The fact that $S_0^2 = (1 - n^{-1})S_n^2 + (\bar{X}_n - \mu_0)^2$ then yieldsa

$$D^*(\mathbf{X}, \tau_{m,n}) = \frac{(m+n)n(\bar{X}_n - \mu_0)^2}{m\tau^2 + (n-1)S_n^2 + n(\bar{X}_n - \mu_0)^2} = \frac{(m+n)T^2(\mathbf{X})}{m\lambda_n + (n-1) + T^2(\mathbf{X})}, \tag{8.11}$$

where $\lambda_n = \tau^2/S_n^2$ compares our prior estimate of σ^2 with that from the data. A great deal of the behavior about *ppp* can be learned from this example by examining how $p(\mathbf{X}; \eta)$ varies as a function of η for given $\mathbf{X}$, and how $p(\mathbf{X}; \eta)$ behaves stochastically for fixed η.

First, if we take $m = 0$, which corresponds to the common improper constant prior for $\log(\sigma^2)$ (e.g., Rubin, 1984; Bayarri and Berger, 2000), then (8.10)–(8.11) implies that

$$p(\mathbf{X};\eta) = \Pr\left(\frac{F_{1,n-1}}{(1-n^{-1})+n^{-1}\tilde{F}_{1,n-1}} \geq \frac{T^2(\mathbf{X})}{(1-n^{-1})+n^{-1}T^2(\mathbf{X})}\right), \tag{8.12}$$

where $F_{1,n-1}$ and $\tilde{F}_{1,n-1}$ are two dependent $F(1, n-1)$ variables such that $F_{1,n-1} = (n-1)$ χ_1^2/χ_{n-1}^2, $\tilde{F}_{1,n-1} = (n-1)\tilde{\chi}_1^2/\chi_{n-1}^2$, and $\{\chi_1^2, \tilde{\chi}_1^2, \chi_{n-1}^2\}$ are mutually independent. From this expression, we see that if χ_1^2 were the same as $\tilde{\chi}_1^2$, then the *ppp* $p(\mathbf{X};\eta)$ would be identical to the classical *p*-value from the *t*-test. The minor difference is due to whether the information $\mu = \mu_0$ is used or not when we derive the posterior of σ^2. As shown in Meng (1994), if we replace the null prior for μ, namely, $\mu = \mu_0$, by the standard improper constant prior for μ, then the resulting *ppp* will be $\Pr\left(F_{1,n-1} \geq T^2(\mathbf{X})\right)$, which recovers the classical *p*-value based on the *t* test.

Second, for any finite $m \geq 1$, we see from (8.10) to (8.11) that $p(\mathbf{X};\eta)$ is a monotonic increasing function of $\lambda_n = \tau^2/S_n^2$. Worse, for any given data set $\mathbf{X}$ (and hence fixed n and S_n^2), $p(\mathbf{X};\eta) \nearrow 1$ because $D^*(\mathbf{X}, \tau_{m,n}) \searrow 0$, as $\tau \nearrow \infty$. This implies that $p(\mathbf{X};\eta)$ cannot be an U-value, and that it can be arbitrarily large even when H_0 is clearly violated. This is the phenomenon for which *ppp* has been heavily criticized, and a typical reason given is that *ppp* uses the data twice, and a number of methods have been proposed to address this issue (Bayarri and Berger, 2000; Robins et al., 2000; Hjort et al., 2006). However, the current example indicates that the "double dipping" reason is not inherent, or at least not universal. For example, neither the partial nor the U-conditional posterior predictive approaches designed to remove 'double dipping' (Bayarri and Berger, 2000) can eliminate the impact of the hyper-parameter τ because they only alter the likelihood. (Of course, in the asymptotic regime, if we do not allow the hyper-parameters to depend on the size of n, then the prior impact will vanish by design; but see Reimherr et al. (2021) for a non-standard asymptotic regime for investigating the impact of prior asymptotically.)

Indeed, changes of $p(\mathbf{X};\eta)$ with the hyper-parameter τ reveal the cause of the behavior of *ppp*. As long as $m > 0$, we see from (8.8) that the larger τ, the less extreme the observed discrepancy $n(\bar{X}_n - \mu_0)^2$ compared to $\tau_{m,n}^2(\mathbf{X})$. This is intuitive because our effective marginal data model is $t_p(\mu_0, \tau)$. Therefore, when we adopt a large value of τ, the *ppp* will take that into account in deciding whether an observed $(\bar{X}_n - \mu_0)^2$ is too large to cast doubt about $H_{B,0}$, which is increasingly more flexible than H_0 as τ increases. Hence $p(\mathbf{X};\eta)$ is faithful to its design – it locates the value of $(\bar{X}_n - \mu_0)^2$ in its posterior predictive distribution, under the prescribed $H_{B,0}$.

Third, setting $m = \infty$ is the same as declaring $\sigma^2 = \tau^2$ with certainty. In such a case, we see from (8.8) that $\tau_{m,n}^2(\mathbf{X}) = \tau^2$, and then from (8.5) that $p(\mathbf{X};\eta) = \Pr(\chi^2 \geq D^*(\mathbf{X};\tau))$, the *p*-value from the classic (two-sided) Z-test, assuming that τ is correctly specified. If $\lambda \equiv \tau^2/\sigma_0^2 \neq 1$, where σ_0^2 is the correct variance of X_i under null, then $p(\mathbf{X};\eta) = \Pr(\lambda\chi^2 \geq D^*(\mathbf{X};\sigma_0^2))$. This *p*-value will be smaller or larger than the *p* value from the Z test, $p = \Pr(\chi^2 \geq D^*(\mathbf{X};\sigma_0^2))$, depending on, respectively, whether $\lambda < 1$ or $\lambda > 1$. The failure of $p(\mathbf{X};\eta)$ being a U-value then is due to the incorrect specification of the hyper-parameter. Whereas a large observed discrepancy $|\lambda_n - 1|$ – recall $\lambda_n = \tau^2/S_n^2$ – should compel us to rethink if we have made a sensible choice of prior for σ^2, we should not attribute our failure in assessing σ^2 as an inherent problem of the Z-test itself. All probability calculations are vulnerable to the assumptions that render the calculations possible; *ppp* or any probabilistically derived Bayesian *p*-values are no exception.

8.6 A Fiducialistic Encounter: A BFFer's Taste of Trinity

8.6.1 A Multi-Perspective Prelude: Integration or Optimization?

A hallmark of the Bayesian paradigm is its inherent coherence, a consequence of adhering to strict probabilistic reasoning and manipulations (Lin, 2022). The fact that it uses only probabilistic calculus also renders an operational coherence. That is, regardless of the nature of an unknown quantity, whether missing data, latent variable, or a nuisance parameter – to marginalize it out exactly is always done via integration (with respect to a dominating measure on a properly defined joint space). This operational coherence, however, is not automatic for other inferential frameworks, even if we do not question their theoretical or logical coherence (albeit one could argue that the lack of operational coherence is a symptom of more malignant incoherence).

For example, for likelihood inference, missing data and latent variables are marginalized out via integration, just as with Bayesian methods, since they possess proper probabilistic distributions. However, there is no principled and unified way of marginalizing out any other nuisance objects (e.g., parameters or shapes) for a likelihood function. A common approach is to profile out the nuisance parameters via maximization (Cole et al., 2014), which can be done even when the entire likelihood parameter is nuisance, as with likelihood inference for Monte Carlo integration (Kong et al., 2003, 2006). Integration and optimization both are workhorses for scientific computation. Statistically however they represent two distinctive mentalities for dealing with variations. Integration demands proportional representation, while optimization implements a winner-takes-all strategy. This stark difference should warn us that these approaches can produce markedly different results.

Broadly speaking, this contrast reflects a main difference between the statistical culture and the machine learning culture (and its kin), to borrow the two-culture dichotomy of Breiman (2001). Statistics, being the scientific discipline devoted to the study of variations, is inherently about proportional representation, in its descriptive and prescriptive endeavors. Machine learning has been dominated by optimal predictions and pattern seeking, and hence inevitably it is anointed by the winner-takes-all mentality. Within statistics, the preference for integration over optimization is reflected by attempts to bring Bayesian-esque integration into likelihood inference, more so than efforts in the opposite direction. Notable examples include the partially Bayes approach of Cox (1975) and McCullagh (1990), which introduces a prior for the nuisance parameter only and hence it can be integrated out, and the integrated likelihood approach investigated by Berger et al. (1999), which also reviewed many methods for handling nuisance parameters.

In contrast, the only 'opposite-direction' attempt that has been noted – or rather has received some dose of notoriety – is the h-likelihood idea of Lee and Nelder (1996), expanded in Lee et al. (2018). My learning about this approach started when I was asked to discuss Lee and Nelder (2009), an excursion which was my nearest-neighbor imputation of Dan Brown's thrill of deciphering the *Da Vinci Code* (Brown, 2004). For any reader who wish to fully experience the simulated thrill that I shall provide a trailer for in Sections 8.6.2–8.6.3, I'd encourage a reading of Section 1.3.1 of Berger et al. (1999), *What is the likelihood function?*, as a warm-up preparation. That section reveals that the definition of likelihood function to be less precise than textbooks imply, particularly in models with unobserved variables like missing data, latent variables, or future realizations.

In particular, Berger et al. (1999) constructs a likelihood by putting all random variables on the left side of the bar, regardless of whether they are observed or not. Here the term 'bar' refers to the divider in the notation of $f(y, u|\lambda, \psi)$, where y represents the observed data, u unobserved variables, with $f(\cdot|\lambda, \psi)$ being their joint probability distribution or density, which depends on the parameter of interest, λ, and the nuisance parameter, ψ. When we fix the value of $\{y, u\}$ and consider $f(\cdot|\lambda, \psi)$ as a function only of $\theta = \{\lambda, \psi\}$, it gives us the *complete-data* or *augmented-data* likelihood. Obtaining the observed-data likelihood for θ seems to be a trivial matter, since we can marginalize out u via integration to obtain $L(\theta|y) \equiv f(y|\theta) = \int f(y, u|\theta)\mu(du)$. However, there are plenty of cases where this integration is hard or impractical. This practical difficulty has encouraged many approximations such as numerical approximations and Monte Carlo integration. Somewhat unexpectedly, Lee and Nelder (1996) embarked on a journey of replacing the integration by maximization over u. And this is where the thrill and trouble started.

8.6.2 Seeking the *Da Vinci* Code for the *H*-likelihood

Using the notation from Section 8.6.1, the initial definition of the h-loglikelihood by Lee and Nelder (1996) is simply $h(\theta, u|y) = \log f(y, u|\theta)$, which treats the unobserved variable u on equal footing with the parameter θ. Viewed this way, it does not seem to be illogical to maximize (θ, u) jointly. However, as the discussants of Lee and Nelder (1996) pointed out, this seemingly coherent approach is generally invalid and can produce meaningless estimates for both θ and u.

Consider the example where the data are an i.i.d sample $\vec{y} = \{y_1, \ldots, y_n\}$ from an exponential distribution with mean λ. We take $u = y_{n+1}$, a future realization, which yields the h-loglikelihood

$$h(\lambda, u; \vec{y}) = -(n+1)\log\lambda - \frac{n\bar{y}_n + u}{\lambda}, \qquad \text{for } \lambda > 0,\ u \geq 0, \tag{8.13}$$

where $\bar{y}_n$ is the sample mean. But this is a decreasing linear function of u for any given λ, and hence the maximizer for u will be $u = 0$ for any $\bar{y}_n$ and $\lambda > 0$, a clearly senseless predictor.

Given $u = 0$, the maximizer for λ from (8.13) is $\frac{n}{n+1}\bar{y}_n$, which differs only slightly from the MLE for λ, $\bar{y}_n$. However, we can define h-likelihood for arbitrarily many future realizations, say $\vec{u} = \{y_{n+1}, \ldots, y_{n+m}\}$. Following the same maximizing approach over $\vec{u}$ would result from the MHLE (maximum h-likelihood estimator), $\hat{\lambda}_{\text{MHLE}} = \frac{n}{n+m}\bar{y}_n$, which can be arbitrarily small by making m arbitrarily large, clearly indicating that maximizing over nuisance variables is not valid in general.

So why is this not the end of the story? Why would any statistician still push for the idea of maximizing over unobserved variables? This was the puzzle that motivated my decoding h-likelihood pursuit (Meng, 2009). To answer this question, let us consider setting up the h-likelihood with $v = \log(u)$ as unobservable, where u is still the future realization y_{n+1}. Because the distribution for v is given by $f(v|\lambda) = \lambda^{-1}e^v e^{-\lambda^{-1}e^v}$, the resulting h-loglikelihood becomes

$$h(\lambda, v; y) = -(n+1)\log\lambda - \frac{n\bar{y}_n + e^v}{\lambda} + v. \tag{8.14}$$

As shown in Meng (2009), the corresponding MHLEs for both λ and $u = e^v$ are now given by the correct MLE for λ, $\hat{\lambda} = \bar{y}_n$. That is, the MHLE apparently is working on the v scale; and indeed the h-loglikelihood in (8.14) also satisfies the first and second Bartlett identities, and hence it will deliver the correct Fisher information matrix for $\theta = \{\lambda, v\}$ as well (Meng, 2009).

Standard likelihood approaches, whether for testing or estimation, are invariant to log or other one-to-one transformation of the parameters, at least when we do not invoke any approximation in our calculations.[4] But for the h-likelihood, the above example shows that the transformation literally transfers a senseless result to a sensible one. What's the secret here? Does such a transformation always exist? If so, how do we find it?

8.6.3 Following the Frequentist and Bayesian Paths ...

To comprehend why MHLE offers nonsensical estimators for u but logical ones for $v = \log u$, compare the stochastic behaviors of $\hat{u} - u$ with that of $\hat{v} - v$. The sampling distribution of $u - \hat{u} = y_{n+1} - \bar{y}_n$ depends on the unknown scale parameter λ, while $v - \hat{v} = \log(y_{n+1}/\bar{y}_n)$ is independent of λ. Thus, on the v scale, exact prediction intervals can be formed without estimating λ—a nuisance parameter for predicting future observations. The v scale leverages the ratio scale's ability to cancel out λ, making $v - \hat{v}$ a pivotal quantity for prediction. The sampling distribution of $r = y_{n+1}/\bar{y}_n$ exhibits a Pareto-type density (Meng, 2009),

$$f(r) = \left(1 + \frac{r}{n}\right)^{-(n+1)}, \quad r \geq 0. \tag{8.15}$$

However, the above concerns the *marginal* distribution of $v - \hat{v}$ over all potential values of the dataset. This typical frequentist imaginary raises the question: when predicting y_{n+1} from the observed $\bar{y}_n$, why concern ourselves with other potential values of $\bar{y}_n$? Conditioning on $\bar{y}_n$ would address this concern, but then r would be non-pivotal because its conditional distribution, given $\bar{y}_n$, is an exponential distribution with mean $s_n = \lambda/\bar{y}_n$. While s_n converges to 1 asymptotically, for any given $\bar{y}_n$, s_n is a linear function of λ, making our predictions of y_{n+1} vary with the unknown λ.

The Bayesian approach easily addresses this problem[5], at the expense of specifying a prior distribution for λ, $\pi(\lambda)$. Upon adopting a prior, the remaining steps become mostly mathematical or computational. With a constant prior for $\eta = \log(\lambda)$, the *posterior predictive distribution* for y_{n+1} is determined by $r\bar{y}_n$, with the density of r, denoted by $f^B(r)$, taking the same form as in (8.15). Whereas $f(r)$ and $f^B(r)$ are mathematically the same, they represent distinct sets of probabilistic replications for assessing the uncertainty in our predictions.

Probabilistic replications, discussed in Section 8.2.2, are imagined studies resembling the current one. The definition of similarity is determined by extracting characteristics from the present study and constructing analogous ones. For $f(r)$, these characteristics involve (A) n independent observations from a single exponential distribution and (B) a future observation from the same distribution. Realizing this similarity, say through Monte Carlo simulation, is straightforward because (A) does not specify the value of $\lambda > 0$. The simulation however may create values far from the λ that generated the observed $\bar{y}_n$, resulting in uncertain assessments not indicative of the uncertainty in our prediction based on the given value of $\bar{y}_n$.

Limiting (A) to cases sharing the same unknown λ that generated the observed $\bar{y}_n$ requires simulating the uncertainty states of λ given $\bar{y}_n$. Per Section 8.2.2, the Bayesian approach poses a prior distribution for λ, $\pi(\lambda)$, to enable posterior predictive simulation. However, invoking $\pi(\lambda)$ creates vulnerability because we almost never have full and reliable information to pinpoint $\pi(\lambda)$.

8.6.4 .... with a Fall in a Fisher's Pit, but a Gain in My Fiducial Wit ...

Given these challenges, numerous efforts have been made to derive a post-data distribution for parameters like λ without invoking a prior. Such attempts include the fiducial approach and the h-likelihood. The latter, as discussed in Lee and Nelder (2009), uses the 'adjusted profile h-likelihood' (APHL) method to yield an 'h-distribution' for the subject of interest.

Specifically, for the h-likelihood of (8.14) and with the transformation $\eta = \log(\lambda)$, the APHL is defined as

$$\tilde{h}_\eta(v; y) = \left[h(\eta, v; y) - \frac{1}{2}\log\left[-h''_\eta(\eta, v; y)\right]\right]_{\eta=\eta(\nu)}, \tag{8.16}$$

where $\eta(v)$ is the root of $h'_\eta(\eta, v; y) = 0$, with h'_η and h''_η denoting respectively the first and second partial derivatives with respect to η. From (8.14), $\eta(v) = \log(n\bar{y}_n + e^v) - \log(n+1)$, and $h''_\eta(\eta(v), v; y) = -(n+1)$. Consequently,

$$\tilde{h}_\eta(v; y) = -(n+1)\eta(v) + v - 0.5\log(n+1). \tag{8.17}$$

By treating $\tilde{h}_\eta(v; y)$ as an authentic post-data log-density for v, the distributional inferential approach of Lee and Nelder (1996) amounts to setting the predictive density for v as

$$f^H(v|y) \propto e^{\tilde{h}_\eta(v;y)} = e^v \eta^{-n-1}(v) \propto e^v(n\bar{y}_n + e^v)^{-n-1}. \tag{8.18}$$

Because $r = u/\bar{y}_n = e^v/\bar{y}_n$, it follows that the implied post-data h-distribution for r is

$$f^H(r|y) = \left(1 + \frac{r}{n}\right)^{-(n+1)}, \quad r \geq 0, \tag{8.19}$$

which is identical to $f(r)$ in (8.15).

The rare trinity where $f(r) = f^B(r) = f^H(r)$ is significant because the interpretation of $f^H(r)$ given in Lee and Nelder (1996) is fundamentally a fiducial argument. As I explained in Meng (2009), their approach is equivalent to applying Laplace approximation to the marginal distribution of $\nu - \hat{\nu}$, where $\hat{\nu}$ maximizes the log profile likelihood $h(\eta(\nu), \nu; y)$. This approximation as a mathematical tool is neutral, applicable in both Bayesian and frequentist inferences (Kass, 1990; Shun and McCullagh, 1995; Hart, 2009). However, the probabilistic interpretation of the resulting approximation by Lee and Nelder (1996) involves a sleight of hand: they interpreted the joint imaginary from all $\{y, \nu\}$ pairs as the conditional one, with y fixed at its observed value.

This use of parameter-free pre-data distribution as post-data distributional estimate is at the core of the fiducial approach, known as the "continuing to regard" doctrine (Dempster, 1963b). Indeed, as pointed out by Desmond and Xie (2011), this exponential problem was a special case in the chapter of 'Fiducial prediction' of Fisher (1956), where Fisher considered predicting the sum of arbitrarily many future observations. In the case of predicting a single future observation, Fisher's fiducial distribution $f^F(r)$ is identical to $f^H(r)$.

Therefore, my h-likelihood meandering apparently was predesignated to fall into a Fisher's trap, despite the efforts made by all h-likelihoodists, attempting to enjoy the perks of the Bayesian club without paying its entry fee. (Every generation needs their own metaphor to keep Savage's omelette fresh.) However, a fall in Fisher's pit is a gain in my fiducial wit (to paraphrase a Chinese idiom, "A fall in the pit, a gain in your wit", which also captured my pivotal lesson at Fudan).

Before my exploration of h-likelihood, I understood that to align the fiducial distribution with the Bayesian posterior for a real parameter, we must limit ourselves to problems transformable into a location family, and apply a constant prior on the resulting location parameter. This was shown by Lindley (1958)[6], and further investigated by Consonni and Veronese (1993). A generally overlooked result in the latter indicates that if we specify a joint imaginary for $(\mathbf{X}, \theta)$ (as shown in Figure 8.1's rightmost diagram) such that the fiducial distribution of θ aligns with its Bayesian posterior distribution for all $\mathbf{X}$ realizations, then $\mathbf{X} = \theta$ almost certainly. This is due to the impossibility of both $\mathrm{E}[\mathbf{X}|\theta] = \theta$ and $\mathrm{E}[\theta|\mathbf{X}] = \mathbf{X}$ holding simultaneously for any proper joint distribution of two variables $\{\mathbf{X}, \theta\}$ within the same state space, except when $\mathbf{X} = \theta$ almost surely. An intuitive reason for this impossibility is that if both $\mathrm{E}[\mathbf{X}|\theta] = \theta$ and $\mathrm{E}[\theta|\mathbf{X}] = \mathbf{X}$ hold, there would be no regression towards the mean, contradicting this universal phenomenon in stochastic relationships.

In this context, $\mathbf{X} = \bar{y}_n$ and $\theta = \lambda$. Hence, for $f^B(r) = f^F(r)$ to occur, the prior must be improper. The apparent unification $f^B(r) = f^F(r)$ therefore is incorporeal because its empirical demonstration is impossible. However, the frequentist $f(r)$ can be demonstrated through simulation with any proper prior on λ, since $r = y_{n+1}/\bar{y}_n$ is ancillary under the joint distribution of $\{\bar{y}_n, y_{n+1}\}$. Therefore, while the mathematical identity between $f^F(r)$ and $f^B(r)$ doesn't make $f^F(r)$ tangible, the identity between $f^F(r)$ and $f(r)$ does for both $f^F(r)$ and $f^B(r)$. This realization, both in simulation and stimulation, encouraged me to consider if this circuitous route via prediction is Fisher's thought process for providing an operational meaning to his hypothetical fiducial distribution for the model parameter, a nuisance object for prediction. This roundabout route is facilitated by a predictive pivotal quantity, once again underscoring the crucial role of statistical pivotality.

8.7 A Mathematical Enlightenment: A Taste of the Cauchy Power

8.7.1 A Rare Statistically Surprising Conjecture

In my initial learning journey, I assumed pivotal quantities were easy to identify by their appearances. If I couldn't picture how a parameter dependency disappears in a stochastic representation, then it would not be a pivotal quantity. However, this simplicity is rare in complex models. A conversation with my colleague Natesh Pillai in early 2014 reminded me that this rarity could stem from our inability to detect hidden pivotality, even in Normal models.

Specifically, over a dinner for a seminar speaker, Natesh intrigued me with a conjecture. Let $\mathbf{X} = (X_1, \ldots, X_p)^\top$ and $\mathbf{Y} = (Y_1, \ldots, Y_p)^\top$ be two independent $N(0, \Sigma = \{\sigma_{ij}\})$ variables, where $\sigma_{jj} > 0$ for all $1 \le j \le p$. Then Drton and Xiao (2016)[7] conjectured that

$$T_{\vec{w}}(\mathbf{X}, \mathbf{Y}) \equiv \sum_{j=1}^{p} w_j \frac{X_j}{Y_j} \sim \text{Cauchy}(0, 1), \tag{8.20}$$

that is, the standard Cauchy distribution, regardless of the value of Σ or of $\vec{w} = (w_1, \ldots, w_p)^\top$ as long as $\sum_{j=1}^{p} w_j = 1$ and $w_j \ge 0$ (indeed $\vec{w}$ can be random, provided its independence with $\{\mathbf{X}, \mathbf{Y}\}$). Drton and Xiao (2016) proved the conjecture for $p = 2$, and provided empirical evidences for $p > 2$.

It's textbook knowledge that the ratio of two independent $N(0, \sigma^2)$ variables is standard Cauchy(0, 1), and a convex mix of independent Cauchy(0,1)'s remains Cauchy(0, 1). When Σ is diagonal, result (8.20) then is obvious. However, when Σ isn't diagonal, many would expect the distribution of T to be impacted by Σ's correlations. It was, and remains surprising—as noted since the publication of Pillai and Meng (2016) [e.g., Cohen et al. (2020) and Xu et al. (2022)]—that the Drton-Xiao conjecture is indeed a mathematical fact.

Rarely in statistics do we require mathematical proof to dispel collective doubts. Prior to this conjecture, I hadn't experienced a situation where every statistician (I communicated with) demanded a proof for conviction. I strived to build intuition for (8.20). Since the result is trivial in extreme cases—when all $R_j = X_j/Y_j$'s are identical or independent—I attempted a 'sandwich argument' to bracket general cases, hoping to form intuitions or a proof. Sadly, my sandwich was merely a sandwish, and all my other attempts turned out to be either a quicksand or a lasting wish.

I therefore was elated when Natesh sent me a dozen-page proof. However, as the long proof didn't inspire intuition—nor would it receive a full mark from my Fudan teachers—I pushed for a more succinct and intuitive proof. Natesh eventually discovered an elegant result in Williams (1969), stating that for any $\{u_1, \ldots, u_p\}$, where $u_j \in \mathbb{R}$, and $\Theta_1 \sim \text{Unif}(-\pi, \pi]$ independent of $\{w_1, \ldots, w_p\}$ where $w_j \geq 0$ and $\sum_j w_j = 1$, we have

$$\sum_{j=1}^{p} w_j \tan(\Theta_1 + u_j) \sim \text{Cauchy}(0, 1). \tag{8.21}$$

Writing $\{X_j = R_i \cos(\Theta_j), Y_j = R_j \sin(\Theta_j)\}$ re-expresses $T_w(X, Y)$ of (8.20) as $\sum_{j=1}^{p} w_j \tan(\Theta_j)$. Let $\delta_j = \Theta_j - \Theta_1$. Whereas the support for δ_j depends on Θ_1 when $j > 1$, $u_i = \delta_i(mod(2\pi))$ is independent of Θ_1 under the Normal model, as we proved in Pillai and Meng (2016). The conjecture than follows (8.21) because $\tan(\Theta_j) = \tan(\Theta_1 + \delta_j) = \tan(\Theta_1 + u_j)$, for all j.

Proving the conjecture didn't satisfy us entirely, as the intuitive understanding remained elusive. We're not alone, as the disappearance of Σ in the distribution of T is a "mystery" per Xu et al. (2022), who explore our speculation about the overpowering effect of the Cauchy tails on the interdependence among the R_i's. This mystery persists, since Xu et al. (2022) shows that other similarly heavy-tailed margins do not exhibit the same phenomenon, but in examples given in Cohen et al. (2020), the phenomenon did appear. However, the most recent work by Fang et al. (2023) offers an important clue, highlighting a link with the known tail independence of binary Normal with correlation ρ, provided $\rho^2 < 1$.

8.7.2 Cauchy's Rescue to Hotelling (but There Is No Free Lunch)

Drafting this chapter led me to a realization that the mysterious equation (8.20) could provide a new tool to navigate through the curse of dimensionality. Consider the task of constructing an exact confidence region for the mean of a multivariate Normal distribution of arbitrary dimension p, with any covariance Σ and sample size $n \geq 2$. Suppose $\mathbf{Z}_1, \ldots, \mathbf{Z}_n \overset{\text{i.i.d}}{\sim} N(\vec{\mu}, \Sigma)$, with $\mathbf{Z}_i = (Z_{i1}, \ldots, Z_{ip})^\top, i = 1, \ldots, n$, and $\vec{\mu} = (\mu_1, \ldots, \mu_p)^\top$. Let $\mathcal{Z} = [\mathbf{Z}_1, \ldots, \mathbf{Z}_n]$ be the $n \times p$ data matrix, and $\vec{r} = (r_1, \ldots, r_n)^\top$ and $\vec{c} = (c_1, \ldots, c_n)^\top$ be two orthogonal vectors in R^n. Then $\mathcal{Z}^\top \vec{r}$ and $\mathcal{Z}^\top \vec{c}$ are independent due to their joint normality and zero covariance because $\vec{r}^\top \vec{c} = 0$.

To apply this result, let $\mathcal{C} = \{c_{ij}\} \equiv (C_1, C_2, \ldots, C_n)$ be an $n \times n$ orthonormal matrix where the first column $C_1 = n^{-1/2}\mathbf{1}$. Let $\mathbf{Y}_1 = \sum_{i=1}^{n} c_{i1}(\mathbf{Z}_i - \vec{\mu}) = \sqrt{n}(\bar{\mathcal{Z}}_n - \vec{\mu})$ and $\mathbf{Y}_k = \sum_{i=1}^{n} c_{ik}\mathbf{Z}_i, k = 2, \ldots, n$; here we slightly abuse the notation by writing $\bar{\mathcal{Z}}_n$ as the sample mean of the n columns in $\mathcal{Z}$. Then the result in the previous paragraph implies that $\mathbf{Y}_1, \ldots, \mathbf{Y}_n \overset{\text{i.i.d}}{\sim} N(0, \Sigma)$, because they are all independent of each other, and $\mathbb{E}[\mathbf{Y}_k] = [\mathbf{1}^\top C_k]\vec{\mu} = 0$ for $k \geq 2$ since C_k is orthogonal to $C_1 \propto \mathbf{1}$, that is, C_k forms an *orthogonal contrast*. Of course this separation of $\mathbf{Y}_1$ from $\{\mathbf{Y}_2, \ldots, \mathbf{Y}_n\}$ is well-known, with $\{\mathbf{Y}_2, \ldots, \mathbf{Y}_n\}$ explicating the $(n-1)$ degrees of freedom for estimating Σ.

Hence, when $n \leq p$, there will be insufficient degrees of freedom to fully estimate Σ. In such cases, we cannot construct confidence regions using the well-known Hotelling T-square statistic:

$$T_H^2 \equiv n(\bar{\mathcal{Z}}_n - \vec{\mu})^\top S_n^{-1}(\bar{\mathcal{Z}}_n - \vec{\mu}) \sim \frac{p(n-1)}{n-p} F_{p,n-p}. \tag{8.22}$$

where $F_{a,b}$ denotes the usual F distribution with degrees a and b. The zero or negative denominator $n - p$ when $n \leq p$ clearly alerts us that T_H^2 is not a legitimate pivotal quantity for $\vec{\mu}$ when $n \leq p$.

Result (8.20) however reveals that there is always a collection of $n-1$ legitimate pivotal quantities for $\vec{\mu}$ that are independent of each other, conditioning on the MLE for $\vec{\mu}$, that is, $\bar{\mathcal{Z}}_n$. Each collection is given by $\{T_{\vec{w}}(\mathbf{Y}_1, \mathbf{Y}_j), j = 2, \ldots, p\}$, where $T_{\vec{w}}(\mathbf{X}, \mathbf{Y})$ is defined by (8.20). How best to use such collections for constructing exact simultaneous confidence regions is for further research. Here we illustrate the construction of a single region by using any linear combination of $\{\mathbf{Y}_2, \ldots, \mathbf{Y}_n\}$.

Let $\mathbf{B}_\xi = \sum_{k=2}^{n} \xi_k \mathbf{Y}_k$, where $\sum_{k=2}^{n} \xi_k^2 = 1$ and hence $\mathbf{B}_\xi \sim N(0, \Sigma)$. Applying (8.20) then yields

$$T_{\vec{w}}(\mathbf{Y}_1, \mathbf{B}_\xi) = \sqrt{n}(\bar{\mathcal{Z}}_n - \vec{\mu})^\top W \mathbf{B}_\xi^- \sim \text{Cauchy}(0,1), \tag{8.23}$$

where $W = \text{Diag}\{w_1, \ldots, w_p\}$, and $\mathbf{B}_\xi^- = (b_1^{-1}, \ldots, b_p^{-1})^\top$, with b_j being the jth component of $\mathbf{B}_\xi$,

$$b_j = \sum_{k=2}^{n} \xi_k \left[\sum_{i=1}^{n} c_{ik} Z_{ij} \right], \quad j = 1, \ldots, p. \tag{8.24}$$

Here $\mathbf{B}_\xi$ is a contrast-based benchmark for $\sqrt{n}(\bar{\mathcal{Z}}_n - \vec{\mu})$ because the two variables share the same distribution. This leads to the following exact $(1-\alpha)$-level confidence region for $\vec{\mu}$ based on data $\mathcal{Z} = \{Z_{ij}\}_{n \times m}$, where $\bar{Z}_{+j} = \sum_{i=1}^{n} Z_{ij}/n$,

$$\left| \sum_{j=1}^{p} \frac{w_j(\bar{Z}_{+j} - \mu_j)}{b_j} \right| \leq \frac{1}{\sqrt{n}} \cot(\alpha \frac{\pi}{2}), \tag{8.25}$$

because the q-percentile for Cauchy(0,1) is $-\cot(q\pi)$.

To compare T_H with T of (8.23), we take $w_j = p^{-1}$ for all j, which then implies that

$$T_C^2 \equiv n(\bar{\mathcal{Z}}_n - \vec{\mu})^\top \mathbf{B}_\xi^- \mathbf{B}_\xi^{-\top} (\bar{\mathcal{Z}}_n - \vec{\mu}) \sim p^2 F_{1,1}, \tag{8.26}$$

because the square of a Cauchy(0, 1) is $F_{1,1}$. Comparing (8.26) with (8.22) reveals the trade-off made for avoiding estimating Σ. Computationally, T_C^2 avoids matrix inversion by using component-wise inverse, as in $\mathbf{B}_\xi^-$. Inferentially, because $\mathbb{E}(b_j^2) = \sigma_{jj}$, we can view b_j as a 'signed standard error' for $\sqrt{n}(\bar{Z}_{+j} - \mu_j)$. The use of this non-standard error (pun intended) as a benchmark then induces the Cauchyanity, which then eliminates the influence of all σ_{ij} for $i \neq j$.

There is no free lunch of course. When $n > p$, the confidence region from the Hotelling pivotal quantity is a bounded region in R^p, whereas the Cauchy region is bounded only in a one-dimensional subspace because B_ξ is of rank 1. The two are the same in the extreme case with $p = 1$ and $n = 2$, where $T_H^2 = 4(\bar{z} - \mu)^2/(z_1 - z_2)^2 = T_C^2$, with $\bar{z} = (z_1 + z_2)/2$.

The case of $n = 2$ is particularly useful for combining p (possibly) dependent studies for the same quantity θ. For jth study, suppose we randomly split the data into two independent parts with (approximately) the same size, and we assume the two independent random sub-samples can still provide valid estimators $\hat{\theta}_{ij}, i = 1, 2$ for θ, where the assignment of the index value of i is random. Let $\hat{\Theta}_i = \{\hat{\theta}_{i1}, \ldots, \hat{\theta}_{ip}\}, i = 1, 2$. When the studies are of reasonable sizes, it is often acceptable to assume $\hat{\Theta}_i - \theta \mathbf{1}_n \sim N(0, \Sigma), i = 1, 2$. We can then apply (8.25) with $\bar{Z}_{+j} = (\hat{\theta}_{1j} + \hat{\theta}_{2j})/2$ and $b_j = \delta_j/\sqrt{2}$, where $\delta_j = \hat{\theta}_{1j} - \hat{\theta}_{2j}$ (note the probability that $\delta_j = 0$ is zero)

$$\left| \theta - \frac{\sum_{j=1}^{p} (\hat{\theta}_{1j} + \hat{\theta}_{2j}) \delta_j^{-1}}{2 \sum_{j=1}^{p} \delta_j^{-1}} \right| \leq \frac{p \cot\left(\alpha \frac{\pi}{2}\right)}{2 \left| \sum_{j=1}^{p} \delta_j^{-1} \right|}, \tag{8.27}$$

which is an $(1-\alpha)\%$ confidence interval for θ. This procedure will give more weight to those studies for which δ_j's are small, which makes intuitive sense since small δ_j's indicate more accurate studies. This procedure, although exact (when the assumption of Gaussianity for $\hat{\Theta}_i$ holds), may be dominated by an approximate method also based on Cauchyanity, as described below to end this winding chapter with a clear blaze for continuing exploration.

8.7.3 Finally Cauchy May Have Its Moments (and Generate Many Open Problems)

Perhaps it was a divine allusion that Pillai and Meng (2016) was a part of a special theme dedicated to the memory of Peter Hall, arguably the most prolific statistician of his generation, because many articles need to be written to unleash the hidden power of Cauchyanity for managing arbitrary dependence structures in arbitrarily high dimensions. Such potential is already shown in a series of articles, with both methodological development and large-scale applications (e.g., Liu and Xie, 2020; Sun et al., 2020; Liu et al., 2021, 2022; Weber et al., 2021; Li et al., 2022; Fang et al., 2023).

In particular, Liu and Xie (2020) propose to combine d possibly correlated p-values $\{p_1, \ldots, p_d\}$ for testing the same null H_0 via

$$T = \sum_{i=1}^{d} w_i \tan\{(0.5 - p_i)\pi\} = \sum_{i=1}^{d} w_i \cot(p_i \pi). \tag{8.28}$$

Clearly, when the d p-values are mutually independent, $T \sim$ Cauchy(0,1) under H_0, and hence we can compute the combined p-value as $\text{arccot}(T)/\pi$. The resemblance between (8.21) and (8.28) suggests that this Cauchyanity may be robust against possible dependence among the p-values.

Indeed, Liu and Xie (2020) provide theoretical and empirical evidence for this robustness. Their main theoretical result is obtained by converting p-values into 'Z-values' via $p_i = 2[1-\Phi(|Z_i|)]$ (where $\Phi(z)$ is the CDF of the standard Normal variable), and by assuming that (Z_i, Z_j)'s are bivariate Normal with mean zero for any $i \neq j$; otherwise Σ, the covariance matrix of $(Z_1, \ldots, Z_p)$, is arbitrary (but exists). Under this set of assumption, Liu and Xie (2020) establish that

$$\lim_{t\to\infty} \frac{P(T \geq t)}{P(C \geq t)} = 1, \tag{8.29}$$

where $C \sim$ Cauchy(0, 1). That is, T of (8.28) will have a Cauchy-like tail regardless of the value Σ. Most recently, Fang et al. (2023) show that such robustness properties against Σ can be extended to other ways of combining p_i, such as the harmonic mean combination $\sum_i p_i^{-1}$ method of Good (1958) and Wilson (2019).

Adopting the well-known idea of converting a testing procedure into a confidence procedure, we can easily turn (8.28) into a confidence procedure for combining d possibly dependent studies for estimating θ. Specially, suppose we have estimator $\hat{\theta}_j$ from j study, with $\hat{\sigma}_j$ as its variance estimate, $j = 1, \ldots, d$. Assuming $(\hat{\theta}_i, \hat{\theta}_j)$ is approximately bivariate Normal for any $i \neq j$, then a $(1-\alpha)\%$ confidence region for θ can be constructed based on (8.28) by letting $p_i = 2\left[1 - \Phi(|\hat{\theta}_i - \theta|/\hat{\sigma}_i)\right]$, which leads to the confidence region with approximately $(1-\alpha)\%$ confidence (as $\alpha \to 1$):

$$\sum_{i=1}^{d} w_i \cot\left[2\pi\left(1-\Phi\left(\frac{|\hat{\theta}_i-\theta|}{\hat{\sigma}_i}\right)\right)\right] \le \cot(\alpha\pi). \tag{8.30}$$

For $d = 1$, this recovers the usual (Gaussian) confidence interval $[\theta_1 - z_{1-0.5\alpha}\hat{\sigma}_1, \theta_1 + z_{1-0.5\alpha}\hat{\sigma}_1]$.

Many research questions arise immediately: practical, mathematical, and statistical. Is this always an interval for θ? If so, how could we compute the two end points efficiently? How should we choose $w_i's$? Is the natural choice of $w_i = 1/d$ optimal? How good is its coverage? Is the optimal choice admissible among those that can be similarly (and optimally) constructed by inverting the robust class of p-value combinations studied in Fang et al. (2023)? [The answer to the last question is likely to be negative, considering (8.30) contains all $\hat{\theta}_j, j = 1, \ldots, d$.]

My learning journey clearly is far from complete, and indeed the remaining segment just became heavier ...

Acknowledgments

I cannot thank enough my co-editors, especially Min-ge Xie, for their extraordinary patience with my tardiness in completing this chapter. I offer my deep apologies to fellow chapter authors for dragging the whole project due to my poor time management skills. I am grateful to Nancy Reid for meticulous proofreading, and to Min-ge Xie, Ben Brown, Natesh Pillai, and Tyler Liu for very helpful comments and discussions. I thank NSF for (many years of) partial financial support that facilitated my learning journey and the preparation of this chapter.

Notes

1. According to Google's English dictionary by Oxford Languages, a construct is "an idea or theory containing various conceptual elements, typically one considered to be subjective and not based on empirical evidence."
2. In 1982, I took a graduate school entry exam, which was too basic by Fudan's standard. I therefore handed in my answers in less than 50% of the allocated time, but I only scored 60% and hence my application was rejected. Later I found out that I lost 40% mostly on "It is obvious ..." or "It is well known ...", since the skipped steps were not obvious or known to the grader, to whom I'm grateful because otherwise I'd not be here to write this chapter.
3. I certainly have encountered more challenging questions by undergraduate students, such as "in what sense is much of the current *Objective Bayes* objective or Bayes?"
4. This caveat is needed because using large-sample Normal approximations can lead to very different results for different transformations. For example, a direct Normal approximation for a variance parameter σ^2 may lead to a confidence interval containing negative values, a problem easily avoided when we start with $\log \sigma^2$.
5. But it is not the only method available; see for example Lawless and Fredette (2005) and Shen et al. (2018) for non-Bayesian methods such as confidence distributions.
6. Lindley acknowledged an earlier work in Russian (Dynkin, 1951), which was brought to Lindley's attention after his article was written.
7. For future historians: one could work in 2014 on a conjecture published in 2016 thanks to the delay in publication.

Bibliography

Bayarri, M. and Berger, J. O., 2000. P values for composite null models. *Journal of the American Statistical Association*, *95*(452), pp. 1127–1142.

Berger, J. O., Liseo, B. and Wolpert, R. L., 1999. Integrated likelihood methods for eliminating nuisance parameters. *Statistical Science*, *14*, pp. 1–22.

Berger, J. O. and Wolpert, R. L., 1988. *The Likelihood Principle.* Hayward, CA: IMS.

Box, G. E., 1980. Sampling and bayes' inference in scientific modelling and robustness. *Journal of the Royal Statistical Society: Series A (General)*, *143*(4), pp. 383–404.

Breiman, L., 2001. Statistical modeling: The two cultures (with comments and a rejoinder by the author). *Statistical Science*, *16*(3), pp. 199–231.

Brooks, S., Gelman, A., Jones, G., and Meng, X.-L., 2011. *Handbook of Markov Chain Monte Carlo.* CRC Press.

Brown, D. (2004). *The Da Vinci Code.* Random House.

Cohen, J. E., Davis, R. A. and Samorodnitsky, G., 2020. Heavy-tailed distributions, correlations, kurtosis and taylor's law of fluctuation scaling. *Proceedings of the Royal Society A*, *476*(2244), p. 20200610.

Cole, S. R., Chu, H. and Greenland, S., 2014. Maximum likelihood, profile likelihood, and penalized likelihood: A primer. *American Journal of Epidemiology*, *179*(2), pp. 252–260.

Consonni, G. and Veronese, P., 1993. Unbiased bayes estimates and improper priors. *Annals of the Institute of Statistical Mathematics*, *45*, pp. 303–315.

Cox, D., 1975. A note on partially bayes inference and the linear model. *Biometrika*, *62*(3), 651–654.

Craiu, R. V., Gong, R. and Meng, X.-L., 2022. Six statistical senses. *Annual Review of Statistics and Its Application*, *10*, pp. 699–725.

Dahl, F. A., 2006. On the conservativeness of posterior predictive p-values. *Statistics & Probability Letters*, *76*(11), pp. 1170–1174.

Dawid, A. P. and Stone, M., 1982. The functional-model basis of fiducial inference. *The Annals of Statistics*, *10*, pp. 1054–1067.

Dawid, P., 2023. Fiducial inference then and now. In *Handbook on Bayesian, Fiducial and Frequentist (BFF) Inference*, eds J. Berger, X.-L. Meng, N. Reid and M. Xie. New York: Chapman & Hall.

Dempster, A. P., 1963a. Further examples of inconsistencies in the fiducial argument. *The Annals of Mathematical Statistics*, *34*, pp. 884–891.

Dempster, A. P., 1963b. On direct probabilities. *Journal of the Royal Statistical Society: Series B (Methodological)*, *25*(1), pp. 100–110.

Desmond, A. and Xie, C., 2011. Discussion of "what's the H in H-likelihood: A holy grail or an achilles' heel?". In *Bayesian Statistics 9*, eds J. Bernardo, M. Bayarri, J. Berger, A. Dawid, D. Heckerman, A. Smith and M. West, pp. 473–501. Oxford University Press.

Ding, P. and Blitzstein, J. K., 2018. On the gaussian mixture representation of the laplace distribution. *The American Statistician*, *72*(2), pp. 172–174.

Drton, M. and Xiao, H., 2016. Wald tests of singular hypotheses. *Bernoulli*, *22*(1), pp. 38–59.

Dynkin, E. B. 1951. Necessary and sufficient statistics for a family of probability distributions. *Uspekhi Mat. Nauk* *6*(1), 68–90.

Efron, B., 1992. *Bootstrap Methods: Another look at the Jackknife.* Springer.

Fang, Y., Chang, C., Park, Y. and Tseng, G. C., 2023. Heavy-tailed distribution for combining dependent *p*-values with asymptotic robustness. *Statistica Sinica*, *33*, pp. 1115–1142.

Fienberg, S. E. and Lazar, N., 2001. William sealy gosset. In *Statisticians of the Centuries*, eds C. C. Heyde, E. Seneta, P. Crépel, S. E. Fienberg, and J. Gani, pp. 312–317. Springer.

Fisher, R. A., 1935. The fiducial argument in statistical inference. *Annals of Eugenics*, *6*(4), pp. 391–398.

Fisher, R. A., 1956. *Statistical Methods and Scientific Inference.* Hafner Publishing Co.

Fraser, D. and Haq, M. S., 1969. Structural probability and prediction for the multivariate model. *Journal of the Royal Statistical Society: Series B (Methodological)*, *31*(2), pp. 317–331.

Gelman, A., 2005. Comment: Fuzzy and Bayesian *p*-values and *u*-values. *Statistical Science*, *20*, pp. 380–381.

Gelman, A., Meng, X.-L. and Stern, H., 1996. Posterior predictive assessment of model fitness via realized discrepancies (with discussions). *Statistica Sinica*, *6*, pp. 733–807.

Good, I. J., 1958. Significance tests in parallel and in series. *Journal of the American Statistical Association*, *53*(284), pp. 799–813.

Guttman, I., 1967. The use of the concept of a future observation in goodness-of-fit problems. *Journal of the Royal Statistical Society: Series B (Methodological)*, *29*(1), pp. 83–100.

Guttman, I. and Tiao, G. C., 1964. A bayesian approach to some best population problems. *The Annals of Mathematical Statistics*, *35*(2), pp. 825–835.

Hannig, J., Iyer, H., Lai, R. C. and Lee, T. C., 2016. Generalized fiducial inference: A review and new results. *Journal of the American Statistical Association*, *111*(515), pp. 1346–1361.

Hart, J. D., 2009. Frequentist-bayes lack-of-fit tests based on laplace approximations. *Journal of Statistical Theory and Practice*, *3*(3), pp. 681–704.

Hjort, N. L., Dahl, F. A. and Steinbakk, G. H., 2006. Post-processing posterior predictive p values. *Journal of the American Statistical Association*, *101*(475), pp. 1157–1174.

Kass, R. E., 1990. The validity of posterior expansions based on laplace's method. In *Bayesian and Likelihood Methods in Statistics and Econometrics*, eds S. Geisser, J. S. Hodges, J. Press, and A. Zellner, pp. 473–488. Amsterdam: North-Holland.

Kass, R. E. and Wasserman, L., 1996. The selection of prior distributions by formal rules. *Journal of the American statistical Association*, *91*(435), pp. 1343–1370.

Kong, A., McCullagh, P., Meng, X.-L. and Nicolae, D., 2006. Further explorations of likelihood theory for Monte Carlo integration. In *Advances in Statistical Modeling and Inference: Essays in Honor of Kjell A. Doksum*, ed V. Nair, pp. 563–592. World Scientific Press.

Kong, A., McCullagh, P., Meng, X.-L., Nicolae, D. and Tan, Z., 2003. A theory of statistical models for Monte Carlo integration. *Journal of the Royal Statistical Society Series B: Statistical Methodology*, *65*, 585–604.

Lawless, J. F. and Fredette, Marc., 2005. Frequentist prediction intervals and predictive distributions. *Biometrika* *92*(3), pp. 529–542. Oxford University Press.

Lee, Y. and Nelder, J. A., 1996. Hierarchical generalized linear models. *Journal of the Royal Statistical Society: Series B (Methodological)*, *58*(4), pp. 619–656.

Lee, Y. and Nelder, J. A., 2009. Likelihood inference for models with unobservables: Another view. *Statistical Science*, *24*(3), pp. 255–269.

Lee, Y., Nelder, J. A. and Pawitan, Y., 2018. *Generalized Linear Models with Random Effects: Unified Analysis via H-Likelihood.* Chapman and Hall/CRC.

Leonelli, S., 2021, January 29. Data science in Times of Pan(dem)ic. *Harvard Data Science Review*, *3*(1). https://hdsr.mitpress.mit.edu/pub/fi1rol2i.

Li, X. and Meng, X.-L., 2021. A multi-resolution theory for approximating infinite-p-zero-n: Transitional inference, individualized predictions, and a world without bias-variance tradeoff. *Journal of the American Statistical Association*, *116*(533), 353–367.

Li, Y., Zhou, X. and Cao, H., 2022. Statistical analysis of spatially resolved transcriptomic data by incorporating multiomics auxiliary information. *Genetics*, *221*(4), iyac095.

Lin, H., 2022. Bayesian epistemology. In *The Stanford Encyclopedia of Philosophy* (Fall 2022 ed.), eds E. N. Zalta and U. Nodelman, Metaphysics Research Lab. Stanford University.

Lindley, D. V., 1958. Fiducial distributions and bayes' theorem. *Journal of the Royal Statistical Society: Series B (Methodological)*, *20*(1) pp. 102–107.

Liu, D., Liu, R. Y. and Xie, M., 2021. Nonparametric fusion learning for multiparameters: Synthesize inferences from diverse sources using data depth and confidence distribution. *Journal of the American Statistical Association*, *117*, pp. 2086–2104.

Liu, K. and Meng, X.-L., 2016. There is individualized treatment. why not individualized inference? *Annual Review of Statistics and Its Application*, *3*, pp. 79–111.

Liu, Y. and Xie, J., 2020. Cauchy combination test: A powerful test with analytic p-value calculation under arbitrary dependency structures. *Journal of the American Statistical Association*, *115*(529), pp. 393–402.

Liu, Z., Shen, J., Barfield, R., Schwartz, J., Baccarelli, A. A. and Lin, X., 2022. Large-scale hypothesis testing for causal mediation effects with applications in genome-wide epigenetic studies. *Journal of the American Statistical Association*, *117*(537), pp. 67–81.

Lo, A. W., Siah, K. W. and Wong, C. H., 2019, July 1. Machine learning with statistical imputation for predicting drug approvals. *Harvard Data Science Review*, *1*(1). https://hdsr.mitpress.mit.edu/pub/ct67j043.

McCullagh, P., 1990. A note on partially bayes inference for generalized linear models. In *Technical Report 284*, Department of Statistics, University of Chicago Chicago, IL.

Meng, X.-L., 1994. Posterior predictive *p*-values. *The Annals of Statistics*, *22*(3), pp. 1142–1160.

Meng, X.-L., 2009. Decoding the H-likelihood. *Statistical Science*, *24*(3), pp. 280–293.

Meng, X.-L., 2011. What's the H in H-likelihood: A holy grail or an achilles' heel? (with discussions and rejoinder). In *Bayesian Statistics 9*, eds J. Bernardo, M. Bayarri, J. Berger, A. Dawid, D. Heckerman, A. Smith and M. West, pp. 473–501. Oxford University Press.

Pillai, N. S. and Meng, X.-L., 2016. An unexpected encounter with Cauchy and Lévy. *The Annals of Statistics*, *44*(5), pp. 2089–2097.

Pitman, E. and Williams, E., 1967. Cauchy-distributed functions of Cauchy variates. *The Annals of Mathematical Statistics*, *38*(3), pp. 916–918.

Reimherr, M., Meng, X.-L. and Nicolae, D. L., 2021. Prior sample size extensions for assessing prior impact and prior-likelihood discordance. *Journal of the Royal Statistical Society: Series B (Statistical Methodology)*, *83*(3), pp. 413–437.

Robins, J. M., van der Vaart, A. and Ventura, V., 2000. Asymptotic distribution of p values in composite null models. *Journal of the American Statistical Association*, *95*(452), pp. 1143–1156.

Rubin, D. B., 1984. Bayesianly justifiable and relevant frequency calculations for the applied statistician. *The Annals of Statistics*, *12*(4), pp. 1151–1172.

Savage, L. J., 1961. *The Foundations of Statistics Reconsidered.* University of California Press.

Shun, Z. and McCullagh, P., 1995. Laplace approximation of high dimensional integrals. *Journal of the Royal Statistical Society: Series B (Methodological)*, *57*(4), pp. 749–760.

Sun, S., Zhu, J. and Zhou, X., 2020. Statistical analysis of spatial expression patterns for spatially resolved transcriptomic studies. *Nature Methods*, *17*(2), pp. 193–200.

Wallace, D. L., 1980. The behrens-fisher and fieller-creasy problems. In *RA Fisher: An Appreciation*, eds S. E. Fienberg and D. V. Hinkley, pp. 119–147. Springer.

Weber, M., Apel, S. and Siegmund, N., 2021. White-box performance-influence models: A profiling and learning approach. In *2021 IEEE/ACM 43rd International Conference on Software Engineering (ICSE)*, pp. 1059–1071. IEEE.

Williams, E., 1969. Cauchy-distributed functions and a characterization of the Cauchy distribution. *The Annals of Mathematical Statistics*, *40*(3), pp. 1083–1085.

Wilson, D. J., 2019. The harmonic mean p-value for combining dependent tests. *Proceedings of the National Academy of Sciences*, *116*(4), pp. 1195–1200.

Xu, H., Cohen, J. E., Davis, R. A. and Samorodnitsky, G., 2022. Cauchy, normal and correlations versus heavy tails. *Statistics & Probability Letters*, *186*, p. 109489.

9

Bayesian Neural Networks and Dimensionality Reduction

Deborshee Sen
University of Bath

Theodore Papamarkou
The University of Manchester

David Dunson
Duke University

9.1 Introduction

Deep learning (DL) has emerged as one of the most successful tools in image and signal processing. Excellent predictive performance in various high-profile applications has led to an enormous increase in focus on DL methods in a wide variety of fields, including artificial intelligence, astrophysics, chemistry and material science, healthcare, and manufacturing. The DL paradigm relies on multi-layer neural networks, which have been around for decades, with recent success due mostly to advances in computing power, algorithms, and the availability of enormous datasets for training. This has been propelled by the rise of graphical processing units (GPUs). Many dedicated toolboxes have been developed and continue to be developed, including Apache MXNet, Caffe, the Microsoft Cognitive Toolkit, PyTorch, and TensorFlow. Many thousands of papers are published every year on deep learning.

The meteoric rise in the field of DL in the last decade started with a key paper by Geoffrey Hinton and collaborators showing that multi-layer convolutional neural networks could be tuned to imaging data to obtain a large improvement in classification performance (Krizhevsky et al., 2012). Convolutional neural networks consist of convolutional layers. A convolutional layer performs a convolution, which is a linear operation between neural network parameters and patches of input data (or their transformations). Convolutions learn local features of the data, thus enhancing the predictive capacity of DL. This stimulated a flood of researchers into the field, particularly in the imaging and signal processing communities, with a more recent spillover into statistics and other fields. The vast majority of the intellectual energy in studying DL approaches has been focused on numerical optimization and engineering methods – trying to modify the neural network architecture, tune hyper-parameters, and automate optimization algorithms to obtain good performance on test data in a variety of settings.

 DOI: 10.1201/9780429341731-9

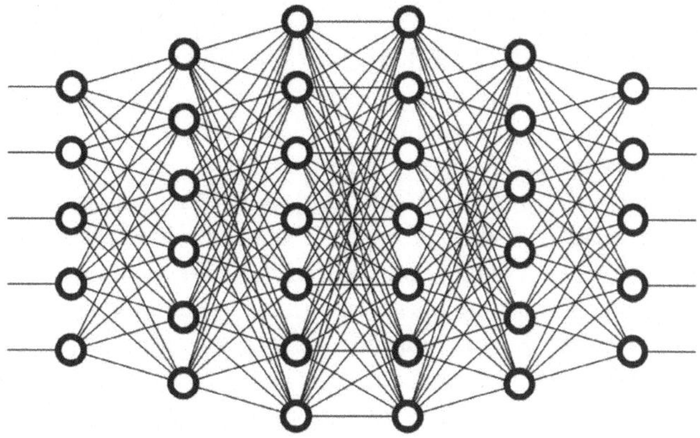

FIGURE 9.1
A neural network with four hidden layers. The left most layer is the input layer with five neurons, the right most layer is the output layer with five neurons, and the four hidden layers have six, seven, seven, and six neurons from left to right, respectively. In other words, we have $q = 5$, $h_1 = 6$, $h_2 = 7$, $h_3 = 7$, $h_4 = 6$, and $p = 5$.

A neural network is a non-linear function $f : \mathbb{R}^q \to \mathsf{Y} \subseteq \mathbb{R}^p$ consisting of an input layer, one or more hidden layers, and an output layer (Figure 9.1). In addition, a non-linear *activation function* is applied element-wise to each input of each layer. Mathematically, this can be written as

$$Y^i = f(z^i) = W_5\varphi\left(W_4\varphi\left[W_3\varphi\left\{W_2\varphi(W_1z^i + b_1) + b_2\right\} + b_3\right] + b_4\right) + b_5 + \varepsilon^i, \qquad (9.1)$$

where $W_1 \in \mathbb{R}^{q\times h_1}, W_2 \in \mathbb{R}^{h_1\times h_2}, W_3 \in \mathbb{R}^{h_2\times h_3}, W_4 \in \mathbb{R}^{h_3\times h_4}$, and $W_5 \in \mathbb{R}^{h_4\times p}$ are *weights*, $b_1 \in \mathbb{R}^{h_1}, b_2 \in \mathbb{R}^{h_2}, b_3 \in \mathbb{R}^{h_3}, b_4 \in \mathbb{R}^{h_4}$, and $b_5 \in \mathbb{R}^p$ are *biases*, $\varphi : \mathbb{R} \to \Omega \subseteq \mathbb{R}$ is the non-linear activation function, $\varepsilon^i \in \mathbb{R}^p$ denotes an error term, and i indexes the observations. Here, x_i denotes an input (for example, a set of features describing an image), and Y^i denotes a prediction made using input z^i (for instance, classification probabilities of the image belonging to different classes). For ease of exposition, equation (9.1) describes a neural network with four hidden layers; this can be easily modified to include any number of hidden layers. Common activation functions include sigmoid $\varphi(z) = \{1 + \exp(-z)\}^{-1}$, the rectified linear unit (ReLU) $\varphi(z) = z\,\mathbb{I}(z \geq 0)$, and the leaky ReLu $\varphi(z) = 0.01z\,\mathbb{I}(z < 0) + z\,\mathbb{I}(z \geq 0)$, where $\mathbb{I}$ is the indicator function. The universal approximation theorem for neural networks (Cybenko, 1989) guarantees that the function f can be made arbitrarily flexible by adjusting the number of layers of the neural network and the number of hidden neurons per layer.

In what follows, we let $\theta = (W_1, b_1, W_2, b_2, \dots)$ denote all the parameters of the neural network; this includes the weights and biases. In the sequel, we use the notation $f_\theta(\cdot)$ in place of $f(\cdot)$ in order to make the function f's dependence on θ explicit. Given training data $\{(z^i, y^i)\}_{i=1}^N$, neural network training typically involves optimizing a loss function $L(\theta)$ as follows:

$$\theta_\star = \underset{\theta}{\operatorname{argmin}} \left\{ L(\theta) = \frac{1}{N}\sum_{i=1}^{N}\{y^i - f_\theta(z^i)\}^2 + p_0(\theta) \right\}, \qquad (9.2)$$

where $p_0(\theta)$ denotes a regularization term. The optimization in equation (9.2) is usually achieved through a technique known as *backpropagation* (Rumelhart et al., 1986), which allows gradients with respect to θ of the loss function $L(\theta)$ to be calculated. The number of observations N is usually very large, and stochastic gradient-based algorithms are used for finding local optima of the non-convex loss function $L(\theta)$. Clever variants to usual stochastic gradient descent have been devised such as including momentum (Sutskever et al., 2013) and moment-based algorithms (Kingma and Ba, 2015).

Dimensionality reduction is very important in making sense of the very high-dimensional and complex data routinely collected in modern scientific and technological applications. High-dimensional data tend to be highly structured and can often be accurately summarized with much lower-dimensional latent variables. Linear dimensionality reduction methods produce a lower-dimensional linear mapping of the original high-dimensional dataset. This includes techniques such as principal component analysis, factor analysis, linear multidimensional scaling, Fisher's linear discriminant analysis, etc. We refer the reader to Cunningham and Ghahramani (2015) for a detailed discussion. While linear methods have advantages in terms of interpretability and simplicity, they are limited in their ability to concisely represent complex data. This has motivated a rich literature on non-linear dimensionality reduction, including popular manifold learning algorithms Isomap (Tenenbaum et al., 2000) and locally linear embeddings (Roweis and Saul, 2000). These approaches learn a mapping from the high-dimensional observation space to a low-dimensional space that preserves certain desirable aspects of the original data such as pairwise distances.

An alternate generative approach for modeling high-dimensional data assumes that the data are generated by transforming a lower-dimensional latent variable X^i to a higher-dimensional observation Y^i through a function g as $Y^i = (X^i, \varepsilon^i)$, where ε^i is a random error component. For example, a linear latent factor model is obtained when $g(X^i, \varepsilon^i) = AX^i + \varepsilon^i$, with $\varepsilon^i \sim \text{Normal}(\mu, \Sigma)$ for a factor loadings matrix A. More generally, the function g can be a non-linear function of X; particular instances include Gaussian processes (Lawrence, 2004) and neural networks; we are interested in the latter approach in this text. Variational Bayes algorithms have been devised for neural networks with latent variables, and such methods are known as variational auto-encoders (VAEs; Kingma and Welling, 2014).

VAEs are a popular generative modeling tool, which (in their most basic form) assumes that model (9.1) is of the form

$$Y^i = f_\theta(X^i) + \varepsilon^i, \quad X^i \sim \lambda, \quad \varepsilon^i \sim \nu,$$

where f_θ is a neural network (called the decoder) with θ denoting all the weights and biases in it, $\theta = (W_1, b_1, W_2, b_2, \dots)$. The latent distribution λ is usually assumed to be $\text{Normal}(0, I)$, although extensions ranging from mixture distributions (Dilokthanakul et al., 2016) to sequential models (Chung et al., 2015) to incorporating geometric aspects (Mathieu et al., 2019) have been considered. VAEs have been used successfully in a wide range of contexts, including music generation (Roberts et al., 2017), image generation (Dosovitskiy and Brox, 2016), forecasting (Walker et al., 2016), text generation (Semeniuta et al., 2017), and clustering (Jiang et al., 2017). We refer the reader to Kingma and Welling (2019) for an excellent introduction to VAEs.

VAEs employ a variational approximation to the distribution of $X^i \mid Y^i$. In particular, this is approximated by $\text{Normal}(\mu_\phi(Y^i), \Sigma_\phi(Y^i))$, where $\mu_\phi(\cdot)$ and $\Sigma_\phi(\cdot)$ are estimated through a neural network parameterized by ϕ; this is known as the encoder. This is a restrictive assumption, as the actual distribution of $X \mid Y$ can be far from a Gaussian distribution. Moreover, a practical problem with VAEs is reproducibility, which is crucial in scientific applications. VAEs can be used as a dimension reduction technique to obtain lower dimensional representations $X^1, \ldots, X^N$ of observations $Y^1, \ldots, Y^N$, respectively. However, this is typically not reproducible: if the VAE is trained again to obtain $X_\prime^1, \ldots, X_\prime^N$, these will typically be quite different from $X^1, \ldots, X^N$. Variational approximations also lack theoretical guarantees in general and lead to inaccurate uncertainty quantification. Motivated by this, we adopt a fully Bayesian approach in this chapter.

The rest of this chapter is organized as follows. We provide background on Bayesian parametric techniques for neural networks in Section 9.2. Section 9.3 is devoted to the main theme of this chapter – Bayesian neural networks for dimensionality reduction; this includes techniques to reduce identifiability issues for such problems. Section 9.4 and Section 9.5 provide experimental assessment of our techniques based on simulated and real data, respectively. Finally, Section 9.6 concludes the chapter and discusses open questions and future research directions.

9.2 Background on Bayesian Neural Networks

Variational approximations and Markov chain Monte Carlo algorithms provide alternative approaches for Bayesian inference in neural networks. The former approach dominates the current literature, and MCMC implementations are relatively rare, likely due to the tendency of standard MCMC approaches to have poor performance. We provide a brief review of both strategies in this section.

9.2.1 Variational Inference

Variational Bayes methods enable approximate Bayesian inference for challenging probability densities. The idea behind most variational inference algorithms is to posit a parametric family of densities and then to find via optimization the member of the family that is closest to the target density. Closeness is evaluated using the Kullback-Leibler (KL) divergence or some distance in the space of probability densities. We refer the reader to Blei et al. (2017) for an introduction to variational inference.

Designing effective variational approximations for Bayesian neural network (BNN) posteriors is a non-trivial task. For instance, it has been shown that more expressive variational families can yield significantly worse predictions in comparison to predictions obtained from less expressive variational families (Trippe and Turner, 2018). Regularization techniques can improve the predictive generalization of variational inference in BNNs. This includes pruning (Graves, 2011) or an appropriate choice of priors, such as horseshoe shrinkage (Ghosh et al., 2018) or hierarchical priors (Wu et al., 2019). Esmaeili et al. (2019); Huang et al. (2019); Titsias and Ruiz (2019) developed variational families with hierarchical structure to approximate BNN posteriors. Such hierarchical variational models induce correlation among parameters to approximate the posterior covariance structure. Liu and Wang (2016) introduced Stein variational gradient descent and applied it to BNNs. This is a variational inference framework that relies on a functional form of stochastic gradient descent to approximate the target density by iteratively transporting a set of particles. This links the derivative of KL divergence with Stein's identity and with the kernel Stein discrepancy. Further, Shi et al. (2018) proposed a spectral Stein gradient estimator using Stein's identity, and Sun et al. (2019) harnessed it to develop a functional form of variational inference for BNNs.

9.2.2 Markov Chain Monte Carlo

Markov chain Monte Carlo (MCMC) methods tend to be slower to implement than variational algorithms, but have the advantage of (conceptually) providing posterior summaries that converge to the true values as the number of Monte Carlo samples increases. However, MCMC algorithms for BNNs face several challenges (Papamarkou et al., 2019). Firstly, the huge size of both the parameter space and input data space of BNNs can render MCMC algorithms computationally infeasible. Secondly, weight symmetries in ANNs create identifiability issues and yield multimodal weight posteriors (Pourzanjani et al., 2017): MCMC algorithms spend valuable computational time exploring posterior modes associated with equivalent solutions in the output space (Nalisnick, 2018). Third, BNNs are complex non-linear hierarchical models, and such models are known to cause problems with MCMC (Zhang and Sutton, 2014). Fourth, it is not clear how to choose good default priors for BNNs (Lee, 2004, 2005; Vladimirova et al., 2019). Several types of priors have been studied, including truncated flat priors (Lee, 2005), restricted flat priors (Lee, 2003), Jeffreys priors (Lee, 2007), smoothing priors (de Freitas, 1999), Laplace priors (Williams, 1995) and approximate reference priors (Nalisnick, 2018), but it remains to find good default priors.

These challenges have led to limited literature on MCMC algorithms for BNNs. Sequential Monte Carlo and reversible jump MCMC were applied to multilayer perceptrons (MLPs) and radial basis function networks as early as two decades back (Andrieu et al., 1999; de Freitas, 1999; Andrieu et al., 2000; de Freitas et al., 2001), but have not seen much development since. Geometric Langevin Monte Carlo and population MCMC (power posterior sampling) were also applied to MLPs (Papamarkou et al., 2019). To the best of our knowledge, there are not any instances of exact MCMC algorithms for convolutional neural networks (CNNs) in the literature.

To scale MCMC methods with big data, a log-likelihood can be evaluated on a subset (minibatch) of the data rather than on the entire dataset. The notion of minibatch was employed by Welling and Teh (2011) to develop a stochastic gradient Monte Carlo algorithm. Chen et al. (2014) introduced stochastic gradient Hamiltonian Monte Carlo and applied it to BNNs. Gong et al. (2019) proposed a stochastic gradient MCMC scheme that generalizes Hamiltonian dynamics with state-dependent drift and diffusion, and demonstrated the performance of this scheme on CNNs and on recurrent neural networks.

9.3 Latent Factor Models through Neural Networks

9.3.1 General Model

Consider the situation where we have observed data $Y^1, \ldots, Y^N \in \mathbb{R}^p$, and assume that these lie near a lower-dimensional manifold. We consider latent factor models of the form

$$X^i \overset{\text{i.i.d}}{\sim} g(\cdot), \quad Y^i \mid X^i \mathbf{1}_{\text{Normal}}\{f_\theta(X^i), \tau^2\}, \quad i = 1, \ldots, N, \tag{9.3}$$

with latent factors $X^i \in \mathbb{R}^q$ for $q < p$ and $i = 1, \ldots, N$. In addition, we let f_θ be a neural network with θ denoting all the weights and biases in the network. Let $\mathbf{Y}^N = (Y^1, \ldots, Y^N)$ and $\mathbf{X}^N = (X^1, \ldots, X^N)$ for compact notation. We place a prior $p_0(\theta)$ on θ and are interested in the posterior $p(\theta, \mathbf{X}^N \mid \mathbf{Y}^N)$.

For simplicity, consider a neural network with a single hidden layer with h neurons. This is also known as a multilayer perceptron (MLP). Following equation (9.1), the function $f_\theta : \mathbb{R}^q \mapsto \mathbb{R}^p$ can be written as

$$f_\theta(X^i) = W_2\varphi(W_1X^i + b_1) + b_2 \in \mathbb{R}^p, \tag{9.4}$$

for some $h \times q$ matrix W_1 and $p \times h$ matrix W_2, leading to the log-likelihood

$$\begin{aligned}\ell(\mathbf{Y}^N \mid \mathbf{X}^N, \theta) &\propto -\frac{1}{\tau^2}\sum_{i=1}^{N}\{f_\theta(X^i) - Y^i\}^2 \\ &= -\frac{1}{\tau^2}\sum_{i=1}^{N}\{W_2\varphi(W_1X^i + b_1) + b_2 - Y^i\}^2.\end{aligned} \tag{9.5}$$

In general, the model specified by equation (9.3) is not identifiable; different values $\mathbf{X}^N$ and θ give the same likelihood of $\mathbf{Y}^N$. Instead of focusing on identifiability of $\mathbf{X}^N$ and θ, we focus on identifiability of pairwise distances between the latent X^is. This is akin to dimension reduction techniques that aim to preserve pairwise distances between the latent variables, such as Isomap (Tenenbaum et al., 2000), locally linear embeddings (Roweis and Saul, 2000), and t-distributed stochastic neighbor embedding (Maaten and Hinton, 2008).

Let $X^i = (X^i_1, \ldots, X^i_q)^T \in \mathbb{R}^q$, and $W_1 \in \mathbb{R}^{h\times q}$ be written in a column-wise fashion as $W_1 = (w_{11}, \ldots, w_{1q})$, where w_{1j} is a h-dimensional column vector for $j = 1, \ldots, q$. Then we have $W_1X^i = \sum_{j=1}^q w_{1j}X^i_j$, which is invariant up to a permutation of $\{1, \ldots, q\}$: if $\{\sigma(1), \ldots, \sigma(N)\}$ denotes a permutation of $\{1, \ldots, N\}$, then $\sum_{j=1}^q w_{1j}X^i_j = \sum_{j=1}^q w_{1\sigma(j)}X^i_{\sigma(j)}$. This does not pose a problem since we are interested only in pairwise distances between the X^is. In the next three Sections 9.3.2–9.3.4, we consider more carefully the identifiability issue.

9.3.2 Identifiability

For vectors $a = (a_1, \ldots, a_p), b = (b_1, \ldots, b_p) \in \mathbb{R}^p$ we use the notation the notation $\langle a, b\rangle = \sum_{j=1}^p a_jb_j$ for Euclidean inner product and the notation $\|a\| = \langle a, a\rangle^{1/2}$ for Euclidean norm. Formally, having an identifiable model would mean that

$$\begin{aligned}&\ell(\mathbf{Y}^N \mid \mathbf{X}^N, \theta, \tau^2) = \ell\{\mathbf{Y}^N \mid \mathbf{X}^N_\prime, \theta_\prime, \tau^2_\prime\}\ \forall\ \mathbf{Y}^N \implies \\ &\mathbf{X}^N = \mathbf{X}^N_\prime,\quad \theta = \theta_\prime,\quad \tau^2 = \tau^2_\prime.\end{aligned} \tag{9.6}$$

Equation (9.6) implies

$$\frac{1}{\tau^2}\sum_{i=1}^{N}\|f_\theta(X^i) - Y^i\|^2 = \frac{1}{\tau^2_\prime}\sum_{i=1}^{N}\|f_{\theta_\prime}(X^i_\prime) - Y^i\|^2 \implies$$

$$\frac{1}{\tau^2}\sum_{i=1}^{N}\left[\|f_\theta(X^i)\|^2 + \|Y^i\|^2 - 2\langle f_\theta(X^i), Y^i\rangle\right] = \frac{1}{\tau^2_\prime}\sum_{i=1}^{N}\left[\|f_{\theta_\prime}(X^i_\prime)\|^2 + \|Y^i\|^2 - 2\langle f_{\theta_\prime}(X^i_\prime), Y^i\rangle\right],$$

which in turn implies

$$\tau^2 = \tau^2_\prime,\quad f_\theta(X^i) = f_{\theta_\prime}(X^i_\prime),\quad \text{and}\quad \sum_{i=1}^{N}\|f_\theta(X^i)\|^2 = \sum_{i=1}^{N}\|f_{\theta_\prime}(X^i_\prime)\|^2; \tag{9.7}$$

The third equality of equation (9.7) follows from the second, and so we focus only on the second equality. We now isolate some factors that cause identifiability issues.

Since we are only interested in pairwise distances among the X^is, location shifts do not matter: $\mathsf{dist}(X^i, X^j) = \mathsf{dist}(X^i + c, X^j + c)$ for any $c \in \mathbb{R}^q$, where dist denotes the Euclidean distance. Moreover, the model given by equation (9.4) is not invariant to scale changes in the columns of W_1 as $\sum_{j=1}^q w_{1j} X_j^i = \sum_{j=1}^q c_j w_{1j} X_j^i / c_j$ for any $c_1, \ldots, c_q \in \mathbb{R}$. We address this by imposing a norm constraint on the columns of W_1 as follows:

$$\|w_{1j}\|^2 = \sum_{k=1}^{h} w_{1jk}^2 = 1, \quad j = 1, \ldots, q. \tag{9.8}$$

This has the effect of fixing the scale of the latent variables. Typically, constraints are included in Bayesian inferences by choosing priors to have constrained support. However, such direct approaches can be computationally challenging, motivating recent proposals including constraint relaxation (Duan et al., 2020) and posterior projections (Patra et al., 2020). Fortunately, we can straightforwardly impose equation (9.8) by re-parameterizing equation (9.4) as

$$f_\theta(X^i) = W_2 \varphi \left\{ \left(w_{11}/\|w_{11}\| \quad \cdots \quad w_{1q}/\|w_{1q}\| \right) X^i + b_1 \right\} + b_2. \tag{9.9}$$

Remark 9.1. *The leaky ReLU activation function is linear; therefore, the rows of W_2 have to be taken to have unit length according to $\|w_{2\cdot i}\|^2 = \sum_{j=1}^p w_{2ij}^2 = 1$, $i = 1, \ldots, h$, if using it as the activation function. In this text, we consider the hyperbolic tangent activation function $\varphi(x) = \{\exp(x) - \exp(-x)\}/\{\exp(x) + \exp(-x)\}$, and so we do not need to impose this constraint on the rows of W_2.*

9.3.3 Anchor Points

The idea of *anchor points* has been used in the context of Gaussian mixture models to address the label switching problem (Kunkel and Peruggia, 2020). The main idea behind this is to "anchor" a certain number of latent variables, while conducting a full Bayesian inference for the remaining latent variables. This can be viewed as placing a highly informative point mass prior for the anchored variables. We use the idea of anchor points to improve identifiability for our model.

We fix N_{ref} anchor points out of a total N latent variables as follows. We start by assuming that we know the "true" latent representations of $\{Y^i\}_{i \in \mathcal{A}_{\text{ref}}}$ to be $\{X^i\}_{i \in \mathcal{A}_{\text{ref}}}$, respectively, where $\mathcal{A}_{\text{ref}} \subset \{1, \ldots, N\}$ denotes indices of anchor points, with $|A_{\text{ref}}| = N_{\text{ref}}$. Let $\mathcal{A}_{\text{ref}}^c = \{1, \ldots, N\} \setminus \mathcal{A}_{\text{ref}}$ denote indices of points other than the anchor points. We define the *anchored posterior* as

$$\pi(\{X^i\}_{i \in \mathcal{A}_{\text{ref}}^c}, \theta \mid \mathbf{Y}^N) \propto \prod_{i \in \mathcal{A}_{\text{ref}}^c} \{p(Y^i \mid X^i, \theta)\, p_0(X^i)\} \prod_{i \in \mathcal{A}_{\text{ref}}} p(Y^i \mid X^i, \theta) \times p_0(\theta). \tag{9.10}$$

The posterior for the non-anchored latent variables quantifies uncertainty in the latent representation. Recall that our focus is in Bayesian inference on the pairwise distances, and not the latent variables directly. In our experiments in Sections 9.4 and 9.5, we choose N_{ref} to be less than 6% of N, which means that only 0.36% of pairwise distances are pre-specified.

The question now is how to obtain the anchored $\{X^i\}_{i \in \mathcal{A}_{\text{ref}}}$. Any one of a number of nonlinear dimension reduction techniques can be used for this. For instance, one can use locally linear embeddings (LLEs; Roweis and Saul, 2000), Isomap (Tenenbaum et al., 2000), t-SNE (Maaten and Hinton, 2008), et cetera. In this article, we use LLEs; we are motivated by the fact that there is a literature carefully studying LLEs theoretically (for example,

Algorithm 1 Obtaining anchor points while imposing constraint (9.8)

Input: Observations $Y^1, \ldots, Y^N$.

Choose $\mathcal{A}_{\text{ref}}$ randomly without replacement from $\{1, \ldots, N\}$.
Let $\{Y^i_{\text{lle}}\}_{i \in \mathcal{A}_{\text{ref}}}$ = locally linear embedding($\{Y^i\}_{i \in \mathcal{A}_{\text{ref}}}$).
Let $W_1^{\text{optim}}, b_1^{\text{optim}}, \ldots = \text{argmin}_{W_1, b_1, \ldots} \sum_{i \in \mathcal{A}_{\text{ref}}} \{W_2 \varphi(W_1 Y^i_{\text{lle}} + b_1) + b_2 - Y^i\}^2$.
For $i \in \mathcal{A}_{\text{ref}}$, let $X^i_j = Y^i_{\text{lle},j} \|w_{1j}^{\text{optim}}\|$, $j = 1, \ldots, q$.

Output: Anchor points $\{(X^i, Y^i)\}_{i \in \mathcal{A}_{\text{ref}}}$.

Wu and Wu, 2018). This approximately preserves pairwise distances among the lower-dimensional representation of the observations. More precisely, we consider the locally linear embeddings of the observations $\{Y^i\}_{i=1}^N$. Let these lower-dimensional points be $\{Y^i_{\text{lle}}\}_{i=1}^N$. We let $\mathcal{A}_{\text{ref}}$ be chosen randomly without replacement from $\{1, \ldots, N\}$.

9.3.4 Combining Parameter Constraints with Anchor Points

As discussed in Section 9.3.4, we also impose constraint (9.8) on the columns of W_1. The experiments of Section 9.4 demonstrate empirically that anchor points together with parameter constraints improve the mixing properties of MCMC sampling. In this case, fixing the reference points to be $\{Y^i_{\text{lle}}\}_{i \in \mathcal{A}_{\text{ref}}}$ will cause a scaling issue. To see this, note that if all the latent variables $\{X^i\}_{i=1}^N$ were unrestricted, imposing constraint (9.8) leaves the model unchanged as discussed in the previous Section 9.3.2. However, since we have some fixed anchor points $\{Y^i_{\text{lle}}\}_{i \in \mathcal{A}_{\text{ref}}}$, this changes the model as we are now forcing the neural network to map *fixed* $\{Y^i_{\text{lle}}\}_{i \in \mathcal{A}_{\text{ref}}}$ to $\{Y^i\}_{i \in \mathcal{A}_{\text{ref}}}$ while imposing constraint (9.8), and mapping unrestricted $\{X^i\}_{i \in \mathcal{A}^c_{\text{ref}}}$ to $\{Y^i\}_{i \in \mathcal{A}^c_{\text{ref}}}$. Having a constraint on the weights reduces the ability of a neural network to map $\{Y^i_{\text{lle}}\}_{i \in \mathcal{A}_{\text{ref}}}$ to $\{Y^i\}_{i \in \mathcal{A}_{\text{ref}}}$.

To address this issue, we train an initial neural network using stochastic gradient descent with $\{Y^i_{\text{lle}}\}_{i \in \mathcal{A}_{\text{ref}}}$ as input and $\{Y^i\}_{i \in \mathcal{A}_{\text{ref}}}$ as output. Let the weights and biases of the trained neural network be W_1^{optim}, b_1^{optim}, etc. We then multiply $\{Y^i_{\text{lle}}\}_{i \in \mathcal{A}_{\text{ref}}}$ by the norms of the columns of W_1^{optim} to obtain the anchored points $\{X^i\}_{i \in \mathcal{A}_{\text{ref}}}$; that is, let $X^i_j = Y^i_{\text{lle},j} \|w_{1j}^{\text{optim}}\|$, $j = 1, \ldots, q$, $i \in \mathcal{A}_{\text{ref}}$, where $X^i = (X^i_1, \ldots, X^i_q)$, $Y^i_{\text{lle}} = (Y^i_{\text{lle},1}, \ldots, Y^i_{\text{lle},q})$, and $W_1^{\text{optim}} = (w_{11}^{\text{optim}}, \ldots, w_{1q}^{\text{optim}})$ is written in a column-wise fashion as in equation (9.9). This leads to the anchor points that we use and is summarized in Algorithm 1.

9.4 Simulation Study for Hyper-sphere

9.4.1 Model and Sampling Details

We conduct a simulation study in which we have $N = 640$ three-dimensional noisy observations lying on a hypersphere. The observations are plotted in Figure 9.2. The dimension of the latent manifold is two in this case.

We use a multilayer perceptron with a single hidden layer with $h = 10$ neurons and the hyperbolic tangent activation function. We place independent Normal$(0, \sigma^2)$ priors on each component of θ (that is, on each component of W_1, b_1, et cetera). We place a half-Cauchy(5) prior on τ^2 and a half-Cauchy(5) hyperprior on σ^2.

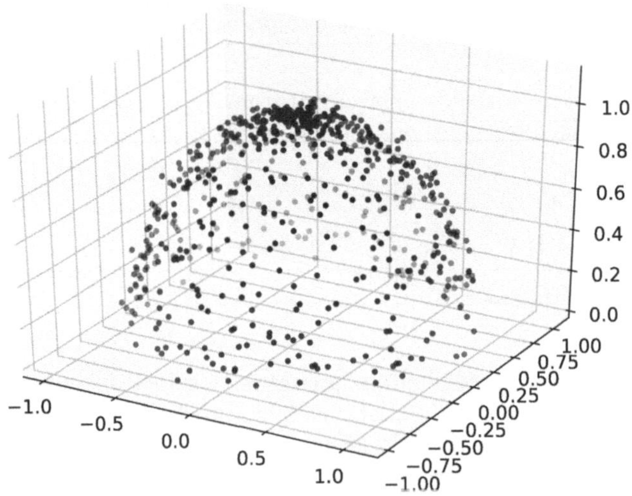

FIGURE 9.2
Noisy observations on hypersphere.

As discussed in Section 9.2, sampling from the posterior of the parameters of a neural network is a highly challenging problem in itself, and this is only exacerbated by the fact that we place a distribution on the input (latent) variables as well. For this reason, we do not focus on designing sampling algorithms in this chapter as this is a daunting task; instead, we use Hamiltonian Monte Carlo (HMC; Duane et al., 1987). In particular, we consider the no-U-turn sampler (Hoffman and Gelman, 2014) as implemented in Stan (Gelman et al., 2015) as this provides a popular and automated approach for choosing the hyperparameters of HMC. The initial value of θ for the Markov chain is set randomly using Stan's `init="random"` command.

9.4.2 Results

We use traceplots of the posterior samples of pairwise distances as a proxy for their identifiability. Badly mixing traceplots may indicate that the model suffers from identifiability issues. In a first experiment, we consider only constraint (9.8) on the weights W_1 without anchor points. We randomly pick some pairs of latent variables and display the traceplots of pairwise distances in Figure 9.3; these figures show very poor mixing/convergence consistent with a lack of identifiability. In a second experiment, we choose $N_{\text{ref}} = 40$ anchor points and do not impose constraint (9.8) on the weights W_1. The anchored points constitute about 6% of all latent variables. The resulting posterior samples of pairwise distances of some randomly chosen pairs of latent variables are plotted in Figure 9.4. We observe that while the traceplots indicate better mixing in comparison to Figure 9.3, mixing remains poor. We increase the number of anchor points to $N_{\text{ref}} = 120$ and display the traceplots of the pairwise distances in Figure 9.5, which still exhibit slow mixing. This demonstrates that using even a relatively large number of anchor points (almost 20% of the observations) is still not sufficient to ensure good mixing for the Markov chains of pairwise distances.

Finally, we consider both $N_{\text{ref}} = 40$ anchor points as well as constraint (9.8) on the weights W_1; in this case, the anchor points are obtained as described in Algorithm 1.

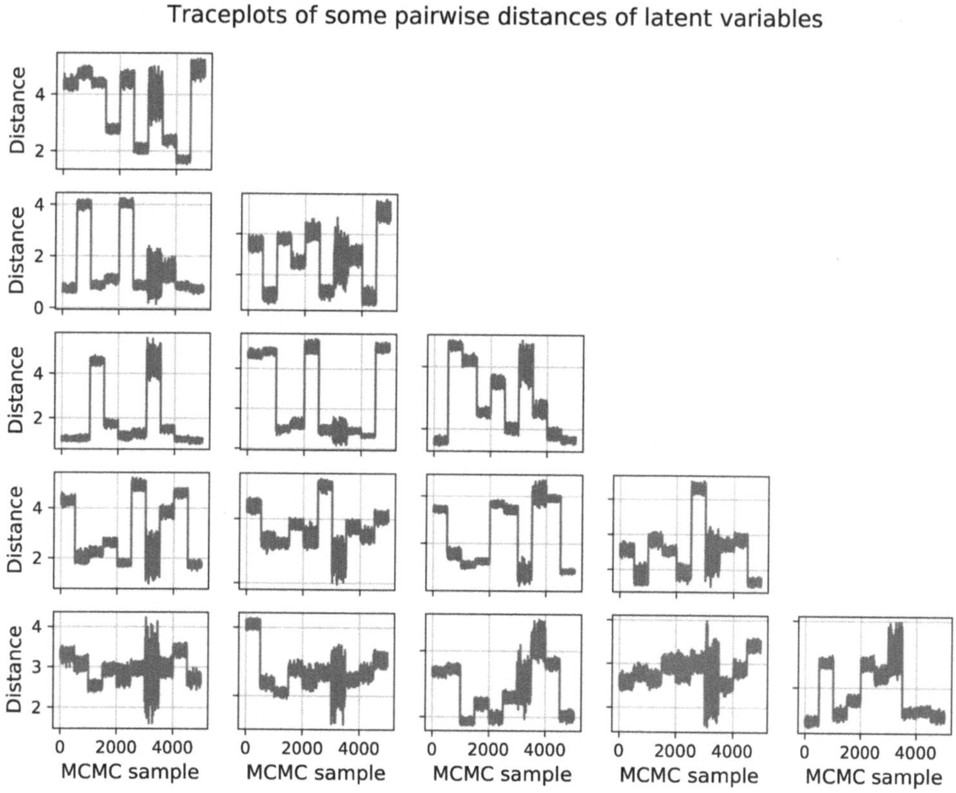

FIGURE 9.3
Traceplots of posterior samples of pairwise distances for randomly chosen latent variables when imposing constraint (9.8) on the weights W_1 and without anchor points for the hypersphere example.

The resulting pairwise distances are plotted in Figure 9.6, where we observe that the traceplots of pairwise distances attain significantly improved mixing (though there is still clear room for improvement). This provides evidence that the identifiability issue has been addressed, and using the posterior samples of such pairwise distances allows us to infer structure in the data, as will be done in the real data examples of the following Section 9.5.

We also compare the pairwise distances for the posterior samples with the pairwise distances of the true latent variables; this ground truth is known since we have simulated it. Let D be the $N \times N$ matrix of the true pairwise distances, and let $D^{(k)}$ be the $N \times N$ matrix of the pairwise distances of the kth posterior sample; here $k = 1, \ldots, K$ indexes the posterior samples. We plot the error $(1/N)\|D^{(k)} - D\| = (1/N)[\sum_{i=1}^{N} \sum_{j=1}^{N} \{D^{(k)ij} - D^{ij}\}^2]^{1/2}$ for $k = 1, \ldots, K$ with constraint (9.8) and without and with anchor points, respectively, in Figure 9.7. We have used ten chains in Stan. We note that the pairwise distances are of a similar order, which is expected as Stan concentrates posterior samples in regions where the target density is high and we have fixed the scale of the latent representation by imposing constraint (9.8). However, we note that when we do not use anchor points, the errors in pairwise distances are all clearly different for the ten chains (indeed, most of them are non-overlapping), which suggest identifiability issues: there are different ways of having a lower-dimensional latent representation.

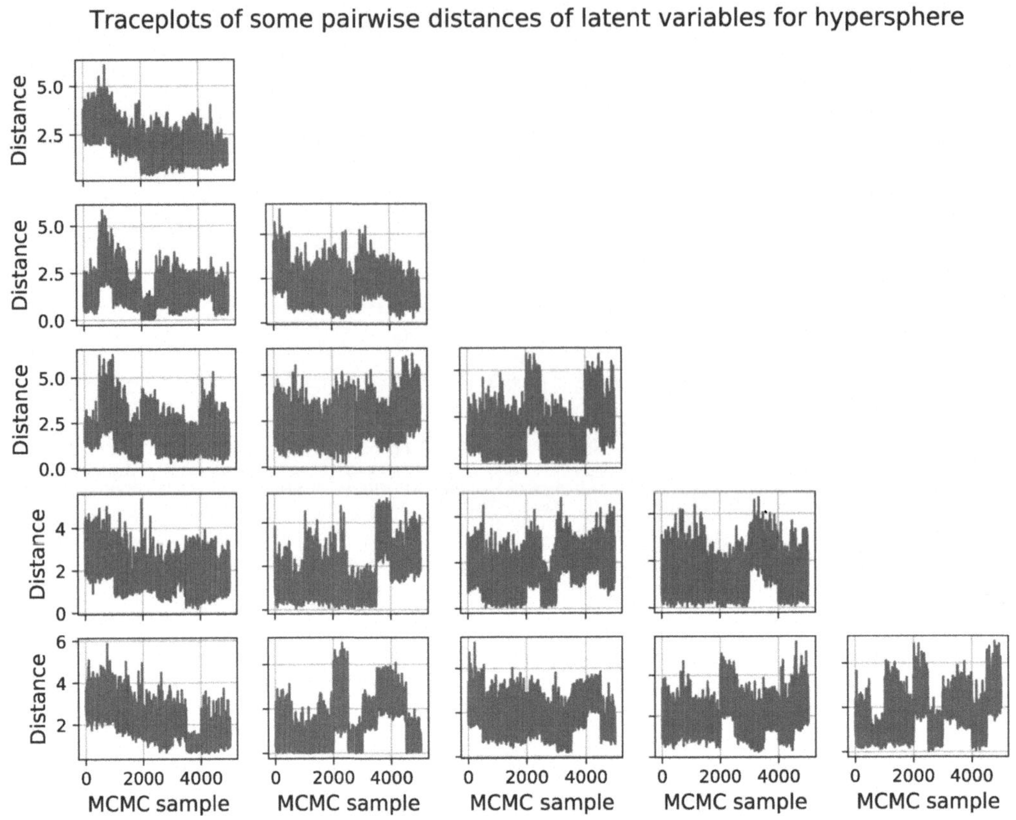

FIGURE 9.4
Traceplots of posterior samples of pairwise distances for randomly chosen latent variables without imposing constraints on the weights W_1 and with around 6% anchor points for the hypersphere example.

9.5 Real Data Applications

All datasets were obtained from the UCI machine learning repository (Dua and Graff, 2017). For each application in this section, the prior for θ and initial values of the Markov chains are the same as described in Section 9.4.1.

9.5.1 E. coli Data

We consider data on Escherichia coli (E. coli) proteins (Horton and Nakai, 1996). Each observation consists of a classification label for the protein (with eight classes in total) and has seven attributes; we refer the reader to Horton and Nakai (1996) for details. The number of observations is $N = 336$. We model the seven-dimensional attributes as lying on a one-dimensional latent manifold, and we use a multilayer perceptron with one hidden layer with five neurons.

In the previous Section 9.4, we observed that imposing constraint (9.8) on the weights W_1 and tying down around 6% of the observations as anchor points lead to adequate MCMC

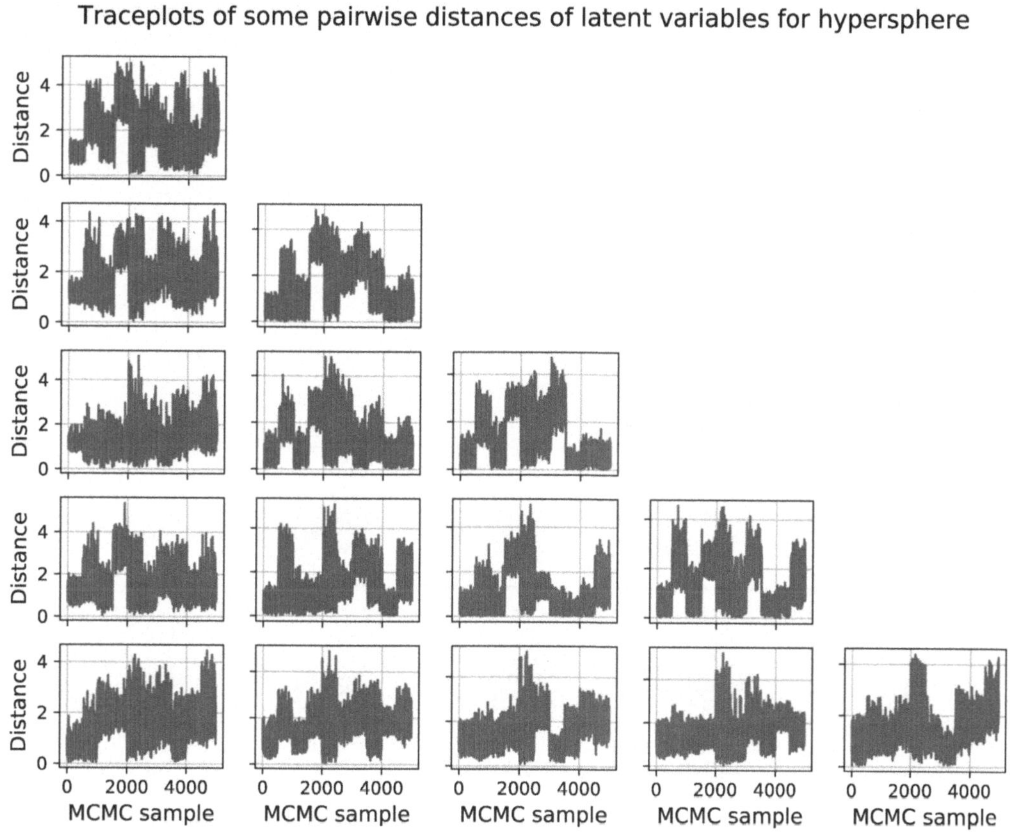

FIGURE 9.5
Traceplots of posterior samples of pairwise distances for randomly chosen latent variables without imposing constraints on the weights W_1 and with almost 20% anchor points for the hypersphere example.

mixing and good performance in estimating the pairwise distances. We therefore impose the constraint and consider 20 anchor points as described in Section 9.3.3. We plot traceplots of posterior samples of pairwise distances for some randomly chosen pairs of latent variables in Figure 9.8 and note that they exhibit relatively good mixing (compared to analyses that do not include constraints – improvements on the HMC algorithm being used may lead to further gains).

We use the posterior samples of the pairwise distances to uncover some structure in the dataset. In particular, we use spectral clustering (Spielman and Teng, 1996), as implemented in Python's `sklearn.cluster` command, to obtain a clustering of the latent variables for every posterior sample; we refer the reader to Von Luxburg (2007) for a detailed introduction to spectral clustering. Based on this clustering, we form an $N \times N$ clustering matrix $C^{(k)}$ for each posterior sample $\mathbf{X}^{N(k)} = (X^{1(k)}, \ldots, X^{N(k)})$, where

$$C_{ij}^{(k)} = \begin{cases} 1 & \text{if } X^{i(k)} \text{ and } X^{j(k)} \text{ are in the same cluster,} \\ 0 & \text{otherwise;} \end{cases}$$

here $k = 1, \ldots, K$ indexes the posterior samples. The posterior mean $(1/K)\sum_{k=1}^{K} C_{ij}^{(k)}$ estimates the probability that observations i and j belong to the same cluster. These are

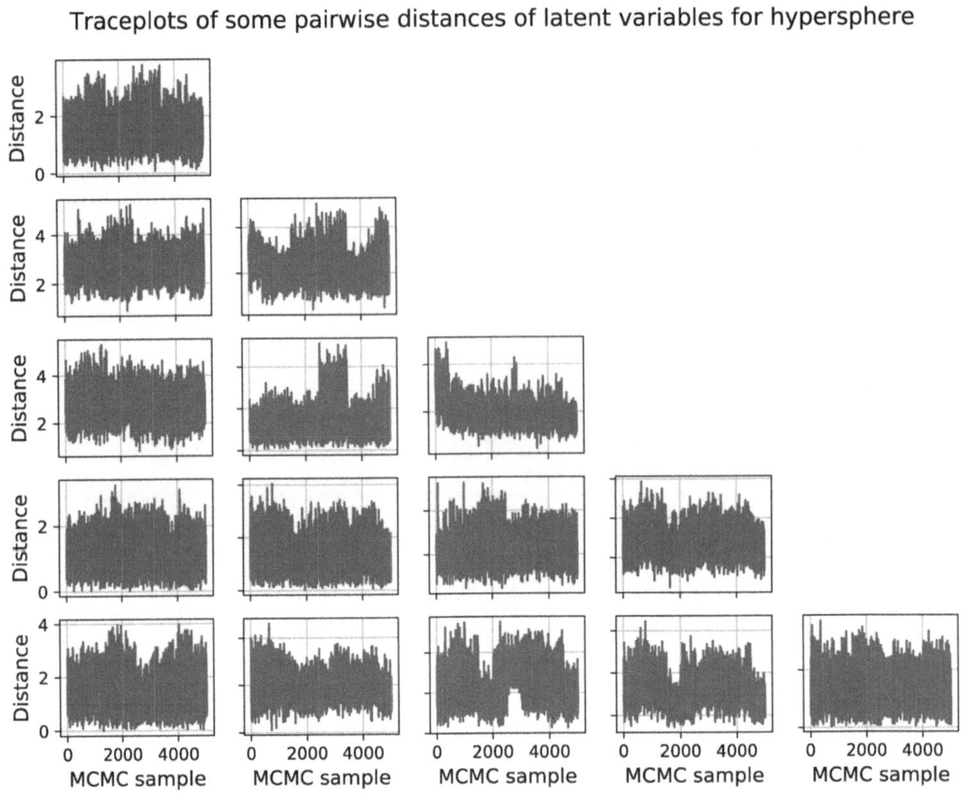

FIGURE 9.6
Traceplots of posterior samples of pairwise distances for randomly chosen latent variables when imposing constraint (9.8) on the weights W_1 and with around 6% anchor points for the hypersphere example.

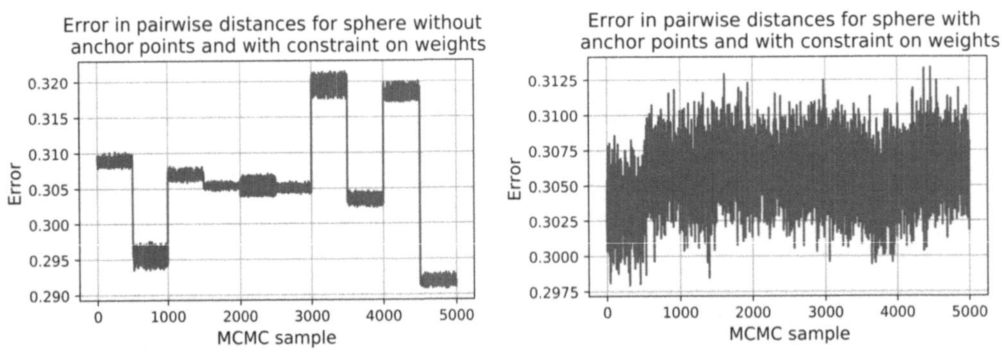

FIGURE 9.7
Errors in estimating the pairwise distances for the hypersphere example.

plotted in the left panel of Figure 9.9, where the observations are sorted by their true class labels. Moreover, using these probabilities as a pairwise affinity matrix and applying the least squares clustering method of Dahl (2006) leads to an estimated clustering of the data samples. This is plotted in the right panel of Figure 9.9. Some structure is visible here, with a few classes being distinguishable.

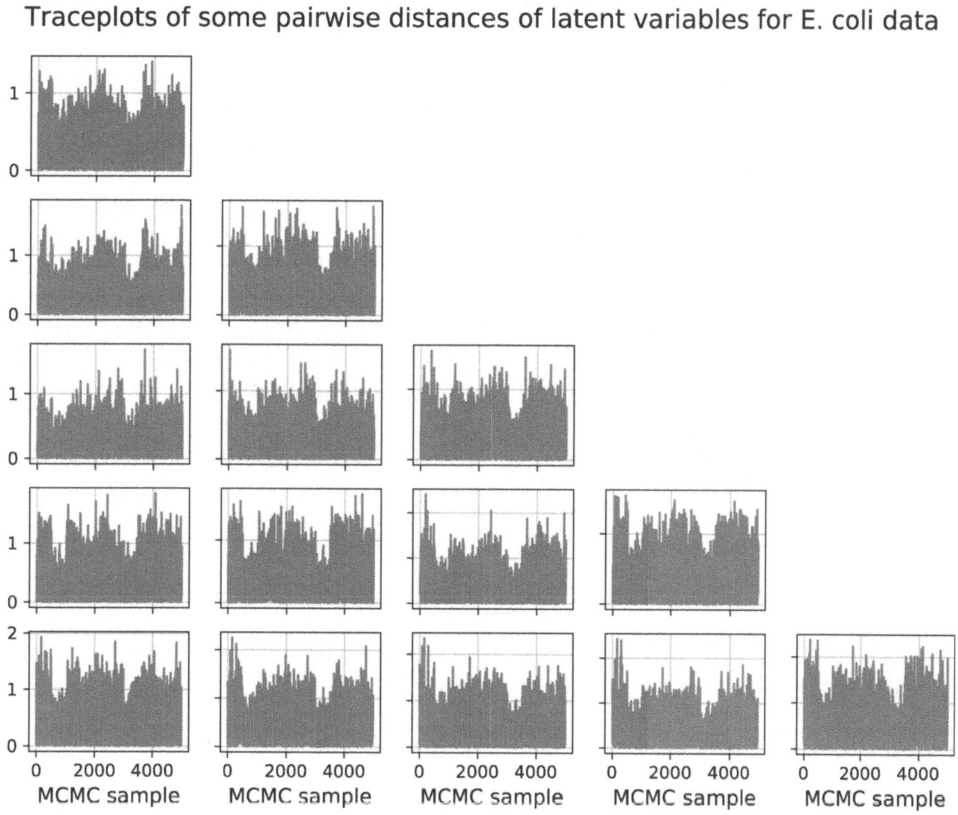

FIGURE 9.8
Traceplots of posterior samples of pairwise distances for randomly chosen latent variables when imposing constraint (9.8) on the weights W_1 and with anchor points for the E. coli dataset.

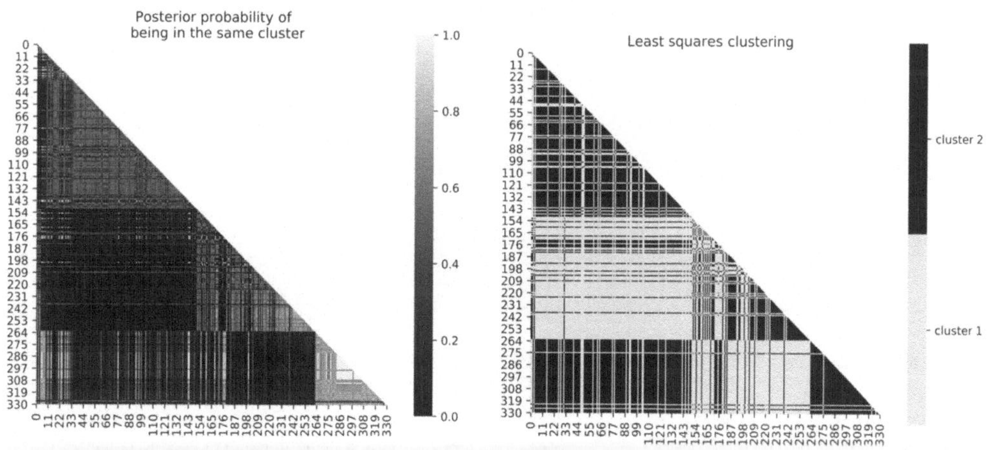

FIGURE 9.9
Posterior probability of pairs of observations belonging to the same cluster (left figure) and least squares clustering of Dahl (2006) (right figure) for the E. coli dataset.

9.5.2 User Knowledge Data

In a second application, we consider the user knowledge data (Kahraman et al., 2013). This consists of five dimensional attributes, including study time for goal object materials and for related objects materials. The number of observations is $N = 257$, with each observation being additionally classified into one of four classes indicating the knowledge level of user. We model the attributes as lying on a one-dimensional latent manifold. We use a multilayer perceptron with one hidden layer consisting of $h = 10$ neurons. As in the E. coli data (Section 9.5.1), we consider constraint (9.8) and additionally tie down $N_{\text{ref}} = 15$ anchor points. Traceplots of posterior samples of pairwise distances of some some randomly chosen pairs of latent variables, plotted in Figure 9.10, demonstrate relatively good mixing (again with substantial room for improvement). As in Section 9.5.1, we also plot the posterior probabilities of pairs of observations being in the same cluster and well as the least squares cluster in Figure 9.11, where the observations are sorted by their true class labels. This figure indicates that the uncertainty in clustering using the lower-dimensional representation of the attributes is high.

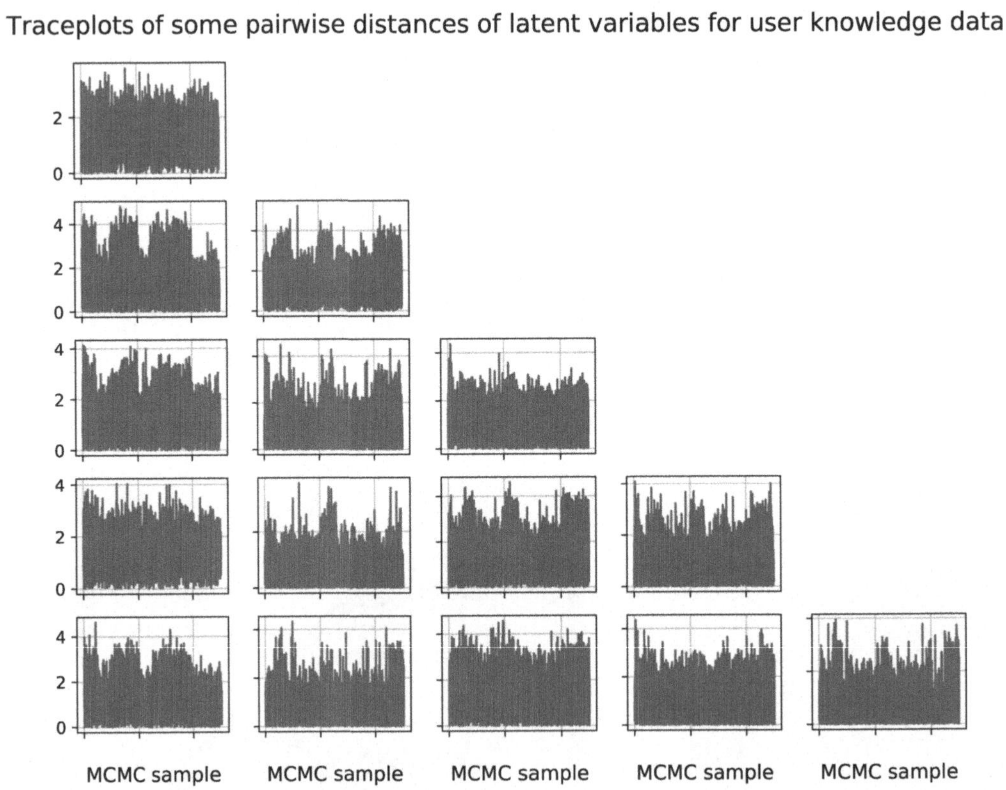

FIGURE 9.10
Traceplots of posterior samples of pairwise distances for randomly chosen latent variables when imposing constraint (9.8) on the weights W_1 and with anchor points for user knowledge dataset.

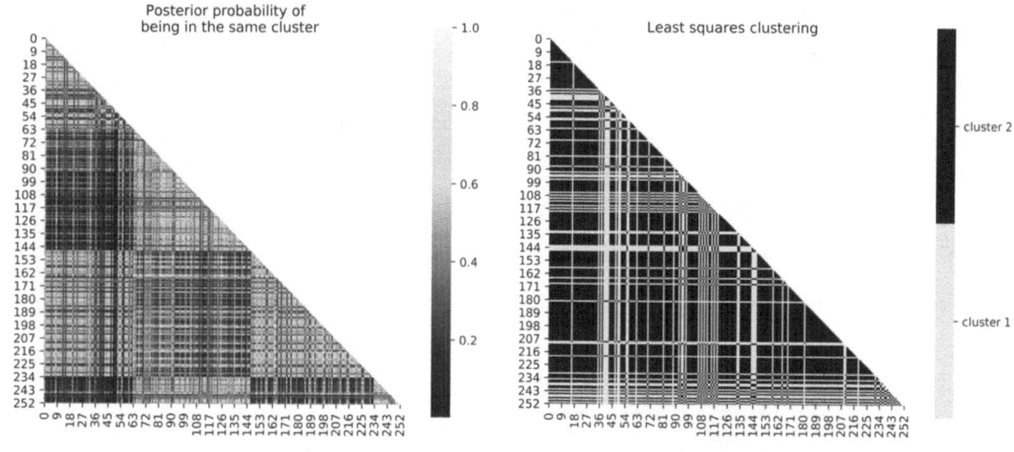

FIGURE 9.11
Posterior probability of pairs of observations belonging to the same cluster (left figure) and least squares clustering of Dahl (2006) (right figure) for the user knowledge dataset.

9.5.3 Banknote Data

We consider data consisting of four-dimensional attributes which were extracted from images that were taken from genuine and forged banknote-like specimens. Wavelet transform tools were used to extract four features from images. Each of $N = 1371$ observations is further classified into genuine and fake. We model the four-dimensional attributes as lying on a two-dimensional latent manifold, and we use a multilayer perceptron with one hidden layer with $h = 20$ neurons. We impose constraint (9.8) and consider eighty anchor points as described in Section 9.3.3. We plot some pairwise distances of posterior samples of the latent variables in Figure 9.12, where we note that the Markov chains of pairwise distances are mixing adequately. As in Section 9.5.1, we also plot the posterior probabilities of pairs of observations being in the same cluster as well as the least squares cluster in Figure 9.13, where the observations are sorted by their true class labels. The uncertainty in clustering using the lower-dimensional representation is very high in this case.

9.6 Open Questions and Ongoing Research

Efficient Bayesian inference for the posterior of the parameters of a neural network remains an open question. Neural networks have routinely been used in many tech industries, and a recent paper (Wenzel et al., 2020) has made the point that as of mid-2020, there are no publicized deployments of Bayesian neural networks in industry. They further argue that sampling the posterior predictive yields worse predictions as compared to simpler methods such as stochastic gradient descent. As we discussed in Section 9.2.2, choosing priors for the parameters of a neural network remains an open question as well.

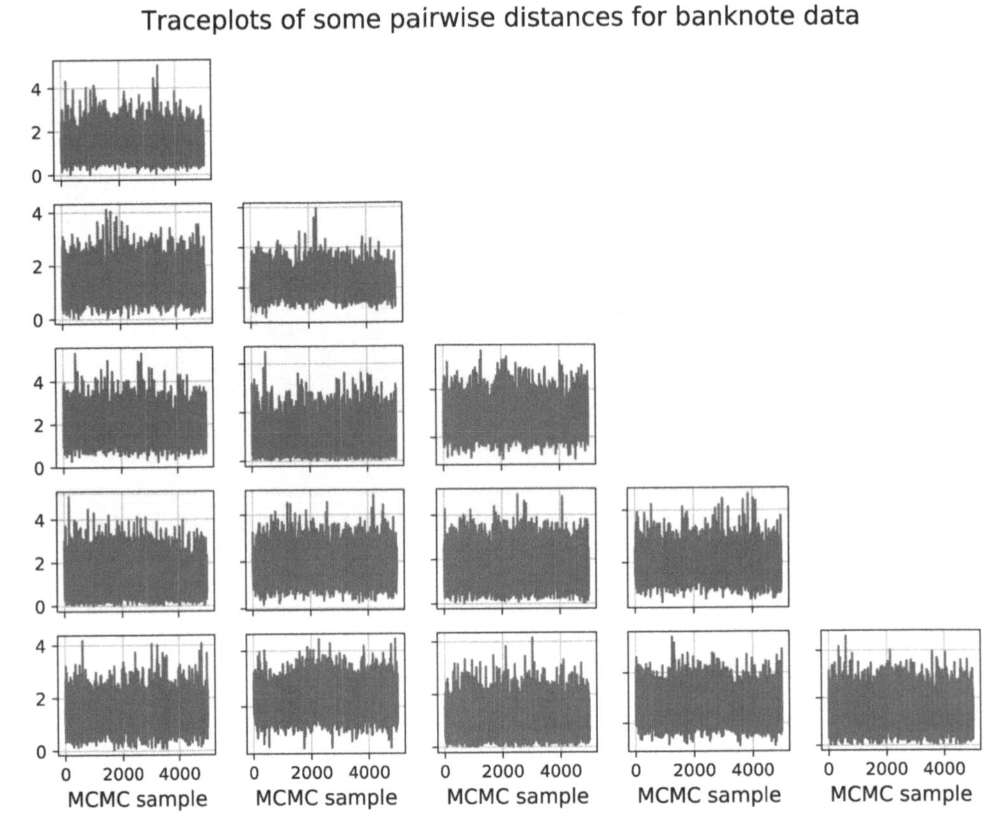

FIGURE 9.12
Traceplots of posterior samples of pairwise distances for randomly chosen latent variables when imposing constraint (9.8) on the weights W_1 and with anchor points for the banknote dataset.

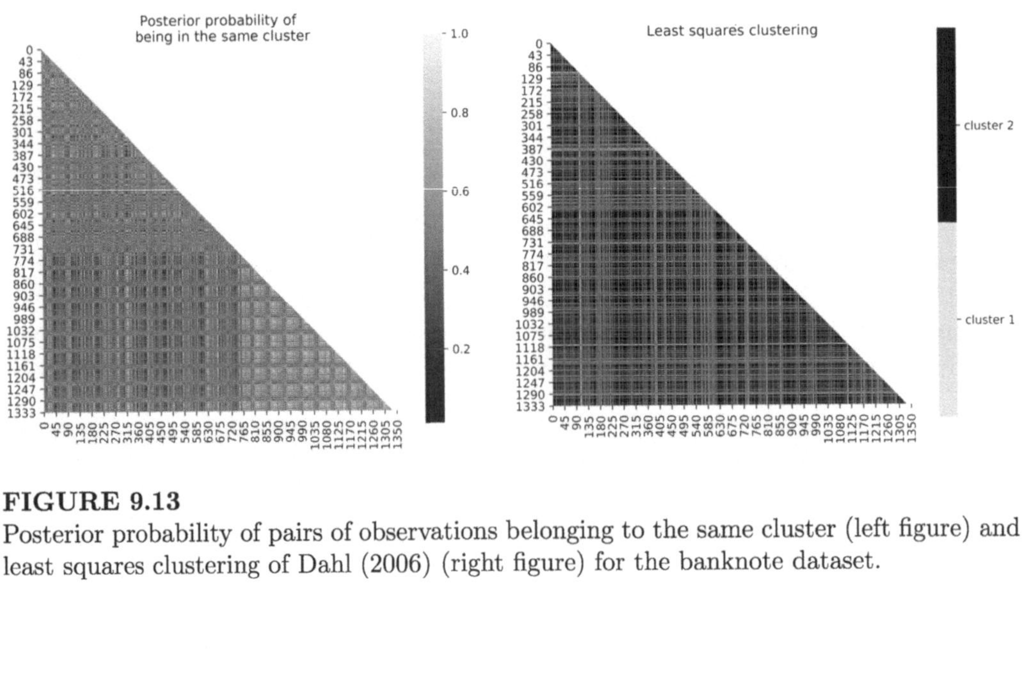

FIGURE 9.13
Posterior probability of pairs of observations belonging to the same cluster (left figure) and least squares clustering of Dahl (2006) (right figure) for the banknote dataset.

Bayesian inference for neural networks based on MCMC methods has not flourished due to various challenges posed by neural networks, such as the high number of their parameters, their hierarchical model structure and non-trivial likelihood landscape, and their likelihood function complexity and associated identifiability and label switching issues. In this text, we demonstrate that parameter constraints can alleviate identifiability and anchor points can mitigate the label switching problem for neural networks, thus yielding MCMC sampling with much better (but still not ideal) mixing for statistics of interest, such as pairwise distances for latent factors. A potentially promising avenue for future MCMC research for neural networks may thus focus on estimating lower dimensional summaries of the parameters by inducing parameter constraints and by exploiting re-labelling algorithms. Such summaries may involve pairwise distances of latent variables or the posterior covariance structure of the parameters. Alternatively, new classes of MCMC algorithms that can more adeptly explore and mix across the complex posterior landscape need to be developed.

Acknowledgments

DS acknowledges support from grant DMS-1638521 from SAMSI. DD acknowledges National Institutes of Health (NIH) grant R01ES027498 and Office of Naval Research (ONR) grant N00014-17-1-2844.

This manuscript has been authored by UT-Battelle, LLC, under contract DE-AC05-00OR22725 with the US Department of Energy (DOE). The US government retains and the publisher, by accepting the article for publication, acknowledges that the US government retains a nonexclusive, paid-up, irrevocable, worldwide license to publsh or reproduce the published form of this manuscript, or allow others to do so, for US government purposes. DOE will provide public access to these results of federally sponsored research in accordance with the DOE Public Access Plan (http://energy.gov/down-loads/doe-public-access-plan). This research was sponsored by the Laboratory Directed Research and Development Program of Oak Ridge National Laboratory, managed by UT-Battelle, LLC, for the US Department of Energy under contract DE-AC05-00OR22725.

Bibliography

Andrieu, C., de Freitas, J. and Doucet, A., 1999. Sequential Bayesian estimation and model selection applied to neural networks. *Citeseer.*

Andrieu, C., de Freitas, N. and Doucet, A., 2000. Reversible jump MCMC simulated annealing for neural networks. In *Proceedings of the Sixteenth Conference on Uncertainty in Artificial Intelligence*, pp. 11–18. San Francisco, CA: Morgan Kaufmann Publishers Inc.

Blei, D. M., Kucukelbir, A. and McAuliffe, J. D., 2017. Variational inference: A review for statisticians. *Journal of the American Statistical Association, 112*(518), pp. 859–877.

Chen, T., Fox, E. and Guestrin, C., 2014. Stochastic gradient Hamiltonian Monte Carlo. In *Proceedings of the 31st International Conference on Machine Learning*, volume 32 of Proceedings of Machine Learning Research, pp. 1683–1691. PMLR.

Chung, J., Kastner, K., Dinh, L., Goel, K., Courville, A. C. and Bengio, Y., 2015. A recurrent latent variable model for sequential data. *Advances in Neural Information Processing Systems*, pp. 2980–2988. Montreal, Canada.

Cunningham, J. P. and Ghahramani, Z., 2015. Linear dimensionality reduction: Survey, insights, and generalizations. *The Journal of Machine Learning Research*, *16*(1), pp. 2859–2900.

Cybenko, G., 1989. Approximation by superpositions of a sigmoidal function. *Mathematics of Control, Signals and Systems*, *2*(4), pp. 303–314.

Dahl, D. B., 2006. Model-based clustering for expression data via a Dirichlet process mixture model. *Bayesian Inference for Gene Expression and Proteomics*, *4*, pp. 201–218.

de Freitas, N., 1999. *Bayesian Methods for Neural Networks*. PhD thesis, Trinity College, University of Cambridge.

de Freitas, N., Andrieu, C., Hojen-Sorensen, P., Niranjan, M. and Gee, A., 2001. *Sequential Monte Carlo Methods for Neural Networks*, pp. 359–379. New York: Springer.

Dilokthanakul, N., Mediano, P. A., Garnelo, M., Lee, M. C., Salimbeni, H., Arulkumaran, K. and Shanahan, M., 2016. Deep unsupervised clustering with Gaussian mixture variational autoencoders. *arXiv preprint*, arXiv:1611.02648.

Dosovitskiy, A. and Brox, T., 2016. Generating images with perceptual similarity metrics based on deep networks. In *Advances in Neural Information Processing Systems*, pp. 658–666. Barcelona Spain.

Dua, D. and Graff, C., 2017. UCI Machine Learning Repository.

Duan, L. L., Young, A. L., Nishimura, A. and Dunson, D. B., 2020. Bayesian constraint relaxation. *Biometrika*, *107*(1), pp. 191–204.

Duane, S., Kennedy, A. D., Pendleton, B. J. and Roweth, D., 1987. Hybrid Monte Carlo. *Physics Letters B*, *195*(2), pp. 216–222.

Esmaeili, B., Wu, H., Jain, S., Bozkurt, A., Siddharth, N., Paige, B., Brooks, D. H., Dy, J. and Meent, J.-W., 2019. Structured disentangled representations. In *Proceedings of Machine Learning Research*, volume 89 of Proceedings of Machine Learning Research, eds K. Chaudhuri and M. Sugiyama, pp. 2525–2534. Naha, Okinawa, Japan: PMLR.

Gelman, A., Lee, D. and Guo, J., 2015. Stan: A probabilistic programming language for Bayesian inference and optimization. *Journal of Educational and Behavioral Statistics*, *40*(5), pp. 530–543.

Ghosh, S., Yao, J. and Doshi-Velez, F., 2018. Structured variational learning of Bayesian neural networks with horseshoe priors. In *Proceedings of the 35th International Conference on Machine Learning*, volume 80 of Proceedings of Machine Learning Research, eds J. Dy and A. Krause, pp. 1744–1753. PMLR.

Gong, W., Li, Y. and Hernández-Lobato, J. M., 2019. Meta-learning for stochastic gradient MCMC. In *International Conference on Learning Representations*, New Orleans, LA, USA.

Graves, A., 2011. Practical variational inference for neural networks. In *Advances in Neural Information Processing Systems 24*, eds J. Shawe-Taylor, R. S. Zemel, P. L. Bartlett, F. Pereira and K. Q. Weinberger, pp. 2348–2356. Granada, Spain: Curran Associates, Inc.

Hoffman, M. D. and Gelman, A., 2014. The no-U-turn sampler: Adaptively setting path lengths in Hamiltonian Monte carlo. *Journal of Machine Learning Research, 15*(1), pp. 1593–1623.

Horton, P. and Nakai, K., 1996. A probabilistic classification system for predicting the cellular localization sites of proteins. In *ISMB*, volume 4, pp. 109–115, St. Louis, Missouri, USA.

Huang, C.-W., Sankaran, K., Dhekane, E., Lacoste, A. and Courville, A., 2019. Hierarchical importance weighted autoencoders. In *Proceedings of the 36th International Conference on Machine Learning*, volume 97, pp. 2869–2878. PMLR.

Jiang, Z., Zheng, Y., Tan, H., Tang, B. and Zhou, H., 2017. Variational deep embedding: An unsupervised and generative approach to clustering. In *Proceedings of the 26th International Joint Conference on Artificial Intelligence, IJCAI'17*, pp. 1965–1972. AAAI Press.

Kahraman, H. T., Sagiroglu, S. and Colak, I., 2013. The development of intuitive knowledge classifier and the modeling of domain dependent data. *Knowledge-Based Systems, 37*, pp. 283–295.

Kingma, D. P. and Ba, J., 2015. Adam: A method for stochastic optimization. In *Third International Conference on Learning Representations*, San Diego, CA, USA.

Kingma, D. P. and Welling, M., 2014. Stochastic gradient VB and the variational auto-encoder. In *Second International Conference on Learning Representations*, Banff, Canada.

Kingma, D. P. and Welling, M., 2019. An introduction to variational autoencoders. *Foundations and Trends in Machine Learning, 12*(4), 307–392.

Krizhevsky, A., Sutskever, I. and Hinton, G. E., 2012. Imagenet classification with deep convolutional neural networks. In *Advances in Neural Information Processing Systems*, pp. 1097–1105, Lake Tahoe, NV, USA.

Kunkel, D. and Peruggia, M., 2020. Anchored Bayesian Gaussian mixture models. *Electronic Journal of Statistics, 14*(2), pp. 3869–3913.

Lawrence, N. D., 2004. Gaussian process latent variable models for visualisation of high dimensional data. In *Advances in Neural Information Processing Systems*, pp. 329–336, Vancouver, Canada.

Lee, H. K., 2003. A noninformative prior for neural networks. *Machine Learning, 50*(1), pp. 197–212.

Lee, H. K., 2004. Priors for neural networks. In *Classification, Clustering, and Data Mining Applications*, eds D. Banks, F. R. McMorris, P. Arabie and W. Gaul, pp. 141–150. San Diego, California: Springer.

Lee, H. K., 2005. Neural networks and default priors. In *Proceedings of the American Statistical Association, Section on Bayesian Statistical Science*, volume 4.

Lee, H. K., 2007. Default priors for neural network classification. *Journal of Classification, 24*(1), pp. 53–70.

Liu, Q. and Wang, D., 2016. Stein variational gradient descent: A general purpose Bayesian inference algorithm. In *Advances in Neural Information Processing Systems 29*, eds D. D. Lee, M. Sugiyama, U. V. Luxburg, I. Guyon and R. Garnett, pp. 2378–2386. Barcelona, Spain: Curran Associates, Inc.

Maaten, L. V. D. and Hinton, G., 2008. Visualizing data using t-SNE. *Journal of Machine Learning Research*, *9*(11), pp. 2579–2605.

Mathieu, E., Le Lan, C., Maddison, C. J., Tomioka, R. and Teh, Y. W., 2019. Continuous hierarchical representations with Poincaré variational auto-encoders. In *Advances in Neural Information Processing Systems 32*, eds H. Wallach, H. Larochelle, A. Beygelzimer, F. d'Alché Buc, E. Fox and R. Garnett, pp. 12565–12576. Vancouver, Canada: Curran Associates, Inc.

Nalisnick, E. T., 2018. *On Priors for Bayesian Neural Networks.* PhD thesis, UC Irvine.

Papamarkou, T., Hinkle, J., Young, M. T. and Womble, D., 2019. Challenges in Markov chain Monte Carlo for Bayesian neural networks. *arXiv preprint*, arXiv:1910.06539.

Patra, S., Sen, D. and Dunson, D. B., 2020. Constrained Bayesian inference through posterior projections. *arXiv preprint*, arXiv:1812.05741.

Pourzanjani, A. A., Jiang, R. M. and Petzold, L. R., 2017. Improving the identifiability of neural networks for Bayesian inference. In *NIPS Workshop on Bayesian Deep Learning*, Long Beach, CA, USA.

Roberts, A., Engel, J. and Eck, D., 2017. Hierarchical variational autoencoders for music. In *NIPS Workshop on Machine Learning for Creativity and Design*, Long Beach, CA, USA.

Roweis, S. T. and Saul, L. K., 2000. Nonlinear dimensionality reduction by locally linear embedding. *Science*, 290(5500), 2323–2326.

Rumelhart, D. E., Hinton, G. E. and Williams, R. J., 1986. Learning representations by back-propagating errors. *Nature*, *323*(6088), pp. 533–536.

Semeniuta, S., Severyn, A., and Barth, E., 2017. A hybrid convolutional variational autoencoder for text generation. *arXiv preprint*, arXiv:1702.02390.

Shi, J., Sun, S. and Zhu, J., 2018. A spectral approach to gradient estimation for implicit distributions. In *Proceedings of the 35th International Conference on Machine Learning*, volume 80 of Proceedings of Machine Learning Research, pp. 4644–4653. PMLR.

Spielman, D. A. and Teng, S.-H., 1996. Spectral partitioning works: planar graphs and finite element meshes. In *Proceedings of 37th Conference on Foundations of Computer Science*, pp. 96–105. IEEE.

Sun, S., Zhang, G., Shi, J. and Grosse, R., 2019. Functional variational Bayesian neural networks. In *International Conference on Learning Representations*, New Orleans, LA, USA.

Sutskever, I., Martens, J., Dahl, G. and Hinton, G., 2013. On the importance of initialization and momentum in deep learning. In *Proceedings of the 30th International Conference on Machine Learning*, volume 28 of Proceedings of Machine Learning Research, eds S. Dasgupta and D. McAllester, pp. 1139–1147. PMLR.

Tenenbaum, J. B., De Silva, V. and Langford, J. C. 2000. A global geometric framework for nonlinear dimensionality reduction. *Science*, *290*(5500), pp. 2319–2323.

Titsias, M. K. and Ruiz, F., 2019. Unbiased implicit variational inference. In *Proceedings of Machine Learning Research*, volume 89 of Proceedings of Machine Learning Research, eds K. Chaudhuri and M. Sugiyama, pp. 167–176. PMLR.

Trippe, B. and Turner, R., 2018. Overpruning in variational Bayesian neural networks. *arXiv preprint*, arXiv:1801.06230.

Vladimirova, M., Verbeek, J., Mesejo, P. and Arbel, J., 2019. Understanding priors in Bayesian neural networks at the unit level. In *Proceedings of the 36th International Conference on Machine Learning*, volume 97 of Proceedings of Machine Learning Research, eds K. Chaudhuri and R. Salakhutdinov, pp. 6458–6467. PMLR.

Von Luxburg, U. (2007). A tutorial on spectral clustering. *Statistics and Computing*, *17*(4), pp. 395–416.

Walker, J., Doersch, C., Gupta, A. and Hebert, M., 2016. An uncertain future: Forecasting from static images using variational autoencoders. In *European Conference on Computer Vision*, pp. 835–851. Springer.

Welling, M. and Teh, Y. W., 2011. Bayesian learning via stochastic gradient Langevin dynamics. In *Proceedings of the 28th International Conference on International Conference on Machine Learning*, ICML'11, pp. 681–688. Omnipress.

Wenzel, F., Roth, K., Veeling, B. S., Światkowski, J., Tran, L., Mandt, S., Snoek, J., Salimans, T., Jenatton, R. and Nowozin, S., 2020. How good is the Bayes posterior in deep neural networks really? *arXiv preprint*, arXiv:2002.02405.

Williams, P. M., 1995. Bayesian regularization and pruning using a Laplace prior. *Neural Computation*, *7*(1), pp. 117–143.

Wu, A., Nowozin, S., Meeds, E., Turner, R. E., Hernandez-Lobato, J. M. and Gaunt, A. L., 2019. Deterministic variational inference for robust Bayesian neural networks. In *International Conference on Learning Representations*, New Orleans, LA, USA.

Wu, H.-T. and Wu, N., 2018. Think globally, fit locally under the manifold setup: Asymptotic analysis of locally linear embedding. *The Annals of Statistics*, *46*(6B), pp. 3805–3837.

Zhang, Y. and Sutton, C., 2014. Semi-separable Hamiltonian Monte Carlo for inference in Bayesian hierarchical models. In *Advances in Neural Information Processing Systems 27*, pp. 10–18. Curran Associates, Inc.

10

The Tangent Exponential Model

A. C. Davison

Ecole Polytechnique Fédérale de Lausanne

N. Reid

University of Toronto

10.1 Introduction

In a series of papers starting in the late 1980s, D. A. S. Fraser, N. Reid and coworkers developed the tangent exponential model for higher-order likelihood inference. This chapter aims to explain the motivation and justification for this model and to describe how it is used to compute accurate approximations. The literature on this is not entirely transparent, as the argument evolved over numerous articles (Fraser, 1988, 1990, 1991, 2004; Fraser and Reid, 1988, 1993, 1995, 2001; Cakmak et al., 1994; Fraser et al., 1999). We give a heuristic account of this construction, for the most part skating over the technical details, and provide an annotated bibliography as a road map through the literature. The high accuracy of the resulting approximations has been verified empirically both in numerous articles and in books such as Brazzale et al. (2007), Chapter 8 of which overlaps with the account here.

Approximations based on the tangent exponential model build on the theory of conditional and marginal inference, and on Laplace and saddlepoint approximations. Chapter 12 of Davison (2003) defines some basic notions and derives some of the results presented here. Section 11.3.1 of that book contains an account of the Laplace method for integrals and related approximations for cumulative distribution functions. Fuller accounts may be found in Barndorff-Nielsen and Cox (1989, 1994), McCullagh (1987) and Severini (2000). Jensen (1995) and Butler (2007) provided comprehensive accounts of saddlepoint approximations and their many applications in analysis, probability, and statistics, and a helpful derivation is given in Kolassa (2006).

To fix notation and provide some building blocks that will be useful later, we start by summarising first-order inferential approximations related to the normal distribution and the central limit theorem. We then outline how the principles of sufficiency and ancillarity lead to consideration of significance functions, which are central in this and other frameworks for statistical inference, before discussing how the likelihood function can be viewed as a pivot. This leads to two density approximations, the p^* approximation of Barndorff-Nielsen (1983) and the tangent exponential model density approximation of Fraser (1988). Higher-order approximations to significance functions are then developed; first for linear exponential family models, where the link to saddlepoint approximations is most explicit, and then for general models, where the tangent exponential model plays a crucial role. Some extensions and generalisations are sketched in the concluding section.

 DOI: 10.1201/9780429341731-10

10.2 Likelihood and Significance Functions

10.2.1 Background

We consider a vector $Y = (Y_1, \ldots, Y_n)^{\mathrm{T}}$ of continuous responses and a statistical model for Y with joint density function $f(y;\theta)$ that depends on a parameter $\theta \in \Theta \subset \mathbb{R}^p$. Vectors throughout are column vectors, and the superscripts ^{T} and $^{\circ}$ denote transpose and a quantity evaluated at the observed data, so y denotes a generic response vector and y° its observed value, and $\widehat{\theta}$ a maximum likelihood estimator and $\widehat{\theta}^{\circ}$ the maximum likelihood estimate computed from y°. Continuity of the response distribution plays a crucial role in the general development; some comments on extending the tangent exponential model for discrete responses are given in Section 10.5.2.

The likelihood function is proportional to the density of the observed data regarded as a function of the unknown parameter θ, i.e.,

$$L(\theta) = L(\theta; y^{\circ}) \propto f(y^{\circ};\theta),$$

and measures the relative plausibility of different values of θ as explanations of y°. Weighted by prior information, the likelihood is a main ingredient in Bayesian approaches to inference, and, unweighted, it is key to the "pure likelihood" approach (Edwards, 1972; Royall, 1997). A central issue in using the likelihood function for inference is the distribution of $L(\theta; y)$ in repeated sampling under $f(y;\theta)$. This is needed in order to *calibrate* inferences based on the likelihood function, i.e., to ensure that their stated properties are correct under repeated sampling from the model. Calibration is essential to give inferences objective validity, ideally while respecting basic principles of inference, such as conditionality and sufficiency. One important approach to calibration is through the notion of a significance function, to be developed in Section 10.2.2.

We denote the log likelihood by $\ell(\theta) = \log L(\theta)$, and derivatives by subscripts, such as $\ell_\theta(\theta) = \partial \ell(\theta)/\partial\theta$ and $\ell_{\theta\theta}(\theta) = \partial^2\ell(\theta)/\partial\theta\partial\theta^{\mathrm{T}}$, which are respectively a column vector and a matrix. The observed information function is $\jmath(\theta) = -\ell_{\theta\theta}(\theta)$. The classical asymptotic theory for likelihood-based inference is derived under the following smoothness conditions on the model (Davison, 2003, Section 4.4.2):

i. the true value of θ is interior to the parameter space Θ;
ii. the densities $\{f(y;\theta) : \theta \in \Theta\}$ are distinct and have common support;
iii. there is a neighbourhood $\mathcal{N}$ of the true value of θ within which the first three derivatives of $\ell(\theta)$ with respect to θ exist, and for $j, k, l = 1, \ldots, p$, $n^{-1}\mathrm{E}\{|\ell_{\theta_j\theta_k\theta_l}(\theta)|\}$ is uniformly bounded for $\theta \in \mathcal{N}$;
iv. the expected Fisher information matrix $I(\theta) = \mathrm{E}\{\jmath(\theta)\}$ is finite and positive definite, and $I(\theta) = \mathrm{E}\{\ell_\theta(\theta)\ell_\theta^{\mathrm{T}}(\theta)\}$.

Chapter 16 of van der Vaart (1998) gives weaker conditions for the limiting distributional results now described.

When θ is a scalar, this classical theory provides three basic distributional approximations for inference:

$$
\begin{aligned}
s = s(\theta; y) &= \ell_\theta(\theta)\jmath^{-1/2}(\widehat{\theta}) \mathrel{\dot\sim} N(0,1), &(10.1)\\
t = t(\theta; y) &= (\widehat{\theta} - \theta)\jmath^{1/2}(\widehat{\theta}) \mathrel{\dot\sim} N(0,1), &(10.2)\\
r = r(\theta; y) &= \mathrm{sign}(\widehat{\theta} - \theta)[2\{\ell(\widehat{\theta}) - \ell(\theta)\}]^{1/2} \mathrel{\dot\sim} N(0,1), &(10.3)
\end{aligned}
$$

where the maximum likelihood estimator, $\widehat{\theta}$, is assumed to satisfy $\ell_\theta(\widehat{\theta}) = 0$. The approximations (10.1)–(10.3) are derived from the central limit theorem for $\ell_\theta(\theta)$, which is $O_p(n^{1/2})$ in independent sampling, as $\mathrm{E}_\theta\{\ell_\theta(\theta)\} = 0$ and $\mathrm{var}_\theta\{\ell_\theta(\theta)\} = O(n)$ under conditions (i)–(iv). Their derivations also rely on the consistency of $\widehat{\theta}$ for θ, and the convergence of the observed Fisher information to its expectation.

Each of s, t and r depends on both θ and the data y. The distributional approximations in (10.1)–(10.3) are all with reference to the distribution of y under the model $f(y;\theta)$, and $s(\theta;y)$, $t(\theta;y)$ and $r(\theta;y)$ are approximate pivots, i.e., functions of the data and a parameter whose distribution is known, at least approximately. Thus if y is fixed at y^{o}, then $s(\theta;y^{\mathrm{o}})$, $t(\theta;y^{\mathrm{o}})$ and $r(\theta;y^{\mathrm{o}})$ can be used to obtain the significance functions $\Phi(s)$, $\Phi(q)$, or $\Phi(r)$, where $\Phi(\cdot)$ is the distribution function for a standard normal random variable.

Versions of (10.1), (10.2) and (10.3) are also available for vector-valued θ. We write $\theta = (\psi, \lambda)$, where ψ is a scalar parameter of interest and λ is a nuisance parameter. The profile log likelihood $\ell_{\mathrm{p}}(\psi) = \ell(\psi, \widehat{\lambda}_\psi)$ can be used to define analogous pivotal quantities,

$$\begin{aligned} s &= \ell_{\mathrm{p}}'(\psi) \jmath_{\mathrm{p}}^{-1/2}(\widehat{\psi}) \stackrel{\cdot}{\sim} N(0,1), \\ t &= (\widehat{\psi} - \psi) \jmath_{\mathrm{p}}^{1/2}(\widehat{\psi}) \stackrel{\cdot}{\sim} N(0,1), \\ r &= \mathrm{sign}(\widehat{\psi} - \psi)[2\{\ell_{\mathrm{p}}(\widehat{\psi}) - \ell_{\mathrm{p}}(\psi)\}]^{1/2} \stackrel{\cdot}{\sim} N(0,1), \end{aligned} \tag{10.4}$$

where the prime denotes differentiation with respect to ψ, $\widehat{\lambda}_\psi$ is the constrained maximum likelihood estimator for λ for ψ fixed, and

$$\jmath_{\mathrm{p}}(\psi) = \jmath_{\psi\psi}(\widehat{\theta}_\psi) - \jmath_{\psi\lambda}(\widehat{\theta}_\psi) \jmath_{\lambda\lambda}^{-1}(\widehat{\theta}_\psi) \jmath_{\lambda\psi}(\widehat{\theta}_\psi).$$

We use the shorthand $\widehat{\theta}_\psi = (\psi, \widehat{\lambda}_\psi^{\mathrm{T}})^{\mathrm{T}}$, and partition the observed information matrix as

$$\jmath(\theta) = \begin{pmatrix} \jmath_{\psi\psi}(\theta) & \jmath_{\psi\lambda}(\theta) \\ \jmath_{\lambda\psi}(\theta) & \jmath_{\lambda\lambda}(\theta) \end{pmatrix},$$

with $\jmath_{\psi\lambda}(\theta) = -\partial^2 \ell(\theta)/\partial\psi\partial\lambda^{\mathrm{T}}$, and so forth.

The normal approximations to the distributions of s, t and r are equivalent to first order, but that for r respects any asymmetry of the log likelihood about its maximum, whereas that for t does not, which suggests that for complex problems these approximations may be inadequate. Their generally poor performance in models with many nuisance parameters has been borne out in empirical work.

The approximations (10.1)–(10.4) are usually derived from Taylor series expansion of the score equation $\ell_\theta(\widehat{\theta}) = 0$ that defines the maximum likelihood estimate. A more elegant, but more difficult, approach shows that the log likelihood, as a function of θ, converges to the log likelihood for a normal distribution. See, for example Fraser and McDunnough (1984) for a Taylor-series type approach to this and LeCam (1960) or van der Vaart (1998, Ch. 6) for a related approach that LeCam called "local asymptotic normality". We will return to this in describing the p^* approximation in Section 10.2.3.

In some settings we may be able to construct exact significance functions, using principles of sufficiency and ancillarity, which we now describe.

10.2.2 Significance Functions

There are two ideal settings for exactly calibrating inference on a scalar parameter θ. In the first, Y can be reduced to a scalar minimal sufficient statistic S, giving the *significance function*, sometimes also known as a *p-value function*,

$$p^{\mathrm{o}}(\theta) = \mathrm{pr}(S \leq s^{\mathrm{o}}; \theta), \tag{10.5}$$

which without loss of generality we suppose to be decreasing in θ. The quantity $P^{\mathrm{o}}(\theta)$ obtained by treating s^{o} as random has a uniform distribution under sampling from the model $f(y;\theta)$, and this allows the calibration of inferences described in Section 10.2.1. For example, the limits of a $(1-2\alpha)$ confidence interval $(\theta_\alpha, \theta_{1-\alpha})$ for θ are the solutions of equations $p^{\mathrm{o}}(\theta_\alpha) = 1-\alpha$ and $p^{\mathrm{o}}(\theta_{1-\alpha}) = \alpha$. Likewise, the evidence against the hypothesis that $\theta = \theta_0$ with alternative $\theta \neq \theta_0$ may be summarised by the p-value $2\min\{p^{\mathrm{o}}(\theta_0), 1-p^{\mathrm{o}}(\theta_0)\}$. The exact uniform distribution of $P^{\mathrm{o}}(\theta)$ under repeated sampling from $f(y;\theta)$ implies that both of these inferential summaries are perfectly calibrated, in the sense that they have exactly their stated properties: the confidence interval contains the true θ with probability $1-2\alpha$ and the p-value is uniformly distributed.

In the second ideal setting there is a known transformation from Y to a minimal sufficient statistic (S, A), where S is scalar and A is *ancillary*, i.e., its distribution does not depend on θ. The role of A can be understood by noting that the density $f(y;\theta)$ factorises as $f(s \mid a;\theta)f(a)$. We can therefore envisage the data as arising first by generating A and then generating S conditional on A. But as the first step does not depend on θ, the relevant subset of the sample space for inference about θ fixes the observed value a^{o} of A (Cox, 1958). Thus we should base inference for θ on the significance function

$$p^{\mathrm{o}}(\theta; a^{\mathrm{o}}) = \mathrm{pr}(S \leq s^{\mathrm{o}} \mid A = a^{\mathrm{o}}; \theta), \tag{10.6}$$

which provides perfectly calibrated inferences, conditionally on the observed value of the ancillary statistic A. Now the limits of an exact $(1-2\alpha)$ confidence interval for θ, the solutions of equations $p^{\mathrm{o}}(\theta; a^{\mathrm{o}}) = \alpha$ and $p^{\mathrm{o}}(\theta; a^{\mathrm{o}}) = 1-\alpha$, depend on a^{o}, thus emphasising how conditioning on the ancillary statistic A affects the precision of inferences.

Expressions (10.5) and (10.6) are functions of θ, evaluated only at the observed data.

Example 10.1

Suppose that Y_1/θ and $Y_2\theta$ are independent gamma variables with unit scale and shape parameter n; their joint density function is

$$f(y_1, y_2; \theta) = \frac{(y_1 y_2)^{n-1}}{\{\Gamma(n)\}^2} \exp\left(-y_1/\theta - y_2\theta\right), \quad y_1, y_2 > 0, \theta > 0,$$

and the minimal sufficient statistic is (Y_1, Y_2). We set $S = (Y_1/Y_2)^{1/2}$ and $A = (Y_1 Y_2)^{1/2}$, which is ancillary, and note that since $Y_1 = AS$ and $Y_2 = A/S$, we have $y^{\mathrm{T}} = (y_1, y_2) = (as, a/s)$, so $|\partial(y_1, y_2)/\partial(a, s)| = 2a/s$ and

$$f(s \mid a; \theta) = \frac{1}{sI(a)} \exp\left\{-a(s/\theta + \theta/s)\right\}, \quad a, s > 0, \theta > 0,$$

where $I(a) = \int_{-\infty}^{\infty} \exp(-2a\cosh u)\,du$ is a normalising constant. The significance function (10.6) is readily obtained by numerical integration of $f(s \mid a^{\mathrm{o}}; \theta)$ over the interval $(0, s^{\mathrm{o}})$.

The log likelihood for this model can be written as

$$\ell(\theta; s, a) \equiv -a(s/\theta + \theta/s), \quad \theta > 0,$$

so the maximum likelihood estimator is $\widehat{\theta} = s$ and the observed information is $\jmath(\widehat{\theta}) = a/s^2$. Larger values of the ancillary a yield more precise inferences, because the standard error for $\widehat{\theta}$, $\jmath^{-1/2}(\widehat{\theta}) = s/a^{1/2}$, diminishes as a increases.

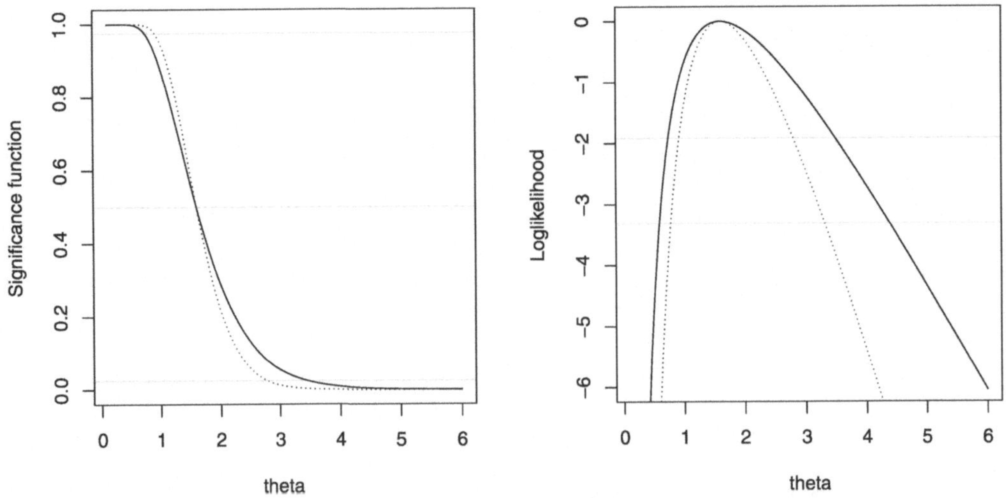

FIGURE 10.1
Inference for gamma example. Significance functions $\text{pr}(s \leq s^{\text{o}} \mid a; \theta)$ (left) and log likelihoods (right) for $a = 3$ (solid) and $a = 6$ (dots). The horizontal lines in the left-hand panel are at 0.025, 0.5 and 0.975; the intersections of the highest and lowest of these with the significance functions give the limits of exact 95% confidence intervals. The horizontal lines in the right-hand panel are at -1.92 and -3.32, and their intersections with the log likelihoods show the limits of 95% and 99% confidence intervals based on the approximation (10.3).

Figure 10.1 shows the significance and log likelihood functions when $s^{\text{o}} = 1.6$ and $a^{\text{o}} = 3, 6$. They are more concentrated when $a^{\text{o}} = 6$, resulting in shorter confidence intervals. Both log likelihood functions show clear asymmetry, suggesting that normal approximation based on t is badly calibrated and would provide poor inferences; indeed, the symmetric approximation (10.2) could produce negative confidence limits.

Though rarely met in practice, the settings above provide blueprints for more complex situations, in which several complications may arise:

- θ consists of a parameter ψ of interest, often scalar, and a vector λ of nuisance parameters. Even if there is a direct analogue of S, the significance functions (10.5) and (10.6) will typically depend on the unknown λ;
- the reduction to a minimal sufficient statistic of dimension p applies principally in linear exponential family models. No mapping $y \mapsto (s, a)$ can be found in general, so exact inferences are mostly unavailable; and
- the interest parameter ψ may be a vector. We shall not consider this situation here, but Section 10.7.2 has pointers to related approaches.

Significance functions were emphasised as a primary tool for inference in Fraser (1991). Fraser (2019) described many summaries that are then directly available: confidence limits and p-values, as described above, as well as aspects of fixed-level testing. Power, for example, is reflected in the "steepness" of the function. There is a close connection between significance functions and confidence distributions (Cox, 1958; Efron, 1993; Xie and Singh, 2013).

10.2.3 Likelihood as Pivot

A fruitful approach to improved approximations is to use the log likelihood more directly—Hinkley (1980) referred to the likelihood function itself as a pivot. One reason this provides more precise calibration of our inferences is that it is closely related to a saddlepoint approximation (Daniels, 1954), which can be uncannily accurate. The saddlepoint approximation can be applied to the density of the maximum likelihood estimator, leading to the p^* *approximation*, also called "Barndorff-Nielsen's formula" (Barndorff-Nielsen, 1980):

$$p^*(\widehat{\theta} \mid a; \theta) = c|\jmath(\widehat{\theta})|^{1/2} \exp\{\ell(\theta) - \ell(\widehat{\theta})\}, \tag{10.7}$$

where $\ell(\theta) = \ell(\theta; \widehat{\theta}, a)$, with $(\widehat{\theta}, a)$ a transformation of the sample $y = (y_1, \ldots, y_n)$ such that a is ancillary. Pivoting this approximate density (10.7), which is supported on $\mathbb{R}^p$, provides confidence intervals or regions for θ, and since a is ancillary no information about the parameter is lost by conditioning. However to compute the approximation requires calculation of the transformation from y to $(\widehat{\theta}, a)$, which can be very difficult. Moreover, computation of a significance function requires only a good approximation at the observed data point, whereas using (10.7) would require high accuracy for all $(\widehat{\theta}, a)$.

An approach intermediate between using the limiting normal form for the log likelihood and using (10.7) was introduced in Fraser (1988). His *tangent exponential model* approximation to the density $f(y; \theta)$, defined for $y \in \mathbb{R}^n$ and $\theta \in \mathbb{R}^p$, is a model on $\mathbb{R}^p$ that implements conditioning on an approximately ancillary statistic a. Its expression is

$$f_{\text{TEM}}(s \mid a; \theta) = \exp[s^{\text{T}}\varphi(\theta) + \ell\{\theta(\varphi); y^{\text{o}}\}]h(s). \tag{10.8}$$

This has the structure of a linear exponential family model for a constructed sufficient statistic $s \in \mathbb{R}^p$ and constructed canonical parameter $\varphi(\theta) \in \mathbb{R}^p$, with $-\ell(\theta; y^{\text{o}})$ playing the role of the cumulant generator. If the underlying density is in the exponential family, then $\varphi(\theta)$ is simply the canonical parameter. In more general models the canonical parameter $\varphi(\theta)$ may depend on the data y^{o}, and is constructed using principles of approximate ancillarity, as described in Section 10.5.1.

In the next section, we show how this tangent exponential model is built on the exact significance functions of Section 10.2.2, but incorporates aspects of direct approximation of the log likelihood.

10.3 Approximate Conditional Inference

10.3.1 Ancillary and Sufficient Directions

Below we describe a general approach to approximate but accurate inference when the dimension p of the parameter vector θ is less than the dimension n of the data y. We assume that the components of y are independent and that the regularity conditions outlined in Section 10.2.1 hold. The tangent exponential model is often said to have "asymptotic properties", which essentially is shorthand for the assumption that the expansions in Section 10.6 have the behaviour in n indicated there, i.e., the log-density is differentiable in both θ and y up to quartic terms, and the standardised derivatives decrease in powers of $n^{-1/2}$.

For a heuristic development we suppose initially that there is a smooth bijection between y and (s, a), where s and a have respective dimensions p and $n - p$, and a is ancillary. The conditionality principle implies that inference should be based on the conditional density $f(s \mid a^{\circ}; \theta)$, so the reference set for frequentist inference is $\mathcal{A}^{\circ} = \{y \in \mathbb{R}^n : a(y) = a^{\circ}\}$, i.e., the p-dimensional manifold of the sample space on which the ancillary statistic equals its observed value a°. The bijection between y and (s, a) implies that $\mathcal{A}^{\circ}$ can be parametrised in terms of s, at which point its tangent plane $\mathcal{T}_s$ is determined by the columns of the $n \times p$ matrix

$$\frac{\partial y(s, a^{\circ})}{\partial s^{\mathrm{T}}}. \tag{10.9}$$

In particular, the tangent plane $\mathcal{T}^{\circ}$ to $\mathcal{A}^{\circ}$ at s° is determined by

$$V = \left. \frac{\partial y(s, a^{\circ})}{\partial s^{\mathrm{T}}} \right|_{s=s^{\circ}}. \tag{10.10}$$

The columns of V have been called ancillary directions, because they are derived from the ancillary manifold $\mathcal{A}^{\circ}$ at $s = s^{\circ}$, but one can argue that this is a misnomer: the p columns of V determine how y changes in the direction of s locally at s°, so they might better be called *sufficient directions*, and we shall use this term below. The ancillary statistic itself varies locally at a° in the $n - p$ directions orthogonal to the columns of V.

Suppose for simplicity that θ and S are scalar. Then

$$\mathrm{pr}(S \leq s^{\circ} \mid A = a^{\circ}; \theta) \propto \int_{-\infty}^{s^{\circ}} f(s, a^{\circ}; \theta)\, \mathrm{d}s. \tag{10.11}$$

The log likelihood $\ell(\theta; s, a) = \log f(s, a; \theta)$ is a sum of n contributions and therefore has order n. A change of variables can be used to ensure that s is $O_p(1)$ and then, for $s = s^{\circ} + O_p(n^{-1/2})$, Taylor series expansion gives

$$\ell(\theta; s, a^{\circ}) = \ell(\theta; s^{\circ}, a^{\circ}) + (s - s^{\circ}) \left. \frac{\partial \ell(\theta; s, a^{\circ})}{\partial s} \right|_{s=s^{\circ}} + \cdots, \tag{10.12}$$

where the first term on the right-hand side is of order n, the second is the product of a term of order $n^{-1/2}$ with one of order n and is therefore of order $n^{1/2}$, and the remainder is $O(1)$. Standard first-order results, such as that leading to a significance function from applying the asymptotic approximation (10.3) to the signed likelihood root

$$r^{\circ}(\theta) = \mathrm{sign}(\widehat{\theta}^{\circ} - \theta) \left[2 \left\{ \ell(\widehat{\theta}^{\circ}; s^{\circ}, a^{\circ}) - \ell(\theta; s^{\circ}, a^{\circ}) \right\} \right]^{1/2},$$

use only the first term on the right of (10.12), and have error of order $n^{-1/2}$ for one-sided confidence intervals. We hope to reduce this error to order n^{-1}, so-called second-order inference, by including the second term. Although $r(\theta)$ involves only the value of the log likelihood, an approximation based on both terms in (10.12) also requires the derivative of ℓ with respect to s, a so-called sample space derivative. Approximation (10.12) yields a version of the tangent exponential model (10.8), here specialised to the case where we can identify (s, a) directly.

In Section 10.4, we describe the building blocks for exponential family models, but we first return to the example in Section 10.2.2.

Example 10.1 (ctd)

We previously saw that $y^{\mathrm{T}}(s,a) = (y_1, y_2) = (as, a/s)$, yielding

$$\frac{\partial y(s,a)}{\partial s} = \begin{pmatrix} a \\ -a/s^2 \end{pmatrix}, \quad V = \begin{pmatrix} a^{\mathrm{o}} \\ -a^{\mathrm{o}}/s^{\mathrm{o}2} \end{pmatrix}, \quad \frac{\partial \ell(\theta; s, a)}{\partial s} = -a(\theta^{-1} - \theta s^{-2}).$$

The left-hand panel of Figure 10.2 shows $f(y_1, y_2; \theta)$, the conditional sample space $\mathcal{A}^{\mathrm{o}}$ and its tangent space $\mathcal{T}^{\mathrm{o}}$ for $a^{\mathrm{o}} = 3$ and $s^{\mathrm{o}} = 1.6$. The right-hand panel shows the logarithm of the conditional density $f(s \mid a^{\mathrm{o}}; \theta)$ and its tangents at s^{o} for θ equal to 1, to $\widehat{\theta}^{\mathrm{o}} = 1.6$ and to 2.2. The filled circles show $f(s^{\mathrm{o}} \mid a^{\mathrm{o}}; \theta)$ for these values of θ, i.e., the corresponding likelihood values. Notice that $\theta = \varphi b + \{\varphi^2 b^2 + (s^{\mathrm{o}})^2\}^{1/2}$, where $b = (s^{\mathrm{o}})^2/(2a^{\mathrm{o}})$, can be expressed in terms of $\varphi(\theta) = \partial \ell(\theta; s^{\mathrm{o}}, a^{\mathrm{o}})/\partial s$, thus parametrising the model in terms of the slope of the tangent to the log density at y^{o}; in this model, this is a data-dependent parametrisation, to which we return below. □

10.3.2 Computation of V

At first sight the definition of V at (10.10) suggests that the mapping $y \mapsto (s, a)$ must be known. In fact this is not the case, as we see if we write

$$V = \left.\frac{\partial y}{\partial s^{\mathrm{T}}}\right|_{y=y^{\mathrm{o}}} = \left.\frac{\partial y}{\partial \theta^{\mathrm{T}}}\right|_{y=y^{\mathrm{o}}} \times \left.\left(\frac{\partial s}{\partial \theta^{\mathrm{T}}}\right)^{-1}\right|_{y=y^{\mathrm{o}}}. \tag{10.13}$$

The second matrix on the right has dimension $p \times p$ and is invertible, so the column space of V is also the column space of the first matrix on the right; both define the same space of

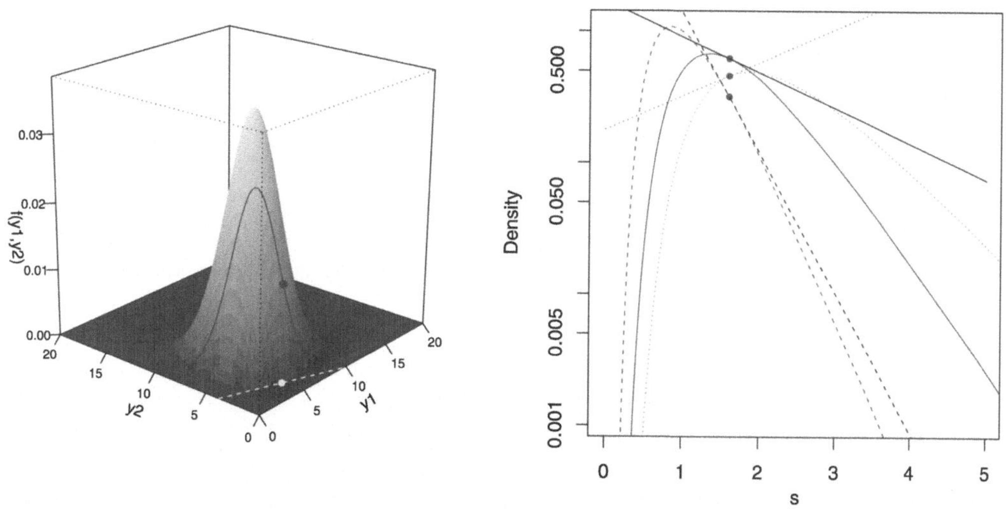

FIGURE 10.2

Conditional inference for gamma example. Left: joint density $f(y;\theta)$ with $\theta = 1$, $y^{\mathrm{o}} = (4.8, 1.875)$, giving $a^{\mathrm{o}} = 3$ and $s^{\mathrm{o}} = 1.6$. The white filled circle shows y^{o} and the black filled circle shows $f(y^{\mathrm{o}};\theta)$. The conditional reference set $\mathcal{A}^{\mathrm{o}}$ is shown by the dashed white line and the density $f\{y(s, a^{\mathrm{o}}); \theta\}$ on $\mathcal{A}^{\mathrm{o}}$ is shown by the solid line. The tangent plane $\mathcal{T}^{\mathrm{o}}$ is shown by the dotted white line. Right: Log conditional density $f(s \mid a^{\mathrm{o}}; \theta)$ (curved) and its tangents at $\log f(s^{\mathrm{o}} \mid a^{\mathrm{o}}; \theta)$ (straight) for $\theta = \widehat{\theta}^{\mathrm{o}}$ (solid), $\theta = 1$ (dashes) and $\theta = 2.2$ (dots). The filled circles are at $f(s^{\mathrm{o}} \mid a^{\mathrm{o}}; \theta)$.

sufficient directions, but $\partial y/\partial\theta^{\mathrm{T}}$ does not require y to be expressed in terms of (s, a). Hence V could be right-multiplied by any invertible $p \times p$ matrix of constants without changing the span of the sufficient directions. Below we shall generally take V to be $\partial y/\partial\theta^{\mathrm{T}}$, evaluated at $y = y^{\mathrm{o}}$ and $\theta = \widehat{\theta}^{\mathrm{o}}$.

Although it is not customary to express y as a function of θ, it is natural to do so when considering how a dataset would be simulated. In a regression model, for example, we can write $\theta = (\beta, \sigma)$ and

$$y(\theta) = X\beta + \sigma\varepsilon, \tag{10.14}$$

with the design matrix X and error vector ε fixed, giving

$$V = \left.\left(\frac{\partial y}{\partial\beta^{\mathrm{T}}}, \frac{\partial y}{\partial\sigma}\right)\right|_{y=y^{\mathrm{o}}} = (X, \varepsilon)|_{y=y^{\mathrm{o}}} = (X, (y^{\mathrm{o}} - X\hat{\beta}^{\mathrm{o}})/\hat{\sigma}^{\mathrm{o}}), \tag{10.15}$$

where $\widehat{\theta}^{\mathrm{o}} = (\hat{\beta}^{\mathrm{o}}, \hat{\sigma}^{\mathrm{o}})$ are maximum likelihood estimates computed from the data y^{o}. One way to think about (10.14) is as a quantile function, or structural equation, whereby changes in θ are reflected in changes to y for fixed ε.

An alternative approach to deriving V is to note that if the y_j are independent and have distribution functions $F_j(\cdot;\theta)$, then the pivotal equation $F(y_j;\theta) = u_j$ for some fixed u_j implicitly defines how y_j depends on θ. Total differentiation of this equation with respect to θ^{T} yields

$$0 = \frac{\mathrm{d}u_j}{\mathrm{d}\theta^{\mathrm{T}}} = \frac{\partial y_j}{\partial\theta^{\mathrm{T}}}\frac{\partial F(y_j;\theta)}{\partial y_j} + \frac{\partial F(y_j;\theta)}{\partial\theta^{\mathrm{T}}},$$

which implies that

$$\frac{\partial y_j}{\partial\theta^{\mathrm{T}}} = -\left\{\frac{\partial F_j(y_j;\theta)}{\partial y_j}\right\}^{-1}\frac{\partial F_j(y_j;\theta)}{\partial\theta^{\mathrm{T}}}. \tag{10.16}$$

In the case of a regression model, we have

$$F_j(y_j;\beta,\sigma) = F\{(y_j - x_j^{\mathrm{T}}\beta)/\sigma\},$$

where x_j^{T} is the jth row of X and $\varepsilon_1, \ldots, \varepsilon_n \stackrel{\text{iid}}{\sim} F$, which gives (10.15) when evaluated at y^{o}.

In group transformation models, there is no need to invoke the distribution function F_j, because one can write $y_j = g_j(\theta, \varepsilon_j)$ for a known function g_j and an error ε_j whose distribution does not depend on θ. Then V can be computed as $\partial y_j/\partial\theta$ for fixed ε_j, or equivalently

$$\frac{\partial y_j}{\partial\theta^{\mathrm{T}}} = -\left\{\frac{\partial\varepsilon_j(y_j;\theta)}{\partial y_j}\right\}^{-1}\frac{\partial\varepsilon_j(y_j;\theta)}{\partial\theta^{\mathrm{T}}}.$$

This construction is used in the bivariate normal example of Section 10.5.2. The derivation of the sufficient directions in Fraser and Reid (1995) and Fraser and Reid (2001) builds on the local location model of Fraser (1964); see Section 10.7.3.

The construction in terms of a structural equation such as (10.14) does not apply to discrete models, which require a slightly different treatment. We discuss the construction of the sufficient directions V further in Section 10.5.2.

In the next section, we outline the accurate approximation of significance functions in continuous exponential families.

10.4 Exponential Family Inferences

10.4.1 One-Parameter Case

Consider a continuous one-parameter exponential family with a scalar parameter θ, expressed as a tilted version of a baseline density $f_0(y)$,

$$f(y;\theta) = f_0(y) \exp\left[n\left\{s(y)\theta - \kappa(\theta)\right\}\right], \tag{10.17}$$

with canonical statistic $ns(y)$, canonical parameter θ and cumulant generator $n\kappa(\theta)$. The marginal density of the sufficient statistic s is of the same form,

$$f(s;\theta) = h_0(s) \exp\left[n\left\{s\theta - \kappa(\theta)\right\}\right], \tag{10.18}$$

where $h_0(s) = \int 1\{y \in \mathbb{R}^n : s(y) = s\} f_0(y)dy$. Here s is assumed to be an average of n independent observations, so its variation around its mean $\kappa_\theta(\theta)$ is $O_p(n^{-1/2})$. Writing densities explicitly in terms of the sample size n, as in (10.17) or (10.18), is a technical device that helps in keeping track of powers of n in theoretical development, but it does not change the quality of subsequent approximations and is unnecessary when applying them.

Although the integral defining h_0 is not usually available in closed form, the saddlepoint approximation can be used to approximate $f(s;\theta)$ very accurately. And as $\widehat{\theta}$ satisfies $s = \kappa_\theta(\widehat{\theta})$, this gives an approximation to the density of $\widehat{\theta}$ (e.g., Davison, 2003, equation (12.32)):

$$f(\widehat{\theta};\theta) = c\,\jmath^{\,1/2} \exp\left\{\ell(\theta;\widehat{\theta}) - \ell(\widehat{\theta};\widehat{\theta})\right\}\left\{1 + O(n^{-1})\right\}, \tag{10.19}$$

where $\jmath = n\kappa_{\theta\theta}(\widehat{\theta})$ is the observed information evaluated at $\theta = \widehat{\theta}$, and the normalising constant c ensures that (10.19) has unit integral. Our goal below is to use (10.19) to obtain a convenient and general expression for the significance function $\mathrm{pr}(\widehat{\theta} \le \widehat{\theta}^{\mathrm{o}};\theta)$, where $\widehat{\theta}^{\mathrm{o}}$ is the observed value of $\widehat{\theta}$.

We first make a monotone change of variable $\widehat{\theta} \mapsto r(\theta)$, where

$$r(\theta) = \mathrm{sign}(\widehat{\theta} - \theta)\left[2\left\{\ell(\widehat{\theta};\widehat{\theta}) - \ell(\theta;\widehat{\theta})\right\}\right]^{1/2}$$

The Jacobian of this transformation may be obtained from the derivative

$$\begin{aligned} r(\theta)\frac{\partial r(\theta)}{\partial\widehat{\theta}} &= \left.\frac{\partial \ell(\theta;\widehat{\theta})}{\partial\theta}\right|_{\theta=\widehat{\theta}} + \left.\frac{\partial \ell(\theta;\widehat{\theta})}{\partial\widehat{\theta}}\right|_{\theta=\widehat{\theta}} - \frac{\partial \ell(\theta;\widehat{\theta})}{\partial\widehat{\theta}}, \\ &= \ell_{;\widehat{\theta}}(\widehat{\theta};\widehat{\theta}) - \ell_{;\widehat{\theta}}(\theta;\widehat{\theta}), \end{aligned}$$

where here and below the appearance of a variable after a subscripted semi-colon indicates a sample-space derivative with respect to that variable. Differentiation with respect to $\widehat{\theta}$ is necessitated by the change of variable $\widehat{\theta} \mapsto r(\theta)$. For the exponential model, we have

$$r(\theta)\frac{\partial r(\theta)}{\partial\widehat{\theta}} = n\kappa_{\theta\theta}(\widehat{\theta})(\widehat{\theta} - \theta) = \jmath(\widehat{\theta} - \theta),$$

and (10.19) becomes

$$f\{r(\theta);\theta\} = c\frac{r(\theta)}{q(\theta)} \exp\left\{-r(\theta)^2/2\right\}, \tag{10.20}$$

where the Wald statistic for θ,

$$q(\theta) = \hat{\jmath}^{\,1/2}(\widehat{\theta} - \theta) = \hat{\jmath}^{-1/2}\left\{\ell_{;\widehat{\theta}}(\widehat{\theta};\widehat{\theta}) - \ell_{;\widehat{\theta}}(\theta;\widehat{\theta})\right\},$$

has an asymptotic standard normal distribution; see (10.2), where q is called t.

As $\theta \to \widehat{\theta}$ we have $r(\theta), q(\theta) \to 0$, but it is possible to show that, using an abbreviated notation, $q = r + a_1 r^2/n^{1/2} + a_2 r^3/n + \cdots$, so there is no singularity in (10.20). Taylor series expansion of the logarithm yields

$$\frac{1}{r}\log\left(\frac{q}{r}\right) = a_1 n^{-1/2} + (a_2 - a_1^2/2) r n^{-1} + O(n^{-3/2}), \tag{10.21}$$

so the square of (10.21) is of order n^{-1}. This is useful in the next step.

It follows from (10.21) that a second change of variable

$$r(\theta) \mapsto r^*(\theta) = r(\theta) + \frac{1}{r(\theta)}\log\left\{\frac{q(\theta)}{r(\theta)}\right\} \tag{10.22}$$

has Jacobian $1+(a_2-a_1^2/2)/n+O(n^{-3/2}) = 1+O(n^{-1})$. That the coefficient of the $1/n$ term is constant in r is important for renormalisation, discussed in Section 10.4.4. Although the transformation (10.22) may not be strictly monotonic over its entire range, it is monotone to the order considered here. After this transformation and a little more algebra, (10.20) becomes

$$f\{r^*(\theta);\theta\} = c\exp\left\{-r^*(\theta)^2/2\right\}\left\{1+O(n^{-1})\right\}. \tag{10.23}$$

We deduce that $c = (2\pi)^{-1/2}\{1+O(n^{-1})\}$, so the significance function may be written

$$\begin{aligned} \mathrm{pr}(\widehat{\theta} \le \widehat{\theta}^{\mathrm{o}};\theta) &= \mathrm{pr}\left\{r^*(\theta) \le r^{*\mathrm{o}}(\theta);\theta\right\} \\ &= \Phi\left\{r^{*\mathrm{o}}(\theta)\right\}\left\{1+O(n^{-1})\right\}, \end{aligned} \tag{10.24}$$

where $r^{*\mathrm{o}}(\theta)$ is the value of $r^*(\theta)$ actually observed, i.e., with y and $\widehat{\theta}$ replaced by y^{o} and $\widehat{\theta}^{\mathrm{o}}$. The discussion in Section 10.5.1 then allows inference on θ by treating $r^{*\mathrm{o}}(\theta)$ as a realisation of a standard normal variable. An alternative to (10.24) with the same order of asymptotic error is the Lugannani–Rice (1980) formula

$$\mathrm{pr}\left(\widehat{\theta} \le \widehat{\theta}^{\mathrm{o}};\theta\right) \doteq \Phi\left\{r^{\mathrm{o}}(\theta)\right\} + \left\{\frac{1}{r^{\mathrm{o}}(\theta)} - \frac{1}{q^{\mathrm{o}}(\theta)}\right\}\phi\left\{r^{\mathrm{o}}(\theta)\right\}, \tag{10.25}$$

where ϕ denotes the standard normal density function and $r^{\mathrm{o}}(\theta)$ and $q^{\mathrm{o}}(\theta)$ are the ingredients to $r^{*\mathrm{o}}(\psi)$. Both (10.24) and (10.25) are typically highly accurate, with neither systematically better than the other in applications, but they can become numerically unstable for θ near $\widehat{\theta}$ and dealing with this may require some careful programming. In Section 10.4.4 we show that when the density is renormalised to integrate to unity the relative error in (10.24) and (10.25) becomes $O(n^{-3/2})$, as is that in (10.23).

Expression (10.21) shows that the second term of $r^*(\theta)$ is an $O(n^{-1/2})$ correction to the $O(1)$ quantity $r(\theta)$; recall the second term on the right-hand side of (10.12).

10.4.2 Linear Exponential Family

The previous argument generalises to a linear exponential family in which $\theta^{\mathrm{T}} = (\psi, \lambda^{\mathrm{T}})$ consists of a scalar parameter ψ of interest and a nuisance parameter λ of dimension $p-1$. In this case,

$$f(s;\theta) = h(s)\exp\left[n\left\{s^{\mathrm{T}}\theta - \kappa(\theta)\right\}\right],$$

where $s(y)^{\mathrm{T}} = (s_1, s_2)$ is partitioned conformably with θ. In this model the conditional density of s_1 given s_2 does not depend on λ, and the ratio of the saddlepoint approximations to the densities of (s_1, s_2) and of s_2 yields the approximation

$$f(s_1 \mid s_2; \psi) \doteq c \left\{ \frac{|\jmath_{\lambda\lambda}(\widehat{\theta}_\psi)|}{|\jmath(\widehat{\theta})|} \right\}^{1/2} \exp\left\{ \ell(\widehat{\theta}_\psi) - \ell(\widehat{\theta}) \right\}, \tag{10.26}$$

where $\widehat{\theta}_\psi = (\psi, \widehat{\lambda}_\psi^{\mathrm{T}})^{\mathrm{T}}$ denotes the maximum likelihood estimator for fixed ψ and $\jmath_{\lambda\lambda}(\theta)$ is the sub-matrix of $\jmath(\theta)$ corresponding to λ. A calculation similar to that leading to (10.23) establishes that apart from a relative error of order n^{-1}, the resulting significance function for inference on ψ, $\mathrm{pr}(S_1 \leq s_1^{\mathrm{o}} \mid S_2 = s_2; \psi)$ is again of form (10.24), now with

$$r(\psi) = \operatorname{sign}(\widehat{\psi} - \psi) \left[2 \left\{ \ell(\widehat{\theta}) - \ell(\widehat{\theta}_\psi) \right\} \right]^{1/2}, \tag{10.27}$$

$$q(\psi) = (\widehat{\psi} - \psi) \left\{ \frac{|\jmath(\widehat{\theta})|}{|\jmath_{\lambda\lambda}(\widehat{\theta}_\psi)|} \right\}^{1/2} \tag{10.28}$$

evaluated at the observed data y^{o} and maximum likelihood estimate $\widehat{\theta}^{\mathrm{o}}$. Note that

$$q(\psi) = t(\psi)\rho(\psi, \widehat{\psi}), \quad \rho(\psi, \widehat{\psi}) = \left\{ \frac{|\jmath_{\lambda\lambda}(\widehat{\theta})|}{|\jmath_{\lambda\lambda}(\widehat{\theta}_\psi)|} \right\}^{1/2},$$

with $t(\psi)$ the Wald statistic based on the profile log likelihood defined in (10.4). For derivations of these results see Fraser and Reid (1993) or Davison (2003, Section 12.3.3), for example.

Although expression (10.24) was derived by approximating a conditional distribution, it is also an approximation to the marginal distribution of $r^*(\psi)$, because the normal distribution of $r^*(\psi)$ in (10.24) does not depend on the conditioning variable S_2.

If the parameter of interest is a linear function of the natural parameter θ of the exponential model, say $\psi = C_1^{\mathrm{T}}\theta$ for some known $p \times 1$ vector C_1, then we can set $\lambda = C_2^{\mathrm{T}}\theta$, so that $\varphi^{\mathrm{T}} = (\psi, \lambda^{\mathrm{T}}) = \theta^{\mathrm{T}}(C_1, C_2) = \theta^{\mathrm{T}}C$, say, where the $p \times p$ matrix C is invertible, express the exponential family in terms of

$$s_*(y) = C^{-1}s(y), \quad \varphi(\theta) = C^{\mathrm{T}}\theta, \quad \kappa_*(\varphi) = \kappa(C^{-\mathrm{T}}\varphi) = \kappa(\theta),$$

and finally apply approximation (10.24) in this reparametrised model.

10.4.3 General Exponential Family

We now extend the normal approximation to the distribution of r^*, developed in the previous section, to the general exponential family

$$f(s; \theta) = h(s) \exp\left[n \left\{ s^{\mathrm{T}}\varphi(\theta) - \kappa(\theta) \right\} \right], \tag{10.29}$$

where $\varphi(\theta)$ may be a nonlinear function of (ψ, λ). For this we shall need the analogues of $r(\psi)$ and $q(\psi)$ of (10.27) and (10.28). The likelihood is invariant to reparametrisation, so $r(\psi)$ is unchanged, but as ψ is no longer a component of the canonical parameter φ, we need a new form for $q(\psi)$ using a surrogate for ψ. Taylor series expansion for small $\widehat{\theta} - \widehat{\theta}_\psi$ gives

$$\varphi(\widehat{\theta}) - \varphi(\widehat{\theta}_\psi) = \frac{\partial \varphi(\widehat{\theta}_\psi)}{\partial \theta^{\mathrm{T}}} (\widehat{\theta} - \widehat{\theta}_\psi) + \cdots,$$

so if the matrix on the right-hand side is invertible at $\widehat{\theta}_\psi$, which is a condition for θ to be identifiable, then

$$\widehat{\theta} - \widehat{\theta}_\psi = \left\{ \frac{\partial \varphi(\widehat{\theta}_\psi)}{\partial \theta^{\mathrm{T}}} \right\}^{-1} \left\{ \varphi(\widehat{\theta}) - \varphi(\widehat{\theta}_\psi) \right\} + \cdots, \tag{10.30}$$

and as the partial derivatives satisfy

$$I_p = \frac{\partial \theta}{\partial \varphi^{\mathrm{T}}} \frac{\partial \varphi}{\partial \theta^{\mathrm{T}}} = \begin{pmatrix} \partial \psi / \partial \varphi^{\mathrm{T}} \\ \partial \lambda / \partial \varphi^{\mathrm{T}} \end{pmatrix} \begin{pmatrix} \partial \varphi / \partial \psi & \partial \varphi / \partial \lambda^{\mathrm{T}} \end{pmatrix} = \begin{pmatrix} \frac{\partial \psi}{\partial \varphi^{\mathrm{T}}} \frac{\partial \varphi}{\partial \psi} & \frac{\partial \psi}{\partial \varphi^{\mathrm{T}}} \frac{\partial \varphi}{\partial \lambda^{\mathrm{T}}} \\ \frac{\partial \lambda}{\partial \varphi^{\mathrm{T}}} \frac{\partial \varphi}{\partial \psi} & \frac{\partial \lambda}{\partial \varphi^{\mathrm{T}}} \frac{\partial \varphi}{\partial \lambda^{\mathrm{T}}} \end{pmatrix}, \tag{10.31}$$

the first row of the inverse on the right-hand side of (10.30) is $\partial \psi / \partial \varphi^{\mathrm{T}}$, yielding

$$\widehat{\psi} - \psi = \frac{\partial \psi(\widehat{\theta}_\psi)}{\partial \varphi^{\mathrm{T}}} \left\{ \varphi(\widehat{\theta}) - \varphi(\widehat{\theta}_\psi) \right\} + \cdots .$$

Hence $q(\psi)$ should be based on a local departure of $\widehat{\theta}$ from $\widehat{\theta}_\psi$, or equivalently of $\hat{\varphi}$ from $\hat{\varphi}_\psi$, which can be measured through the constructed parameter $\chi(\theta)$ given by

$$\chi(\theta) = u^{\mathrm{T}} \varphi(\theta), \quad u = \frac{\partial \psi(\widehat{\theta}_\psi)/\partial \varphi}{\left\| \partial \psi(\widehat{\theta}_\psi)/\partial \varphi \right\|},$$

i.e., the orthogonal projection of $\varphi(\theta)$ onto a unit vector u parallel to $\partial \psi(\widehat{\theta}_\psi)/\partial \varphi$; note that $\chi(\theta)$ depends on the data through $\widehat{\theta}_\psi$. The Wald-type measure q from (10.28) for $\chi(\theta)$ is

$$q(\psi) = \mathrm{sign}(\widehat{\psi} - \psi) \left| \chi(\widehat{\theta}) - \chi(\widehat{\theta}_\psi) \right| \left\{ \frac{|\jmath(\hat{\varphi})|}{|\jmath_{(\lambda\lambda)}(\hat{\varphi}_\psi)|} \right\}^{1/2}, \tag{10.32}$$

where the determinants are computed in the φ parametrisation,

$$|\jmath(\hat{\varphi})| = \left| \jmath(\widehat{\theta}) \right| \left| \frac{\partial \varphi(\widehat{\theta})}{\partial \theta^{\mathrm{T}}} \right|^{-2}, \quad |\jmath_{(\lambda\lambda)}(\hat{\varphi}_\psi)| = \left| \jmath_{\lambda\lambda}(\widehat{\theta}_\psi) \right| \left| \frac{\partial \varphi^{\mathrm{T}}(\widehat{\theta}_\psi)}{\partial \lambda} \frac{\partial \varphi(\widehat{\theta}_\psi)}{\partial \lambda^{\mathrm{T}}} \right|^{-1}, \tag{10.33}$$

and the second factor on the right-hand side of the second expression here stems from the "area formula" (see, for example, Krantz and Parks, 2008, Lemma 5.1.4). An equivalent expression for $q(\psi)$ is obtained by substituting these expressions and simplifying, yielding

$$q(\psi) = \frac{|\varphi(\widehat{\theta}) - \varphi(\widehat{\theta}_\psi) \quad \partial \varphi(\widehat{\theta}_\psi)/\partial \lambda^{\mathrm{T}}|}{\left| \partial \varphi(\widehat{\theta})/\partial \theta^{\mathrm{T}} \right|} \times \left\{ \frac{|\jmath(\widehat{\theta})|}{|\jmath_{\lambda\lambda}(\widehat{\theta}_\psi)|} \right\}^{1/2}. \tag{10.34}$$

If $\varphi^{\mathrm{T}} = (\psi, \lambda^{\mathrm{T}})$, then (10.34) reduces to (10.28).

The equivalence of (10.32) and (10.34) follows by noting that the determinant of the $p \times p$ matrix

$$\left(\varphi(\widehat{\theta}) - \varphi(\widehat{\theta}_\psi) \quad \partial \varphi(\widehat{\theta}_\psi)/\partial \lambda^{\mathrm{T}} \right)$$

is the signed volume $l_1 v_2$ of the parallelepiped generated by its p columns, with l_1 the length of the component of its first column in the direction orthogonal to its other columns and v_2 the volume of the $(p-1)$-dimensional parallelepiped these last columns generate (Fraser et al., 1999; Skovgaard, 1996), which is

$$v_2 = \left| \frac{\partial \varphi^{\mathrm{T}}(\widehat{\theta}_\psi)}{\partial \lambda} \frac{\partial \varphi(\widehat{\theta}_\psi)}{\partial \lambda^{\mathrm{T}}} \right|^{1/2}.$$

As the vector $\partial \psi(\widehat{\theta}_\psi)/\partial \varphi$ is orthogonal to $\partial \varphi(\widehat{\theta}_\psi)/\partial \lambda^{\mathrm{T}}$, we have $l_1 = \left| \chi(\widehat{\theta}) - \chi(\widehat{\theta}_\psi) \right|$; the sign of $\widehat{\psi} - \psi$ supplies the sign of $l_1 v_2$.

Example 10.2

To illustrate the development above, consider independent exponential variables y_1 and y_2 with rate parameters $\lambda\psi$ and λ; here $\theta = (\psi, \lambda)^{\mathrm{T}}$. The corresponding log likelihood is

$$\ell(\psi, \lambda) = 2\log\lambda + \log\psi - \lambda(\psi y_1 + y_2), \quad \psi, \lambda > 0,$$

leading to $\widehat{\lambda}_\psi = 2/(\psi y_1 + y_2)$, $\widehat{\lambda} = 1/y_2$, $\widehat{\psi} = y_2/y_1$, $|\jmath(\widehat{\theta})| = 1/(\widehat{\lambda}\,\widehat{\psi})^2$ and $|\jmath_{\lambda\lambda}(\widehat{\theta}_\psi)| = 2/\widehat{\lambda}_\psi^2$.

The elements of (10.29) are

$$\begin{aligned} \varphi &= \begin{pmatrix} \varphi_1 \\ \varphi_2 \end{pmatrix} = -\begin{pmatrix} \lambda\psi \\ \lambda \end{pmatrix}, \\ s(y) &= \begin{pmatrix} y_1 \\ y_2 \end{pmatrix}, \\ \kappa(\varphi) &= -\log(-\varphi_1) - \log(-\varphi_2), \quad \varphi_1, \varphi_2 < 0, \end{aligned}$$

so $\psi = \varphi_1/\varphi_2$, and fixing ψ is equivalent to forcing φ to lie on a line through the origin of gradient $1/\psi$. Now

$$\frac{\partial\varphi}{\partial\lambda} = -\begin{pmatrix} \psi \\ 1 \end{pmatrix}, \quad \frac{\partial\psi}{\partial\varphi} = \begin{pmatrix} 1/\varphi_2 \\ -\varphi_1/\varphi_2^2 \end{pmatrix} = \frac{1}{\lambda}\begin{pmatrix} -1 \\ \psi \end{pmatrix}.$$

Hence the unit vector in the direction $\partial\psi/\partial\varphi$ is $(1+\psi^2)^{-1/2}(-1, \psi)^{\mathrm{T}}$, and u is obtained from this by evaluating it at $\varphi(\widehat{\theta}_\psi)$; this is orthogonal to $\partial\varphi/\partial\lambda$ by construction.

To be concrete, let $y_1 = 1$ and $y_2 = 2$, and consider computing the significance function at $\psi = 1$. In this case $\widehat{\theta}_\psi = (1, 2/3)^{\mathrm{T}}$, $\widehat{\theta} = (2, 1/2)^{\mathrm{T}}$, $\varphi(\widehat{\theta}_\psi) = -(2/3, 2/3)^{\mathrm{T}}$, $\varphi(\widehat{\theta}) = -(1, 1/2)^{\mathrm{T}}$ and $u = (-1/\sqrt{2}, 1/\sqrt{2})^{\mathrm{T}}$.

Figure 10.3 shows the log likelihoods, the maximum likelihood estimates and the partial maximum likelihood estimates in the θ and φ parametrisations. The construction of $\chi(\theta) = u^{\mathrm{T}}\varphi(\theta)$ in terms of u and $\varphi(\theta)$, shown in the right-hand panel, yields $\chi(\widehat{\theta}) = 1/\sqrt{8}$ and $\chi(\widehat{\theta}_\psi) = 0$. The difference $\chi(\widehat{\theta}) - \chi(\widehat{\theta}_\psi)$ in the constructed parameter is a signed distance along the dashed line. As ψ varies, the grey line $\varphi_1 = \psi\varphi_2$ and the vector u, which is orthogonal to that line, also vary. Increasing ψ starting from $\psi = 1$ would move $\varphi(\widehat{\theta}_\psi)$ along the dotted line closer to $\varphi(\widehat{\theta})$, inclining the vector u and the dashed line closer to vertical and reducing significance by decreasing $\chi(\widehat{\theta}) - \chi(\widehat{\theta}_\psi)$. When the grey line passes though the black square representing $\varphi(\widehat{\theta})$, it coincides with the black circle and the open square and then $\chi(\widehat{\theta}) - \chi(\widehat{\theta}_\psi) = 0$. Decreasing ψ would move $\varphi(\widehat{\theta}_\psi)$ along the black dotted line away from $\varphi(\widehat{\theta})$, inclining the vector u and the dashed line further from the vertical, and increasing $\chi(\widehat{\theta}) - \chi(\widehat{\theta}_\psi)$ and thus the significance. □

10.4.4 Renormalisation

Although the error in (10.24) is ostensibly $O(n^{-1})$, it is actually $O(n^{-3/2})$ in wide generality after renormalisation of the approximation to ensure that it integrates to 1. While in practice this would be done numerically, we note that if the error term in (10.23) takes the form B/n where B does not depend on the variable of integration, then

$$1 = \int f\{r^*(\theta); \theta\}\mathrm{d}r^* = \int c\sqrt{(2\pi)}\phi\{r^*(\theta)\}\{1 + B/n + O(n^{-3/2})\}\mathrm{d}r^*$$

implies

$$c\sqrt{(2\pi)} = 1 - B/n + O(n^{-3/2}),$$

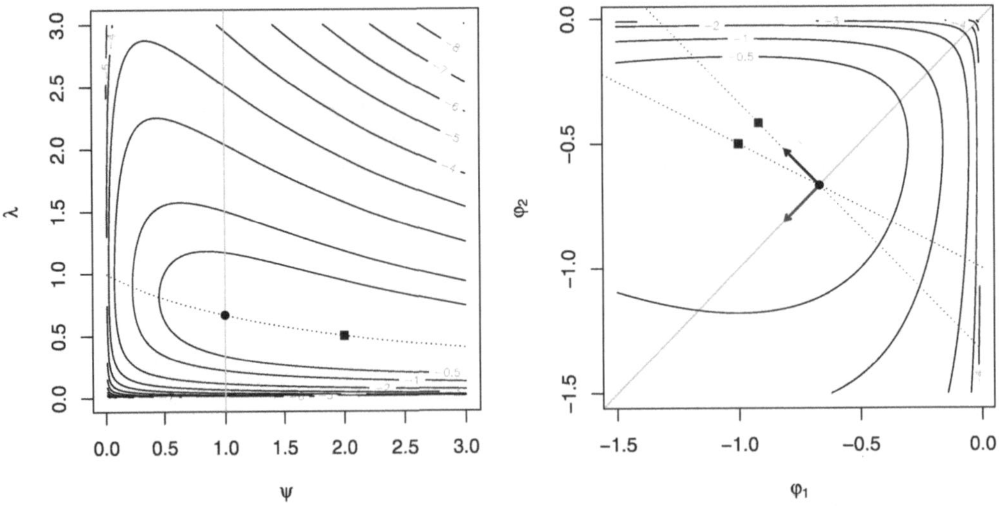

FIGURE 10.3
Log likelihoods for exponential example in θ parametrisation (left) and φ parametrisation (right). The solid grey lines show $\psi = 1$ or equivalently $\varphi_1 = \varphi_2$. The overall maximum likelihood estimates $\widehat{\theta}$ and $\varphi(\widehat{\theta})$ are shown by black squares, and the partial maximum likelihood estimates $\widehat{\theta}_\psi$ and $\varphi(\widehat{\theta}_\psi)$, with $\psi = 1$, by black circles. The dotted lines show $\widehat{\theta}_\psi$ and $\varphi(\widehat{\theta}_\psi)$ as functions of ψ. The arrows in the right-hand panel show the directions of the vectors $\partial\varphi(\widehat{\theta}_\psi)/\partial\lambda$ (light) and $\partial\psi(\widehat{\theta}_\psi)/\partial\varphi$ (heavy). The difference $\chi(\widehat{\theta}) - \chi(\widehat{\theta}_\psi)$ appearing in $q(\psi)$ is a distance along the dotted line, between $\chi(\widehat{\theta}_\psi)$ (black filled circle) and $\chi(\widehat{\theta})$ (open square).

so (10.23) becomes

$$f(r^*;\theta) = (1 - B/n)\phi\{r^*(\theta)\}\{1 + B/n + O(n^{-3/2})\} = \phi\{r^*(\theta)\}\{1 + O(n^{-3/2})\}.$$

As noted below (10.22), in the linear exponential family with $p = 1$ we indeed have B free of r^*. In work as yet unpublished Y. Tang has verified that B is also constant in r^* for the multi-parameter linear exponential family treated in Section 10.4.2, and for regression-scale models (10.14).

If the $O(n^{-1})$ error term depends on $r^*(\theta)$, then under mild conditions

$$B\{r^*(\theta)\} = B(0) + r^*(\theta)B'(s), \quad 0 < |s| < |r^*(\theta)|,$$

and the term $B(0)/n$ cancels with the normalizing constant as above, so we only need consider

$$c\surd(2\pi)\phi\{r^*(\theta)\}r^*(\theta)B'(s).$$

If $B'(s)$ is constant, then this term integrates to 0 and the norming constant is as before. If $B'(s)$ depends on r^*, it would have to to rise very rapidly for the error term to be unbounded, as the normal density function drops rapidly as $|r^*(\theta)|$ increases. Slower, but not constant, dependence of B' on r^* would lead to relative error just $O(1/n)$, not improved by renormalization.

Mild regularity conditions (Daniels, 1956) ensure the validity of this argument, which also applies to Laplace and similar approximations, including those in Sections 10.4.2 and 10.4.3. A similar analysis verifies that the relative error in the Lugannani–Rice (1980) approximation (10.25) is also $O(n^{-3/2})$.

10.5 General Likelihood

10.5.1 Basic Approximation

Section 10.4 described approximations useful for inference in exponential families. We now outline how these may be extended to general models, using the tangent exponential model (10.8). As noted there, the tangent exponential model has the structure of an exponential family model, with canonical parameter φ and cumulant generator $-\ell(\varphi; y^{\text{o}})$. To compare (10.8) to the general exponential family density (10.29), we take the baseline density $h(s)$ to be $h(s; \widehat{\theta}^{\text{o}})$, which effectively centers the score variable s so that $s^{\text{o}} = 0$. Thus we have

$$f_{\text{TEM}}(s \mid a; \theta) = \exp\{s^{\text{T}}\varphi(\theta; y^{\text{o}}) + \ell(\theta; y^{\text{o}})\}h(s; \widehat{\theta}^{\text{o}}), \tag{10.35}$$

with canonical variable s and canonical parameter $\varphi(\theta) \equiv \varphi(\theta; y^{\text{o}})$ defined locally at y^{o}, as we now describe.

In an ordinary exponential family, the canonical parameter can be obtained (up to linear transformations) from $\partial\ell(\theta; s)/\partial s$. In a general model, $\varphi(\theta)$ is also constructed by differentiating in the sample space, now using the $n \times p$ matrix V of sufficient directions of Section 10.3. As shown there, projecting onto the space spanned by these directions implements conditioning on an approximate ancillary statistic without requiring an explicit form for that ancillary. In more detail,

$$\varphi(\theta; y^{\text{o}}) = \ell_{;V}(\theta; y^{\text{o}}), \tag{10.36}$$

where

$$\ell_{;V}(\theta; y^{\text{o}}) = \frac{\text{d}}{\text{d}t}\ell(\theta; y^{\text{o}} + Vt)\bigg|_{t=0} = V^{\text{T}}\left.\frac{\partial\ell(\theta; y)}{\partial y}\right|_{y=y^{\text{o}}} = \sum_{j=1}^{n} V_j^{\text{T}}\frac{\partial\ell(\theta; y_j^{\text{o}})}{\partial y_j},$$

with $t = (t_1, \ldots, t_p)^{\text{T}}$; the last equality applies when the y_j make independent contributions $\ell(\theta; y_j)$ to the log likelihood. Note that both $\ell(\theta; y^{\text{o}}) = \log f(y^{\text{o}}; \theta)$ and $\varphi(\theta; y^{\text{o}})$ are computed from the original model $f(y; \theta)$.

The validity of this local approximation can be established by Taylor series approximation, as described in Section 10.6. Taking (10.35) as our starting point, we simply apply the formulae derived for the general exponential family model in Section 10.4.3, computing $r(\psi)$ at (10.27) and $q(\psi)$ at (10.34). The approximate significance function for an interest parameter ψ is $\Phi\{r^*(\psi)\}$, using (10.22).

The tangent model approximates the original density in a neighbourhood of the observed data point y^{o}; more precisely the original model and the tangent exponential model have the same observed log-likelihood function, and the same sample-space derivative of the log-likelihood function. This turns out to be enough to ensure that the significance function is accurate to $O(n^{-3/2})$.

The saddlepoint approximation to (10.35) is

$$f_{\text{TEM}}(s \mid a; \theta) \doteq c|j(\hat{\varphi})|^{-1/2}\exp\left[s^{\text{T}}\{\varphi(\theta) - \varphi(\hat{\theta}^{\text{o}})\} + \ell(\theta; y^{\text{o}}) - \ell(\hat{\theta}^{\text{o}}; y^{\text{o}})\right], \tag{10.37}$$

where $j(\varphi)$ is the observed Fisher information computed in the φ parametrisation as at (10.33), and $\ell(\theta; y^{\text{o}}) = \ell\{\theta(\varphi); y^{\text{o}}\}$. The validity of this saddlepoint approximation is verified by Taylor-series expansion of the log likelihood $\ell(\theta; s)$ about $\widehat{\theta}^{\text{o}}$ and s^{o}, described in Section 10.6.

10.5.2 Generalising V

The matrix V of sufficient directions is central to approximations based on the tangent exponential model. Our previous discussion has taken V to equal $\partial y/\partial\theta^{\mathrm{T}}$, evaluated at the observed values y^{o} of the data y and the maximum likelihood estimate $\widehat{\theta}^{\mathrm{o}}$, but this expression presupposes that the individual observations y_j are scalar and can be differentiated with respect to θ; the latter is not the case for discrete responses. In this section we outline how V can be generalised.

As a preliminary remark, notice that the span of V is invariant to smooth invertible reparametrisation $\theta \mapsto \theta'$, which has the effect of right-multiplying V by an invertible $p \times p$ matrix of constants; see (10.13). It is readily checked that such a multiplication leaves (10.34) unchanged, so V can be computed in whatever parametrisation is simplest.

When the y_j are continuous and independent vectors of possibly different dimensions d_j we can write

$$\varphi(\theta) = \sum_{j=1}^{n} \left.\frac{\partial y_j^{\mathrm{T}}}{\partial\theta}\right|_{y_j=y_j^{\mathrm{o}},\theta=\widehat{\theta}^{\mathrm{o}}} \times \left.\frac{\partial\ell(\theta;y_j)}{\partial y_j}\right|_{y_j=y_j^{\mathrm{o}}} = \sum_{j=1}^{n} V_j^{\mathrm{T}} \left.\frac{\partial\ell(\theta;y_j)}{\partial y_j}\right|_{y_j=y_j^{\mathrm{o}}},$$

say, where V_j has dimension $d_j \times p$. This effectively replaces the matrix V by a tensor.

Example 10.3

If $\{(y_{1j}, y_{2j}), j = 1,\ldots,n\}$ are independent pairs from a bivariate normal distribution with zero means, unit variances and covariance θ, V can be contructed using the pivotal quantities $z_{1j} = (y_{1j}+y_{2j})^2/\{2(1+\theta)\}$ and $z_{2j} = (y_{1j}-y_{2j})^2/\{2(1-\theta)\}$, leading to

$$V_j = \left(\frac{y_{2j}^{\mathrm{o}} - \widehat{\theta}^{\mathrm{o}} y_{1j}^{\mathrm{o}}}{1-\widehat{\theta}^{\mathrm{o}2}}, \frac{y_{1j}^{\mathrm{o}} - \widehat{\theta}^{\mathrm{o}} y_{2j}^{\mathrm{o}}}{1-\widehat{\theta}^{\mathrm{o}2}}\right), \quad j = 1,\ldots,n,$$

and thus to

$$\varphi(\theta) = \ell_{;V}(\theta) = \frac{\theta(t^{\mathrm{o}} - \widehat{\theta}^{\mathrm{o}} s^{\mathrm{o}}) - (s^{\mathrm{o}} - \widehat{\theta}^{\mathrm{o}} t^{\mathrm{o}})}{(1-\theta^2)(1-\widehat{\theta}^{\mathrm{o}2})},$$

where $t = \sum(y_{1j}^2 + y_{2j}^2)/2$ and $s = \sum y_{1j}y_{2j}$. The sufficient statistics (s,t) emerge naturally in the construction of $\varphi(\theta)$. If a preliminary reduction to sufficiency is made, the resulting V is a 2×1 vector instead of a $2n \times 1$ vector as above, though $\varphi(\theta)$ is unchanged. The sample space contours determined by V are illustrated in Reid (2003), and the accuracy of the normal approximation to the distribution of r^* is illustrated in Reid (2005). □

Computing V for discrete responses is more awkward. In Section 10.3.2 we saw that in the continuous case, total differentiation of the pivot $F_j(y_j;\theta)$ led to the expression (10.16) for V_j, but $\partial F_j(y;\theta)/\partial y = 0$ almost everywhere in the discrete case. To deal with this, note that in a continuous exponential family model with canonical observation y_j, we can write $\partial\ell(\theta;y_j)/\partial y_j = \alpha_j(\theta)$, say, and observe that as

$$\begin{aligned} n^{-1}\varphi(\theta) &= n^{-1}\sum_{j=1}^{n}\frac{\partial y_j^{\mathrm{T}}}{\partial\theta}\alpha_j(\theta) \\ &= n^{-1}\sum_{j=1}^{n}\mathrm{E}\left(\frac{\partial y_j^{\mathrm{T}}}{\partial\theta}\right)\alpha_j(\theta) + n^{-1}\sum_{j=1}^{n}\left\{\frac{\partial y_j^{\mathrm{T}}}{\partial\theta} - \mathrm{E}\left(\frac{\partial y_j^{\mathrm{T}}}{\partial\theta}\right)\right\}\alpha_j(\theta), \end{aligned}$$

and each term in the final average has mean zero, that term is $O_p(n^{-1/2})$. Thus if the order of integration and differentiation can be interchanged, $n^{-1}\varphi(\theta)$ can be replaced with

$$n^{-1}\tilde{\varphi}(\theta) = n^{-1}\sum_{j=1}^{n} \left.\frac{\partial \mathrm{E}(y_j^{\mathrm{T}})}{\partial \theta}\right|_{\theta=\widehat{\theta}^{\mathrm{o}}} \alpha_j(\theta),$$

at the expense of introducing an $O_p(n^{-1/2})$ error. Hence using $\tilde{\varphi}(\theta)$ in (10.34) does not change the $O(n^{-1})$ error in (10.24). The use of $\partial \mathrm{E}(y_j^{\mathrm{T}})/\partial\theta$ replaces the sufficient directions with their expectation, which is only tangential to $\mathcal{A}^{\mathrm{o}}$ on average, so renormalisation no longer reduces the order of error to $n^{-3/2}$: inference accurate to third order is unavailable.

As the expectation of a discrete response is typically continuous in the parameters, this approach can be applied to discrete exponential family models such as the Poisson, binomial and multinomial. Extension to more general discrete response distributions entails replacing $\alpha_j(\theta)$. Davison et al. (2006) show that one can use a locally-defined score variable w_j, and

$$w_j = \left.\frac{\partial \ell(\theta; y_j)}{\partial \theta}\right|_{\theta=\widehat{\theta}^{\mathrm{o}}}, \quad V_j = \left.\frac{\partial \mathrm{E}(w_j;\theta)}{\partial \theta^{\mathrm{T}}}\right|_{\theta=\widehat{\theta}^{\mathrm{o}}}, \quad \varphi(\theta) = \sum_{j=1}^{n} V_j^{\mathrm{T}} \frac{\partial \ell(\theta; y_j)}{\partial w_j}.$$

Here w_j has dimension $p \times 1$, so V_j is a $p \times p$ matrix that is easily seen to be the contribution from y_j to the expected information matrix, evaluated at $\widehat{\theta}^{\mathrm{o}}$. The derivative $\partial\ell(\theta; y_j)/\partial w_j$ is most easily computed as $\partial\ell(\theta; y_j)/\partial y_j \times (\partial w_j/\partial y_j)^{-1}$.

Skovgaard (1996) derived a version of r^* that replaces q in (10.22) with a quantity that is computed entirely from cumulants of the log likelihood. The resulting approximation has a relative error that is $O(n^{-1})$ in a large deviation region about the maximum likelihood estimator. This provides highly accurate results far out into the tails of its distribution, which Skovgaard argues may be of more practical value than higher, $O(n^{-3/2})$, accuracy near its mean. Reid and Fraser (2010) show that Skovgaard's approximation can be related to a tangent exponential model with canonical parameter determined from the derivative of $I(\theta; \widehat{\theta}^{\mathrm{o}}) = \int \ell(\theta; y) f(y; \widehat{\theta}^{\mathrm{o}})dy$. It can also be used for discrete models, and gives the same approximation as Davison et al. (2006) in curved exponential families, but not more generally.

10.6 Derivation of the Tangent Exponential Model

10.6.1 Preliminary Remarks

The expression for the tangent exponential model, (10.8), is concise and emphasises the connection to exponential family models and the role of φ as a canonical parameter, but does not lend itself to ready understanding. It can be derived using Taylor series approximations that can be given explicitly when y and θ are scalar and provide some theoretical illumination. The development for higher dimensions is similar but much more laborious and does not yield additional insights.

We have seen above that the tangent exponential model requires computation of a first derivative in the sample space. We can think of this as assessing how the log likelihood $\ell(\theta; y^{\mathrm{o}})$ changes not only as a function of θ, as is standard in both likelihood and Bayesian inference, but also as a function of y, in a small neighbourhood of the observed data point y^{o}.

Employing this first derivative probes $\ell(\theta; y)$ more deeply than simply using the observed log likelihood function $\ell(\theta; y^o)$, but does not involve computing the log likelihood function on its entire domain, i.e., for $(\theta, y) \in \mathbb{R}^p \times \mathbb{R}^n$.

10.6.2 No Nuisance Parameters

Suppose that we have a model $f(s;\theta)$, $s \in \mathbb{R}$ and $\theta \in \mathbb{R}$, and that there is an implicit dependence on n, in the sense that $\ell(\theta; s) = \log f(s;\theta)$ is $O_p(n)$. We will address how to get this reduction in general at the end of this section, but this would be the case for example if s was the sufficient statistic based on a random sample from a linear exponential family model, and it is also the case for models like the Cauchy, where s is the maximum likelihood estimator or any other location-equivariant estimator of θ, based on a sample of size n, and the distribution is conditional on the $(n-1)$-dimensional ancillary statistic a.

We first expand the log likelihood $\log f(s;\theta)$ in a Taylor series in both s and θ, about the fixed points s^o and $\widehat{\theta}^o$, giving

$$\begin{aligned}
\ell(\theta; s) &= \log f(s;\theta) \\
&= \ell(\widehat{\theta}^o; s^o) + (s - s^o)\ell_s(\widehat{\theta}^o; s^o) + (\theta - \widehat{\theta}^o)\ell_\theta(\widehat{\theta}^o; s^o) \\
&\quad + \frac{1}{2}(s - s^o)^2 \ell_{ss}(\widehat{\theta}^o; s^o) + (s - s^o)(\theta - \widehat{\theta}^o)\ell_{\theta s}(\widehat{\theta}^o; s^o) \\
&\quad + \frac{1}{2}(\theta - \widehat{\theta}^o)^2 \ell_{\theta\theta}(\widehat{\theta}^o; s^o) + \cdots \\
&= \sum_{i,j=0}^{\infty} \frac{1}{i!j!}(\theta - \widehat{\theta}^o)^i (s - s^o)^j b_{ij}, \qquad (10.38)
\end{aligned}$$

say, where $b_{ij} = \partial^i \partial^j \ell(\widehat{\theta}^o; s^o)/\partial\theta^i \partial s^j$ for $i, j = 0, 1, \ldots$; note that $b_{10} = 0$.

Now consider how expansion (10.38) would differ if $f(s;\theta)$ were an exponential family model with canonical parameter θ and sufficient statistic s. In this case

$$\log f(s;\theta) = \ell(\theta; s) = \theta s - \kappa(\theta) - d(s),$$

so the coefficients b_{i0} would be those for a Taylor series expansion of $-d(s)$, the b_{0j} would be those for a Taylor series expansion of $-\kappa(\theta)$, and the only other non-zero term would be $b_{11} = 1$.

Andrews et al. (2005) showed that for any continuously differentiable model $f(s;\theta)$, there exists a transformation $x = x(s)$ and $\varphi = \varphi(\theta)$ such that the expansion of $\log f(x;\varphi)$ has the coefficient array starting with $i = j = 0$ at the top left and terms b_{ij} in row i and column j, given by

$$\begin{pmatrix}
b + \dfrac{3\alpha_4 - 5\alpha_3^2 - 12\gamma}{24n} & \dfrac{-\alpha_3}{2n^{1/2}} & -\left(1 + \dfrac{\alpha_4 - 2\alpha_3^2 - 5\gamma}{2n}\right) & \dfrac{\alpha_3}{n^{1/2}} & \dfrac{\alpha_4 - 3\alpha_3^2 - 6\gamma}{n} \\
0 & 1 & 0 & 0 & \\
-1 & 0 & \dfrac{\gamma}{n} & & \\
-\dfrac{\alpha_3}{n^{1/2}} & 0 & & & \\
-\dfrac{\alpha_4}{n} & & & &
\end{pmatrix}. \qquad (10.39)$$

Only terms up to $O(n^{-1})$ are shown: the terms in the blank spaces are $O(n^{-3/2})$ or smaller, as are the terms implicitly omitted. The constants α_3, α_4 and γ are the derivatives of $\log f(x;\varphi)$ at $x=0$ and $\varphi=0$:

$$\begin{aligned}\alpha_3 &= -\left.\frac{\partial^3 \log f(x;\varphi)}{\partial\varphi^3}\right|_{x=0,\varphi=0},\\ \alpha_4 &= -\left.\frac{\partial^4 \log f(x;\varphi)}{\partial\varphi^4}\right|_{x=0,\varphi=0},\\ \gamma &= \left.\frac{\partial^4 \log f(x;\varphi)}{\partial x^2\partial\varphi^2}\right|_{x=0,\varphi=0},\end{aligned}$$

and γ is related to the exponential curvature of the model (Efron, 1975). Ignoring terms of $O(n^{-3/2})$, the expansion (10.39) in terms of x and φ is almost that of an exponential family model; the only additional coefficient is the $(2,2)$ entry γ/n, which adds a term $\gamma\varphi^2x^2/(4n)$ to the log likelihood expansion.

The variables s and θ are both scaled and centered as part of the transformation to x and φ: note that the observed information $-\partial^2 \log f(0;0)/\partial\varphi^2 = -b_{20} = 1$. The point $(x,\varphi) = (0,0)$ corresponds to the original point of expansion $(s^o, \widehat{\theta}^o)$, where s^o is the observed value and $\widehat{\theta}^o = \widehat{\theta}(s^o)$ is the corresponding value of the maximum likelihood estimator.

Another way to write the model given by (10.39) is

$$\log f(x;\varphi) = b_{00} + P_{1n}(x) + P_{2n}(\varphi) + x\varphi + \gamma x^2\varphi^2/(4n) + O(n^{-3/2}),$$

where $P_{1n}(x)$ is given by the first row of the array and $P_{2n}(\varphi)$ by its first column, each omitting b_{00}. On examining the elements of (10.39) we see that we can write

$$f(x;\varphi) \propto \phi(x-\varphi)\left\{1 + \frac{a_1(x,\varphi)}{n^{1/2}} + \frac{a_2(x,\varphi)}{n} + \gamma\frac{a_3(x,\varphi)}{n} + O(n^{-3/2})\right\}, \qquad (10.40)$$

in terms of the standard normal density function ϕ and suitable polynomials a_1, a_2 and $a_3(x,\varphi) = (x^2\varphi^2 - x^4 + 5x^2 - 2)/4$. Equation (10.40) can be integrated term by term with respect to x; an explicit array for the resulting approximate distribution function $F(x;\varphi)$ is given in Andrews et al. (2005). Remarkably, although $F(x;\varphi)$ depends on γ in general, $F(0;\varphi)$ does not depend on γ, because $\phi(x-\varphi)a_3(x;\varphi)$ has integral zero over the negative half-line. As $x=0$ corresponds to $s=s^o$, the significance function $F(s^o;\theta)$ does not depend on γ and can be computed using the exponential family version of (10.39), in which $\gamma=0$.

The tangent exponential model (10.8) is just an invariant version of the simplified expansion (10.40): s is now the transformed variable called x here, and is the corresponding score variable, and φ is by definition $\partial \log f_{\text{TEM}}(0;0)/\partial x$, the canonical parameter of the exponential model approximation.

The steps in going from the original model to (10.39) are outlined in Andrews et al. (2005) and Cakmak et al. (1998).

The reduction to a single variable s in a scalar parameter model is straightforward if, for example, $y = (y_1,\ldots,y_n)$ is a sample from an exponential family model, with density function (10.17), as the log likelihood $\ell(\theta;s)$ then equals $\exp\{\varphi(\theta)s - nc(\theta)\}$, where $s = \sum_{i=1}^n s(y_i)$, and has the dependence on n summarised in (10.39), with $\gamma=0$.

Similarly, in the case of a sample $y = (y_1,\ldots,y_n)$ from a location model, the exact distribution of any location-invariant estimator, say s, of the location parameter θ given the location ancillary statistic $a = (y_1 - s,\ldots,y_n - s)$ is

$$f(s \mid a;\theta) = \exp\{\ell(\theta;s,a)\} \Big/ \int_t \exp\{\ell(t;s,a)\}\,dt,$$

and (10.39) is equivalent to the density approximation arising when Laplace's method is applied to the denominator integral.

For more general models, the discussion in Section 10.3 establishes the existence of a conditional distribution on $\mathbb{R}$ that can be determined by finding the $n \times 1$ vector V of sufficient directions. The arguments above show that this conditional distribution, which now has a scalar variable and scalar parameter, is effectively an exponential family for the purpose of approximating the significance function.

10.6.3 Nuisance Parameters

The expansion in (10.39) can be generalised to vector parameters, as in Cakmak et al. (1994) and Fraser and Reid (1993), but the notation is cumbersome, and the various multi-dimensional analogues to $\alpha_3, \alpha_4, \gamma$ are not explicitly available. However, the expansion verifies that the coordinate-free version of the tangent exponential model has the form given at (10.8) and (10.35), with saddlepoint approximation (10.37).

This gives a tangent exponential model on $\mathbb{R}^p$ for inference about θ, which is implicitly conditioned on an approximate ancillary statistic through the use of the $n \times p$ matrix V, and this is now the full model used to obtain an approximate significance function for a scalar parameter of interest ψ.

In this full model, consider fixing ψ, and constructing a new tangent exponential model on $\mathbb{R}^{p-1}$ with parameter λ. We can write, suppressing the conditioning on a,

$$f_{\mathrm{TEM}}(s;\theta) = f_{1,\mathrm{TEM}}(s_\psi \mid \tilde{a}_\psi;\lambda) f_2(\tilde{a}_\psi), \tag{10.41}$$

where $\tilde{a}_\psi$ is a new approximate ancillary statistic for the model with ψ held fixed. This gives us a one-dimensional distribution for inference about ψ,

$$f_2(\tilde{a}_\psi) = f_{\mathrm{TEM}}(s;\theta)/f_{1,\mathrm{TEM}}(s_\psi \mid \tilde{a}_\psi;\lambda),$$

and as we know the left-hand side is free of both s_ψ and λ, we can choose $s_\psi = 0$ and $\lambda = \widehat{\lambda}_\psi$. Using the saddlepoint form (10.37) of the tangent exponential model in the numerator and denominator yields a model on $\mathbb{R}$ of the form

$$h(s;\psi) = c\exp\{\ell(\hat{\varphi}_\psi;s) - \ell(\hat{\varphi};s)\}|\jmath_{\varphi\varphi}(\hat{\varphi})|^{-1/2}|\jmath_{(\lambda\lambda)}(\hat{\varphi}_\psi)|^{1/2}, \quad s \in \mathcal{L}_\psi, \tag{10.42}$$

where $\ell(\varphi;s) = s^{\mathrm{T}}\varphi + \ell(\varphi;y^{\mathrm{o}})$, and $\mathcal{L}_\psi$ is a line in the sample space corresponding to fixing $\widehat{\lambda}_\psi$ (and a). Expressing the result with a constraint on s avoids explicit identification of $\tilde{a}_\psi$: it is enough to know that it exists. As in Section 10.4.3 the information determinants are computed in the φ parameterization; see (10.33).

The right-hand-side of (10.42) has the form of our original tangent exponential model (10.37), with an adjustment factor in the ratio of determinants; note also the similarity to the approximate conditional density (10.26) for linear exponential families. As a result, the approximate significance function is the same as that for general exponential families outlined in Section 10.4.3, with the significance function as in (10.24) or (10.25), with r defined in (10.27), and q defined in (10.34). Once the tangent exponential approximation to the original model has been established, the exponential model formulas apply directly.

It would be natural to partition s into a component related to ψ and one related to λ, and this is how the result is presented in Fraser and Reid (1995, Sec. 6). In later work (Reid and Fraser, 2010; Fraser et al., 2016b) the simpler notation of (10.42) is preferred, with a constraint on s to emphasise that the density is for a variable of the same dimension as the parameter of interest ψ.

10.7 Concluding Remarks

10.7.1 Summary

The tangent exponential model and associated significance function implement inference conditional on an approximate ancillary statistic, followed by marginalization to a pivotal quantity, $r^*(\psi; y)$, for a scalar parameter of interest. This pivot is readily computed using only $\ell(\theta; y^o)$ and $\varphi(\theta; y^o) = \ell_{;V}(\theta; y^o)$, and the full and constrained maximum likelihood estimators $\widehat{\theta}$ and $\widehat{\theta}_\psi$. Fraser et al. (1999) and Reid (2003) present this "inference algorithm" as two dimension-reduction steps: from a model $f(y; \theta)$ on $\mathbb{R}^n$ to a model $f(s \mid a; \theta)$ on $\mathbb{R}^p$, by conditioning, and from this model to another on $\mathbb{R}$, by marginalizing. The model on $\mathbb{R}$ can be approximated by a simple standard normal distribution for the pivotal quantity $r^*(\psi; y)$, and in continuous models the approximation to the significance function based on $r^*(\psi; y^o)$ has relative error $O(n^{-3/2})$.

The final approximation step is somewhat separate from the development of the model on $\mathbb{R}$, and follows closely the derivation of the r^* approximation in Barndorff-Nielsen (1986). It can also be applied in other contexts, and in particular to approximation of a Bayesian posterior survivor function, starting from the the Laplace approximation to the posterior marginal density (Tierney and Kadane, 1986).

As our focus here is on the steps leading to the tangent exponential model and their implications for inference, we have not included numerical work indicating the accuracy of the approximations. There are many examples and exercises in Brazzale et al. (2007), in the literature referred to there and in Brazzale and Davison (2008).

There is a close relation between the r^* approximation and the parametric bootstrap; the higher-order properties of the latter are investigated in DiCiccio and Young (2008), Lee and Young (2005), and DiCiccio et al. (2015). To achieve the same order of accuracy it is necessary to bootstrap under the constrained maximum likelihood estimate $(\psi, \widehat{\lambda}_\psi)$, which increases the computational burden. Fraser and Rousseau (2008) also consider the relationship between significance functions based on the parametric bootstrap, on r^*, and on Bayesian versions of predictive p-values.

10.7.2 Extensions

If the parameter of interest is a vector, a significance function is not easily obtained unless one can construct a scalar measure of departure such as the log likelihood ratio statistic $w(\psi) = 2\{\ell(\widehat{\theta}) - \ell(\widehat{\theta}_\psi)\}$, Wald statistic $(\widehat{\psi} - \psi)^{\mathrm{T}} j^{\psi\psi}(\widehat{\theta})(\widehat{\psi} - \psi)$, or score statistic, each to first order approximately distributed as χ^2_d. Davison et al. (2014) and Fraser et al. (2016b) use the tangent exponential model as the building block for a directional approach to inference for a d-dimensional parameter ψ which creates a univariate summary, by considering the magnitude of ψ conditional on its direction from a null value ψ_0. The saddlepoint approximation to $f_{\mathrm{TEM}}(s \mid a; \theta)$ on this line in the sample space forms the basis for inference. A new scalar-parameter exponential family is constructed from the multi-parameter exponential family model or the approximating tangent exponential model.

Our discussion has presumed that the underlying data are independent, but the geometric motivation in Section 10.3 suggests that the approach should provide improved accuracy more generally. Belzile and Davison (2022) adapt the approach for discrete responses to the inhomogeneous Poisson process, but this is a special case owing to its independence properties. The main difficulty in broader settings is to compute V, and from this the constructed parameter $\varphi(\theta)$. In a time series setting, a series of pivotal quantities may be generated from the predictive distributions $F(y_j \mid y_{j-1},\ldots,y_1;\theta)$ for $j = 2,\ldots,n$, using martingale differences or a lower triangular square root of the covariance matrix for the response (Fraser et al., 2005; Lozada-Can and Davison, 2010). It is not yet clear whether other decompositions of the covariance matrix would lead to asymptotically equivalent results.

There is an r^* approximation for Bayesian inference, readily obtained from the Laplace approximation, as mentioned above; see also Fraser et al. (1999). This provides a route to examining the discrepancy between posterior survivor functions and significance functions. Equating the two versions of r^* leads to a data-dependent prior that ensures agreement of the significance and survivor functions up to terms of $O(n^{-1})$. The former was emphasised in Fraser (2011) and Fraser et al. (2016a); the latter formed the basis for a discussion of default priors in Fraser et al. (2010b).

10.7.3 A Brief Historical Note

Fraser viewed the dimension-reduction steps in Section 10.7.1 as essentially unique, and consequently the pivotal quantity r^*_ψ not as an arbitrary choice among several potential pivotal quantities, but as the only route to higher-order approximation for a scalar parameter in the presence of nuisance parameters:

> This ancillary density is uniquely determined by steps that retain continuity of the model in the derivation of the marginal distribution. It thus provides the unique null density for assessing a value $\psi = \psi_0$, and anyone suggesting a different null distribution would need to justify inserting discontinuity where none was present (Fraser, 2017, Section 4).

The continuity referred to there is the presumption that changes in θ are smoothly related to changes in y and vice-versa, as in a pure location model $f(y - \theta)$. The vectors V determining the tangent plane to the ancillary surface are based on the local location model defined in Fraser (1964). Suppose y_i has density $f(y_i;\theta)$ and cumulative distribution function $F(y_i;\theta), \theta \in \mathbb{R}$. Define a transformation $y_i \mapsto x_i$ by setting

$$x_i = \int^{y_i} -\frac{F_\theta(y;\theta_0)}{F_y(y;\theta_0)}\,\mathrm{d}y,$$

where θ_0 is some fixed value, and write the density of x_i as $g(x_i;\theta)$. Then $g\{x_i-(\theta-\theta_0);\theta_0\}$ coincides with the true density at θ_0, as does its derivative $g_x(x_i;\theta_0)$. For a sample $y_1,\ldots,y_n$, this local location model has an ancillary statistic $(x_1-\bar{x},\ldots,x_n-\bar{x})$, and sufficient direction $(1,\ldots,1)$, which transforms back to the sufficient direction

$$V^{\mathrm{T}} = -\left(\frac{F_\theta(y_1;\theta_0)}{F_y(y_1;\theta_0)},\ldots,\frac{F_\theta(y_n;\theta_0)}{F_y(y_n;\theta_0)}\right)$$

in terms of $y_1,\ldots,y_n$; see (10.16). As noted in Fraser and Reid (2001), this construction does not give a local location model for the full sample $y_1,\ldots,y_n$ because the vector field $V(y)$ is not guaranteed to be integrable. But the expansions in that paper verify that the approximations derived from the tangent exponential model are still valid, as the sufficient directions V describe the same tangent plane as a second-order ancillary statistic that exists under mild regularity conditions. Fraser and Reid (2001, Section 10.3) promised that "the integrability of the $V(y)$ to the required order will be examined elsewhere", and this was fulfilled in Fraser et al. (2010a).

Fraser viewed as intrinsically linked the construction of the tangent exponential model, the application of the saddlepoint approximation and the construction of significance functions as key inferential summaries. A first, lengthy, paper written shortly after the simpler developments in Fraser (1988, 1990, 1991) included all these pieces, and was met with some puzzlement by editors and reviewers: one reviewer advised "it should probably be several papers" — a reaction that might be rather unusual nowadays. This led to the asymptotic expansions being the focus of Fraser and Reid (1993), although much of the original draft was published in Fraser and Reid (1995). That latter paper derived the tangent exponential approximation to general models, derived the directional vectors V from a local location model, showed the existence of a second-order ancillary statistic with the same directional vectors, verified that the dimension of this ancillary is fixed as $n \to \infty$, and derived the r^* approximation in its general form. The construction of the directional vectors was discussed in more detail in Fraser and Reid (2001), which is confusingly referred to in some of his papers as Fraser and Reid (1999).

The annotations in the bibliography below attempt to provide a road map through the most relevant of these papers. Copies of the less readily accessible ones are posted at

`https://utstat.toronto.edu/reid/fraser-papers.html`

Acknowledgements

The work was supported by the Swiss National Science Foundation and the Natural Sciences and Engineering Council of Canada. We thank Léo Belzile, Yanbo Tang and two anonymous referees for helpful comments.

Bibliography

Andrews, D. A., Fraser, D. A. S. and Wong, A., 2005. Computation of distribution functions from likelihood information near observed data. *Journal of Statistical Planning and Inference, 134*, pp. 180–193. *Detailed derivation of Taylor series expansions for density and distribution functions. Corrections to some formulae available at https://utstat.toronto.edu/dfraser/documents/210Typo-corr.pdf.*

Barndorff-Nielsen, O. E., 1980. Conditionality resolutions. *Biometrika, 67*, pp. 293–310.

Barndorff-Nielsen, O. E., 1983. On a formula for the distribution of the maximum likelihood estimator. *Biometrika, 70*, pp. 343–365.

Barndorff-Nielsen, O. E., 1986. Inference on full or partial parameters based on the standardized signed log likelihood ratio. *Biometrika, 73*, pp. 307–322.

Barndorff-Nielsen, O. E. and Cox, D. R., 1989. *Asymptotic Techniques for Use in Statistics.* London: Chapman & Hall.

Barndorff-Nielsen, O. E. and Cox, D. R., 1994. *Inference and Asymptotics.* London: Chapman & Hall.

Belzile, L. R. and Davison, A. C., 2022. Improved inference on risk measures for univariate extremes. *Annals of Applied Statistics, 16*, pp. 1524–1549.

Brazzale, A. R. and Davison, A. C., 2008. Accurate parametric inference for small samples. *Statistical Science, 23*, pp. 465–484.

Brazzale, A. R., Davison, A. C. and Reid, N., 2007. *Applied Asymptotics: Case Studies in Small Sample Statistics.* Cambridge: Cambridge University Press.

Butler, R. W., 2007. *Saddlepoint Approximations with Applications.* Cambridge: Cambridge University Press.

Cakmak, S., Fraser, D. A. S., McDunnough, P., Reid, N. and Yuan, X., 1998. Likelihood centered asymptotic model: Exponential and location model versions. *Journal of Statistical Planning and Inference, 66*, pp. 211–222. *One of a series of papers giving expansions of likelihood quantities in various versions; here for a scalar parameter and scalar response.*

Cakmak, S., Fraser, D. A. S. and Reid, N., 1994. Multivariate asymptotic model: Location and exponential approximations. *Utilitas Mathematica, 46*, pp. 21–31. *Extends the expansion given in 10.6.1 above to p-dimensional parameter and p-dimensional response. A location model version is also developed, which was used in the development of so-called default priors.*

Cox, D. R., 1958. Some problems connected with statistical inference. *Annals of Mathematical Statistics, 29*, pp. 357–372.

Daniels, H. E., 1954. Saddlepoint approximations in statistics. *Annals of Mathematical Statistics, 25*, pp. 631–650.

Daniels, H. E., 1956. The approximate distribution of serial correlation coefficients. *Biometrika, 43*, pp. 169–185.

Davison, A. C., 2003. *Statistical Models.* Cambridge: Cambridge University Press.

Davison, A. C., Fraser, D. A. S. and Reid, N., 2006. Improved likelihood inference for discrete data. *Journal of the Royal Statistical Society: Series B, 68*, pp. 495–508. *Derives an approximation to V based on the derivative of the expected value of the response, as described in Section 10.5.2.*

Davison, A. C., Fraser, D. A. S., Reid, N. and Sartori, N., 2014. Accurate directional inference for vector parameters in linear exponential families. *Journal of the American Statistical Association , 109*, pp. 302–314. *Uses the tangent exponential model and a conditional argument to construct approximations to P-values for assessing a vector parameter of interest.*

DiCiccio, T. J., Kuffner, T. A., Young, G. A. and Zaretzki, R., 2015. Stability and uniqueness of p-values for likelihood-based inference. *Statistica Sinica, 25*, pp. 1355–1376.

DiCiccio, T. J. and Young, G. A., 2008. Conditional properties of unconditional parametric bootstrap procedures for inference in exponential families. *Biometrika, 95*, pp. 747–758.

Edwards, A. W. F., 1972. *Likelihood.* Cambridge: Cambridge University Press.

Efron, B., 1975. Defining the curvature of a statistical problem (with applications to second order efficiency). *Annals of Statistics, 3*, pp. 1189–1242.

Efron, B., 1993. Bayes and likelihood calculations from confidence intervals. *Biometrika, 80*, pp. 3–26.

Fraser, A. M., Fraser, D. A. S. and Staicu, A.-M., 2010a. Second order ancillary: A differential view from continuity. *Bernoulli, 16*, pp. 1208–1223. *Resolves concerns about non-uniqueness of the approximate ancillary determined using the tangent exponential model.*

Fraser, D. A. S., 1964. Local conditional sufficiency. *Journal of the Royal Statistical Society: Series B, 26*, pp. 52–62.

Fraser, D. A. S., 1988. Normed likelihood as saddlepoint approximation. *Journal of Multivariate Analysis, 27*, pp. 181–193. *Studies the relationship of the* p^* *and saddlepoint approximations for the distribution of the score function. Introduction of the tangent exponential model for scalar* y *and scalar* θ.

Fraser, D. A. S., 1990. Tail probabilities from observed likelihoods. *Biometrika, 77*, pp. 65–76. *Shows the relationship of the* p^* *and saddlepoint approximations in exponential families, and discusses how this generalizes. Gives an expression for the tangent exponential model when* y *and* θ *have the same dimension.*

Fraser, D. A. S., 1991. Statistical inference: Likelihood to significance. *Journal of the American Statistical Association, 86*, pp. 258–265. *Based on Fisher Lecture at 1990 Joint Statistical Meetings; proposes significance functions as an encompassing inferential method.*

Fraser, D. A. S., 2004. Ancillaries and conditional inference (with discussion). *Statistical Science, 19*, pp. 333–369. *Detailed discussion of conditioning as the main focus for inference, with many examples of exact ancillary statistics and a discussion of the approximate ancillarity underlying the tangent exponential model.*

Fraser, D. A. S., 2011. Is Bayes posterior just quick and dirty confidence? *Statistical Science, 26*, pp. 299–316. *Argues that the use of Bayesian inference with convenience priors could be adequate to first order but potentially misleading in finite samples.*

Fraser, D. A. S., 2017. p-values: The insight to modern statistical inference. *Annual Review of Statistics and Its Application, 4*, pp. 1–14. *Fraser continued to simplify and refine his approach, here giving a concise but clear account of the material covered in this chapter.*

Fraser, D. A. S., 2019. The p-value function and statistical inference. *The American Statistician, 73*, pp. 135–147. *This is in a special issue on the role of "*$p < 0.05$*" in contributing to a lack of replicability of scientific work. This paper emphasizes the significance, or p-value, function as preferable to a single p-value.*

Fraser, D. A. S., Bédard, M., Wong, A., Lin, W. and Fraser, A. M., 2016a. Bayes, reproducibility and the quest for truth. *Statistical Science, 31*, pp. 578–590.

Fraser, D. A. S. and McDunnough, P., 1984. Further remarks on asymptotic normality of likelihood and conditional analyses. *Canadian Journal of Statistics, 12*, pp. 183–190.

Fraser, D. A. S. and Reid, N., 1988. On conditional inference for a real parameter: A differential approach on the sample space. *Biometrika, 75*, pp. 251–264.

Fraser, D. A. S. and Reid, N., 1993. Third order asymptotic models: Likelihood functions leading to accurate approximations to distribution functions. *Statistica Sinica, 3*, pp. 67–82. *Taylor expansion of §10.6.1 first published; also a multivariate version. In the scalar parameter case the* r^* *approximation based on the tangent exponential model established. Builds on Fraser (1990) where the* p^* *and tangent exponential model approximations are* *connected, for scalar y and scalar θ. Nuisance parameters tackled in linear exponential and location models; suggests using r^* based on either approximate conditional log likelihood or approximate marginal log-likelihood.*

Fraser, D. A. S. and Reid, N., 1995. Ancillaries and third order significance. *Utilitas Mathematica, 47,* pp. 33–53. *Most of the main theoretical results first appear here, re-worked in later papers. Uses the Taylor expansion as in Fraser and Reid (1993), §10.6.1 and Andrews et al. (2005) to derive the saddlepoint version of the tangent exponential model. Shows that only second-order ancillarity is needed for third-order inference, following Skovgaard (1986); that V gives a tangent plane for a fixed ancillary to $O(n^{-1/2})$, but can be 'bent' to get ancillary to $O(n^{-1})$ without changing V, i.e., there exists an $O(n^{-1})$ ancillary W, but V is adequate. Shows the relation to local location model in (Fraser, 1964). An elegant argument in Section 3 shows that the dimension of the ancillary statistic does not grow with sample size.*

Fraser, D. A. S. and Reid, N., 2001. Ancillary information for statistical inference. In *Empirical Bayes and Likelihood Inference*, eds S. E. Ahmed and N. Reid, pp. 185–207. New York: Springer. *Clarifies the development of approximate ancillarity in Fraser and Reid (1995), later put on a more rigorous footing in Fraser et al. (2010a).*

Fraser, D. A. S., Reid, N., Marras, E. and Yi, G. Y., 2010b. Default priors for Bayesian and frequentist inference. *Journal of the Royal Statistical Society: Series B, 72,* pp. 631–654.

Fraser, D. A. S., Reid, N. and Sartori, N., 2016b. Accurate directional inference for vector parameters. *Biometrika, 103,* pp. 625–639.

Fraser, D. A. S., Reid, N. and Wu, J., 1999. A simple general formula for tail probabilities for frequentist and Bayesian inference. *Biometrika, 86,* pp. 249–264. *Spells out the dimension reduction from n to p to 1; compares the r^* approximation using the tangent exponential model to the approach of Barndorff-Nielsen (1986). Derives the Bayesian version of r^*. Also considers models where the parameter of interest is defined implicitly via a constraint on the full parameter θ. Provides several numerical examples.*

Fraser, D. A. S., Rekkas, M. and Wong, A., 2005. Highly accurate likelihood analysis for the seemingly unrelated regression problem. *Journal of Econometrics, 127,* pp. 17–33.

Fraser, D. A. S. and Rousseau, J., 2008. Studentization and deriving accurate p-values. *Biometrika, 95,* pp. 1–16.

Hinkley, D. V., 1980. Likelihood as approximate pivotal distribution. *Biometrika, 67,* pp. 287–292.

Jensen, J. L., 1995. *Saddlepoint Approximations.* Oxford: Oxford University Press.

Kolassa, J., 2006. *Series Approximation Methods in Statistics* (3rd ed.). New York: Springer.

Krantz, S. G. and Parks, H. R., 2008. *Geometric Integration Theory.* Boston: Birkhaüser.

LeCam, L., 1960. Locally asymptotically normal families of distributions. *University of California Public Statistics, 3,* pp. 27–98.

Lee, S. M. S. and Young, G. A., 2005. Parametric bootstrapping with nuisance parameters. *Statistics & Probability Letters, 71,* pp. 143–153.

Lozada-Can, C. and Davison, A. C., 2010. Three examples of accurate likelihood inference. *American Statistician, 64,* pp. 131–139.

Lugannani, R. and Rice, S., 1980. Saddlepoint approximation for the distribution of the sum of independent random variables. *Advances in Applied Probability, 12*, pp. 475–490.

McCullagh, P., 1987. *Tensor Methods in Statistics.* London: Chapman & Hall.

Reid, N., 2003. Asymptotics and the theory of inference. *Annals of Statistics, 31*, pp. 1695–1731.

Reid, N., 2005. Asymptotics and the theory of statistics. In *Celebrating Statistics: Papers in Honour of D.R. Cox*, eds A. C. Davison, Y. Dodge and N. Wermuth, pp. 73–88. Oxford: Oxford University Press.

Reid, N. and Fraser, D. A. S., 2010. Mean loglikelihood and higher-order approximations. *Biometrika, 97*, pp. 159–170. *Compares inference based on* r^* *using* $\varphi(\theta) = \ell_{;V}(\theta; y^o)$ *to the method proposed by (Skovgaard, 1996), which can be formulated as the same approximation using a parametrization* $\bar{\varphi}(\theta)$ *that is constructed from likelihood cumulants. The Appendix attempts to clarify the derivation of the tangent exponential model.*

Royall, R. M., 1997. *Statistical Evidence: A Likelihood Paradigm.* London: Chapman & Hall/CRC.

Severini, T. A., 2000. *Likelihood Methods in Statistics.* Oxford: Clarendon Press.

Skovgaard, I. M., 1986. Successive improvement of the order of ancillarity. *Biometrika, 73*, pp. 516–519.

Skovgaard, I. M., 1996. An explicit large-deviation approximation to one-parameter tests. *Bernoulli, 2*, pp. 145–165.

Tierney, L. and Kadane, J. B., 1986. Accurate approximations for posterior moments and marginal densities. *Journal of the American Statistical Association, 81*, pp. 82–86.

van der Vaart, A. W., 1998. *Asymptotic Statistics.* Cambridge: Cambridge University Press.

Xie, M. and Singh, K., 2013. Confidence distributions: The frequentist distribution estimator of a parameter. *International Statistical Review, 81*, pp. 3–39.

11

Data Integration and Model Fusion in the Bayesian and Frequentist Frameworks

Emily C. Hector
North Carolina State University

Lu Tang
University of Pittsburgh

Ling Zhou
Southwestern University of Finance and Economics

Peter X.-K. Song
University of Michigan

11.1 Introduction

Current statistical practice is greatly challenged by rapid advances in technologies that enable practitioners to frequently collect massive and complex data to study various problems of practical importance. Scalability, studied and implemented in the divide-and-conquer framework, is a key feature of statistical methods developed to address the computational needs of modern data analysis [27]. Divide-and-conquer strategies generally proceed in three steps: first, the whole data is divided into smaller and more manageable subsets, or blocks, of data; then each data block is analyzed in parallel; finally, results from data block analyses are combined under some optimality criterion. In contrast, many practical studies, such as genetic studies, attempt to combine datasets collected under similar study protocols to form a mega dataset in order to increase the sample size and variety of features for high statistical power and generalizability. In both cases, the essential statistical difficulty concerns data integration and model fusion.

In this chapter, our primary goal is to introduce and review statistical methods related to data integration and model fusion in the context of regression analysis. Regression analysis has played a central role in studying relationships between variables in practice. Data integration and model fusion methods have become very popular in the recent statistical literature, and our review is not meant to be exhaustive; rather, we aim to provide a reader with enough background and insight to formulate and evaluate existing methods in the Bayes and Frequentist (BF) paradigm. Specifically, in this chapter, data integration refers to the process of combining multiple datasets that are generated from a class of homogeneous statistical models, such as regression models. By model fusion, we mean specifically a process of merging models with both homogeneous and heterogeneous parameters that generate

 DOI: 10.1201/9780429341731-11

individual nonhomogeneous datasets; the fusion technique enables the integration of relevant information to yield more efficient statistical estimation and inference for homogeneous parameters.

To be precise, let us introduce notation. We consider a collection of D data sources $\{\boldsymbol{Y}_{i,d}, \boldsymbol{X}_{i,d}\}_{i,d=1}^{n_d,D}$, $D \in \mathbb{N}$, where $\boldsymbol{Y}_{i,d} = (Y_{i1,d}, \dots, Y_{iM_d,d}) \in \mathbb{R}^{M_d}$ is the ith response and $\boldsymbol{X}_{i,d} \in \mathbb{R}^{M_d \times p}$ the corresponding deterministic covariate matrix in data source $d \in \{1, \dots, D\}$. The primary goal is to combine these data sources to create a mega dataset for an integrative analysis or a meta-analysis. In specifying a regression model, we assume that the random component of the model is given by $\boldsymbol{Y}_{i,d}|\boldsymbol{X}_{i,d} \overset{iid}{\sim} f(\boldsymbol{y}|\mathbf{x}, \boldsymbol{\beta}_d), i = 1, \dots, n_d$, for $d = 1, \dots, D$, where f is a certain M_d-dimensional density function, such as an M_d-variate normal distribution for the linear model or M_d-variate dispersion distribution generated from a copula for the vector generalized linear model [48]. We highlight the differences between data integration and model fusion:

i. Data integration refers to the task of estimation and inference on the homogeneous model parameter $\boldsymbol{\beta} :\equiv \boldsymbol{\beta}_d, d = 1, \dots, D$ with the merged data. Although data integration is similar to the classical meta-analysis, the difference between these two types of analyses is clear. In the integrative data analysis, presumably one can access the raw data of individual datasets, whereas only summary statistics are accessible in a meta-analysis.

ii. Model fusion (sometimes also known as data harmonization) can be broadly categorized as fusion learning or model parameter fusion. Typically, dataset-specific parameters are partitioned into two or more subvectors, where a fusion procedure first evaluates the compatibility of model parameters across different studies and then fuses similar parameters to reach model parsimony for better interpretations. As an example, let us consider a simple partition into two subvectors $\boldsymbol{\beta}_d = (\boldsymbol{\beta}_{1d}, \boldsymbol{\beta}_2)$, where the first parameter subvector $\boldsymbol{\beta}_{1d}$ remains heterogeneous while the second parameter subvector $\boldsymbol{\beta}_2$ is assumed to be the same across the D datasets for the purpose of dimension reduction. Model fusion consists of two steps: a clustering analysis of parameters, and parameter estimation. Frequently in practice, homogeneity of the second subvector $\boldsymbol{\beta}_2$ holds only in some of the D datasets, rather than in all the D datasets. Thus, a technical challenge in fusion learning is to identify the subset of the D datasets over which parameter homogeneity is present and then estimate $\boldsymbol{\beta}_2$ by combining the corresponding datasets to reach higher statistical power than an analysis based on individual datasets. In a general model fusion analysis, both the size of the subvector and the homogeneous members of the parameter vector are unknown and have to be learned from the available datasets by clustering analysis. Due to this need, this model fusion analysis is sometimes also called *fusion learning*. Unsurprisingly, this parameter subgroup fusion should be conducted for each element of $\boldsymbol{\beta}$, which can be very computationally expensive when the dimension of $\boldsymbol{\beta}$ is not small. For more details, refer to [51, 53, 44], among others, and additional references therein.

The statistical methodology discussed in this chapter describes an integrative data analysis in the situation where the D data sources are the result of the first step of a divide-and-conquer procedure applied to one large dataset or correspond to datasets from a multi-center study. Clearly, $M_d = 1$ corresponds to a univariate outcome and $M_d > 1$ to a multivariate outcome. For simplicity, we consider the marginal linear regression model; that is, in data source d, the regression model takes the form: $E(\boldsymbol{Y}_{i,d}) = \boldsymbol{X}_{i,d}\boldsymbol{\beta}_d$ for subject $i = 1, \dots, n_d$

in data source d. The analytic goal is to integrate estimators of $\boldsymbol{\beta}_d$ from each data source in a suitable way to increase statistical power. Optimizing a fusion procedure depends on the characteristics of the data sources, such as homogeneity/heterogeneity patterns and the dependence structure of the data sources. In the case of the linear model with a one-dimensional outcome ($M_d = 1$), we typically estimate $\boldsymbol{\beta}_d$ by the ordinary least-squares estimator $\widehat{\boldsymbol{\beta}}_d = (\boldsymbol{X}_d^T \boldsymbol{X}_d)^{-1} \boldsymbol{X}_d \boldsymbol{Y}_d$, where $\boldsymbol{X}_d = (\boldsymbol{X}_{1,d}^T, \ldots, \boldsymbol{X}_{n_d,d}^T) \in \mathbb{R}^{p \times d}$. In the case of the multivariate linear model for correlated data ($M_d > 1$), several estimators of $\boldsymbol{\beta}_d$ are available, such as the generalized estimating equations (GEE) estimator $\widehat{\boldsymbol{\beta}}_d$ [32], which solves the estimating equation:

$$\boldsymbol{\Psi}(\boldsymbol{\beta}_d) = \sum_{i=1}^{n_d} \boldsymbol{X}_{i,d}^T \boldsymbol{V}_i^{-1} (\boldsymbol{Y}_{i,d} - \boldsymbol{X}_{i,d} \boldsymbol{\beta}_d) = 0, \tag{11.1}$$

where $\boldsymbol{V}_i = \boldsymbol{A}_i^{1/2} \boldsymbol{R}(\alpha) \boldsymbol{A}_i^{1/2}$, $\boldsymbol{R}(\alpha)$ a working correlation matrix. When the data are assumed to be multivariate Normally distributed, the estimating function $\boldsymbol{\Psi}(\cdot)$ in Equation (11.1) is the score function. The solution of the GEE $\boldsymbol{\Psi}(\boldsymbol{\beta}) = 0$, however, does not require the outcomes to be normally distributed according to the more general pseudo-likelihood estimation framework. In this chapter, whenever applicable, we use the notation $\mathcal{N}$ to denote the univariate Normal distribution and $\mathcal{N}_k$ to denote the k-variate Normal distribution.

The remainder of this chapter is organized as follows, as shown in Figure 11.1. Readers may refer to Figure 11.1 to reach a particular topic of interest. In Section 11.2, we begin with a historical review of data integration methods that focuses on meta-analysis from the BF perspective. We then discuss the current state of the data integration and model fusion literature categorized by analytic goals and assumptions: Section 11.3 reviews methods for data integration with independent data sources, Section 11.4 for data integration with dependent data sources, and Section 11.5 for model fusion with independent data sources. We briefly review existing software implementations in Section 11.6. We conclude with a discussion and summary of future work in Section 11.7.

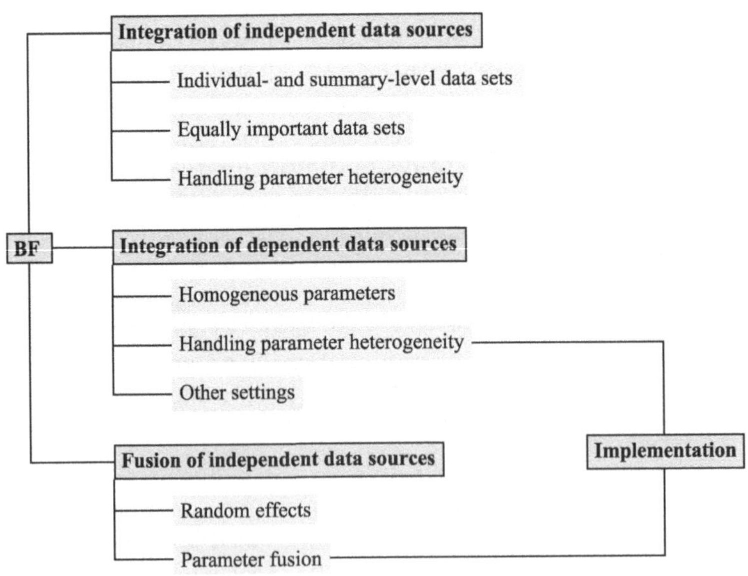

FIGURE 11.1
Organizational chart of this chapter.

11.2 Bayesian and Frequentist Perspectives

We begin with the familiar linear regression setting to illustrate the differences between the meta-estimation of regression coefficients within the Bayesian and frequentist approaches. The basic insights established in this section are helpful to understand more sophisticated data integration and model fusion techniques discussed in the subsequent sections. For ease of exposition, we focus on the meta-estimation of one regression coefficient in a one-dimensional regression model ($M_d = 1$) with Normally distributed errors.

Meta-analysis [18] describes a general methodology that appears as early as [9] and [10]. Meta-analysis assumes model parameters $\boldsymbol{\beta}_d$ are homogeneous across data sources and combines the results from individual data sources to provide one unified estimator $\widehat{\boldsymbol{\beta}}_{meta}$ for the regression coefficient vector $\boldsymbol{\beta}$. Recalling our motivating marginal linear regression example, this homogeneity assumption manifests itself as $\boldsymbol{\beta}_d \equiv \boldsymbol{\beta}$ for $d = 1, \ldots, D$. Meta-analysis does not require access to individual data sources, instead using only suitable summary statistics from each data source. This data-sharing advantage, coupled with its relative ease of computational implementation, partly explains why meta-analysis has been extensively studied in the statistical literature, why it remains incredibly popular in applied scientific domains and the proliferation of generalizations and extensions for different applications. Meta-analysis also generally produces estimates that are robust to outliers since it aggregates information from many data sources, a phenomenon sometimes called outlier dilution. Generally, both independence of the data sources $\{Y_{i,d}, \boldsymbol{X}_{i,d}\}_{i,d=1}^{n_d,D}$ and homogeneity of model parameters are key assumptions necessary for a meta-analysis to provide valid inference, in the sense that the meta-estimator is consistent and efficient for $\boldsymbol{\beta}$. We describe three approaches to meta-analysis from the Frequentist and Bayes perspectives and highlight the challenges when generalizing these perspectives to more complex settings.

Frequentist Fusion Learning. The Best Linear Unbiased Estimator (BLUE) is one of the earliest data integration solutions from the frequentist perspective when data sources are independent. The BLUE is derived from the Gauss-Markov estimation paradigm. For convenience, let us consider one element (β_j) of the parameter vector $\boldsymbol{\beta}$, denoted by a generic notation β, with suppression of subscript j unless otherwise noted. Suppose that the data source-specific estimators $\hat{\beta}_d, d = 1, \ldots, D$, of β are unbiased or asymptotically unbiased (i.e. consistent) as $n_d \to \infty$, and the variance is $Var(\hat{\beta}_d) = \sigma_d^2$. The variance σ_d^2 measures the precision of the estimator, and a smaller value of σ_d^2 corresponds to higher precision. The objective is to obtain a combined estimator that is (asymptotically) unbiased and has minimal variance, leading to efficient inference. The BLUE of β is

$$\begin{aligned} \hat{\beta}_{\text{BLUE}} &= \arg \min_{c_1, \ldots, c_D} Var(c_1\hat{\beta}_1 + \cdots + c_D\hat{\beta}_D), \\ &\text{subject to} \quad c_d \geq 0, d = 1, \ldots, D, \quad c_1 + \cdots + c_D = 1. \end{aligned} \tag{11.2}$$

The above constraint $c_1 + \cdots + c_D = 1$ ensures the (asymptotic) unbiasedness of the resulting meta-estimator $\hat{\beta}_{\text{BLUE}}$. It is easy to see that the above BLUE has a closed form expression given by

$$\hat{\beta}_{\text{BLUE}} = \frac{\sigma_1^{-2}\hat{\beta}_1 + \cdots + \sigma_D^{-2}\hat{\beta}_D}{\sigma_1^{-2} + \cdots + \sigma_D^{-2}}. \tag{11.3}$$

In the literature, the estimator in Equation (11.3) is also called the inverse variance weighted estimator. The weighting scheme in the BLUE is intuitively appealing. The weight for the d-th estimator $\hat{\beta}_d$ in the BLUE is $\sigma_d^{-2}/(\sigma_1^{-2} + \cdots + \sigma_D^{-2})$, which reflects the proportion of the contribution from the estimator $\hat{\beta}_d$ from the d-th data source. Thus, the influence of

$\hat{\beta}_d$ is down-weighted in the meta-estimator if it is a poor estimator with large variance or low precision. In this integration procedure, the raw data from individual data sources are not involved; rather, only the data source-specific summary statistics, including estimates and variances, are needed in the construction of the BLUE in Equation (11.3). Therefore, the BLUE can be easily implemented in a parallelized computing architecture using, for example, the MapReduce paradigm. The BLUE gives rise to a form of distributed fusion learning that is scalable to an arbitrary number of data sources. It is not straightforward, however, to generalize this BLUE to the case where the data sources are correlated, which will be discussed in Section 11.4.

Bayesian Fusion Learning. Bayesian fusion learning allows us to incorporate dependence and heterogeneity between data sources. Consider a hierarchical regression model that specifies a prior distribution for the model parameters of interest, similar to the use of random effects. That is, assume the data are generated from the following model:

$$\beta_j \sim \mathcal{N}(a_j, \tau_j^2), \quad j = 1, \ldots, p,$$
$$Y_{i,d}|\boldsymbol{\beta} \overset{ind}{\sim} \mathcal{N}(\boldsymbol{X}_{i,d}\boldsymbol{\beta}, \eta^2), \quad i = 1, \ldots, n_d, \; d = 1, \ldots, D,$$

and denote the prior distribution $f(\boldsymbol{\beta}) = \prod_j f(\beta_j)$. Each prior distribution describes the heterogeneity of the respective regression parameter β_j, $j = 1, \ldots, p$. Denote the classical OLS estimate of $\boldsymbol{\beta}$ from the d-th data source by $\hat{\boldsymbol{\beta}}_d = (\boldsymbol{X}_d^T\boldsymbol{X}_d)^{-1}\boldsymbol{X}_d^T\boldsymbol{Y}_d, d = 1, \ldots, D$, a sufficient statistic of $\boldsymbol{\beta}$. It follows that

$$\hat{\boldsymbol{\beta}}_d = (\hat{\beta}_{1,d}, \ldots, \hat{\beta}_{p,d})^T \sim \mathcal{N}_p(\boldsymbol{\beta}, \eta(\boldsymbol{X}_d^T\boldsymbol{X}_d)^{-1}), \quad d = 1, \ldots, D. \tag{11.4}$$

For a given one-dimensional parameter β_j of $\boldsymbol{\beta}$, $j = 1, \ldots, p$, there are D independent samples $\hat{\beta}_j$ that follow univariate Normal distributions with a common mean parameter β_j and variances $\sigma_{j,d}^2$, respectively. Here $\sigma_{j,d}^2$ is the j-th diagonal element of the covariance matrix $\eta(\boldsymbol{X}_d^T\boldsymbol{X}_d)^{-1}$. Thus, the posterior of β_j is proportional to

$$f(\beta_j|\hat{\boldsymbol{\beta}}_j) \propto -\sum_{d=1}^{D}\frac{1}{2\sigma_{j,d}^2}(\hat{\beta}_{j,d} - \beta_j)^2 - \frac{1}{2\tau_j^2}(\beta_j - a_j)^2.$$

It follows that the Bayes meta-estimator of β_j is

$$\begin{aligned}\hat{\beta}_{j,\text{bayes}} &= \frac{\sigma_{j,1}^{-2}\hat{\beta}_{j,1} + \cdots + \sigma_{j,D}^{-2}\hat{\beta}_{j,D} + \tau_j^{-2}a_j}{\sigma_{j,1}^{-2} + \cdots + \sigma_{j,D}^{-2} + \tau_j^{-2}} \\ &\overset{a_j=0}{=} \frac{\sigma_{j,1}^{-2}\hat{\beta}_{j,1} + \cdots + \sigma_{j,D}^{-2}\hat{\beta}_{j,D}}{\sigma_{j,1}^{-2} + \cdots + \sigma_{j,D}^{-2} + \tau_j^{-2}}.\end{aligned}$$

Clearly, this Bayesian approach to data integration can also be implemented in the distributed MapReduce framework. In this solution, the well-known shrinkage phenomenon is a consequence of forcing heterogeneous regression parameters toward the common hyper mean a_j in the prior controlled by the nonzero hyper variance parameter τ_j^2. When $\tau_j^2 \to \infty$, the heterogeneity diffuses with little concentration on the prior mean parameter a_j, and the Bayesian meta-estimator converges to the frequentist BLUE given in Equation (11.3). Moreover, the D data sources are implicitly assumed to be dependent, since they share a common random variable $\beta_j \sim N(a_j, \tau_j^2)$. This provides an intuition for the utility of random effects models in meta-analysis. When the number D of data sources is small, estimation of τ_j^2 may not be reliable because the inter-batch variability τ_j^2 may not be estimated properly.

Confidence Distribution Fusion Learning. The confidence distribution approach [64] provides a viable solution to data integration. As in the Frequentist Fusion Learning, for convenience, we consider one element (β_j) of the parameter vector $\boldsymbol{\beta}$, denoted by a generic notation β, with suppression of subscript j unless otherwise noted. From Equation (11.4), we have $\hat{\beta}_d \overset{ind.}{\sim} \mathcal{N}(\beta, \sigma_d^2)$, where we treat σ_d^2 as known (a suitable estimator may be substituted otherwise), $d = 1, \ldots, D$. Following [14], a confidence distribution for β is given by

$$f_{CD}(\beta) = \prod_{d=1}^{D} \phi(\hat{\beta}_d; \beta, \sigma_d^2) \propto -\sum_{d=1}^{D} \frac{1}{2\sigma_d^2}(\hat{\beta}_d - \beta)^2,$$

where $\phi(\cdot)$ denotes the standard Normal density function. [37] define the confidence distribution meta-estimator as the maximizer of the product of the Normal CD densities $f_{CD}(\beta)$:

$$\hat{\beta}_{CD} = \arg\max_{\beta} \prod_{d=1}^{D} \phi(\hat{\beta}_d; \beta, \sigma_d^2) = \arg\min_{\beta} \sum_{d=1}^{D} \frac{1}{2\sigma_d^2}(\hat{\beta}_d - \beta)^2. \tag{11.5}$$

It follows that

$$\hat{\beta}_{CD} = \frac{\sigma_1^{-2}\hat{\beta}_1 + \cdots + \sigma_D^{-2}\hat{\beta}_D}{\sigma_1^{-2} + \cdots + \sigma_D^{-2}}, \tag{11.6}$$

which is the same as the BLUE meta-estimator in Equation (11.3). Despite having the same form as (11.3), the two estimators are derived using different principles of data integration with different purposes. Using only the first two moments of the estimators, the BLUE estimator aims to optimize statistical power by minimizing the variance of the resulting meta-estimator. In contrast, relying on the inference distributions, the confidence distribution approach aims to integrate individual inferential estimators to yield an integrative inferential instrument. Using a similar formulation to that of maximum likelihood, the confidence distribution approach is conceptually more appealing and methodologically more flexible. For example, regularization can be introduced in the confidence distribution fusion learning approach, and the confidence distributions may take certain empirical distributions by bootstrap procedures. Besides the approach of multiplying CDs, which closely resembles multiplying likelihood functions in parametric inference, the confidence distribution fusion approach can also be used to define a general framework for combining p-value functions [46]; see for example [65] and [67], among many others. This second approach does not require parametric assumptions, hence can be applied to CDs obtained from nonparametric approaches, such as bootstrap. Recently, [35] considered a broader setting than that discussed in [36] and extended this framework to cover multiparameters.

We make a connection between the confidence distribution and fiducial inference originally proposed by [16] in an attempt to provide an inverse probability interpretation of parameters based on the observed data. Letting $\Phi(\cdot)$ the standard Normal cumulative distribution function, we define

$$U_d = \Phi\left\{\sigma_d^{-1}(\widehat{\beta}_d - \beta)\right\}.$$

By the probability integral transformation, U_d are independent and follow the Uniform distribution on the interval $(0, 1)$, and may therefore be regarded as pivotal quantities. From the perspective of generalized fiducial inference [19], the confidence distribution in fact provides a generator for the $\hat{\beta}_d$ using the data-generating equation $\hat{\beta}_d = G(U_d, \beta)$, with data-generating function $G(\cdot, \cdot)$ of the form

$$G(U_d, \beta) = \sigma_d \left\{ \Phi^{-1}(U_d) + \beta \right\}.$$

Thus, in principle, the output from the confidence distribution fusion learning, $\hat{\beta}_{CD}$ in equation (11.6), may also be formulated in the fiducial paradigm. We expect that the above connection may be generalized to the family of location-scale distributions, including the Normal distribution as a special case. We do not explore this connection formally in this chapter, but the theoretical link between confidence distribution and fiducial inference in fusion learning is worth further exploration, especially in the context of estimating functions, where $\hat{\beta}_b$ may be replaced by an estimating function to establish alternative fusion learning methods dealing with more complex data structures.

Remarks. We highlight the limitations of the above BF perspectives on meta-analysis that have motivated various recent advances in the BF approaches.

i. In reality, data accessibility may appear in different forms. For example, in a meta-analysis, part of the data may contain person-level information and the other part of the data may be available in the form of summary statistics. Data sources of varying resolutions call for specific extensions of the BF approaches. Section 11.3.1 is devoted to an overview of data integration methods that handle data of mixed resolutions.

ii. Fast technological advances have made the collection of high-dimensional data easy. The classical BF paradigm does not provide any principled ways of integrating high-dimensional data. Specifically, in the setting of regression analysis, when the number of predictors p is bigger than the sample size n_d in each data source, an extended BF framework is needed to address the challenge of large p small n_d. Section 11.3.2 presents some recent developments for distributed regularized estimation of regression coefficients, including the distributed Lasso through an extension of the confidence distribution approach.

iii. Violation of the assumption of independent data sources is pervasive in practice, for example with spatial and imaging data. The confidence distribution perspective for handling dependent data sources is discussed in detail in Section 11.4. The key insight is that we may rewrite the right-hand side of equation (11.5) as a quadratic function:

$$\hat{\beta}_{j,CD} = \arg\min_{\beta_j} 0.5(\hat{\beta}_{j,1}, \ldots, \hat{\beta}_{j,D})^T \text{diag}(\sigma_{j,1}^2, \ldots, \sigma_{j,D}^2)^{-1}(\hat{\beta}_{j,1}, \ldots, \hat{\beta}_{j,D}),$$

where the matrix of variances is diagonal due to the independence of $\hat{\boldsymbol{\beta}}_d, d = 1, \ldots, D$. This diagonal matrix may be replaced by the covariance matrix of the data source-specific estimators $\hat{\boldsymbol{\beta}}_b$ with nonzero off-diagonal entries when data sources are correlated.

iv. The homogeneity assumption on model parameters is required in meta-analysis and can easily be invalid. In practice, however, both the size and the homogeneous members of the parameter vector are unknown. These have to be learned from the available data sources as part of parameter estimation. Section 11.5 provides an overview of several existing frequentist methods to identify parameter heterogeneity and associated distinct parameter clusters.

11.3 Integration of Independent Data Sources

This section presents several important extensions of the data integration techniques for independent data sources discussed in Section 11.2. To improve power and efficiency of integration methods, it is critical to develop tailored strategies that take specific study designs, data types and structures into consideration. We focus on two major types of data to be integrated: (i) individual-level information (usually small in size and termed "internal data") and summary-level information (usually big in size and termed "external data"); (ii) equally important but separate individual-level data sets, like multi-source or multi-modality data. Integration methods that are built upon summary-level information, as opposed to individual-level information, are particularly appealing for practical reasons such as data sharing, storage, and computing. In addition, ethical reasons, such as maintenance of the privacy of the study subjects and protection of the future research interests of data-generating institutions and investigators, present big hurdles in integrative data analyses.

11.3.1 Integration of Individual- and Summary-Level Information

For data of type (i) consisting of an internal dataset of individual-level data and an external dataset of summary -level data, properly integrating data can improve statistical inference in terms of efficiency even if external information is not detailed [5]. More specifically, we have two data sources, $\{Y_{i,1}, \boldsymbol{X}_{i,1}, \boldsymbol{Z}_{i,1}\}_{i=1}^{n_1}$ and $\{Y_{i,2}, \boldsymbol{X}_{i,2}\}_{i=1}^{n_2}$, where the former is defined as "internal data", and the latter is "external data". Here the covariates $\boldsymbol{Z}_{i,1}$ represent extra covariate information included only in the internal data. The likelihood is given by

$$L_{\beta,F} = \prod_{i=1}^{n_1} f_\beta(Y_{i,1} \mid \boldsymbol{X}_{i,1}, \boldsymbol{Z}_{i,1}) dF(\boldsymbol{X}_{i,1}, \boldsymbol{Z}_{i,1}).$$

For the external data, suppose a model $g_\theta(Y \mid \boldsymbol{X})$ is built, and $U(Y \mid \boldsymbol{X}; \theta) = \partial \log\{g_\theta(Y \mid \boldsymbol{X})\}/\partial\theta$, with $\mathbb{E}\{U(Y \mid \boldsymbol{X}, \theta^*)\} = 0$, where θ^* is the underlying true parameter. The goal is to maximize $\log\{L_{\beta,F}\}$ with respect to β and $F(\cdot)$ with the following key assumption satisfied

$$\int_{\boldsymbol{X},\boldsymbol{Z}} \left\{ \int_Y U(Y \mid \boldsymbol{X}, \theta^*) f_{\beta_0}(Y \mid \boldsymbol{X}, \boldsymbol{Z}) dY \right\} dF(\boldsymbol{X}, \boldsymbol{Z}) = 0, \tag{11.7}$$

where β_0 is the true value of β. Clearly, the above equation (11.7) could be obtained by rewriting the equation $\mathbb{E}\{U(Y \mid \boldsymbol{X}, \theta^*)\} = 0$ and changing the order of integrals. If the internal data and the external data include the same information, it is easy to see that $\theta^* \equiv \beta_0$ and $f_\beta \equiv g_\theta$. Equation (11.7) essentially converts the external information to a set of constraints, which may be used for internal data to improve the efficiency of parameter estimates and generalizability of the models; refer to [5] for details.

Recently, [68] considered the estimation of causal effects by integrating small validation data $\mathcal{O}_2$ containing supplementary information on confounders and bigger data $\mathcal{O}_1$ containing unmeasured confounders. In their setting, the causal effect $\boldsymbol{\tau}$ is estimated consistently based on the validation data $\mathcal{O}_2$, but inefficiently without using the big data $\mathcal{O}_1$. The big data $\mathcal{O}_1$, however, does not contain important confounders U, which causes the corresponding estimators to be inconsistent for $\boldsymbol{\tau}$ and thus error-prone in general. Without imposing additional modeling assumptions linking U and the validation data, they use the correlation of estimators to improve estimation efficiency. Specifically, they apply the same error-prone procedure to both the big and the validation data. Their key insight is that the difference between the two error-prone estimates is consistent for 0 and can be used to improve the efficiency of the initial estimator due to its association with $\hat{\boldsymbol{\tau}}_2$, the estimator based on $\mathcal{O}_2$. That is,

$$n_2^{1/2}\begin{pmatrix} \hat{\boldsymbol{\tau}}_2 - \boldsymbol{\tau} \\ \hat{\boldsymbol{\tau}}_{2,ep} - \hat{\boldsymbol{\tau}}_{1,ep} \end{pmatrix} \xrightarrow{d} \mathcal{N}\left\{\mathbf{0}, \begin{pmatrix} \boldsymbol{v}_2 & \boldsymbol{\Gamma}^T \\ \boldsymbol{\Gamma} & \boldsymbol{V} \end{pmatrix}\right\},$$

as $n_2 \to \infty$, where n_2 is the sample size in $\mathcal{O}_2$. Using the conditional distribution

$$\begin{aligned} & n_2^{1/2}(\hat{\boldsymbol{\tau}}_2 - \boldsymbol{\tau}) \mid n_2^{1/2}(\hat{\boldsymbol{\tau}}_{2,ep} - \hat{\boldsymbol{\tau}}_{1,ep}) \\ & \qquad \sim \mathcal{N}\left\{n_2^{1/2}\boldsymbol{\Gamma}^T\boldsymbol{V}^{-1}(\hat{\boldsymbol{\tau}}_{2,ep} - \hat{\boldsymbol{\tau}}_{1,ep}), \boldsymbol{v}_2 - \boldsymbol{\Gamma}^T\boldsymbol{V}^{-1}\boldsymbol{\Gamma}\right\}, \end{aligned}$$

we set $n_2^{1/2}(\hat{\boldsymbol{\tau}}_2 - \boldsymbol{\tau})$ to equal its estimated conditional mean $n_2^{1/2}\boldsymbol{\Gamma}^T\boldsymbol{V}^{-1}(\hat{\boldsymbol{\tau}}_{2,ep} - \hat{\boldsymbol{\tau}}_{1,ep})$, leading to an estimating equation for $\boldsymbol{\tau}$. Solving this equation for $\boldsymbol{\tau}$ yields the integrated estimator

$$\hat{\boldsymbol{\tau}} = \hat{\boldsymbol{\tau}}_2 - \hat{\boldsymbol{\Gamma}}^T\hat{\boldsymbol{V}}^{-1}(\hat{\boldsymbol{\tau}}_{2,ep} - \hat{\boldsymbol{\tau}}_{1,ep}),$$

where $\hat{\boldsymbol{\Gamma}}$ and $\hat{\boldsymbol{V}}$ are consistent estimators for $\boldsymbol{\Gamma}$ and $\boldsymbol{V}$. The updated estimator $\hat{\boldsymbol{\tau}}$ is more efficient than $\hat{\boldsymbol{\tau}}_2$ with the use of information from the big data $\mathcal{O}_1$.

11.3.2 Integration of High-Dimensional Data Sources

For data of type (ii), since different modalities could provide complementary information, statistical prediction methods using multimodality data could deliver better prediction performance than using single modality data. When multiple data sources satisfy certain homogeneous assumptions, parallel or distributed methods are routinely used in integration procedures. Specifically, two strategies are commonly used: one-shot approaches and communication-efficient methods.

One-shot methods separately analyze each data source and then merge results together. This includes classical meta-analysis and the confidence distribution approaches. [40] describe an exact, unconditional, non-randomized procedure for producing confidence intervals for the grand mean in a normal-normal random effects meta-analysis. This type of method is also referred to as "embarrassingly parallel" because only one round of communication between data source analyses and integration is involved. In the frequentist setting, most one-shot approaches rely on (weighted) averaging [76, 54], where the integrated estimator is the (weighted) average of the data source estimators. Though computationally fast, these average-based, one-shot communication approaches suffer from several drawbacks. To achieve the minimax rate of convergence and obtain asymptotically unbiased results, the order of the weighted averaging estimator should be much smaller than $\sqrt{N}$, a highly restrictive assumption. [33] developed a new method through a residual-adjustment composition estimate to relax this strong assumption and achieve theoretical optimality. Specifically, for each data source, the debiased lasso estimator takes the form

$$\hat{\boldsymbol{\beta}}_d^{DBL} = \hat{\boldsymbol{\beta}}_d^{\text{Lasso}} + \frac{1}{n_d}\hat{\boldsymbol{\Theta}}_d\boldsymbol{X}_d^T(\boldsymbol{y}_d - \boldsymbol{X}_d\hat{\boldsymbol{\beta}}_d^{\text{Lasso}}).$$

To obtain an integrated estimator, the one-shot approach takes the form

$$\bar{\boldsymbol{\beta}}^{GDBL} = \frac{1}{D}\sum_{d=1}^{D}\hat{\boldsymbol{\beta}}_d^{DBL}.$$

In [33], they rewrite the above debiased lasso estimator by a residual-adjustment lasso estimator:

$$\hat{\boldsymbol{\beta}}_d^{DBL} = \hat{\boldsymbol{\beta}}_d^{\text{Lasso}} + \frac{1}{n_d}\hat{\boldsymbol{\Theta}}_d \boldsymbol{X}_d^T \boldsymbol{X}_d(\boldsymbol{\beta}_d - \hat{\boldsymbol{\beta}}_d^{\text{Lasso}}) + \hat{\boldsymbol{\varepsilon}}_d,$$

where $\hat{\boldsymbol{\varepsilon}}_d = (1/n_d)\hat{\boldsymbol{\Theta}}_d \boldsymbol{X}_d^T \boldsymbol{\varepsilon}_d$. Multiplying a p-dimensional random vector $\boldsymbol{\eta}_d$ on both sides and combining similar terms, we have

$$\hat{\mathbf{Z}}_d^{\text{Lasso}} = \hat{\mathbf{U}}_d^T \boldsymbol{\beta} + \hat{\boldsymbol{\varepsilon}}_d, \quad d = 1, \ldots, D, \tag{11.8}$$

where $\hat{\mathbf{U}}_d = (1/n_d)\boldsymbol{X}_d^T \boldsymbol{X}_d \hat{\boldsymbol{\Theta}}_d^T \boldsymbol{\eta}_d$ and $\hat{\mathbf{Z}}_d^{Lasso} = \boldsymbol{\eta}_d^T(\hat{\boldsymbol{\beta}}_d^{DBL} - \hat{\boldsymbol{\beta}}_d^{Lasso}) + \hat{\mathbf{U}}_d^T \hat{\boldsymbol{\beta}}_d^{Lasso}$. Then, since data sources are independent and $\mathbb{E}(\hat{\boldsymbol{\varepsilon}}_d \mid \boldsymbol{X}_d) = \mathbf{0}$, (weighted) least-square methods can be used to obtain an integrated estimator of $\boldsymbol{\beta}$ based on the above model (11.8). That is,

$$\tilde{\boldsymbol{\beta}}^{\text{GRAL}} = \left(\sum_{d=1}^{D} \hat{\sigma}_d^{-2} \hat{\mathbf{U}}_d \hat{\mathbf{U}}_d^T\right)^{-1} \left(\sum_{d=1}^{D} \hat{\sigma}_d^{-2} \hat{\mathbf{U}}_d \hat{\mathbf{Z}}_d^{Lasso}\right),$$

where $\hat{\sigma}_d^2 = (1/n_d)\boldsymbol{\eta}_d^T \hat{\boldsymbol{\Theta}}_d \boldsymbol{X}_d^T \boldsymbol{X}_d \hat{\boldsymbol{\Theta}}_d^T \boldsymbol{\eta}_d$. For details on the choice of $\boldsymbol{\eta}_d$ and other details we refer the reader to [33].

Communication-efficient methods are developed to yield unbiased results, relax strong assumptions, and achieve the same rate of convergence as approaches that use all data sources simultaneously. [28] present a communication-efficient surrogate likelihood framework for solving distributed statistical inference problems, which can be used for low-dimensional estimation, high-dimensional regularized estimation and Bayesian inference. Let $N = \sum_{d=1}^{D} n_d$ and define the local and global loss functions as

$$\mathcal{L}_d(\boldsymbol{\beta}) = \frac{1}{n_d}\sum_{i=1}^{n_d} \mathcal{L}(\boldsymbol{\beta}; \mathbf{Z}_{id}), \quad d = 1, \ldots, D$$

$$\mathcal{L}_N(\boldsymbol{\beta}) = \frac{1}{N}\sum_{i=1}^{n_d}\sum_{d=1}^{D} \mathcal{L}(\boldsymbol{\beta}; z_{ik}) = \frac{1}{D}\sum_{d=1}^{D} \mathcal{L}_d(\boldsymbol{\beta})$$

respectively. Here, $\mathcal{L}_d(\boldsymbol{\beta})$ is the loss function evaluated at $\boldsymbol{\beta}$ by using the local data source d, $d = 1, \ldots, D$. The key insight of [28]'s method is motivated by the Taylor series expansion of $\mathcal{L}_N$:

$$\mathcal{L}_N(\boldsymbol{\beta}) = \mathcal{L}_N(\bar{\boldsymbol{\beta}}) + \langle \nabla \mathcal{L}_N(\bar{\boldsymbol{\beta}}), \boldsymbol{\beta} - \bar{\boldsymbol{\beta}}\rangle + \sum_{j=2}^{\infty} \frac{1}{j!}\nabla^j \mathcal{L}_N(\bar{\boldsymbol{\beta}})(\boldsymbol{\beta} - \bar{\boldsymbol{\beta}})^{\otimes j}.$$

Here $\bar{\boldsymbol{\beta}}$ is any initial estimator of $\boldsymbol{\beta}$, say, the local empirical loss minimizer $\arg\min_{\boldsymbol{\beta}} \mathcal{L}_1(\boldsymbol{\beta})$ in the first data source. The most time-consuming part is to evaluate the higher-order derivatives $\nabla^j \mathcal{L}_N(\bar{\boldsymbol{\beta}})(j > 1)$. Thus, [28] propose to replace the global higher-order derivatives with local derivatives, leading to an approximation of $\mathcal{L}_N$:

$$\tilde{\mathcal{L}}(\boldsymbol{\beta}) = \mathcal{L}_N(\bar{\boldsymbol{\beta}}) + \langle \nabla \mathcal{L}_N(\bar{\boldsymbol{\beta}}), \boldsymbol{\beta} - \bar{\boldsymbol{\beta}}\rangle + \sum_{j=2}^{\infty} \frac{1}{j!}\nabla^j \mathcal{L}_1(\bar{\boldsymbol{\beta}})(\boldsymbol{\beta} - \bar{\boldsymbol{\beta}})^{\otimes j}.$$

Clearly, after some calculations, the approximation error is of order:

$$\tilde{\mathcal{L}}(\boldsymbol{\beta}) - \mathcal{L}_N(\boldsymbol{\beta}) = O\left(\frac{1}{\sqrt{n_1}}\|\boldsymbol{\beta} - \bar{\boldsymbol{\beta}}\|_2^2 + \|\boldsymbol{\beta} - \bar{\boldsymbol{\beta}}\|_2^3\right).$$

Replacing the infinite sum of high-order derivatives yields

$$\tilde{\mathcal{L}}(\boldsymbol{\beta}) = \mathcal{L}_N(\bar{\boldsymbol{\beta}}) + \langle \nabla \mathcal{L}_N(\bar{\boldsymbol{\beta}}), \boldsymbol{\beta} - \bar{\boldsymbol{\beta}} \rangle + \mathcal{L}_1(\boldsymbol{\beta}) - \mathcal{L}_1(\bar{\boldsymbol{\beta}}) - \langle \nabla \mathcal{L}_1(\bar{\boldsymbol{\beta}}), \boldsymbol{\beta} - \bar{\boldsymbol{\beta}} \rangle.$$

Thus, they suggest to use the following surrogate loss function to generate global estimators:

$$\tilde{\mathcal{L}}(\boldsymbol{\beta}) := \mathcal{L}_1(\boldsymbol{\beta}) - \langle \boldsymbol{\beta}, \nabla \mathcal{L}_1(\bar{\boldsymbol{\beta}}) - \nabla \mathcal{L}_N(\bar{\boldsymbol{\beta}}) \rangle.$$

In other work, [7] proposed a computationally efficient method for inference in quantile regression under a limited memory constraint. Later, [6] studied distributed estimation and support recovery for high-dimensional linear regression with heavy-tailed noise. [47] present a general-purpose framework for distributed computing environments, CoCoA, that has an efficient communication scheme and is applicable to a wide variety of problems in machine learning and signal processing.

11.3.3 Integration of Homogeneous Parameters in the Presence of Heterogeneous Parameters

For multiple large datasets, it may be not realistic to assume that the parameter of interest is homogeneous across data sources, especially for high-dimensional datasets. Some covariates may have different effects on the response across various data sources. Ignoring this heterogeneity can lead to bias and incorrect conclusions [26]. Under the framework of confidence distributions, [8] proposed a new and simple resampling method to construct confidence interval estimators for parameters regardless of the presence or absence of parameter heterogeneity. [13] developed a distributed algorithm to account for the heterogeneous distributions by allowing site-specific nuisance parameters. Their method extended [28]'s surrogate likelihood approach to the heterogeneous setting by applying a novel density ratio tilting method to the efficient score function. [3] developed a new integration method, aggregating and debiasing local estimators, to protect individual data through a summary statistics-based integrating procedure. Their approach also accommodates between source heterogeneity in both the covariate distribution and model parameters, and attains consistent variable selection.

11.4 Integration of Dependent Data Sources

Data integration of dependent data sources remains relatively unexplored in comparison to the wealth of methods available for independent data sources. Clearly, incorporating the dependence structure in the integration procedure is not necessary, and a working independence assumption will lead to consistent estimators of model parameters. Inference can be incorrect, however, depending on how severely the working independence assumption is violated. For the purposes of scientific discovery, incorporating the dependence structure of the data sources is crucial to making efficient use of the data [17]. Particular challenges surround the estimation of this dependence structure, particularly in distributed settings such as meta-analysis where it can not be directly evaluated.

11.4.1 Integration of Homogeneous Parameters

Consider the simple setting from Section 11.1 where the dependent data sources correspond to dependent observations on the same independent individuals. We observe data

$\{\boldsymbol{Y}_{i,d}, \boldsymbol{X}_{i,d}\}_{i,d=1}^{N,D}$ from D data sources, where $\boldsymbol{Y}_{i,d} = (Y_{i1,d}, \ldots, Y_{iM_d,d}) \in \mathbb{R}^{M_d}$ is the ith response and $\boldsymbol{X}_{i,d} \in \mathbb{R}^{M_d \times p}$ the corresponding deterministic covariate matrix in data source $d \in \{1, \ldots, D\}$. We aim to integrate homogeneous parameters $\boldsymbol{\beta}$ from marginal mean models $E(Y_{ij,d}) = \boldsymbol{X}_{ij,d}\boldsymbol{\beta}$ across data sources. The Frequentist, Bayesian and confidence distribution estimators $\widehat{\boldsymbol{\beta}}_{\text{BLUE}}$, $\widehat{\boldsymbol{\beta}}_{\text{Bayes}}$ and $\widehat{\boldsymbol{\beta}}_{CD}$ proposed in Section 11.1, while all consistent, will be inefficient because they fail to account for the dependence within and between $\boldsymbol{Y}_{i,d}$.

[23, 24] recently showed how to obtain a communication-efficient meta-estimator similar in spirit to the estimators in Section 11.1 that incorporates dependence between data sources for improved statistical efficiency. For $d = 1, \ldots, D$, let

$$\boldsymbol{\Psi}_d(\boldsymbol{\beta}; \boldsymbol{y}_d, \boldsymbol{\gamma}_d) = \frac{1}{N} \sum_{i=1}^{N} \boldsymbol{\psi}_{i,d}(\boldsymbol{\beta}; \boldsymbol{y}_{i,d}, \boldsymbol{\gamma}_d)$$

be an additive, unbiased estimating function for $\boldsymbol{\beta}$ based solely on observations in data source d, with $\boldsymbol{\gamma}_d$ potential nuisance parameters, such that $\boldsymbol{\Psi}_d(\boldsymbol{\beta}; \boldsymbol{y}_d, \boldsymbol{\gamma}_d)$ has a unique $\mathbf{0}$ at the true value of $\boldsymbol{\beta}$ and $\boldsymbol{\gamma}_d$. For example, $\boldsymbol{\Psi}_d(\boldsymbol{\beta}; \boldsymbol{y}_d)$ can correspond to the score function if the multivariate likelihood of $\boldsymbol{Y}_{i,d}$ is available, the pairwise composite likelihood score function [34, 57]

$$\boldsymbol{\Psi}_d(\boldsymbol{\beta}; \boldsymbol{y}_d, \boldsymbol{\gamma}_d) = \frac{1}{N} \sum_{i=1}^{n_k} \sum_{r=1}^{M_d-1} \sum_{t=r+1}^{M_d} \nabla_{\boldsymbol{\beta}} f_d(y_{ir,d}, y_{it,d}; \boldsymbol{\beta}, \boldsymbol{\gamma}_d, \boldsymbol{X}_{i,d})$$

if bivariate likelihoods are available, or the generalized estimating equation [32] if the functional form of the full or composite likelihood is not available for modeling or computational reasons. Solving $\boldsymbol{\Psi}_d(\boldsymbol{\beta}, \boldsymbol{y}_d, \boldsymbol{\gamma}_d) = \mathbf{0}$ yields an estimator $\widehat{\boldsymbol{\beta}}_d$ of $\boldsymbol{\beta}$ for each data source $d = 1, \ldots, D$. Estimators $\widehat{\boldsymbol{\gamma}}_d$ of $\boldsymbol{\gamma}_d$ may be obtained through a similar unbiased additive estimating function with unique $\mathbf{0}$ at the true value of $\boldsymbol{\gamma}_d$ and allow for heterogeneous modeling of the dependence structure of the outcome.

The key for making efficient inference on $\boldsymbol{\beta}$ is to account for the correlation between data sources when integrating estimators $\widehat{\boldsymbol{\beta}}_d$ into one combined estimator of $\boldsymbol{\beta}$. To this end, denote $\widehat{\mathbf{S}}_d = \nabla_{\boldsymbol{\beta}} \boldsymbol{\Psi}_d(\boldsymbol{\beta}; \boldsymbol{y}_d, \boldsymbol{\gamma}_d)\Big|_{\boldsymbol{\beta}=\widehat{\boldsymbol{\beta}}_d, \boldsymbol{\gamma}=\widehat{\boldsymbol{\gamma}}_d}$. Let

$$\widehat{\boldsymbol{V}}_N = \sum_{i=1}^{N} \left\{ \left(\boldsymbol{\psi}_{i,d}(\widehat{\boldsymbol{\beta}}_d; \boldsymbol{y}_{i,d}, \widehat{\boldsymbol{\gamma}}_d) \right)_{d=1}^{D} \right\}^{\otimes 2}$$

be the sample covariance matrix of the estimating functions $\boldsymbol{\Psi}_d(\boldsymbol{\beta}; \boldsymbol{y}_d, \boldsymbol{\gamma}_d)$, where $\boldsymbol{a}^{\otimes 2}$ denotes the outer product of a vector $\boldsymbol{a}$ with itself. Let $[\widehat{\boldsymbol{V}}_N]^{i,j}$ be the rows corresponding to data source i and columns corresponding to data source j of matrix $\widehat{\boldsymbol{V}}_N^{-1}$. An integrated estimator of $\boldsymbol{\beta}$ can be defined as

$$\widehat{\boldsymbol{\beta}}_{DIMM} = \left(\sum_{i,j=1}^{D} \widehat{\mathbf{S}}_i^T \left[\widehat{\boldsymbol{V}}_N \right]^{i,j} \widehat{\mathbf{S}}_j \right)^{-1} \sum_{i,j=1}^{D} \widehat{\mathbf{S}}_i^T \left[\widehat{\boldsymbol{V}}_N \right]^{i,j} \widehat{\mathbf{S}}_j \widehat{\boldsymbol{\beta}}_j. \tag{11.9}$$

The asymptotic properties and numerical performance of $\widehat{\boldsymbol{\beta}}_{DIMM}$ in Equation (11.9) are studied in detail in [23]. The derivation of $\widehat{\boldsymbol{\beta}}_{DIMM}$ originates from an extension of the confidence distribution, termed confidence estimating function, which has close ties to the Generalized Method of Moments of [20]. The estimator $\widehat{\boldsymbol{\beta}}_{DIMM}$ in Equation (11.9) is also computationally attractive, communication-efficient and privacy-preserving since it does not require access to individual-level data and weights estimators from each data source according to within- and between- data source dependence.

[23] show this distributed approach to model building is generally more efficient than a whole data approach because it builds a better model of the multi-level nested dependence structure of the observations combined over dependent data sources. To see how $\widehat{\boldsymbol{V}}_N$ accounts for dependence between data sources, take $\boldsymbol{\Psi}_d(\boldsymbol{\beta}; \boldsymbol{y}_d, \boldsymbol{\gamma}_d)$ to be the score function for $\boldsymbol{\beta}$ in data source d:

$$\boldsymbol{\Psi}_d(\boldsymbol{\beta}; \boldsymbol{y}_d, \boldsymbol{\gamma}_d) = \sum_{i=1}^{N} \boldsymbol{X}_{i,d}^T \boldsymbol{\Sigma}_d^{-1} (\boldsymbol{Y}_{i,d} - \boldsymbol{\mu}_{i,d}), \quad \boldsymbol{\Sigma}_j = Var(\boldsymbol{Y}_{i,d}).$$

Letting $d_1, d_2 \in \{1, \ldots, D\}$,

$$\begin{aligned} &Cov\left\{\boldsymbol{X}_{i,d_1}^T \boldsymbol{\Sigma}_{d_1}^{-1} (\boldsymbol{Y}_{i,d_1} - \boldsymbol{\mu}_{i,d_1}), \boldsymbol{X}_{i,d_2}^T \boldsymbol{\Sigma}_{d_2}^{-1} (\boldsymbol{Y}_{i,d_2} - \boldsymbol{\mu}_{i,d_2})\right\} \\ &\quad = \boldsymbol{X}_{i,d_1}^T \boldsymbol{\Sigma}_{d_1}^{-1} E\left[\left(\boldsymbol{Y}_{i,d_1} - \boldsymbol{\mu}_{i,d_1}\right)\left(\boldsymbol{Y}_{i,d_2} - \boldsymbol{\mu}_{i,d_2}\right)^T\right] \boldsymbol{\Sigma}_{d_2}^{-1} \boldsymbol{X}_{i,d_2} \\ &\quad = \boldsymbol{X}_{i,d_1}^T \boldsymbol{\Sigma}_{d_1}^{-1} Cov(\boldsymbol{Y}_{i,d_1}, \boldsymbol{Y}_{i,d_2}) \boldsymbol{\Sigma}_{d_2}^{-1} \boldsymbol{X}_{i,d_2}. \end{aligned}$$

The covariance of the estimating functions is a scaled form of the covariance of the original response vectors. The former, however, can be computed without access to the original data as the outer product of the estimating function, and can thus be computed in a communication-efficient and privacy-preserving distributed framework.

11.4.2 Integration of Homogeneous Parameters in the Presence of Heterogeneous Parameters

Consider a more complex setting in which we observe data $\{\boldsymbol{Y}_{i,jk}, \boldsymbol{X}_{i,jk}\}_{i,j,k=1}^{N_k,J,K}$ from $D = J \times K$ data sources, where $\boldsymbol{Y}_{i,jk} = (Y_{i1,jk}, \ldots, Y_{iM_j,jk}) \in \mathbb{R}^{M_j}$ is the ith response and $\boldsymbol{X}_{i,jk} \in \mathbb{R}^{M_j \times p}$ the corresponding deterministic covariate matrix in data source (j,k), $j \in \{1, \ldots, J\}$, $k \in \{1, \ldots, K\}$. Data sources $(1,k), \ldots, (J,k)$ are assumed dependent, and data sources $(j,1), \ldots, (j,K)$ are assumed independent. In other words, dependent observations from data sources $(1,k), \ldots, (J,k)$ are also dependent, and independent observations from data sources $(j,1), \ldots, (j,K)$ are also independent. We aim to integrate homogeneous parameters $\boldsymbol{\beta}$ from marginal mean models $E(Y_{ij,jk}) = \boldsymbol{X}_{ij,jk}(\boldsymbol{\beta}^T, \boldsymbol{\alpha}_{jk}^T)^T$ across data sources, where $\boldsymbol{\alpha}_{jk}$ corresponds to heterogeneous fixed mean effects. When $K = 1$ and $\boldsymbol{\alpha}_{jk}$ is empty, this setting clearly reduces to the setting in Section 11.4.1 as a special case. We again show how to obtain a communication-efficient meta-estimator similar in spirit to the estimators in Section 11.1.

Following the framework proposed in [24], let

$$\begin{aligned} \boldsymbol{\Psi}_{jk}(\boldsymbol{\beta}; \boldsymbol{y}_{jk}, \boldsymbol{\gamma}_{jk}) = \frac{1}{N_k} \sum_{i=1}^{N_k} \boldsymbol{\psi}_{i,jk}(\boldsymbol{\beta}; \boldsymbol{y}_{i,jk}, \boldsymbol{\alpha}_{jk}, \boldsymbol{\gamma}_{jk}), \\ j \in \{1, \ldots, J\}, \quad k \in \{1, \ldots, K\}, \end{aligned}$$

be an additive unbiased estimating function for $\boldsymbol{\beta}$ based solely on observations in data source (j,k), with $\boldsymbol{\gamma}_{jk}$ potential nuisance parameters, such that $\boldsymbol{\Psi}_{jk}(\boldsymbol{\beta}; \boldsymbol{y}_{jk}, \boldsymbol{\alpha}_{jk}, \boldsymbol{\gamma}_{jk})$ has a unique $\boldsymbol{0}$ at the true values of $\boldsymbol{\beta}$, $\boldsymbol{\alpha}_{jk}$ and $\boldsymbol{\gamma}_{jk}$. As in the case of $K = 1$, $\boldsymbol{\Psi}_{jk}(\boldsymbol{\beta}; \boldsymbol{y}_{jk}, \boldsymbol{\alpha}_{jk}, \boldsymbol{\gamma}_{jk})$

can correspond, for example, to the score function, pairwise composite score function or generalized estimating equation. Estimators of $\widehat{\boldsymbol{\beta}}_{jk}$, $\widehat{\boldsymbol{\alpha}}_{jk}$ and $\widehat{\boldsymbol{\gamma}}_{jk}$ can be computed by solving appropriate estimating equations. As before, an integrated estimator of $\boldsymbol{\beta}$ that accounts for dependence between data sources is defined as

$$\widehat{\boldsymbol{\beta}}_{DDIMM} = [\widehat{C}]^{\boldsymbol{\beta}} \sum_{k=1}^{K} \sum_{j=1}^{J} n_k^2 \widehat{\boldsymbol{C}}_{k,j} \begin{pmatrix} \widehat{\boldsymbol{\beta}}_{jk} \\ \widehat{\boldsymbol{\alpha}}_{\text{list}} \\ \widehat{\boldsymbol{\gamma}}_{\text{list}} \end{pmatrix} \tag{11.10}$$

where $[\widehat{\boldsymbol{C}}]^{\boldsymbol{\beta}}$ corresponds to the rows of $(\sum_{k=1}^{K} \sum_{j=1}^{J} n_k^2 \widehat{\boldsymbol{C}}_{k,j})^{-1}$ corresponding to parameter $\boldsymbol{\beta}$, $\widehat{\boldsymbol{\alpha}}_{\text{list}} = (\widehat{\boldsymbol{\alpha}}_{jk})_{j,k=1}^{J,K}$, $\widehat{\boldsymbol{\gamma}}_{\text{list}} = (\widehat{\boldsymbol{\gamma}}_{jk})_{j,k=1}^{J,K}$, and $\widehat{\boldsymbol{C}}_{k,j}$ is a function of sample variability and sensitivity matrices and data source-specific estimators $\widehat{\boldsymbol{\beta}}_{jk}$, $\widehat{\boldsymbol{\alpha}}_{jk}$ and $\widehat{\boldsymbol{\gamma}}_{jk}$ defined in detail in [24]. By incorporating the sample variability of the estimating functions $\boldsymbol{\Psi}_{jk}$ in the weighting of $\widehat{\boldsymbol{\beta}}_{jk}$, $\widehat{\alpha}_{jk}$ and $\widehat{\boldsymbol{\gamma}}_{jk}$, the estimator $\widehat{\boldsymbol{\beta}}_{DDIMM}$ in Equation (11.10) non-parametrically accounts for dependence between data sources in a similar way to (11.9).

11.4.3 Data Integration in Other Settings

Data integration of dependent datasets has received attention in [4] and in [30]. [4] develop a Bayesian calibration model for computer model calibration using the block composite likelihood approach based on partitioning the spatial field. They assume conditional independence between outcomes in different blocks (subsets of the spatial field) given the block means, and model the dependence between block means directly. [30] considers a sequential split-and-conquer approach based on the confidence distribution to combine information from dependent datasets. The proposed method sequentially updates the analysis from dependent data sources by removing the dependency.

When conducting data integration, one special challenge arises from variable missingness. Variable information may be completely missing, for example because of dropouts, because there is no need to collect these variables, or because it is infeasible to collect certain sources of information given other known variables. To address this difficulty, [66] proposed a multiple block-wise imputation approach to integrate information from the multi-source observed data to achieve efficient estimators, and [73] proposed a direct sparse regression procedure using covariance from multimodality data.

11.5 Fusion of Parameters with Independent Data Sources

Data integration is often undertaken when data from multiple sources are jointly considered in modeling. Besides the earlier mentioned divide-and-conquer scenario where we randomly divide large-scale data into manageable partitions and integrate the result from each, data integration also considers the more challenging case when a data division already exists and cannot be altered. For example, electronic health record data across multiple hospital systems, insurance claims data across multiple providers, and protected social security data in different states. In all these cases, pooling of individual-level data to construct a centralized repository is often prohibited due to data-sharing barriers and privacy concerns. Even if these regulatory barriers are overcome, the amount of effort required to combine databases is still prohibitively large. To jointly analyze such passively distributed data requires us to acknowledge and account for data heterogeneity.

11.5.1 Random Effects Approach

Formally, data heterogeneity is the difference in the joint distributions across multiple data sources. It may be partially explained by measured variables, but more often than not by unmeasured factors, and is eventually reflected in the heterogeneity of parameters of interest. The common theme of past data integration approaches has been to conveniently assume systematic biases in the parameters of interest across studies. Typically the normal assumption is imposed on random effects across data sources (see Section 11.2). For example, the true effect of covariate $\boldsymbol{X}_{j,d}$ in data source d, $\beta_{j,d}$, is assumed to be independent across data batches and follows a normal distribution around the global average, i.e.,

$$\beta_{j,d} = \beta_j + U_{j,d}, \quad d = 1, \dots, D,$$

where $U_{j,d} \sim \mathcal{N}(0, \tau_j^2)$ is known as the random slope (or random intercept). Note the distinction between this and the Bayesian Fusion Learning in Section 11.2. Despite marginally following similar normal distributions, $\beta_{j,d}, d = 1, \dots, D$ in the Bayesian hierarchical model are assumed common across data sources, and the randomness results from a prespecified normal prior. In a random effect model, the true parameters are different across data sources with an unknown randomness to be estimated.

Estimation and inference in random effects meta-analysis can be carried out using a likelihood-based approach. Despite its popularity and well-studied theory (see a summary in [12]), the method has a few limitations. First, by assuming a random effect across data sources, the method only attempts to *describe* data heterogeneity, as opposed to *leverage* multiple studies for higher statistical power. One may argue that the power for inference for the average effect, i.e., β_j, is superior in meta-analysis than in individual data batches. But if we believe heterogeneity exists, the average effect is less meaningful in the traditional sense. The wide prediction confidence intervals (i.e., combining between and within study randomness, $Var(\beta_j) + Var(\hat{\beta}_j)$) usually lead to low power for inference when τ^2 is large, which is counterintuitive to the notion of data integration. Second, the requirement that random effect terms representing data heterogeneity must be uncorrelated with independent variables is restrictive in many settings, despite being a mathematical convenience. A typical example is an unobserved confounding variable of $\boldsymbol{Y}$ that is also correlated with any of the independent variables $\boldsymbol{X}$. Third, the interpretation of results of a random effects meta-analysis, specifically, the use of prediction confidence intervals, is indirect and typically difficult to understand.

11.5.2 Fusion of Parameters

A new class of approaches relying on parameter fusion has appeared in the recent literature for heterogeneous data integration problems. This type of approach is given the name fusion learning (see for example, [65, 37, 51, 44], as it exploits the notion of similarity across data sources to achieve grouping. A main contrast between these new approaches and classic random effects models is on the assumption of heterogeneity and how it is handled. Fusion approaches assume a clustered structure in the heterogeneity across data sources. In a sense, $U_{j,d}$ in $\beta_{j,d} = \beta_j + U_{j,d}$, follows distribution $p_j(u)$, where $p_j(u)$ is allowed to be a mixture-type distribution with grouping on top of data sources or can take a finite number of unique values. In some recent work based on regularization techniques, a distribution assumption for the heterogeneity component $p_j(u)$ is not required, hence partially model free. The grouping can potentially interact with independent covariates through this setup to study grouping effects across data sources.

One of the earliest methods for fusion learning in heterogeneous data integration appears in the context of ANOVA, where the differences among the levels in factors are assumed to be sparse and form some low dimensional subgroups [2]. This method is useful when the number of levels is large, otherwise an exhaustive pairwise comparison will suffice. If we view the levels defined by a factor as a stratifying variable, i.e., data sources are defined by levels of this factor, then we can view this problem as a data integration problem. The paper applied fused lasso [55] to penalize the differences in the mean of the outcome across different levels so that smaller differences are shrunk to zero, corresponding to dropping some of the levels within a stratifying factor. To see this, consider a penalty imposed on the distances among $\beta_{j,d}$,

$$d(\beta_{j,d}, \beta_{j,d'}), \quad d, d' \in \{1, \ldots, D\},$$

or in other words $d(U_{j,d}, U_{j,d'})$ among U after subtracting the average β_j, where $d(a,b)$ is a predefined distance between a and b, for example, the ℓ_1 norm absolute distance. Popular alternatives include SCAD [15] and MCP [74].

Parameter fusion was originally studied outside of the context of data integration. The most well-known fused lasso penalty [55] considers ordered parameters,

$$P(\boldsymbol{\beta}) = \sum_{j=1}^{p-1} d(\beta_j, \beta_{j+1}),$$

where, without loss of generality, $j = 1, \ldots, p$ are presented in a meaningful order (e.g., location on gene). Shrinkage on the adjacent pairs forces the parameter to be estimated in a piecewise constant manner. Another class of methods involves network-type penalties [1, 45, 56, 29] that relax the ordering constraint by considering more extensive pairs of parameters defined by a network,

$$P(\boldsymbol{\beta}) = \sum_{(j,j')\in E} d(\beta_j, \beta_{j'}),$$

where E is the edge set defined by the network with vertices V consisting of indices of all parameters, i.e., $\{1, \ldots, p\}$. Within the context of data integration, fusion learning considers clustering parameters of similar meaning, but from various data sources. In this context, many penalty functions have also been considered, including ordered penalties [60, 51],

$$P(\boldsymbol{\beta}) = \sum_{j=1}^{p} P(\boldsymbol{\beta}_j) = \sum_{j=1}^{p} \sum_{d=1}^{d-1} d(\beta_{j,d}, \beta_{j,d+1}),$$

and network-type penalties [31, 38, 21, 72],

$$P(\boldsymbol{\beta}) = \sum_{j=1}^{p} P(\boldsymbol{\beta}_j) = \sum_{j=1}^{p} \sum_{(d,d')\in E} d(\beta_{j,d}, \beta_{j,d'}),$$

where E is the edge set defined by the network with vertices V consisting of indices of all data sources, i.e., $\{1, \ldots, d\}$.

Penalization methods lend themselves well to pure likelihood-based (Frequentist) data integration methods and confidence distribution-based data integration methods. For example, [51] consider a penalized joint likelihood function across D heterogeneous data sources with an ℓ_1 distance function d by a consistent adaptive weighting scheme [77].

A confidence distribution generalization can be achieved with minimal effort by replacing likelihood functions with confidence distributions. Let $f(\boldsymbol{\beta})$ denote a generic objective function (e.g., likelihood function or confidence distribution function). The strength of the penalty is controlled by a tuning parameter λ in the penalized objective function

$$f(\boldsymbol{\beta}) - \lambda P(\boldsymbol{\beta}),$$

and the choice of λ is decided through some model selection criteria. Additional examples with generalization to generalized linear models, GEE and quadratic inference functions can be found in [60] and [51, 52]. All introduced methods are estimated through direct optimization. On the other hand, under a model-based construct of the unobserved component U, the hierarchical mixture structure lends well to Bayesian data integration methods because it is naturally a two level model (a level on individual-level data, and another level on data sources). Analogous to clustered effects in fusion learning, a mixture distribution on β can be from a mixture of normal (as shown in Section 0.2); see also [22, 63]. Analogous to the need to tune λ, a challenge in applying this Bayesian approach is that it requires specifying the number of subgroups. Some of the fusion learning approaches using penalization can be achieved using Bayesian methods, see for example [49].

11.6 Software Implementations

In this section, we briefly review two software implementations for data integration with dependent data sources and for fusion of heterogeneous parameters with independent data sources.

11.6.1 Integration of Dependent Data Sources

We consider the setting of Section 11.4.2 with $J = 10$ and $K = 5$, $N_k = 1000$ and $M_j = 50$ for $k = 1, \ldots, K$, $j = 1, \ldots, J$. The value of the (homogeneous) regression parameter is set to $\boldsymbol{\beta} = (3, -2, 1)^T$.

```
J <- 10; K <- 5; D <- J*K
n_k <- 1000; m_j <- 50
beta <- c(3, -2, 1)
```

We simulate $D = JK = 50$ datasets with a total sample size of $N = \sum_{k=1}^{K} N_k = 5000$ and total outcome dimension $M = \sum_{j=1}^{J} M_j = 500$. Covariates consist of an intercept and two continuous M-dimensional covariates X_1 and X_2, and the outcome vector for each individual i is a M-dimensional multivariate normal variable from the marginal linear regression model

$$Y_{ij,d} = \boldsymbol{\beta}_1 - \boldsymbol{\beta}_2 X_{ij1,d} + \boldsymbol{\beta}_3 X_{ij2,d} + \epsilon_{i,d}, \quad \epsilon_i \sim \mathcal{N}_M(0, \boldsymbol{\Sigma}).$$

Here $\boldsymbol{\Sigma}$ is a $M \times M$ positive definite matrix with block AR(1) structure.

```
block_AR1 <- 4 * 0.5^abs(outer(1:J, 1:J, "-"))
Sigma <- kronecker(
 clusterGeneration::genPositiveDefMat(
              covMethod = "eigen", dim = m_j
                                   )$Sigma, block_AR1)
```

```
error <- t(MASS::mvrnorm(n = n_k*K, mu = rep(0, m_j*J),
                         Sigma = Sigma))

cov1 <- rnorm(n = n_k*K*m_j*J, mean = 0, sd = 1)
cov2 <- rbinom(n = n_k*K*m_j*J, size = 1, prob = 0.4)
Y <- beta[1] + beta[2]*cov1 + beta[3]*cov2 + c(error)
IDs <- sort( rep( 1:(n_k*K), m_j*J ) )

df <- data.frame(id=IDs, Y = Y, cov1 = cov1, cov2 = cov2)
```

The integrated estimator in Equation (11.10), with data source covariance parameters treated as the heterogeneous parameters and the mean regression parameter as the homogeneous parameter of interest, can be computed using the R package DDIMM from [24]:

```
library(DDIMM)

response_ind <- sort( rep(1:J, m_j) )
subject_ind <- sort (rep(1:K, n_k) )
fit <- dimm(Y ~ cov1 + cov2, df, id="id", response_ind,
        subject_ind, family = "gaussian", corstr="AR1",
        analysis = "GEE", cluster=1)
```

The results are given by:

```
summary(fit)
Call:
dimm(formula = Y ~ cov1 + cov2, data = df, id = "id",
  response_indicator = response_ind,
  subject_indicator = subject_ind,
  family = "gaussian", corstr = "AR1",
  analysis = "GEE", cluster = 1)

              Estimate     StdErr z.value   p.value
(Intercept)  2.9982695  0.0061006  491.47 < 2.2e-16 ***
cov1        -1.9959128  0.0023893 -835.35 < 2.2e-16 ***
cov2         1.0072751  0.0048165  209.13 < 2.2e-16 ***
---
Signif. codes:
0 ‘***’ 0.001 ‘**’ 0.01 ‘*’ 0.05 ‘.’ 0.1 ‘ ’ 1
```

Here, the generalized estimating equations [32] approach is used to obtain data source estimators. The pairwise composite likelihood is also available through the specification `analysis = "PCL"`. Parallelization of data source analyses is controlled by the number `cluster`, here set to 1 for sequential analysis.

11.6.2 Fusion of Heterogeneous Parameters with Independent Data Sources

We demonstrate the use of an existing software package to fuse heterogeneous parameters across independent data sources into subgroups. We consider $n_d = 100$ for $d = 1, \ldots, D$, where $D = 10$ with three continuous covariates X_1, X_2, X_3. The corresponding coefficients, including the intercept, are $\boldsymbol{\beta}_0, \ldots, \boldsymbol{\beta}_3$, where each of $\boldsymbol{\beta}_j$ is either homogeneous across all data sources or heterogeneous across data sources but with a small number of subgroups.

```
nd <- 100; D <- 10
beta <- matrix(c(0,0,0,0,0,0,0,0,0,0,     # Intercept
                 0,0,0,0,0,0,0,0,0,0,     # Coefficient of X1
                 0,0,0,0,0,1,1,1,1,1,     # Coefficient of X2
                 -1,-1,-1,0,0,0,0,1,1,1), # Coefficient of X3
               nrow = D, ncol = 4)
```

In the above R example, `beta` contains the true coefficient values across the 10 sources for each of $\boldsymbol{\beta}_0, \ldots, \boldsymbol{\beta}_3$ where $\boldsymbol{\beta}_0$ is the intercept term. Coefficients $\boldsymbol{\beta}_2$ and $\boldsymbol{\beta}_3$ are heterogeneous as compared to $\boldsymbol{\beta}_0$ and $\boldsymbol{\beta}_1$, which are homogeneous across data sources. A fusion estimate that searches for and recovers such potential underlying subgrouping across data sources can be obtained by the R package metafuse from [51]:

```
library(metafuse)
df <- datagenerator(n = nd, beta0 = beta, family = "gaussian")
y_in <- df[, "y"]
X_in <- df[, c("x1", "x2", "x3")]
d_in <- df[, "group"]
fit <- metafuse(X = X_in, y = y_in, sid = d_in,
                intercept = TRUE, fuse.which = c(0, 1, 2, 3),
                family = "gaussian", criterion = "EBIC")
print(fit$betahat)
         Intercept  x1          x2           x3
study1  0.01342207 0.0261268 -0.07103079 -1.045040093
study2  0.01342207 0.0261268 -0.07103079 -1.045040093
study3  0.01342207 0.0261268 -0.07103079 -1.045040093
study4  0.01342207 0.0261268 -0.07103079  0.008527549
study5  0.01342207 0.0261268 -0.07103079  0.008527549
study6  0.01342207 0.0261268  1.02302426  0.008527549
study7  0.01342207 0.0261268  1.02302426  0.008527549
study8  0.01342207 0.0261268  1.02302426  1.038181233
study9  0.01342207 0.0261268  1.02302426  1.038181233
study10 0.01342207 0.0261268  1.02302426  1.038181233
```

The `group` column identifies the data sources of each individual, which is passed to the `metafuse` function. The estimated values for $\boldsymbol{\beta}$ are listed where the tuning parameter is chosen through the extended BIC criterion.

11.7 Future Directions

Data integration relates to the fundamental issue of information integration, while model fusion relates to the integration of data distributions. To avoid the so-called pitfall of *garbage in, garbage out* in meta-analysis, applying some principled methods for creating a mega dataset and extracting statistical features is of critical importance. The BF framework provides a proper theoretical foundation to formulate, derive and evaluate new approaches to analyzing aggregated data. In particular, the flexibility and utility of the confidence distribution approach have been illustrated throughout this chapter, including the distributed regularized estimation and model fusions with correlated data sources. We also highlighted an interesting conceptual connection between the confidence distribution and Fisher's fiducial inference in fusion learning that merits further exploration.

Our review has only cursorily described Bayesian approaches to data integration and model fusion by focusing on random effects models. The Bayesian paradigm offers a natural framework for incorporating prior knowledge through a two-step process: prior knowledge may be generated from a separate data source in advance, formulated from existing scientific knowledge, or specified for a technical requirement. This prior knowledge is then updated with information from a current data source in a second step. The sequential nature of this approach is not easily amenable to data integration and model fusion approaches that consider all data sources simultaneously. Existing Bayesian approaches for information aggregation primarily fall into two categories: combining Markov Chain Monte Carlo chain samples and integrating multi-platform data in a Bayesian hierarchical model. For the former, interested readers should consider [61, 42, 43, 50, 11]. For the latter, [41, 75] constitute good examples; see also the references therein. [62] proposed a Bayesian approach to combining information from independent surveys. More recently, [58] proposed a Bayesian approach for analysis of partitioned data by using expectation propagation's message passing technique. We anticipate these relatively new areas of investigation will see active development over the next few years.

To review a broad range of methodologies in this chapter, we purposely described an ambiguous setup regarding how data sources are formed. We simply assumed that data are available as separate data sources, whether by study design, such as a multi-center study, or as a result of the divide step in a divide-and-conquer procedure. Our review has thus primarily focused on the mechanics of data integration and model fusion, with limited discussion on how to divide data into pieces or compare available data sources. Data heterogeneity, data dependence and data imbalance are key aspects that arise from integrating multiple data sources. This area of research has recently attracted attention in the literature; see for example [39, 25]. Certainly, data heterogeneity can be substantially reduced by careful study design and data collection. For example, studies that analyze batches of samples as more funds become available should, to the best of their ability, maintain a consistent analysis protocol across batches. Similarly, multi-site studies in consortia should agree on data collection and sample preparation protocols beforehand. It is of great interest to evaluate the robustness of statistical methods to data heterogeneity, not only in terms of estimation efficiency but also in terms of reliability and reproducibility. Substantially erroneous conclusions may ensue from assuming homogeneity where none exists [26]. The challenge of data integration methods is to develop the appropriate quantification of this measure of commonality, whether it be parameter or likelihood homogeneity. Some tests to determine heterogeneity of data sources have been developed [59, 70], as well as methods robust to unmeasured confounding [69, 71].

When data are split into data sources following the first step in a divide-and-conquer procedure, heterogeneity can be controlled directly at the data-splitting stage. In this situation, the data is most often split to reduce the computational burdens of data analysis. Clearly, integration methods must be aligned with the data division strategy. Careful consideration of the trade-offs between computational speed and efficiency must be taken into account, since a computationally efficient analysis of each data source is generally preferred. Future directions include the development of statistical methods for integration of misspecified models with a focus on robustness and efficiency, and the integration of approximate estimators, likelihoods or estimating functions. This last direction raises a particular question: can we combine multiple inconsistent or inefficient estimators to form one consistent or efficient estimator? Some work seems to suggest that this is possible given strong assumptions [68], and links to the machine learning literature on combining weak learners are immediately obvious, but considerable additional study is required.

With the increased interest in data integration and model fusion methods of the past decade, statistical methods for many different types of data integration problems have proliferated. Data integration for causal inference, or with missing or multi-modality data are but a few examples. These areas are certainly worthy of exploration and study. For example, statistical methods for the integration of multi-modality imaging data are becoming increasingly popular as it becomes easier to collect different types of imaging data via various platforms such as EEG, fMRI and PET. Distributed causal inference is also an important area of methodological research, since sharing clinical data sources is very difficult due to unfriendly policies and resources. This field stands to benefit tremendously from new statistical methods for data integration and model fusion. We offer words of caution, however, that we must be sure that we can do data integration in simple situations before tackling more complex problems; it is not obvious to us that this is the case, and much work remains to ascertain the validity of data integration and model fusion approaches. This is clearly a fast developing area of statistical methodology research with many big challenges arising from practical studies. Any solutions to address these challenges should be investigated in a rigorous theoretical framework. We believe the BF framework to be eminently desirable for these developments.

Bibliography

[1] Bondell, H. D. and Reich, B. J., 2008. Simultaneous regression shrinkage, variable selection, and supervised clustering of predictors with OSCAR. *Biometrics, 64*(1), pp. 115–123.

[2] Bondell, H. D. and Reich, B. J., 2009. Simultaneous factor selection and collapsing levels in ANOVA. *Biometrics, 65*(1), pp. 169–177.

[3] Cai, T., Liu, M. and Xia, Y., 2019. Individual data protected integrative regression analysis of high-dimensional heterogeneous data. *arXiv preprint*, arXiv:1902.06115.

[4] Chang, W., Haran, M., Olson, R. and Keller, K., 2015. A composite likelihood approach to computer model calibration with high-dimensional spatial data. *Statistica Sinica, 25*(1), pp. 243–259.

[5] Chatterjee, N., Chen, Y.-H., Maas, P. and Carroll, R. J., 2016. Constrained maximum likelihood estimation for model calibration using summary-level information from external big data sources. *Journal of the American Statistical Association, 111*(513), pp. 107–117.

[6] Chen, X., Liu, W., Mao, X. and Yang, Z., 2019. Distributed high-dimensional regression under a quantile loss function. *arXiv preprint*, arXiv:1906.05741.

[7] Chen, X., Liu, W., Zhang, Y. et al., 2019. Quantile regression under memory constraint. *The Annals of Statistics, 47*(6), pp. 3244–3273.

[8] Claggett, B., Xie, M. and Tian, L., 2014. Meta-analysis with fixed, unknown, study-specific parameters. *Journal of the American Statistical Association, 109*(508), pp. 1660–1671.

[9] Cochran, W. G., 1937. Problems arising in the analysis of a series of similar experiments. *Supplement to the Journal of the Royal Statistical Society, 4*(1), pp. 102–118.

[10] Cochran, W. G. and Carroll, S. P., 1953. A sampling investigation of the efficiency of weighting inversely as the estimated variance. *Biometrics, 9*(4), pp. 447–459.

[11] Dai, H., Pollock, M. and Roberts, G., 2019. Monte carlo fusion. *Journal of Applied Probability, 56*(1), pp. 174–191.

[12] DerSimonian, R. and Laird, N., 1986. Meta-analysis in clinical trials. *Controlled Clinical Trials, 7*(3), pp. 177–188.

[13] Duan, R., Ning, Y. and Chen, Y., 2019. Heterogeneity-aware and communication-efficient distributed statistical inference. *arXiv preprint*, arXiv:1912.09623.

[14] Efron, B., 1993. Bayes and likelihood calculations from confidence intervals. *Biometrika, 80*(1), pp. 3–26.

[15] Fan, J. and Li, R., 2001. Variable selection via nonconcave penalized likelihood and its oracle properties. *Journal of the American Statistical Association, 96*(456), pp. 1348–1360.

[16] Fisher, R. A., 1956. *Statistical Methods and Scientific Inference.* Oxford, England: Hafner Publishing Co.

[17] Fitzmaurice, G. M., Laird, N. M. and Rotnitzky, A. G., 1993. Regression models for discrete longitudinal responses. *Statistical Science, 8*(3), pp. 284–309.

[18] Glass, G. V., 1976. Primary, secondary, and meta-analysis of research. *Educational Researcher, 5*(10), pp. 3–8.

[19] Hannig, J., Iyer, H., Lai, R. C. S. and Lee, T. C. M., 2016. Generalized fiducial inference: A review and new results. *Journal of the American Statistical Association, 515*(111), pp. 1346–1361.

[20] Hansen, L. P., 1982. Large sample properties of generalized method of moments estimators. *Econometrica, 50*(4), pp. 1029–1054.

[21] Hao, B., Sun, W. W., Liu, Y. and Cheng, G., 2018. Simultaneous clustering and estimation of heterogeneous graphical models. *Journal of Machine Learning Research, 18*(217), pp. 1–58.

[22] Hastie, T. and Tibshirani, R., 1996. Discriminant analysis by Gaussian mixtures. *Journal of the Royal Statistical Society: Series B, 58*(1), pp. 155–176.

[23] Hector, E. C. and Song, P. X.-K., 2020a. A distributed and integrated method of moments for high-dimensional correlated data analysis. *Journal of the American Statistical Association, 116*(534), pp. 805–818. https://doi.org/10.1080/01621459.2020.1736082.

[24] Hector, E. C. and Song, P. X.-K., 2020b. Doubly distributed supervised learning and inference with high-dimensional correlated outcomes. *Accepted, Journal of Machine Learning Research, 21*, pp. 1–35.

[25] Hector, E. C. and Song, P. X.-K., 2020. Joint integrative analysis of multiple data sources with correlated vector outcomes. *arXiv preprint*, arXiv:2011.14996.

[26] Higgins, J. P. T. and Thompson, S. G., 2002. Quantifying heterogeneity in a meta-analysis. *Statistics in Medicine, 21*(11), pp. 1539–1558.

[27] Jordan, M. I., 2013. On statistics, computation and scalability. *Bernoulli, 19*(4), pp. 1378–1390.

[28] Jordan, M. I., Lee, J. D. and Yang, Y., 2018. Communication-efficient distributed statistical inference. *Journal of the American Statistical Association, 114*(526), pp. 668–681.

[29] Ke, Z. T., Fan, J. and Wu, Y., 2015. Homogeneity pursuit. *Journal of the American Statistical Association, 110*(509), pp. 175–194.

[30] Li, C., Hung, Y. and Xie, M., 2020. A sequential split-and-conquer approach for the analysis of big dependent data in computer experiments. *The Canadian Journal of Statistics, 48*, pp. 712–730. https://doi.org/10.1002/cjs.11559. .

[31] Li, Y., Nan, B. and Zhu, J., 2015. Multivariate sparse group lasso for the multivariate multiple linear regression with an arbitrary group structure. *Biometrics, 71*(2), pp. 354–363.

[32] Liang, K.-Y. and Zeger, S. L., 1986. Longitudinal data analysis using generalized linear models. *Biometrika, 73*(1), pp. 13–22.

[33] Lin, L. and Lu, J., 2019. A race-dc in big data. *arXiv preprint*, arXiv:1911.11993.

[34] Lindsay, B. G., 1988. Composite likelihood methods. *Contemporary Mathematics, 80*, pp. 220–239.

[35] Liu, D., Liu, R. Y. and Xie, M. G., 2021. Nonparametric fusion learning for multiparameters: Synthesize inferences from diverse sources using data depth and confidence distribution. *Journal of the American Statistical Association, 117*(540), 2086–2104.

[36] Liu, D., Liu, R. Y. and Xie, M.-G., 2014. Exact meta-analysis approach for discrete data and its application to 2 × 2 tables with rare events. *Journal of the American Statistical Association, 109*(508), pp. 1450–1465.

[37] Liu, D., Liu, R. Y. and Xie, M.-G., 2015. Multivariate meta-analysis of heterogeneous studies using only summary statistics: Efficiency and robustness. *Journal of the American Statistical Association, 110*(509), pp. 326–340.

[38] Ma, S. and Huang, J., 2017. A concave pairwise fusion approach to subgroup analysis. *Journal of the American Statistical Association, 112*(517), pp. 410–423.

[39] Maity, S., Sun, Y. and Banerjee, M., 2019. Communication-efficient integrative regression in high-dimensions. *arXiv preprint*, arXiv:1912.11928.

[40] Michael, H., Thornton, S., Xie, M. and Tian, L., 2019. Exact inference on the random-effects model for meta-analyses with few studies. *Biometrics, 75*(2), pp. 485–493.

[41] Morris, J. S. and Baladandayuthapani, V., 2017. Statistical contributions to bioinformatics: design, modelling, structure learning and integration. *Statistical Modelling, 17*(4–5), pp. 245–289.

[42] Neiswanger, W., Wang, C. and Xing, E., 2014. Asymptotically exact, embarassingly parallel MCMC. *arXiv preprint*, arXiv:1311.4780.

[43] Scott, S. L., Blocker, A. W., Bonassi, F. V., Chipman, H. A. and George, E. I., 2016. Bayes and Big Data: The consensus Monte Carlo algorithm. *International Journal of Management Sciences and Engineering Management*, *11*(2), pp. 78–88.

[44] Shen, J., Liu, R. Y. and Xie, M.-G., 2020. i fusion: Individualized fusion learning. *Journal of the American Statistical Association*, *115*(531), pp. 1251–1267.

[45] Shen, X. and Huang, H.-C., 2010. Grouping pursuit through a regularization solution surface. *Journal of the American Statistical Association*, *105*(490), pp. 727–739.

[46] Singh, K., Xie, M. and Strawderman, W. E., 2005. Combining information from independent sources through confidence distributions. *The Annals of Statistics*, *33*(1), pp. 159–183.

[47] Smith, V., Forte, S., Ma, C., Takáč, M., Jordan, M. I. and Jaggi, M., 2017. Cocoa: A general framework for communication-efficient distributed optimization. *The Journal of Machine Learning Research*, *18*(1), pp. 8590–8638.

[48] Song, P. X.-K., 2007. *Correlated Data Analysis: Modeling, Analytics, and Applications.* Springer Series in Statistics, New York: Springer Verlag.

[49] Song, Q. and Cheng, G., 2019. Bayesian fusion estimation via t shrinkage. *Sankhya A*, *82*, pp. 353–385.

[50] Srivastava, S., Li, C. and Dunson, D. B., 2018. Scalable Bayes via Barycenter in Wasserstein space. *Journal of Machine Learning Research*, *19*, pp. 1–35.

[51] Tang, L. and Song, P. X.-K., 2016. Fused lasso approach in regression coefficients clustering – learning parameter heterogeneity in data integration. *Journal of Machine Learning Research*, *17*(113), pp. 1–23.

[52] Tang, L. and Song, P. X.-K., 2020. Poststratification fusion learning in longitudinal data analysis. *Biometrics*, *77*(3), pp. 914–928.

[53] Tang, L., Zhou, L. and Song, P. X.-K., 2019. Fusion learning algorithm to combine partially heterogeneous cox models. *Computational Statistics*, *34*, pp. 395–414.

[54] Tang, L., Zhou, L. and Song, P. X.-K., 2020. Distributed simultaneous inference in generalized linear models via confidence distribution. *Journal of Multivariate Analysis*, *176*, p. 104567.

[55] Tibshirani, R., Saunders, M., Rosset, S., Zhu, J. and Knight, K., 2005. Sparsity and smoothness via the fused lasso. *Journal of the Royal Statistical Society: Series B (Statistical Methodology)*, *67*(1), pp. 91–108.

[56] Tibshirani, R. J., and Taylor, J., 2011. The solution path of the generalized lasso. *The Annals of Statistics*, *39*(3), pp. 1335–1371.

[57] Varin, C., Reid, N. and Firth, D., 2011. An overview of composite likelihood methods. *Statistica Sinica*, *21*(1), pp. 5–42.

[58] Vehtari, A., Gelman, A., Sivula, T., Jylänki, P., Tran, D., Sahai, S., Blomstedt, P., Cunningham, J. P., Schiminovich, D. and Robert, C. P., 2020. Expectation propagation as a way of life: a framework for Bayesian inference on partitioned data. *Journal of Machine Learning Research*, *21*, pp. 1–53.

[59] Wang, F., Wang, L. and Song, P. X.-K., 2012. Quadratic inference function approach to merging longitudinal studies: Validation and joint estimation. *Biometrika, 99*(3), pp. 755–762.

[60] Wang, F., Wang, L. and Song, P. X.-K., 2016. Fused lasso with the adaptation of parameter ordering in combining multiple studies with repeated measurements. *Biometrics, 72*(4), pp. 1184–1193.

[61] Wang, X. and Dunson, D. B., 2014. Parallelizing MCMC via Weierstrass sampler. *arXiv preprint*, arXiv:1312.4605.

[62] Wang, Z. Kim, J. K. and Yang, S., 2018. Approximate Bayesian inference under informative sampling. *Biometrika, 105*(1), pp. 91–102.

[63] Wei, S. and Kosorok, M. R., 2013. Latent supervised learning. *Journal of the American Statistical Association, 108*(503), pp. 957–970.

[64] Xie, M. G. and Singh, K., 2013. Confidence distribution, the frequentist distribution estimator of a parameter: A review. *International Statistical Review, 81*(1), pp. 3–39.

[65] Xie, M., Singh, K. and Strawderman, W. E., 2011. Confidence distributions and a unifying framework for meta-analysis. *Journal of the American Statistical Association, 106*(493), pp. 320–333.

[66] Xue, F. and Qu, A., 2020. Integrating multisource block-wise missing data in model selection. *Journal of the American Statistical Association, 116*(536), pp. 1914–1927. https://doi.org/10.1080/01621459.2020.1751176:1–14.

[67] Yang, G., Liu, D., Wang, J. and Xie, M. G., 2016. Meta-analysis framework for exact inferences with application to the analysis of rare events. *Biometrics, 72*(4), pp. 1378–1386.

[68] Yang, S. and Ding, P., 2019. Combining multiple observational data sources to estimate causal effects. *Journal of the American Statistical Association, 115*(531), pp. 1540–1554.

[69] Yang, S., Kim, J. K. and Song, R., 2020. Doubly robust inference when combining probability and non-probability samples with high dimensional data. *Journal of the Royal Statistical Society: Series B, 82*(2), pp. 445–465.

[70] Yang, S., Zeng, D. and Wang, X., 2020. Elastic integrative analysis of randomized trial and real-world data for treatment heterogeneity estimation. *arXiv preprint*, arXiv:2005.10579.

[71] Yang, S., Zeng, D. and Wang, X., 2020. Improved inference for heterogeneous treatment effects using real-world data subject to hidden confounding. *arXiv preprint*, arXiv:2007.12922.

[72] Yang, X., Yan, X. and Huang, J., 2019. High-dimensional integrative analysis with homogeneity and sparsity recovery. *Journal of Multivariate Analysis, 174*, p. 104529.

[73] Yu, G., Li, Q., Shen, D. and Liu, Y., 2019. Optimal sparse linear prediction for block-missing multi-modality data without imputation. *Journal of the American Statistical Association, 115*(531), pp. 1406–1419.

[74] Zhang, C.-H. et al., 2010. Nearly unbiased variable selection under minimax concave penalty. *The Annals of Statistics, 38*(2), pp. 894–942.

[75] Zhang, Y., Morris, J. S., Aerry, S. N., Rao, A. U. K. and Baladandayuthapani, V., 2019. Radio-iBAG: Radiomics-based integrative Bayesian analysis of multiplatform genomic data. *The Annals of Applied Statistics, 13*(3), pp. 1957–1988.

[76] Zhang, Y., Duchi, J. C. and Wainwright, M. J., 2013. Communication-efficient algorithms for statistical optimization. *The Journal of Machine Learning Research, 14*(1), pp. 3321–3363.

[77] Zou, H., 2006. The adaptive lasso and its oracle properties. *Journal of the American Statistical Association, 101*(476), pp. 1418–1429.

12

How the Game-Theoretic Foundation for Probability Resolves the Bayesian vs. Frequentist Standoff

Glenn Shafer
Rutgers University

12.1 Introduction

The conventional picture of the Bayesian vs. frequentist divide conflates two distinctions. One concerns the meaning of probability — whether it is subjective or objective. The other concerns a choice between two methods of inference — one that relies on the type of reasoning introduced by Thomas Bayes (thus using Bayes's theorem and conditional probability calculations), and one that relies on the type of reasoning introduced earlier by Jacob Bernoulli (thus using significance tests, confidence intervals, etc.). Bayesians are said to believe that probabilities are subjective and that Bayes's theorem usually suffices for inference. Frequentists are said to believe that probabilities are frequencies and that this justifies Bernoullian methods.

This chapter explains how the game-theoretic foundation for probability, which begins with a betting game instead of a mere assignment of probabilities to events, can dissolve the conflation and transform the entire notion of frequentism, generalizing Bernoullian inference so that it stands beside Bayesian inference as a way of using betting.

The project of using betting games as the starting point for mathematical probability dates back to Jean Ville, who used it in 1939 to generalize and criticize Richard von Mises's earlier project of axiomatizing probability as a theory about frequencies. Ville's idea was further developed by Per Martin-Löf, Claus-Peter Schnorr, Phil Dawid, and others, including Vladimir Vovk and myself. But it is only now being deployed in mathematical statistics.[1]

The simplest kind of bet is an all-or-nothing bet. It has only two outcomes: you win or you lose. You put your money on the table, and you get back a larger amount known in advance or you get back nothing. But there are more complicated bets — bets with many possible outcomes. You might put \$10 on the table on the understanding that you will get back either \$0 or \$5 or \$30. When we make an all-or-nothing bet on an event, it is natural to focus our minds on the frequency of the event in repeated trials, even if these repeated trials are entirely imaginary. But when we make bets that are not all-or-nothing, we are taking a first step beyond frequentism. This step, which I discuss in Section 12.2 and in [23], already permits a substantial generalization of the Bernoullian methodology of significance tests and confidence intervals. And it already begins to dissolve the illusion of competing subjective and objective interpretations of probability.

In Section 12.3, I take the next step by pointing out the great generality of testing by betting. We can test any probability forecaster by betting against their forecasts. It does not matter whether the forecaster is a statistical model, a machine-learning method,

 DOI: 10.1201/9780429341731-12

a physical model, or a soothsayer. Neither do the topics of the forecasts matter. The forecaster might shift capriciously from one topic to another. So the betting generalization of Bernoullian methodology is vast.

In Section 12.4, I define a formal game for testing a probability forecaster and explain how this game can be used as a starting point for a generalization of probability theory, thus providing for the betting method the same mathematical grounding that conventional probability theory provides for the familiar Bernoullian methods.

In Section 12.5, I discuss how the same formal game can represent Bayesian inference. The Bayesian picture translates into ways of obtaining and using strategies for the probability forecaster. Other ways of finding such strategies are also available.

I conclude with three sections of a more historical nature. In Section 12.6, I document some precedents for the way the adjective "Bernoullian" is used in this chapter. In Section 12.7, I discuss why testing by betting, in spite of seeming so natural, has previously been so little used in mathematical statistics. In Section 12.8, I discuss connections with better known aspects of game theory and measure theory.

12.2 Beyond All-or-Nothing Bets

The frequentist tests a probabilistic hypothesis by singling out an event to which it gives small probability; the hypothesis is discredited at least to some extent if the event then happens. According to the 19th-century French mathematician Augustin Antoine Cournot, this is the only way to link probability theory with phenomena; this has been called Cournot's principle [25]. Cournot's principle can be considered more fundamental than the identification of probability with frequency. If we rely on Jacob Bernoulli's law of large numbers, which says that the frequency of an event in repeated trials will be as close as desired to its probability *with high probability*, equating probability with frequency requires ruling out events of small probability.

Testing by betting generalizes rather than displaces the notion of testing using events of small probability. Indeed, we can make the singling out an event of small probability p more vivid by pretending that the statistician is betting against the event at odds $p : (1-p)$, so that the discredit is associated with winning $(1-p)/p$ times as much as risked. Once we think about testing in this way, it is natural to generalize by allowing the statistician to buy any nonnegative payoff S for the expected value $\mathbf{E}(S)$ the hypothesis assigns it. The discredit is then measured by the factor $S/\mathbf{E}(S)$ by which the bet multiplies the money it risks. Prediction is similarly generalized; once we have provisionally accepted a probabilistic hypothesis, betting against it is associated with predicting outcomes for which the bet will not multiply by a large factor of the money risked.

Are the probabilities and expected values in this picture subjective or objective? An answer to this question must begin with the fact that a bet involves at least two parties. Usually one party proposes or offers the bet; the other decides whether to take it. Here the first party offers to sell any nonnegative payoff S for $\mathbf{E}(S)$, and the other party decides which payoff S to buy. There may be elements of subjectivity and objectivity on both sides. The expected values $\mathbf{E}(S)$ may represent the opinions of a single person, or they may represent the predictions of a well-established theory. The test S may be chosen on a whim, or it may be a conventional choice corresponding to a serious alternative. In the end, neither the subjectivity or the objectivity can be eliminated entirely from the story. Betting requires an agent, but the idea that probabilities say something true about the world is inherent in the project of testing them.

By using bets that are not all-or-nothing, statisticians can acquire some new and very useful tools. As explained in [23], the factor $S/\mathbf{E}(S)$, considered a measure of the strength of the statistical evidence, is more easily communicated to laypeople than a p-value. We also acquire a logic for sequential testing, because a sequence of bets form a single test when each is allowed to use only the capital remaining from the previous one, the final measure of the evidence against the hypothesis then being the ratio of the final to the initial capital.[2] In this context, moreover, the choice of a test $\mathcal{S}$ implies an alternative hypothesis, thus tightening the logic, linking test outcome to prior expected outcome under an alternative in a way that the standard concepts of p-value and power are not linked.

12.3 Improvisation: Bob Tests Alice

For at least 20 years, since Leo Breiman famously contrasted the culture of modeling with the culture of prediction [5], statisticians have struggled to reconcile the free-wheeling methods of machine learning with the probabilistic vision that still defines mathematical statistics. Even when the predictions produced by dynamic neural networks, for example, take the form of probabilities, these probabilities live outside R. A. Fisher's statistical models and Andrei Kolmogorov's global probability measures [11, 17]. This is also true of the probabilistic predictions made by the physical models that dominate weather prediction. How can we test probabilistic predictions when they are not derived from a statistical model or a probability measure?

This question is easily answered once we accept testing by betting. We can test any probability forecaster by betting against each of its successive predictions, no matter how these predictions are generated and no matter how improvised they are. No model is required.

Suppose Alice announces the odds for sports events. One week she looks at the roster for a tennis tournament and assigns each player a probability of winning. The next week she announces probabilities for the outcome of a game between Real Madrid and Barcelona — probabilities for Real Madrid winning, for Barcelona winning, and for a tie. Then she announces a probability distribution for the winning point spread between the Nets and the 76ers. And so on. Bob can challenge Alice's prowess as a probability forecaster by betting at the odds Alice announces. If Bob begins with \$1, bets each time using only what remains of this initial capital together with subsequent winnings, and walks away with \$100 after a year of betting, he will have put a big dent in Alice's reputation as a forecaster. Alice may plead that she was merely unlucky, but she cannot claim success as a forecaster.

We can glean some general insights from this example.

1. Bob can challenge and discredit Alice without giving alternative probabilities. He does not even need to believe that there are meaningful or reliable probabilities for the events in question.

2. Bob does not need to risk real money. He can bet with play money. His goal is to make a point, not to get rich. When he uses play money, he does not need a counterparty to his bets. So Alice is not risking real money either; she is risking only her reputation as a forecaster.

3. Even if real money is risked by both parties, we can posit that the amounts are too small to matter. At question is the factor by which Bob manages to multiply the money he risks. If Bob risked only his initial \$1 and ended up with \$100,

he could have achieved the same factor of 100 by making bets only 1% as great, risking 1 cent and ending up with $1.

4. Alice may know more about the sports and the competitors than Bob. If Alice has a good reputation for using knowledge available to her to forecast sports outcomes, and yet Bob succeeds in making money on her forecasts, then we may conjecture that Alice's additional information is not very relevant.

5. Bob may know more about the sports and the competitors than Alice. If Alice has a good reputation as a sports forecaster, and yet Bob makes money on her forecasts, then we may conjecture that things known to Bob but not to Alice are relevant.

6. If Bob does not make money betting against Alice's probabilities — if he begins with $1 and ends up with only 10 cents or perhaps $1.10 — then we have no evidence against Alice's probabilities. If we know Bob to be very clever and very knowledgeable about the events in question, then this result may be taken as evidence that Alice is doing her job well.

Bob can also test Alice when she gives repeatedly updated probabilities for the same outcome. Suppose, for example, that every week during the football season in the United States she gives probabilities for which team will emerge as the champion of the National Football League. In this case, Bob can test her by buying and selling weekly. Every time Alice assigns a probability to each team being the final champion, Bob buys a payoff that depends on which is the final champion, then sells it back to her the next week at the prices she gives then. If he uses only an initial stake and subsequent winnings for his betting but multiplies that initial stake substantially, then Alice will be discredited. What other way could she be discredited? What other way could any empirical meaning be assigned to her probabilities?

When a forecaster repeatedly changes their probabilities concerning a future event, it may be possible to test their insight even if the event itself is aborted. In early 2020, www.fivethirtyeight.com was regularly giving new probabilities for the season's championship of the National Basketball Association. The betting by testing method might have been used to discredit (or support) this forecaster even though the championship was never settled. The remainder of the season was canceled on March 12 because of the COVID-19 virus.

Financial markets and the prediction markets that imitate them also provide a setting for testing by betting. In the case of financial markets, it is the efficiency of a market rather than the sagacity of a forecaster that is at stake. This way of testing market efficiency has not yet been widely implemented, but see [26, Chapter 16] and [32]. Both financial and prediction markets are constantly being tested by the participants themselves, but transaction costs and the restrictions the markets impose (limitations on short selling in financial markets and limitations on the amounts bet in prediction markets) leave room for thought experiments in the form of imagined testing by betting.

12.4 Strategies for Bob: Bernoullian Inference via Game Theory

The usual theory of significance testing and confidence intervals is justified by results in mathematical probability — the law of large numbers, the central limit theorem, and all their variations and generalizations. Is there analogous justification for the expansive use of testing by betting advocated here? Yes.

Suppose Alice's announcements always take the form of a probability distribution P on some finite nonempty set $\mathcal{Y}$ and that Bob bets by choosing and buying a payoff $f(y)$, where f is a real-valued function on $\mathcal{Y}$ and $y \in \mathcal{Y}$ is the outcome of the sporting event, not yet known. He pays P's expected value $\mathbf{E}_P(f)$ for this payoff. Alice and Bob alternate moves, say for N rounds. Write $\mathcal{K}_0$ for Bob's initial capital, and $\mathcal{K}_n$ for his capital at the end of the nth round. Write $\mathcal{P}(\mathcal{Y})$ for the set of all probability distributions on $\mathcal{Y}$. Write $\mathbb{N}$ for the natural numbers. With this notation, we can lay out the rules for play in Alice's and Bob's game as follows.

Protocol 12.1 (Testing Alice's probabilistic forecasts).
Parameter: $N \in \mathbb{N}$

$\mathcal{K}_0 = 1$.
FOR $n = 1, 2, \ldots, N$:
Alice announces a finite nonempty set $\mathcal{Y}_n$ *and* $P_n \in \mathcal{P}(\mathcal{Y}_n)$.
Bob announces $f_n : \mathcal{Y}_n \to [0, \infty)$ *such that* $\mathbf{E}_{P_n}(f_n) = \mathcal{K}_{n-1}$.
Reality announces $y_n \in \mathcal{Y}_n$.
$\mathcal{K}_n := f_n(y_n)$.

Here, and in the other protocols discussed in this chapter and in [26], each player sees the other players' moves as they are made; the game is one of *perfect information*. This does not rule out the possibility that players acquire private information — information not available to the other players — at the outset or as play proceeds.

The requirement that Bob's move f_n always be non-negative and have expected value equal to his current capital $\mathcal{K}_{n-1}$ enforces the condition that he risk no more than his initial capital $\mathcal{K}_0$. He is not allowed to borrow or otherwise bring more money into the game when play goes badly for him. Because $\mathcal{K}_0 = 1$, his final capital $\mathcal{K}_N$ is the amount by which he multiplies this initial capital.

Within this setup, the mathematical results of probability theory are theorems of game theory — theorems about whether one or more of the players have strategies that guarantee certain goals. Consider, for example, the goal $\mathcal{K}_N \geq 30$ for Bob. He does not have a strategy guaranteed to achieve this goal, because Reality can keep him from ever increasing his capital. But there are many goals he can achieve. For example, if $E \subseteq \mathcal{Y}$, $\epsilon > 0$, and X is the number of the outcomes $y_1, \ldots, y_N$ in E, then Bob has a strategy that will achieve $\mathcal{K}_N \geq N\epsilon^2$ unless the moves by Alice and Reality satisfy

$$\left| \frac{X}{N} - \frac{\sum_{n=1}^{N} P_n(E)}{N} \right| < \epsilon; \tag{12.1}$$

see [26, Exercise 2.8]. The agreement between Alice and Reality represented by (12.1) is one aspect of the game-theoretic law of large numbers. Probability's other classical limit theorems, the law of the iterated logarithm and the central limit theorem in their myriad guises, also have game-theoretic formulations. We can put these game-theoretic results in a more familiar form if we define an *upper probability* $\overline{\mathbb{P}}(A)$ for each set A of possibilities for the moves $\mathcal{Y}_1, P_1, y_1, \ldots, \mathcal{Y}_N, P_N, y_N$ by Alice and Reality:

$$\overline{\mathbb{P}}(A) := \inf \left\{ \alpha \mid \text{Bob has strategy guaranteeing } \mathcal{K}_N \geq \frac{1}{\alpha} \text{ if } A \text{ happens} \right\}. \tag{12.2}$$

The existence of a strategy for Bob guaranteeing $\mathcal{K}_N \geq N\epsilon^2$ unless Alice and Reality make (12.1) happen can then be written as

$$\overline{\mathbb{P}}\left(\left|\frac{X}{N} - \frac{\sum_{n=1}^{N} P_n(E)}{N}\right| \geq \epsilon\right) \leq \frac{1}{N\epsilon^2}.$$

In general, upper probabilities do not obey the rules for probability.[3] But the definition (12.2) tells us that we can interpret small upper probabilities just as Cournot's principle tells us to interpret small probabilities; because we do not think Bob will multiply his money by a large amount, we do not think an event of small upper probability will happen.

In [26], Vovk and I vary and generalize Protocol 12.1 in a great variety of ways. We imagine that play continues indefinitely instead of ending after N rounds. We suppose that N is a stopping time — a rule that stops play when and if moves so far satisfy some condition. We limit the moves one of the players can make or even replace the player with a fixed strategy. We have Reality announce other new information on each round. We introduce other players. We allow Alice to offer fewer bets, in the spirit of imprecise probability [1]. We allow the $\mathcal{Y}_n$ to be infinite. We even consider continuous time. The definition (12.2) can be adapted to all these generalizations, thus providing a rigorous generalization of probability theory and mathematical statistics.

12.5 Strategies for Alice, Bayesian and Non-Bayesian

Let us simplify Protocol 12.1 by fixing a finite nonempty set $\mathcal{Y}$ and requiring Alice to always choose $\mathcal{Y}_n$ equal to $\mathcal{Y}$. In this case, a probability distribution P for Reality's moves $y_1, \ldots, y_N$ in Protocol 12.1 can serve as a strategy for Alice. Alice simply uses P's conditional probability distribution $P(y_n|y_1, \ldots, y_{n-1})$ as her nth move P_n. This is a very special kind of strategy for Alice; when choosing P_n, it takes into account only Reality's previous moves, ignoring Bob's moves and any other information Alice might have or acquire.

Strictly speaking, of course, a probability distribution P for $y_1, \ldots, y_N$ may not define all the needed conditional probabilities; $P(y_n|y_1, \ldots, y_{n-1})$ is not defined when P gives $y_1, \ldots, y_{n-1}$ probability zero. So a more accurate formulation is that a system of conditional probabilities — a family of probability distributions — can serve as a strategy for Alice.

Write $\mathbb{S}$ for the set of all sequences of elements of $\mathcal{Y}$ of length $N-1$ or less; this includes the "empty sequence", denoted by $\Box$. With this notation, we can say that the strategy for Alice is a family $(P_s)_{s\in\mathbb{S}}$ of probability distributions on $\mathcal{Y}$.

Suppose we announce such a strategy for Alice to all the players at the outset and require Alice to play it. This leaves Alice with no role to play. Removing her, we obtain the following protocol.

Protocol 12.2 (Testing a probability distribution).
Parameters: *$N \in \mathbb{N}$, finite nonempty set $\mathcal{Y}$, family $(P_s)_{s\in\mathbb{S}}$*

$\mathcal{K}_0 = 1$.
FOR $n = 1, 2, \ldots, N$:
Bob announces $f_n : \mathcal{Y} \to [0, \infty)$ *such that* $\mathbf{E}_{P_{y_1,\ldots,y_{n-1}}}(f_n) = \mathcal{K}_{n-1}$.
Reality announces $y_n \in \mathcal{Y}$.
$\mathcal{K}_n := f_n(y_n)$.

Given the family $(P_s)_{s\in\mathbb{S}}$, we can define a global probability distribution P for $y_1, \ldots, y_N$ using the usual the formula for a joint probability as a product of conditional probabilities:

$$P(y_1, \ldots, y_N) = P(y_1)P(y_2|y_1)\ldots P(y_N|y_1, \ldots, y_{N-1}).$$

Alternatively, we can use the rule of iterated expectation to define the global expectation operator:

$$\mathbf{E}(f(y_1, \ldots, y_N)) := \mathbf{E}_{P_\square}(\mathbf{E}_{P_{y_1}}(\cdots \mathbf{E}_{P_{y_1,\ldots,y_{N-1}}}(f(y_1, \ldots, y_N)) \cdots)). \tag{12.3}$$

This manner of defining P corresponds to the reasoning about the value of expectations that we find in the work of Blaise Pascal and other early probabilists (c.f. [26, p. 36]).

Changing your predictions for future events by conditioning a probability distribution on what has happened so far is sometimes considered the essence of Bayesian inference. Our formulation respects, moreover, the conditions of coherence on which Bayesians insist. When Alice prices all payoffs $f_n : \mathcal{Y} \to [0, \infty)$, allowing Bob to buy or sell f_n at the same price, the prices must cohere by all being expected values with respect to a probability distribution. The analogous weaker condition of coherence used in the theory of imprecise probability is required when Alice gives separate buying and selling prices [26, Chapter 6].

Another aspect of Bayesian inference is the idea of averaging probability distributions. Suppose Alice is uncertain about what probability distribution on $\mathcal{Y}^N$ will resist Bob's efforts but feels confident that one of the distributions in a certain class $(P_\theta)_{\theta\in\Theta}$ would do the job. In this case, the Bayesian method is to average these distributions with respect to some prior distribution on Θ and use the result as P. Under certain conditions, we know, the P thus obtained will do asymptotically as well as any P_θ. But this is not, of course, the only way Alice might proceed. Another interesting possibility is to average possible strategies for Bob and play against the average; see [26, Chapter 12].

12.6 Who Said "Bernoullian"?

I have been using the adjective *Bernoullian* rather than *frequentist* to designate statistical methods that follow Jacob Bernoulli's example rather than that of Thomas Bayes. This recognizes Bernoulli as the first to state a theory of direct statistical estimation, just as Bayes was the first to state Bayes's formula. It also allows us to contrast Bernoullian and Bayesian methods without asserting anything about how probabilities are to be interpreted.

I have borrowed this use of "Bernoullian" from some prominent predecessors. Here are some quotations documenting their use of it.

- Francis Edgeworth used *Bernoullian* in this way in 1918, contrasting "the *direct* problem associated with the name of Bernoulli" with "the *inverse* problem associated with the name of Bayes" [10].

- Richard von Mises made a similar remark in German in 1919 ([29], page 5): "Man kann die beiden großen Problemgruppen ... als den Bernoullischen und den Bayesschen Ideenkreis charakterisieren." In English: "We can call the two large groups of problems the Bernoullian and Bayesian circles of ideas."

- A. P. Dempster explicitly advocated the use of "Bernoulian" and "Bayesian" in 1966 [8]. In 1968 [9], in a review of three volumes of collected papers by Jerzy Neyman and E. S. Pearson, Dempster wrote

  > Neyman and Pearson rode roughshod over the elaborate but shaky logical structure of Fisher, and started a movement which pushed the Bernoullian approach to a high-water mark from which, I believe, it is now returning to a more normal equilibrium with the Bayesian view.

- Ian Hacking, probably inspired by Dempster, used *Bernoullian* repeatedly in his 1990 book, *The Taming of Chance* [14]. For example, when discussing Poisson's interest in changes in the chance of conviction by a jury (page 97), he wrote:

> Laplace had two ways in which to address such questions. One is Bernoullian, and attends to relative frequencies; the other is Bayesian, and is usually now interpreted in terms of degrees of belief. Laplace almost invited his readers not to notice the difference.

The use of "Bernoullian" by Edgeworth and von Mises predated the introduction of the term "frequentist" by Ernest Nagel in 1936 [20]. It is also notable that von Mises, generally recognized in the mid-20th century as the leading proponent of "the frequency theory of probability", always contended that Bayes's formula provides a correct method of statistical inference [30, 31].

12.7 Why Does Testing by Betting Seem Novel?

Testing by betting is part of our culture. It is commonplace to challenge a strongly expressed opinion by offering a contrary bet or demanding that the opinionated party offer odds. We know, moreover, that mathematical probability grew out of a calculus for betting. So how can testing by betting seem novel in mathematical statistics? Surely this can only be because mathematicians have deliberately put it out of mind. Suppressed it because it is dangerous. Sometimes too immoral. More often, simply too subjective. We want science, and its handmaiden statistics, to be objective.

The celebrated letters between Blaise Pascal and Pierre Fermat in 1654 were about betting, and Christian Huygens's pamphlet, published a few years later, was about betting — about how stakes should be set for bets in games of chance. Pascal's and Huygens's arguments, moreover, were game-theoretic [22]. Hans Freudenthal and Stephen Stigler have emphasized Huygens's argument for equal chances at a and b being worth $(a+b)/2$ [13, 27]. To have equal chances at a and b, Huygens explains, you may play a game in which you and an opponent both stake $(a+b)/2$, with the winner taking the whole $a+b$ but giving the loser back b. Here equal chances might mean simply that the players are treated equally. Huygens was talking about games of chance, but his reasoning could apply just as well to a game of skill, provided that the two players have agreed to play on even terms.

The concept of probability and even the word "probability" did not appear in Pascal's, Fermat's, and Huygens's writings about their calculus for betting. Many authors did write about probability in the 17th century and earlier, but their probabilities were not numerical and were not modeled on games of chance [12, 16]. This changed only with Bernoulli's *Ars conjectandi*, published in 1713 [2, 3]. To make Pascal's and Huygens's calculus into a theory of probability, Bernoulli replaced Huygens's game theory with calculations based on equally possible cases. Why? One obvious motivation was to make probability numbers appear more objective. Betting has an ineradicably subjective element.

Today, when the objectivity of science is under attack, we may feel more tempted than ever to avoid the notion of betting in discussions of statistical testing. But the raging controversies about the meaning of significance testing suggest that this strategy is failing. As I argue in [23], talking about betting may help us communicate the results of statistical tests in a way that better enables the public to understand both their value and their limitations.

12.8 Game Theory and Measure Theory

As its title advertises, this chapter uses a game-theoretic rather than measure-theoretic understanding of mathematical probability. What aspects of game theory are we using, and what aspects of measure theory are we leaving aside?

To answer the question about game theory, we can emphasize the point made right after Protocol 12.1 in Section 12.4: the players move in order and see each other's moves as they are made. Games of this type, *perfect-information games*, appear frequently in mathematics. A venerable example is the proof by Ernst Zermelo, published in 1913, that in a two-player perfect-information game that must end after some fixed finite number of rounds and cannot end in a draw, one of the players has a winning strategy. In the late 20th century, Donald Martin generalized Zermelo's theorem to infinite horizons; see [24, pp. 96–98]. Most of the theorems proven in [26] can be said to instantiate Martin's theorem, because they use a protocol similar to Protocol 12.1, give Bob a goal (the goal (12.1) is an example), make the protocol into a two-player game by saying that the other players are a team that wins if the goal is not achieved, and show that Bob has a winning strategy.

As the reader will have noticed, probability distributions (also known as probability measures) appear in our games in more than one way.

- In Protocol 12.1, Alice specifies the bets she is offering on each round by announcing a probability distribution. (Protocols that offer fewer bets are not discussed in this chapter, but they are also important; see for example [26, Chapter 1].)
- In Protocol 12.2, a strategy for Alice defines a global probability distribution for Reality's moves.
- In (12.2), upper probabilities for global events (events defined by the moves of Bob's opponents) emerge from the game. When a strategy for Alice defined by a global probability distribution is followed, as in Protocol 12.2, Reality is the only opponent, and the upper probabilities are identical with the probabilities given by Alice's strategy [26, p. 178].

As discussed in [23], a probability distribution for Reality's moves, say Q, also defines a strategy for Bob in Protocol 12.1. This strategy tells Bob to make the move

$$f_n := \frac{Q_n}{P_n}$$

when Alice moves P_n, where Q_n is Q's conditional distribution for y_n given $y_1, \ldots, y_{n-1}$. This makes the payoff of Bob's bet equal to the likelihood ratio that compares Bob's probabilities to Alice's.

Twenty-first-century statisticians sometimes say that probabilities define "data-generating mechanisms". If we use probabilities both to forecast outcomes and to test the forecasts, we may not need the additional relatively mystical notion that chance generates outcomes. Instead of saying that Nature produces outcomes by chance, it may be enough to say that Reality's outcomes are in line with Alice's forecasts. But in many set-ups where probability is used — statistical decision theory, for example, it is easier to talk about Nature producing outcomes by chance than to introduce explicitly players who forecast outcomes and then test the forecasts.

In any case, the perfect-information games used in game-theoretic probability constitute a very small part of game theory. In most of the games that economists, computer scientists, and other applied mathematicians use, players have limited information about other players' moves. Von Neumann's minimax theorem and Nash's more general theory of equilibrium show us how players can deal with this mutual uncertainty by randomizing their moves, thus introducing probability exogenously. In practice, "randomizing" means using algorithms that are so hidden and complex that we can only forecast their outcomes with probabilities, but here again this can usually be left implicit.

Protocol 12.1 can be generalized by replacing each finite set $\mathcal{Y}_n$ with an arbitrary measure space and allowing P_n to be any probability measure on this space. In this respect, game-theoretic probability uses the full resources of measure-theoretic probability. The main feature of measure-theoretic probability left aside is its notion of conditional expectation. This notion does not see much service in applied statistics, because a conditional expectation is not fully defined; it can be changed arbitrarily on any set of measure zero. For the same reason, a conditional expectation cannot serve as a strategy in a game. So instead of pretending that we begin with a probability measure and then find conditional expectations via the Radon-Nikodym theorem, as measure-theoretic probability might have us do, we proceed as explained at the beginning of section 12.5, defining directly all the "conditional probabilities" we might use. In the 1980s, A. Philip Dawid called such a collection of conditional probabilities a *probability forecasting system* [7], but he did not invent the idea; it is merely the way joint probabilities have always been defined in statistics and other fields that use probability.

A theorist might ask whether a probability forecasting system $(P_s)_{s\in\mathbb{S}}$, as defined in Section 12.5, always defines a probability measure on $\mathcal{Y}^N$. It obviously does when $\mathcal{Y}$ is finite. In 1949 [15], Cassius T. Ionescu Tulcea showed that a measurability condition is required in the general case: $P_{y_1\dots y_n}(E)$ must be a measurable function of $(y_1,\dots,y_n)$ for every n and every measurable $E\subseteq\mathcal{Y}$. Given that no one has ever seen a non-measurable function, we need not spend much time worrying about whether this condition is satisfied. Notably, however, it is not required in (12.2), our definition of global upper probability.

Notes

1. For an overview of Ville's influence, see [4]. Key milestones include [28, 18, 19, 21, 7]. For my work with Vovk, see our recent book, *Game-Theoretic Foundations for Probability and Finance* [26], and the working papers posted at www.probabilityandfinance.com. For the application to mathematical statistics, see especially [23].
2. The condition that each bet only risk the capital from the previous one is essential. If you can draw on an unlimited line of credit, then it may not be so hard to multiply the capital you actually use; see [6].
3. They do obey the weaker rules used in the theory of imprecise probabilities. See [1, Chapter 6] and [26, Chapter 6].

Bibliography

[1] Augustin, T., Coolen, F. P. A., de Cooman, G. and Troffaes, M. C. M. editors, 2014. *Introduction to Imprecise Probabilities.* West Sussex: Wiley.

[2] Bernoulli, J., 1713. *Ars Conjectandi.* Basel: Thurnisius.

[3] Bernoulli, J., 2006. *The Art of Conjecturing, Together with Letter to a Friend on Sets in Court Tennis.* Baltimore: Johns Hopkins University Press. Translation of [2] and commentary by Edith Sylla.

[4] Bienvenu, L., Shafer, G. and Shen, A., 2009. On the history of martingales in the study of randomness. *Electronic Journal for History of Probability and Statistics, 5*(1), pp. 1–40.

[5] Breiman, L., 2001. Statistical modeling: The two cultures (comments and a rejoinder by the author). *Statistical Science, 16*(3), pp. 199–231.

[6] Crane, H. and Shafer, G., 2020. Risk is random: The magic of the d'Alembert. Working Paper 56, www.probabilityandfinance.com.

[7] Dawid, A. P., 1984. Statistical theory: The prequential approach (with discussion). *Journal of the Royal Statistical Society: Series A, 147*(2), pp. 278–292.

[8] Dempster, A. P., 1966. Further examples of inconsistencies in the fiducial argument. *Annals of Mathematical Statistics, 34*(3), pp. 884–891.

[9] Dempster, A. P., 1968. Crosscurrents in statistics; Review of *The Selected Papers*, by E. S. Pearson, *Joint Statistical Papers*, by Jerzy Neyman and E. S. Pearson, and *A Selection of Early Statistical Papers*, by J. Neyman. *Science, 160*, pp. 661–663.

[10] Edgeworth, F., 1918. Mathematical representation of statistics: A reply. *Journal of the Royal Statistical Society, 81*(2), pp. 322–333.

[11] Fisher, R. A., 1922. On the mathematical foundations of theoretical statistics. *Philosophical Transactions of the Royal Society of London (A), 222*, pp. 309–368.

[12] Franklin, J., 2001. *The Science of Conjecture: Evidence and Probability before Pascal* (2nd ed.). Baltimore: Johns Hopkins, 2015.

[13] Freudenthal, H., 1980. Huygens' foundations of probability. *Historia Mathematica, 7*, pp. 113–117.

[14] Hacking, I., 1990. *The Taming of Chance.* New York: Cambridge University Press.

[15] Ionescu Tulcea, C. T., 1949. Mesures dans les espaces produits. *Atti della Accademia Nazionale dei Lincei Rendiconti Classe di Scienze Fisiche, Matematiche e Naturali, 7*, pp. 208–211.

[16] Knebel, S. K., 2000. *Wille, Würfel und Wahrscheinlichkeit: Das System der moralischen Notwendigkeit in der Jesuitenscholastik 1550–1700.* Hamburg: Meiner.

[17] Kolmogorov, A. N., 1933. *Grundbegriffe der Wahrscheinlichkeitsrechnung.* Berlin: Springer. An English translation by Nathan Morrison appeared under the title *Foundations of the Theory of Probability* (Chelsea, New York) in 1950, with a second edition in 1956.

[18] Martin-Löf, P., 1966. The definition of random sequences. *Information and Control, 9*(6), pp. 602–619.

[19] Martin-Löf, P., 1969. The literature on von Mises' Kollectivs revisited. *Theoria, 35*, pp. 12–37.

[20] Nagel, E., 1936. The meaning of probability. *Journal of the American Statistical Association, 31*(193), pp. 10–30.

[21] Schnorr, C.-P., 1971. Vorwort und Einleitung. In: *Zufälligkeit und Wahrscheinlichkeit. Eine algorithmische Begründung der Wahrscheinlichkeitstheorie.* Lecture Notes in Mathematics, vol. 218. Berlin: Springer. https://doi.org/10.1007/BFb0112459.

[22] Shafer, G., 2019. Pascal's and Huygens's game-theoretic foundations for probability. *Sartoniana, 32*, pp. 117–145.

[23] Shafer, G., 2021. Testing by betting: A strategy for statistical and scientific communication (with discussion). *Journal of the Royal Statistical Society: Series A, 184*(2), pp. 407–431.

[24] Shafer, G. and Vovk, V., 2001. *Probability and Finance: It's Only a Game!* New York: Wiley.

[25] Shafer, G. and Vovk, V., 2006. The sources of Kolmogorov's *Grundbegriffe. Statistical Science, 21*(1), pp. 70–98.

[26] Shafer, G. and Vovk, V., 2019. *Game-Theoretic Foundations for Probability and Finance.* Hoboken, NJ: Wiley.

[27] Stigler, S. M., 2007. Chance is 350 years old. *Chance, 20*(4), pp. 33–36.

[28] Ville, J., 1939. *Étude critique de la notion de collectif.* Paris: Gauthier-Villars.

[29] von Mises, R., 1919. Grundlagen der Wahrscheinlichkeitsrechnung. *Mathematische Zeitschrift, 5*, pp. 52–99.

[30] von Mises, R. 1941. On the foundations of probability and statistics. *Annals of Mathematical Statistics, 12*(2), pp. 191–205.

[31] von Mises, R., 1942. On the correct use of Bayes' formula. *Annals of Mathematical Statistics, 13*(2), pp. 156–165.

[32] Wu, W. and Shafer, G., 2007. Testing lead-lag effects under game-theoretic efficient market hypotheses. Working Paper 23, www.probabilityandfinance.com.

13

Introduction to Generalized Fiducial Inference

Alexander C. Murph

University of NC at Chapel Hill

Jan Hannig

University of NC at Chapel Hill

Jonathan P. Williams

NC State University

Fiducial inference was introduced in the first half of the 20th century by R.A. Fisher as a means to get a posterior-like distribution for a parameter without having to arbitrarily define a prior [11]. While the method originally fell out of favor due to non-exactness issues in multivariate cases, the method has garnered renewed interest in the last decade. This is partly due to the development of generalized fiducial inference, which is a fiducial perspective on generalized confidence intervals: a method used to find approximate confidence distributions. In this chapter, we illuminate the usefulness of the fiducial philosophy, introduce the definition of a generalized fiducial distribution, and apply it to interesting, non-trivial inferential examples.

13.1 Introduction

Fiducial inference attempts to find a middle ground between the frequentist and Bayesian perspectives. The fiducial argument allows one to fit a posterior-like distribution on a target parameter θ in a way that is entirely data-driven and does not rely on a sometimes-arbitrary prior selection. This argument is based on inverting a *data generating algorithm (DGA)* that associates data $\mathbf{Y}$ to the parameters θ and a random component U with a known distribution F_0, e.g. a vector of independent and identically distributed (iid) standard uniforms or standard Gaussians. This is often expressed as the relation $\mathbf{Y} = A(U, \theta)$. Since the DGA is a function of a random variable with a known distribution, it immediately determines the likelihood $f(\mathbf{y}|\theta)$. By solving the DGA for θ, we get the distribution of our parameters, called the *generalized fiducial distribution* (GFD), that is entirely data-driven and does not require the use of Bayes' Theorem.

While the strengths and limitations of *generalized fiducial inference* (GFI) continue to be explored, its usefulness has already been illustrated in numerous practical applications. Recent work has applied the fiducial ideas to bio-equivalence [25, 14], metrology problems [17, 38, 39, 15, 40], inter-laboratory experiments and international key comparison experiments [22]. It has also been used to tackle statistical problems at the forefront of modern topics in statistical research, such as in wavelet regression [16] and extreme value

 DOI: 10.1201/9780429341731-13

estimation [37], and has recently led to a creative new perspective on linear model selection [41] and vector autoregressive graph selection [42].

As a brief motivating example of the fiducial argument, we define the DGA of a single observation from a normal distribution with known mean, $Y = \mu + U$, where $U \sim \mathcal{N}(0,1)$ is our random quantity with known distribution, assumed to be independent of μ. We ascribe a distribution on the parameter μ by inverting the DGA: $\mu = Y - U \sim \mathcal{N}(Y, 1)$. While inverting the DGA for this simple normal example is algebraically simple, inverting a general DGA could be non-trivial.

Heuristically, a smooth DGA $A(U, \theta)$ taken as a function of θ behaves locally like a linear function near our observed data value $\mathbf{y}$. Therefore, for each realization of the random quantity $U^\star$, there is a well-defined point $\theta^\star$ so that $A(U^\star, \theta^\star)$ is closest to $\mathbf{y}$. The GFD is calculated as the distribution of $\theta^\star$ using the implicit function theorem from the distribution of $U^\star$ conditional on the event that $\{A(U^\star, \theta^\star) \approx \mathbf{y}\}$.

Formally, we define the GFD as a limit in the following way. Defining the pseudo-inverse of the DGA using the optimization problem

$$Q_{\mathbf{y}}(u) = \text{argmin}_{\theta^\star} \|\mathbf{y} - A(U, \theta^\star)\|. \tag{13.1}$$

Typically $\|\cdot\|$ is either ℓ_2 or ℓ_∞ norm. This will serve to find the closest point θ described above. While not technically an inverse, this quantity is always defined. Next, for each small $\epsilon > 0$, define the random variable $\theta^\star_\epsilon = Q_{\mathbf{y}}(U^\star_\epsilon)$, where $U^\star_\epsilon$ has distribution F_0 **truncated** to the set

$$\mathcal{M}_{\mathbf{y},\epsilon} = \{U^\star_\epsilon : \|\mathbf{y} - A(U^\star_\epsilon, \theta^\star_\epsilon)\| = \|\mathbf{y} - A(U^\star_\epsilon, Q_{\mathbf{y}}(U^\star_\epsilon))\| \leq \epsilon\}, \tag{13.2}$$

i.e., having the density $f_U(u^\star) I_{M_{\mathbf{y},\epsilon}}(u^\star) / (\int_{M_{\mathbf{y},\epsilon}} f_U(u)\, du)$, where f_U is the original density of U. Then assuming that the random variable $\theta^\star_\epsilon$ converges in distribution as $\epsilon \to 0$, the GFD is defined as the limiting distribution of $\lim_{\epsilon\to 0} \theta^\star_\epsilon$.

When the sampling distribution of $\mathbf{Y}$ is discrete, we can set $\epsilon = 0$ and no limit is necessary. When the sampling distribution of $\mathbf{Y}$ is continuous, the implicit function theorem is used to calculate the limiting distribution of $\theta^\star_\epsilon$, which leads to the following result.

Theorem 13.1 ([19]) *Under mild conditions (see Assumptions B.1–B.4 in [19]), the limiting distribution above has density*

$$r_{\mathbf{y}}(\theta) = \frac{f(\mathbf{y}|\theta) J(\mathbf{y}, \theta)}{\int f(\mathbf{y}|\theta')\, J(\mathbf{y}, \theta')\, d\theta'}, \tag{13.3}$$

where $J(\mathbf{y}, \theta) = D\left(\nabla_\theta A(u, \theta)|_{u = A^{-1}(\mathbf{y}, \theta)}\right)$. *Here* $\nabla_\theta A(u, \theta)$ *is the gradient matrix computed with respect to* θ, *and* D *is a determinant-like operator that depends on the norm inEquation* (13.1), *e.g., when we use* ℓ_2 *norm* $D(M) = (\det M'M)^{\frac{1}{2}}$.

The generalized fiducial approach communicates a simple algorithm: when possible, define a DGA that expresses the relationship between the data, the parameter, and a random quantity, then invert it. The application of this idea in practice, however, can be nuanced and requires careful thought. Our aim in this chapter is to provide a comprehensive overview of GFI by way of detailed examples. As we will illustrate below, in some examples inverting a DGA does not lead to a single point associated with a given U draw from F_0. In this case, we will find a fiducial distribution on the sets of parameters, cf. Demspter-Shafer theory [6]. In particular, we will treat two instances of multivariate normal data and a binomial distribution with an unknown number of trials. We selected these examples both because they are of interest in their own right, but also because they will allow us to demonstrate how to implement a generalized fiducial solution using modern computational tools. The computer codes are available at a GitHub repository [https://github.com/sirmurphalot/IntroductionGFI]. For the binomial algorithms, this GitHub page also includes full pseudocodes of the implementations.

Throughout this paper, numerous approximate fiducial confidence intervals are reported. Although we use the term *confidence*, our approximate fiducial confidence intervals are closely related to the notion of the Bayesian *credible* interval. Each of these intervals involves defining a probability distribution on a target parameter and using this distribution to calculate a set whose probability mass matches a researcher's desired level of coverage. Rather than call the fiducial intervals credible intervals, which are specifically a Bayesian construct, we refer to them as approximate confidence intervals.

The caveat that these intervals are *approximate* is necessary for numerous reasons. For the continuous multivariate normal problems, it is necessary to perform Markov Chain Monte Carlo (MCMC) sampling, an estimation method, to circumvent the need to calculate the untenable marginal integral in the denominator of Equation (13.3). MCMC sampling is also necessary to explore the sample space for the binomial problem for n and p unknown. In addition, for both of the binomial problems, approximation is necessary to address the issue of an unbounded sample space on the n parameter. While all these sources of uncertainty merit the use of the term "approximate," the primary reason we refer to these intervals in this way is because the coverages need not be exact. This "close but not exact" coverage is analogous to the caveats one imposes on frequentist confidence intervals based on asymptotic normality. We address these multiple sources of approximation in this paper using simulation studies to assess whether the performance of our computational strategies (the *empirical coverage*) achieve our theoretical expectations (the *nominal coverage*). The results of these simulations show that, regardless of these approximations, the fiducial method is justified for practical use.

13.2 Multivariate Normal Distribution

13.2.1 Multivariate Normal Data; μ Unknown, Σ Unknown

The estimation of covariance matrices is a fundamental problem in many multivariate methods. Examples include discriminant data analysis, longitudinal data analysis, time series analysis, and spatial data analysis, just to name a few. However, only recently it was pointed out that in the Bayesian context the most commonly used conjugate inverse Wishart prior may not be the best choice for estimation [2, 3, 44]. In particular, the inverse Wishart posterior has the effect of forcing the eigenvalues of the covariance matrix apart. This "systemic distortion" of the eigenstructure of the covariance matrix has negative effects on estimation and has been studied extensively. For further details on this problem, and a survey of proposed solutions, we suggest the papers by [28] and [44].

Let $\mathbf{Y}_1, \mathbf{Y}_2, \ldots, \mathbf{Y}_m$ be iid $\mathcal{N}_d(\mu, \Sigma)$, where μ is a d-dimensional vector and Σ is a $d \times d$ covariance matrix. Our aim is to perform inference on the covariance matrix Σ with the unknown mean parameter μ. To date, there have been proposed two basic approaches to define the GFD for this model. They both start with the DGA,

$$\mathbf{Y}_i = \mu + B\mathbf{U}_i, \quad i = 1, \ldots, n,$$

where $\mathbf{U}_i$ are iid standard Gaussian vectors of dimension d, but differ in the structure of the matrix B. One example treats B as a lower triangular matrix [35]. This leads to

GFD that depends on the arbitrary order of the coordinates. In contrast, we can use an arbitrary $d \times d$ matrix B, which removes the dependence on the coordinate order but is overparametrized [31]. In this second case, the resulting GFD for the covariance matrix $\Sigma = BB^\top$ belongs to the Wishart family. Several other reasonable choices for B lead to a GFD belonging to the Wishart family: $B = \Sigma^{1/2} = Z\Lambda Z^\top$, where Z is an orthogonal matrix, Λ is a diagonal matrix with positive entries on the diagonal, or the lower triangular B with a DGA for the sufficient statistic, i.e., the sample mean and sample covariance modeled by $\bar{\mathbf{Y}} = \mu + n^{-1}BU_1$, $S^2 = n^{-1}BU_2B^\top$ respectively, where U_1 is d-variate standard normal and U_2 has Wishart distribution $W_d(I, n-1)$.

We propose an alternative DGA that does not lead to the Wishart distribution and is not overparameterized. In particular, consider $B = Z\Lambda$, where Z is an orthogonal matrix, Λ is a diagonal matrix with positive entries on the diagonal. Consequently, the covariance matrix is $\Sigma = Z\Lambda^2Z^\top$. To be able to compute the GFD using (13.3), we will need to reparametrize Z using the *Cayley transformation* (see Theorem 13.2). In particular, we will use the following two facts. For the proof of the first fact see, for instance, [9].

Theorem 13.2 (Cayley transform) *Every real orthogonal matrix Z that does not have -1 as a characteristic root can be expressed as*

$$Z = (I_d - A)(I_d + A)^{-1} = (I_d + A)^{-1}(I_d - A) \tag{13.4}$$

by a suitable choice of a real skew-symmetric matrix, i.e., $A^\top = -A$.

Theorem 13.3 ([27]) *For any orthogonal matrix Z, there must exist a signature matrix D such that ZD does not have -1 as a characteristic root and all elements of corresponding Cayley transform $a_{i,j}$ are such that $|a_{i,j}| \leq 1$ for $1 \leq i \leq j \leq d$.*

Recall that D is a diagonal matrix with entries ± 1. Consequently, $ZD\Lambda^2DZ^\top = Z\Lambda^2Z^\top$ which is invariant to the choice of D. This leads us to propose the following data-generating algorithm

$$\mathbf{Y}_i = \mu - (I_d + A)(I_d + A)^{-1}\Lambda\mathbf{U}_i, \quad i = 1, \ldots, n, \tag{13.5}$$

where A is a skew-symmetric matrix with all entries $|a_{ij}| \leq 1$ and Λ is a diagonal matrix with positive entries $\lambda_i > 0$.

We show in Appendix 13.4 that the Jacobian is $J(\mathbf{y}, \theta) = J^*(\mathbf{y}, A)\prod_{i=1}^d \lambda_i^{-1}$, where $J^*(\mathbf{y}, A)$ does not depend on μ or λ. The form of this Jacobian allows us to simplify the generalized fiducial density from Equation (13.3). In particular, calculations in Appendix 13.4 show that the marginal GFD of *the unique entries of* A is

$$r_{\mathbf{y}}(\text{veck}(A)) \propto J^*(\mathbf{y}, A)\prod_{i=1}^{d}(Z^T nS^2\, Z)_{ii}^{\frac{-(n-1)}{2}}, \tag{13.6}$$

where $S^2 = \frac{1}{n}\sum_{i=1}^n(\mathbf{Y}_i - \bar{\mathbf{Y}})(\mathbf{Y}_i - \bar{\mathbf{Y}})^\top$ and $\bar{\mathbf{Y}} = n^{-1}\sum_{i=1}^n \mathbf{Y}_i$. The veck$(A)$ operation vectorizes the strictly lower triangular elements of the skew-symmetric matrix A, as discussed in [20]. The matrix Z is defined as a function of veck(A) in Equation (13.4). The conditional GFD of the diagonal entries of Λ^{-2} given the veck(A) follow independent gamma distributions, $\lambda_i^{-2} \sim \text{Gamma}\left(\frac{n-1}{2}, \frac{(Z^\top nS^2\,Z)_{ii}}{2}\right)$, $i = 1, \ldots, d$. The conditional GFD of μ given Λ and A is multivariate $\mathcal{N}\left(\bar{\mathbf{Y}}, n^{-1}Z\Lambda^2Z^\top\right)$. The simple form of this distribution allows us to implement sampling from the GFD using STAN [32].

Since A is skew-symmetric, its diagonal terms must be zero, and the matrix is determined by the entries of the strict lower triangle of A. The number of free parameters (A, Λ, μ) in Equation (13.5) is therefore $d(d+3)/2$, which matches the number of free equations in the minimal sufficient statistic $(\bar{\mathbf{Y}}, S^2)$, where there are d free equations in the vector $\bar{\mathbf{Y}}$ and $(d+1)d/2$ free equations from the lower triangle of S^2. Consequently, our DGA is not overparameterized. We do not impose an order restriction on the diagonal entries of Λ. This makes the calculation of the entries λ_i easier, but introduces non-uniqueness. To address this issue, we run parallel chains of our MCMC algorithm each starting with their own random ordering of the singular values, obtained by using *Principle Component Analysis* on the sample covariance matrix.

The STAN software allows us to quickly and easily draw samples from the GFD, which we can then use to obtain approximate fiducial confidence intervals for the true covariance matrix. To build these fiducial confidence intervals and evaluate our method, we consider a number of distance metrics and parameters of interest. For the distance metrics, fiducial confidence intervals are developed by defining a ball around the mean of the GFD that covers $(1-\alpha)\%$ of the GFD. This is done by calculating the distance from the average point for every value sampled from the GFD and defining the ball using a distance cutoff such that $(1-\alpha)\%$ of the sampled values have a distance from the average value that is less the chosen cutoff. For parameters of interest, fiducial confidence intervals are developed much like they would be on the real line. That is, after mapping every matrix from the GFD to the real line, we simply take the center $(1-\alpha)\%$ of the GFD as our fiducial confidence interval.

In particular, we construct the fiducial confidence intervals for the true covariance matrix using the following two distance metrics, two norms, and one parameter of interest. The norms are used as both distance metrics (for matrices $M, N : ||M - N||$) and as parameters of interest (for matrix $M : ||M||$). Let M, N be two positive-definite, symmetric matrices. Then,

1. FM_DISTANCE [12] : $dist(M,N) = \sqrt{\sum_{i=1}^{d} \ln^2 \lambda_i(M,N)}$, where $\lambda_i(M,N)$ are the eigenvalues from the equation $\det(\lambda M - N) = 0$;
2. STEIN_LOSS [23] : $dist(M,N) = \mathrm{tr}(M^{-1}N) - \log(\det(M^{-1}N)) - d$;
3. SPECTRAL_NORM [21] : $norm(M) = \lambda_{(d)}$, the maximum value in the Λ matrix;
4. FROBENIUS_NORM [21] : $norm(M) = \sqrt{\mathrm{tr}(MM^H)}$, where M^H is the conjugate transpose of M;
5. LOGDET [10] : $huer(M) = \log(\det(M))$.

Our simulation study consists of 1000 iterations on a 4×4 multivariate normal $\mathcal{N}_4(\mu, \Sigma)$ where

$$\mu = \begin{pmatrix} 1 \\ 2 \\ 3 \\ 1 \end{pmatrix}, \quad \Sigma = \begin{pmatrix} 4 & 1 & 0 & 0 \\ 1 & 1 & 0 & 1 \\ 0 & 0 & 9 & 1 \\ 0 & 1 & 1 & 4 \end{pmatrix}.$$

At each iteration, we simulate a dataset of $n = 100$ observations from the given multivariate normal distribution. Then, we run 20 MCMC chains and use the simulated values from the GFD of the covariance matrix to construct approximate fiducial confidence intervals around the true covariance matrix Σ.

Figure 13.1 shows the performance of these fiducial confidence intervals for different values of α. The nominal coverage corresponds to the value of α we set and the empirical coverage is the proportion of times $(1-\alpha)$ balls made this way contained the true covariance matrix. As we can see, all of our metrics resulted in a line across the diagonal, communicating that nominal and empirical coverage match. Table 13.1 shows the specific coverage of these fiducial confidence intervals for $\alpha = 0.05$. As we can see, the empirical coverage probabilities of each of these intervals are approximately 95%, as desired.

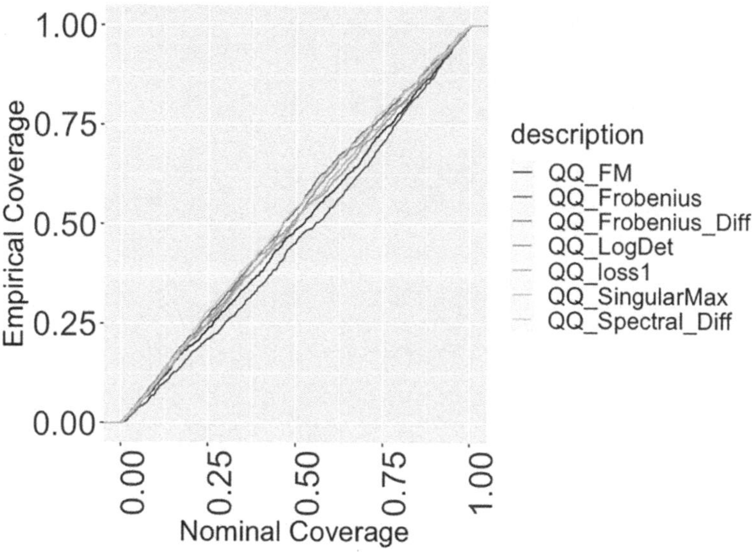

FIGURE 13.1
The coverage of fiducial confidence intervals for the covariance matrix using different notions of distance. This figure is created by setting different levels of coverage in our GFD fiducial confidence intervals and checking the empirical coverage of these intervals over the whole simulation. A perfect diagonal line equates to a match between our observed (empirical) coverage and our theoretical (nominal) coverage. These lines show that our method approximately achieves the desired coverage.

TABLE 13.1
The empirical coverage of fiducial confidence intervals for the covariance matrix using different notions of difference. For the distance metrics, fiducial confidence intervals are developed by defining a ball around the mean of the GFD that covers 95% of the GFD. For parameters of interest, fiducial confidence intervals are developed much like they would be on the real line. That is, after mapping every matrix from the GFD to the real line, we simply take the center 95% of the GFD as our fiducial confidence interval. This table shows how our empirical coverage is approximately equal to our expected nominal coverage of 95%

Distance Metric/Norm	**Empirical Coverage of the 95% Fiducial Confidence Interval**
FM_DISTANCE	0.9505
LOSS1_DISTANCE	0.9463
LOGDET Parameter	0.9589
SPECTRAL_NORM Parameter	0.9568
FROBENIUS_NORM Parameter	0.9547
SPECTRAL_NORM Distance	0.9540
FROBENIUS_NORM Distance	0.9580

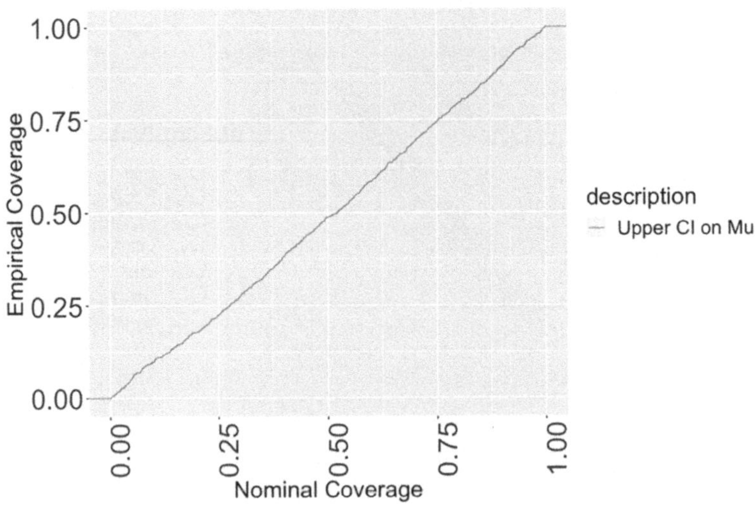

FIGURE 13.2
The coverage of fiducial confidence intervals for the mean vector μ using standard Euclidean distance. This figure was made in the same way as Figure 13.1. This line shows that our method approximately achieves the desired coverage.

The simultaneous fiducial confidence interval for μ is constructed the same way as Σ using the standard Euclidean distance $dist(u, v) = ||u - v||$. Draws from the GFD for μ were calculated at the same time as the draws of Σ in our simulation study. Figure 13.2 shows that nominal coverage matched the empirical coverage, validating our method in the context of this inference problem.

13.2.2 Generalization of One Way Random Effects Model

Continuing to examine the multivariate normal problem, we consider an alternative, more restrictive parameterization, the standard unbalanced one-way random effect model $Y_{i,j} = \mu + \eta_i + \epsilon_{i,j}$, for $\eta_i \sim \mathcal{N}(0, \sigma_a^2)$, $\epsilon_{i,j} \sim \mathcal{N}(0, \sigma_e^2)$, where $i = 1, \ldots, n$ and $j = 1, \ldots, d$, and the η_i and $\epsilon_{i,j}$ terms are independent. The first fiducial solution to inference on this model was given by [7] and used a tailor-made solution to this problem. Let us consider the more general DGA

$$\mathbf{Y} = \mathbb{X}\beta + A\mathbf{U}, \tag{13.7}$$

where β is an unknown vector of fixed effects, $\mathbf{U}$ is standard normal random vector, $\mathbb{X}$ is a fixed effect design matrix and $A = \Sigma^{1/2}$ such that $\Sigma = \sigma_\alpha^2 S_\alpha + \sigma_e^2 I$ and S_α is a matrix of ones and zeros corresponding to the group sizes $n_1, n_2, \ldots, n_m$. Thus, our inferential problem is simplified to providing GFD for β, σ_α^2, and σ_e^2.

While the original study in [7] was able to derive well-performing generalized fiducial intervals for σ_α^2 and σ_e^2, it did so by way of a long calculation and they did not allow for simultaneous inference on the fixed effects β. Using (13.3) the generalized fiducial intervals can be implemented without major computational hassle using the STAN software, while simultaneously obtaining GFD for the fixed effects β that is often of interest [26].

Due to the structure imposed by (13.7), the Jacobian matrix simplifies greatly and does not involve the square root of a complicated matrix. Indeed, the function $J(\mathbf{Y}, S_\alpha, \sigma_\alpha^2, \sigma_e^2, \mathbb{X}\beta)$ is computed from the $n \times (d+2)$ matrix obtained by column concatenation of

$$\nabla_\beta \mathbf{Y} = \mathbb{X}, \quad \frac{\partial \mathbf{Y}}{\partial \sigma_e^2} = (\sigma_\alpha^2 S_\alpha + \sigma_e^2 I)(\mathbf{Y} - \mathbb{X}\beta), \quad \frac{\partial \mathbf{Y}}{\partial \sigma_\alpha^2} = S_\alpha(\sigma_\alpha^2 S_\alpha + \sigma_e^2 I)(\mathbf{Y} - \mathbb{X}\beta).$$

Derivation of the above quantities is outlined in Appendix 13.4.

We will draw samples from our GFD to create fiducial confidence intervals that we will then compare to the truth to assess performance, parallel the simulation study in [7]. In particular, at each instance of our simulation, we generate data using seven different designs of group sizes outlined in Table 13.2 and one of the true parameter values $(\sigma_\alpha^2, \sigma_e^2)$ from the set $\{(.1, 10), (.5, 10), (1, 10), (.5, 2), (1, 1), (2, .5), (5, .2), (10, .1)\}$.

The results of our simulation can be seen in Figure 13.3 and Table 13.3. Across many different parameter choices, it would appear that the generalized fiducial method performs

TABLE 13.2
The different group assignments tested for simulation of mixed models, as suggested by [7]. Here, Φ is an expression of imbalance as defined by [1]. The smaller the value of Φ, the greater the degree of imbalance

Pattern	Φ	n_i					
1	.068	1	1	1	1	1	100
2	.130	2	2	2	2	2	100
3	.187	2	5	60			
4	.410	4	4	4	8	48	
5	.700	5	10	15	20	25	30
6	.807	2	2	4	6		
7	.957	6	6	8	8	10	10

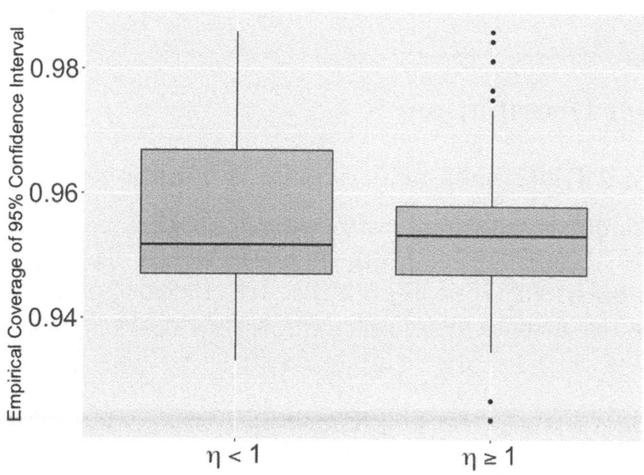

FIGURE 13.3
Empirical coverages for $\eta = \sigma_\alpha^2/\sigma_e < 1$ and $\eta \geq 1$ for both σ_α^2 and σ_e^2 across all parameter choices and all seven group choices. We can see that the generalized fiducial method tends to be more conservative for larger values of η.

TABLE 13.3
Median (IQR) of the CI lengths for all our parameters. Instances of outstanding lengths likely contribute to the conservative empirical coverage we see in Figure 13.3. These results are comparable to the conclusions in [7]

$(\sigma_\alpha^2, \sigma_e^2)\backslash$ Pattern	1	2	3	4
(.1,10)	34.06 (22.54)	19.38 (13.44)	88.16 (93.09)	12.55 (10.49)
(.5,10)	35.84 (27.90)	22.21 (16.12)	102.62 (111.85)	14.99 (12.67)
(1,10)	37.70 (30.91)	23.89 (18.58)	114.65 (123.98)	18.77 (17.68)
(.5,2)	9.26 (7.26)	6.18 (5.28)	28.42 (34.99)	5.38 (5.82)
(1,1)	7.62 (8.05)	6.85 (6.34)	33.84 (49.73)	7.45 (8.04)
(2,.5)	11.33 (11.00)	10.74 (10.05)	52.34 (87.86)	14.83 (15.00)
(5,.2)	24.47 (24.48)	26.08 (23.21)	140.51 (213.16)	34.27 (34.80)
(10,.1)	49.16 (43.76)	47.13 (44.41)	279.18 (419)	69.09 (67.37)
$(\sigma_\alpha^2, \sigma_e^2)\backslash$ Pattern	5	6	7	
(.1,10)	4.00 (3.50)	38.63 (39.64)	7.06 (6.04)	
(.5,10)	5.92 (5.56)	45.66 (44.97)	9.01 (8.18)	
(1,10)	8.63 (8.59)	49.28 (54.27)	10.98 (9.79)	
(.5,2)	3.29 (2.85)	12.88 (15.37)	3.75 (3.23)	
(1,1)	5.58 (4.71)	14.29 (18.18)	5.61 (4.95)	
(2,.5)	10.18 (9.08)	24.27 (28.87)	10.02 (9.80)	
(5,.2)	24.96 (22.16)	54.17 (64.35)	24.79 (22.60)	
(10,.1)	52.37 (42.99)	102.33 (135.72)	47.55 (40.69)	

well, with a slight conservative tendency for certain extreme parameter combinations. Our findings match [7], who found this same conservative tendency for larger values of $\eta = \sigma_\alpha^2/\sigma_e^2$ and greater degrees of imbalance, yet promising performance in all other instances. We conclude by pointing out that we were able to efficiently implement our calculations using RStan which yields accurate results with a frugal use of iterations.

13.3 Binomial Distribution

13.3.1 Binomial Distribution; n Known, p Unknown

When working with discrete distributions, Equation (13.3) can no longer be used, and one must instead invert the DGA directly. To demonstrate how one applies the generalized fiducial method to discrete data, we begin with a well-studied application of the GFI ideas: inference on p from the binomial distribution with n known. The following is a possible data generating algorithm

$$Y = \sum_{i=1}^{n} I_{(0,p)}(U_i), \tag{13.8}$$

where $p \in [0, 1]$ is the unknown parameter and U_i are i.i.d. Uniform$(0, 1)$.

Assume we have an observed realization of our random variable $Y = y$. The uniform random variables U_i that generated the observation y are unknown and therefore we replace them with newly generated U_i^*. When we solve (13.8) for p, we obtain the interval

$(U^*_{(y)}, U^*_{(y+1)}]$, where $U^*_{(y)}$ is the yth order statistic of U^*_i. Recall that $U^*_{(y)}$ follows the $Beta(y, n-y+1)$ distribution.

The above illustrates an important feature of GFI for discrete data: it is based on a random set rather than a random point. This is in contrast to a GFD derived for continuous data using Theorem 13.1. Notions from Dempster-Shafer theory, such as *belief* and *plausibility*, can be used to interpret these sets [6, 8]. Another approach is to reduce the interval to a single distribution. Typically this is defined as a function of the densities of the upper and lower bound of the interval, the density of $Beta(y, n-y+1)$ and $Beta(y+1, n-y)$. Some suggest using the arithmetic mean, while others advocate for the geometric mean [18, 29]. Let $f_{\beta(a,b)}$ be the PDF of a $Beta(a, b)$ distribution. The arithmetic mean leads to $r_{y,n}(p) \propto (p\,(n-2y)+y)\, f_{\beta(y,n-y)}(p)$ for $y \neq 0, n$ while the geometric mean leads to $r_{y,n}(p) \propto f_{\beta(y+\frac{1}{2}, n-y+\frac{1}{2})}(p)$. Thus, simulating values from these GFDs is computationally easy and effective, as they are each derived from Beta density functions.

This result immediately generalizes to multiple independent observations from a binomial distribution. Let $Y_1, \ldots, Y_m$ be i.i.d. $\text{Bin}(n, p)$ where n is known but p is unknown. The sufficient statistic $\sum_{i=1}^m Y_i$ follows $\text{Bin}(nm, p)$ and we can use (13.8) to obtain a GFD.

13.3.2 Binomial Distribution; n Unknown, p Known

With the aim of providing a more nuanced perspective on the generalized fiducial approach for discrete data, we will now consider a more complicated problem: binomial data for n unknown. To the best of our knowledge, the binomial distribution with unknown n has yet to receive a fiducial solution in the literature.

Consider m i.i.d. observations from a $\text{Bin}(n, p)$ distribution, $\mathbf{Y} = (Y_1, \ldots, Y_m)$ where p is *known* but n is unknown, and our objective is inference on n. We begin our solution by choosing a DGA based on the inverse of the distribution function. In particular,

$$Y_i = F^{-1}_{n,p}(U_i), \quad i = 1, \ldots, m, \tag{13.9}$$

where $U_1, \ldots, U_m$ are i.i.d. $Unif(0, 1)$, and $F^{-1}_{n,p}(u) = \inf\{y : F_{n,p}(y) \geq u\}$ with $F_{n,p}$ being the distribution function of $\text{Bin}(n, p)$.

There are two major observations to make about inverting (13.9): first, that $n \geq \max\{\mathbf{Y}\}$; and second, that n does not have an upper bound. Because of this latter observation, we must begin by approximating the GFD by restricting our calculations to a range of reasonable values $\{\max\{\mathbf{Y}\}, \ldots, N-1, N\}$ for some upper bound $N \in \mathbb{N}$. Then, the generalized fiducial probability mass of all relevant subsets in this range is determined by a deterministic algorithm. That is, unlike the usual simulation-based algorithms this algorithm directly assigns meaningful probabilities to sets of possible n values. The precise details of this algorithm, including our proposal for choosing the upper bound N, can be found in Appendix 13.4.

We remark that the difference between GFD and Bayesian posterior using the flat prior is a result of the fiducial probability assigned to non-singleton sets, and if that probability is negligible the GFD would coincide with the Bayesian posterior. In the simulations below, the final GFD on the values of n is obtained by sampling from end-points of all of these sets that have positive fiducial probability.

We illustrate the generalized fiducial solution to this inference problem via a simulation study, comparing the generalized fiducial solution against the Bayesian. For this simulation, we generate data with 10, 50 and 100 observations from the distribution $\text{Bin}(n_0, p)$ where $n_0 = 10$ always and p varies over the set $\{0.01, 0.05, 0.1, 0.2, 0.4, 0.5, 0.6, 0.7, 0.8, 0.9, 0.95, 0.99\}$. For each instance of our binomial data, we draw 1000 values $\hat{n}_B$ from the Bayesian posterior distribution on n as well as 1000 values $\hat{n}_F$ from the GFD on n.

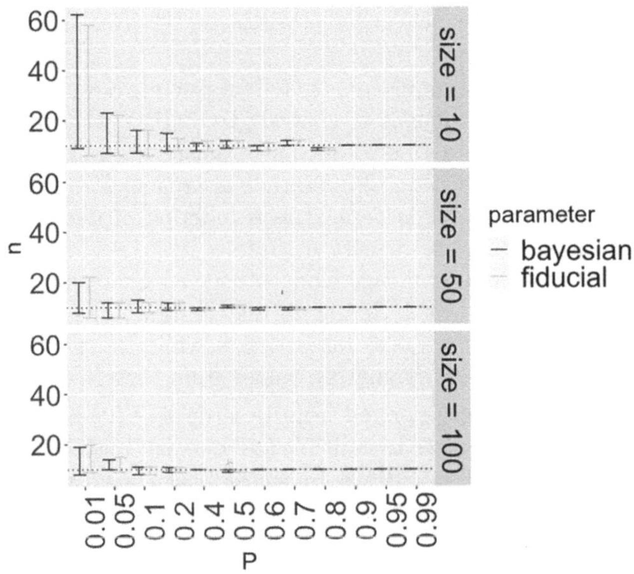

FIGURE 13.4
95% fiducial confidence and credible intervals for each candidate value of p, over different data sizes. This simulation reveals that while the generalized fiducial solution does capture the true value n_0 with an expected amount of regularity, there is not much distinguishing it from the Bayesian solution to the same problem.

Figure 13.4 shows the 95% GFD fiducial confidence intervals and Bayesian credible intervals based on these 1000 draws. This simulation reveals that while the generalized fiducial solution does capture the true value n_0 with an expected amount of regularity, there is not much distinguishing it from the Bayesian solution to the same problem.

To better examine the difference between the generalized fiducial and the Bayesian solutions to this problem, we perform the above simulation 1000 times: each time drawing 1000 values from the GFD and 1000 values from the Bayesian posterior distribution. At each repetition, we record the Mean Absolute Difference (MAD) $\frac{1}{1000}\sum_{i=1}^{1000}|n_0 - \hat{n}_{\eta,i}|$ for both the generalized fiducial and the Bayesian draws, where $\eta \in \{\text{Fiducial, Bayesian}\}$.

Figure 13.5 compares boxplots of each method's 1000 MAD values collected via the described simulation. We can see from this simulation that the generalized fiducial and Bayesian solutions are mostly identical, with some slight variation when the data sizes and values of p are small. We suspect that this slight variation is due to the fact that, in our calculations for these cases, the fiducial method assigned more mass to *sets* of values. This aligns well with the fiducial method's major contrast to the Bayesian method: the former can assign mass to sets of n values while the latter only assigns mass to singleton values.

Figure 13.6 examines the approximate fiducial confidence intervals, built with the upper 95% of the GFD mass, and the Bayesian upper 95% credible intervals, of 1000 simulations of a size 100 draw from a $\text{Bin}(10, \cdot)$ distribution. The horizontal axis of Figure 13.6 is the coverage probability we set for a given interval based on the posterior distribution we derive, while the vertical axis is the observed proportion of time that a given interval captured the true parameter.

Figure 13.6 reaffirms what we saw in Figures 13.4 and 13.5. Namely, the Bayesian solution matches the generalized fiducial, except for some slight variation when p is 0.01 and the data size is small. We also observe that for larger values of p an interval at any coverage

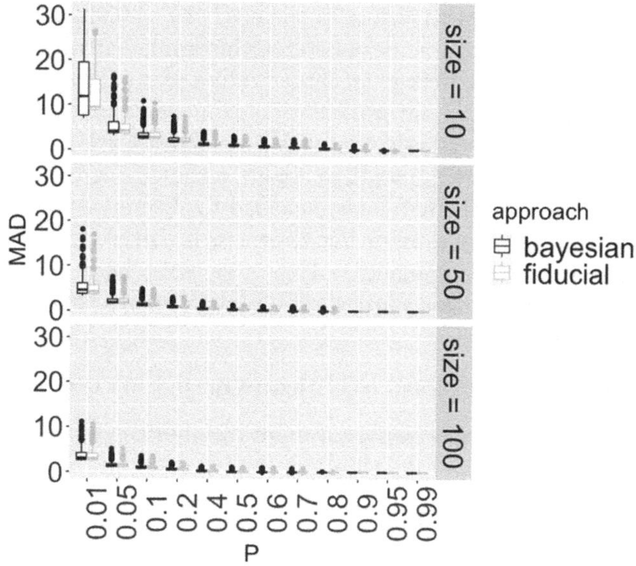

FIGURE 13.5
Boxplots of 1000 MAD calculations over varying values of p and data sizes. There is some slight variation between the fiducial and Bayesian solutions for the smaller data sizes with smaller values of p.

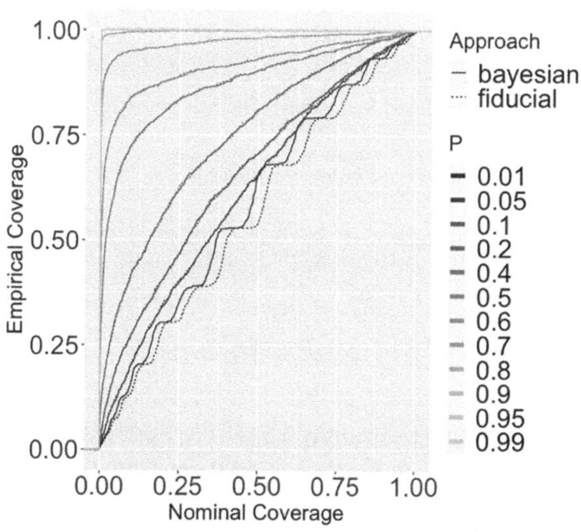

FIGURE 13.6
The coverage of fiducial confidence and credible intervals at varying levels of confidence for the generalized fiducial and Bayesian approaches. This figure was calculated in the same way as Figure 13.1. The Bayesian solution matches the generalized fiducial, except for some slight variation when p is 0.01 and the data size is small.

level will contain the true parameter value. In the case of $p = 0.99$, an interval at any non-zero confidence level captures the true parameter with complete certainty. This relationship between nominal and empirical coverage for larger values of p follows from what we observe in Figure 13.4, which suggests that our posterior distribution is just a single point (the true parameter value n_0), and thus intervals of at any confidence level are simply the set $\{n_0\}$. For smaller values of p, the coverage plots are closer to the 45 degree line indicating a close to correct coverage. The waves in the plot are caused by discretization effects.

Our results show that inference on n for known p is fairly straightforward when the probability of success is high. For smaller values of p, while both the generalized fiducial and Bayesian approaches have less accuracy, they remain comparable.

13.3.3 Binomial Distribution; n Unknown, p Unknown

We will now consider our final non-trivial example through the fiducial lens: simultaneous inference on (n, p) for the binomial distribution. As far as we are aware, this problem has yet to receive a satisfactory solution using any philosophy whatsoever.

We consider a set of m independent binomial data: $\mathbf{Y} = (Y_1, \ldots, Y_m)$ i.i.d. $\text{Bin}(n, p)$, where both p and n are unknown and our objective is inference on the *pair* (n, p). We will use the same DGA (13.9) as in the previous example. Recall the following identity relating to the binomial CDF F and the beta CDF G such that $F_{n,p}(y) = G_{n-y+1,y}(1-p)$ [33]. This allows us to write an equivalent form of (13.9):

$$G^{-1}_{Y_i+1,n-Y_i}(1-U_i) \geq p > G^{-1}_{Y_i,n-Y_i+1}(1-U_i), \quad i = 1, \ldots, m, \tag{13.10}$$

which we can invert to get

$$\hat{p}^{\text{upper}}_{i,n} = G^{-1}_{Y_i+1,n-Y_i}(1-U_i), \quad \hat{p}^{\text{lower}}_{i,n} = G^{-1}_{Y_i,n-Y_i+1}(1-U_i), \quad i = 1, \ldots, m.$$

For any value $\hat{n}$, the set of associated values p that satisfy (13.10) is the interval

$$(\max\{\hat{p}^{\text{lower}}_{1,\hat{n}}, \ldots, \hat{p}^{\text{lower}}_{m,\hat{n}}\}, \min\{\hat{p}^{\text{upper}}_{1,\hat{n}}, \ldots, \hat{p}^{\text{upper}}_{m,\hat{n}}\}].$$

The interval is taken as an empty set if the lower limit is larger than the upper limit.

A generalized fiducial solution to this problem would define a GFD that assigns probability mass to *sets*

$$\bigcup_{\hat{n}} \{\hat{n}\} \times (\max\{\hat{p}^{\text{lower}}_{1,\hat{n}}, \ldots, \hat{p}^{\text{lower}}_{m,\hat{n}}\}, \min\{\hat{p}^{\text{upper}}_{1,\hat{n}}, \ldots, \hat{p}^{\text{upper}}_{m,\hat{n}}\}].$$

These sets contain multiple predicted values $\hat{n}$ that are each associated with a set of values of p. The left panel of Figure 13.7 shows a single random draw from the GFD on the parameter space (n, p). Note that a single draw in the two-dimensional parameter space is a collection of horizontal segments at each $\hat{n}$. Naturally, given a fixed data set, Y, larger and larger values of $\hat{n}$ correspond with smaller values for $\hat{p}$, making them difficult to visualize. For this reason, and because of some desirable properties of the limiting distribution of $\hat{n}\hat{p}$ discussed in Appendix 13.4, we propose to work with a reparameterization (n, μ), where $\mu := np$. This results in a GFD that assigns mass to sets of the form

$$\bigcup_{\hat{n}} \{\hat{n}\} \times (\max\{\hat{\mu}^{\text{lower}}_{1,\hat{n}}, \ldots, \hat{\mu}^{\text{lower}}_{m,\hat{n}}\}, \min\{\hat{\mu}^{\text{upper}}_{1,\hat{n}}, \ldots, \hat{\mu}^{\text{upper}}_{m,\hat{n}}\}],$$

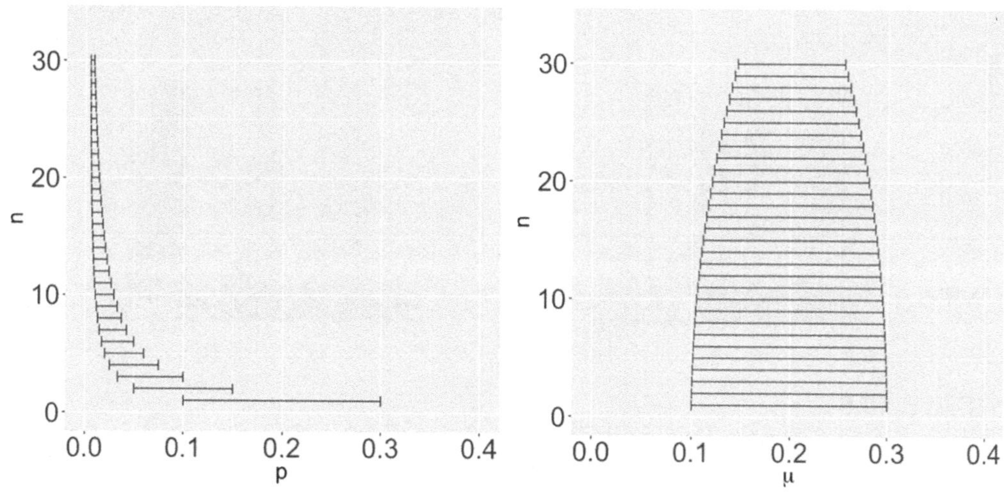

FIGURE 13.7
Two parameterizations for the same single draw from the GFD: using n, p parameterization (on the left) versus using n, μ parameterization (on the right). Note that a draw from this GFD defines a region of values in a joint parameter space that is discrete in n and continuous in p and μ.

where

$$\hat{\mu}_{i,\hat{n}}^{\text{upper}} = \hat{p}_{i,\hat{n}}^{\text{upper}} \hat{n}, \quad \hat{\mu}_{i,\hat{n}}^{\text{lower}} = \hat{p}_{i,\hat{n}}^{\text{lower}} \hat{n}, \quad i = 1, \ldots, m.$$

A draw from this reparameterized GFD can be seen in the right panel of Figure 13.7.

Our algorithm for simulating values from the GFD of (n, μ) uses a Metropolis-within_Gibbs sampler. We start with a set of values $\mathbf{U}$ that give a solution that is not the empty set, we record this solution set, and then we resample the $\mathbf{U}$ values one by one in such a way that the DGA (13.9) is still satisfied. These updates are available in closed form. To improve mixing we also add to each Gibbs scan two additional random walk Metropolis-Hastings steps, one in the n direction and one in the μ direction, so that we can adequately explore our sample space. Further details on the MCMC algorithm can be found in Appendix 13.4.

Figures 13.8 and 13.9 show four implementations of our Gibbs sampler for different true values of n and p. For these figures, we draw 50 observations from the distributions $\text{Bin}(0.1, 15)$, $\text{Bin}(0.1, 75)$, $\text{Bin}(0.9, 15)$, and $\text{Bin}(0.9, 75)$, respectively.

For p small, the binomial distribution approaches a Poisson distribution and results in large sets of $\hat{n}$ values for each draw from our GFD. One can observe this in Figure 13.8, where certain draws from the GFD seem to go upward to infinity in the n direction. Our algorithm senses when a draw from the GFD is reaching its limiting Poisson distribution and stops checking new values for n. The specific details of this stopping criterion are discussed in Appendix 13.4. In contrast, for large p, the algorithm reports a narrower range in n (Figure 13.9). Note that for clarity we jittered the graph in the n direction, especially in the case of Figure 13.9, where most of the sets have a n value concentrated in one area.

Our Gibbs sampler is able to explore both the n space and the μ space simultaneously. We examine the marginal movement in the μ and the n directions in Figures 13.10 and 13.11. To create these figures, we select a single representative μ and n from each draw from the GFD. To select a single μ from a set of values in a single draw from the GFD, we randomly

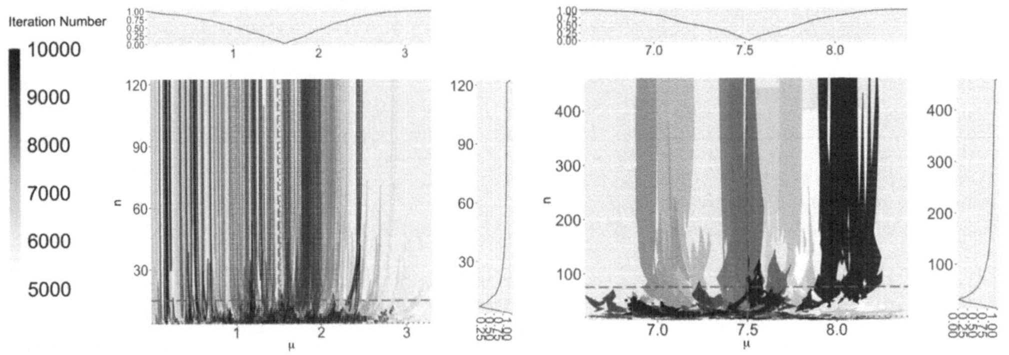

FIGURE 13.8
GFD for Bin(15, 0.1) data (left) and Bin(75, 0.1) data (right). Different shades of color in the center plot correspond to separate draws using a Gibbs sampler on the GFD built on the data. The top and right marginal plots show the confidence curves of μ and n values, respectively, centered at the median. The dashed lines cross at the true (n, μ) value used to generate the distribution and the dotted lines cross at a value of the estimator $(\hat{n}, \bar{\mathbf{Y}})$, developed in [5], that is used to initialize our algorithm.

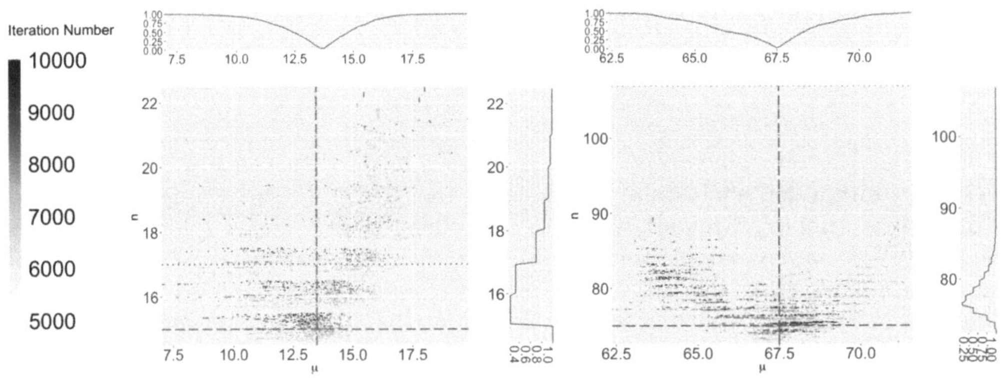

FIGURE 13.9
GFD for Bin(15, 0.9) data (left) and Bin(75, 0.9) (right) visualized in the same way as Figure 13.8. For the Bin(15, 0.9) data, the values are jittered in the n direction for clarity.

choose either the minimum or maximum μ value. We perform the same process in the n direction; for every draw from the GFD, we randomly select either the largest or smallest n value present to be that draw's representative in the trace plot. For each of these selections, the minimum or maximum is chosen with equal probability. The resulting figures show how our algorithm explores the n and μ space around the true parameter values. For smaller true values of p, these plots show how our algorithm visits very large values of n. These draws correspond to the instances where the algorithm begins to capture the limiting Poisson distribution.

We evaluate this method via a simulation study. For each combination $n \in \{15, 75\}$ and $p \in \{0.1, 0.5, 0.9\}$, we draw 100 observations from a Bin(n, p) distribution 300 times. One important thing to note about this solution is that each element of the GFD is a *set* of

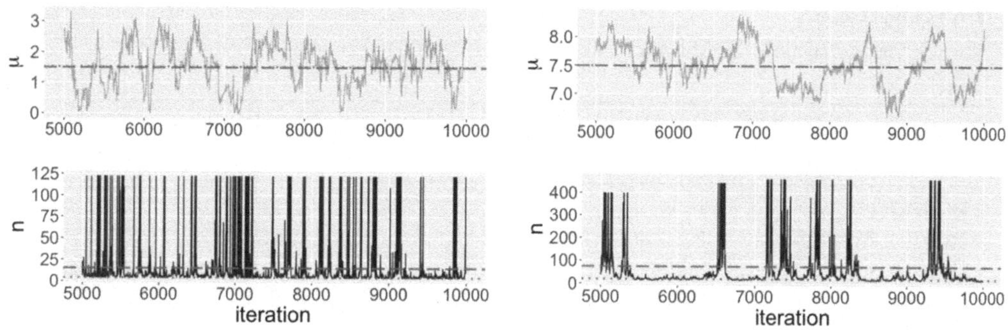

FIGURE 13.10
Marginal trace plots for μ and n for Bin(15, 0.1) data (left) and Bin(75, 0.1) data (right). These plots use the same GFD produced for figures 13.8 and 13.9. The dashed line crosses at the true parameter values used to generate the data and the dotted line crosses at a value of the estimator $(\hat{n}, \bar{\mathbf{Y}})$, developed in [5], that is used to initialize our algorithm.

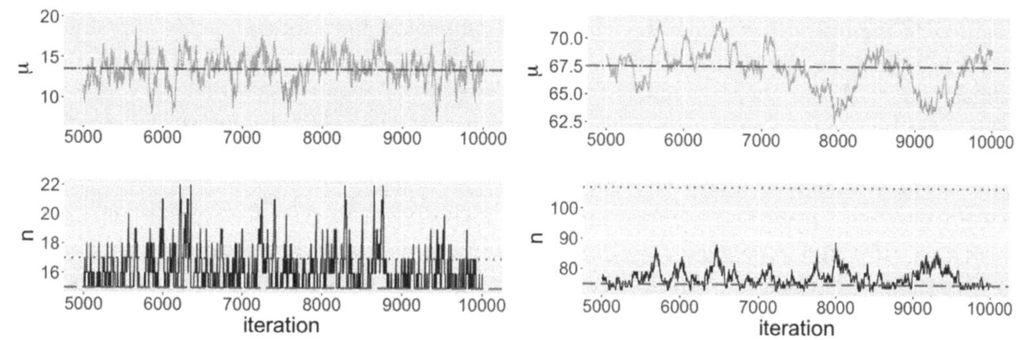

FIGURE 13.11
Marginal trace plots for μ and n for Bin(15, 0.9) data (left) and Bin(75, 0.9) (right) visualized in the same way as Figure 13.10.

values, rather than a singular pair $(\hat{n}, \hat{\mu})$. This complicates the notion of containment ratios slightly. Ideally, for each iteration of our simulation study, we would like to be able to calculate a box of reasonable values that we would expect to contain the true parameter pair (n, μ) 95% of the time. We followed the Dempster-Shafer theory to draw belief and plausibility boxes, which we then used to calculate coverage. In this application, we call the belief box a rectangle centered at the median of each marginal fiducial distribution that fully contained 95% of our fiducial draws, while the plausibility box is a rectangle intersecting 95% of our fiducial draws. One can see the resulting containment ratios using belief and plausibility in Table 13.4. We point out that as expected the belief box is larger than the plausibility box. This is a reflection of the fact that in Dempster-Shafer theory belief of a set is smaller than the plausibility of the same set.

Table 13.4 shows that our method performs mostly conservatively for our choices of parameter values. This is in line with our expectations based on what we observed in Figures 13.8 and 13.9. The belief boxes for these draws from the fiducial posterior would need to be very large, and potentially unbounded, since they need to fully contain 95% of the fiducial samples.

TABLE 13.4
The Coverage of 95% Belief and Plausibility Boxes on Posterior Fiducial Sets of (n, μ).

Distribution	95% Plausibility Containment	95% Belief Containment	μ Marginal Coverage	n Marginal Coverage
Bin(0.1,15)	1	1	0.992	1
Bin(0.1,75)	1	1	0.943	1
Bin(0.5,15)	0.994	0.994	0.961	0.968
Bin(0.5,75)	0.955	0.955	0.961	0.968
Bin(0.9,15)	1	1	1	1
Bin(0.9,75)	0.978	0.986	1	1

13.4 Conclusion and Discussion

GFI provides an alternative perspective on numerous classical inferential problems. Our selection of examples shows how the generalized fiducial framework can derive a meaningful, practically feasible distribution on a target parameter without a need for an arbitrarily defined prior. While there is much to be learned from using this framework on classic theoretical problems, modern research has shown that the fiducial perspective is a strong tool for tackling today's statistical challenges. We provide a short list of such references here for the interested reader: [7, 4, 35, 36, 37, 34, 16, 24].

While the fiducial method has seen much recent success, it has the potential drawback that inverting the DGA for some data applications can be non-trivial. The result by [19], Theorem 13.1, mitigates this issue by developing a general formula for the GFD for continuous data. Of course, Theorem 13.1 replaces the need for a potentially complicated inversion with a need for a potentially complicated differentiation. As a future step towards making fiducial inference accessible to a wide audience, the authors would like to see an autodifferentiator developed to automate the process of inverting a DGA. With such a tool, the fiducial process could be made significantly more accessible.

Acknowledgements

Jan Hannig's research was supported in part by the National Science Foundation under Grant No. IIS-1633074, DMS-1916115, and DMS-2113404.

Appendix

A.1 Derivation of the Multivariate Jacobian Quantity $J(X; \mu, U, \Lambda)$ and Marginal $r(\text{veck}(A))$

Let $J^{i,j}$ be a matrix of all zeros save for a value of 1 at the index (i, j). In general, the form of our Jacobian derivative with respect to any value of the skew-symmetric matrix

A is $J(\mathbf{y}, \mu, A, \Lambda) = \left[\nabla_\mu \mathbf{Y}_i; \frac{\partial \mathbf{Y}_i}{\partial \lambda_j}; \frac{\partial \mathbf{Y}_i}{\partial a_{j,k}}\right]$, where ";" denotes the row concatenation of the following three quantities:

$$\nabla_\mu \mathbf{Y}_i = I_d$$
$$\frac{\partial \mathbf{Y}_i}{\partial \lambda_j} = \lambda_j^{-1}(I_d - A)(I_d + A)^{-1} J^{j,j}[(I_d - A)(I_d + A)^{-1}]^\top (\mathbf{Y}_i - \mu), \quad j = 1, \dots, d$$
$$\frac{\partial \mathbf{Y}_i}{\partial a_{j,k}} = 2(I_d + A)^{-1}\Big(- J^{j,k} + J^{k,j}\Big)(I_d - A)^{-1}(\mathbf{Y}_i - \mu), \quad 1 \le j < k \le d.$$

Using the l_2 norm, the final form of our Jacobian is $D(J(\mathbf{y}, \mu, A, \Lambda))$ where $D(X) = \left|\det\left(\sqrt{X^T X}\right)\right|$. One can show, using the Cauchy-Binnet formula, that the μ vector drops out and that we can factor out the Λ matrix such that $D(J(\mathbf{y}, \mu, A, \Lambda)) = \det(\Lambda)^{-1} D^*(J(\mathbf{y}, A))$. The notation $D^*(J(\mathbf{y}, A))$ is chosen to reiterate that, with the Λ matrix factored out, the remaining Jacobian term relies on neither Λ nor μ. We can thus write the GFD likelihood using (13.3) like so:

$$r_{\mathbf{y}}(\text{veck}(A), \Lambda, \mu) \propto (2\pi)^{\frac{-nd}{2}} |\Sigma|^{-n/2} \exp\left\{ \frac{-1}{2} \operatorname{tr}\left(\Sigma^{-1} \left(\sum_{i=1}^{n} (\mathbf{y}_i - \mu)(\mathbf{y}_i - \mu)^T \right)\right)\right\} D(J).$$

By integrating out μ and rearranging the terms that depend on λ_i we get

$$r_{\mathbf{y}}(\text{veck}(A), \Lambda) \propto D^*(J(\mathbf{y}, A))(2\pi)^{\frac{-d(n-1)}{2}} n^{-d/2} \left[\prod_{i=1}^{d} \left(\frac{1}{\lambda_i^2}\right)^{\frac{n}{2}} \exp\left\{ \frac{-1}{2\lambda_i^2} (Z^T n S^2\ Z)_{ii} \right\} \right],$$

Next, we integrate out Λ to get the marginal GFD of A

$$r_{\mathbf{y}}(\text{veck}(A)) \propto D^*(J(\mathbf{y}, A))(\pi)^{\frac{-d(n-1)}{2}} 2^{-d} n^{\frac{-d}{2}} \Gamma\left(\frac{n-1}{2}\right)^d \prod_{i=1}^{d} (Z^T n S^2\ Z)_{ii}^{\frac{-(n-1)}{2}}.$$

A similar calculation can be done for an alternative data generating algorithm using the minimal sufficient statistic $(\bar{\mathbf{Y}}, S^2)$. The only difference will be the form of $D^*(J(\mathbf{y}, A))$. As a reminder, the veck(A) operation vectorizes the strictly lower triangular elements of the skew-symmetric matrix A, as discussed in [20]. See the STAN implementation on this paper's Github page for details.

A.2 Derivation of the Mixed Models Jacobian Quantity $J(\mathbf{Y}, S_\alpha, \mathbb{X}\beta, \sigma_\alpha, \sigma_e)$

The derivative $\frac{\delta \mathbf{Y}}{\delta \beta} = \mathbb{X}$ is trivial. To solve for the other two terms, note that $\Sigma = AA$ and therefore

$$\frac{\delta \Sigma}{\delta \sigma_\alpha^2} = A \frac{\delta A}{\delta \sigma_\alpha^2} + \frac{\delta A}{\delta \sigma_\alpha^2} A, \text{ where } \frac{\delta A}{\delta \sigma_\alpha^2} = \frac{1}{2} S_\alpha A^{-1},$$

The derivative $\frac{\delta A}{\delta \sigma_\alpha^2}$ commutes with A since

$$A = U\left(\operatorname{diag}\left(\sqrt{\sigma_\alpha^2 \lambda_i + \sigma_e^2}\right)\right) U', \quad \frac{\delta A}{\delta \sigma_\alpha^2} = \frac{1}{2} U\left(\operatorname{diag}\left(\lambda_i \left(\sigma_\alpha^2 \lambda_i + \sigma_e^2\right)^{-1/2}\right)\right) U',$$

and clearly $\frac{\delta A}{\delta_\alpha^2} A = A \frac{\delta A}{\delta_\alpha^2}$. Therefore,

$$\frac{\delta \mathbf{Y}}{\delta \sigma_\alpha^2} = \frac{\delta}{\delta \sigma_\alpha} (\mathbb{X}\beta + A\mathbb{Z})\Big|_{\mathbb{Z} = A^{-1} * (Y - \mathbb{X}\beta)} = \frac{1}{2} S_\alpha \Sigma^{-1} (\mathbf{Y} - \mathbb{X}\beta).$$

(Note that $A^{-1} * (Y - \mathbb{X}\beta)$ is a matrix product, not a realization of a DGA.) Following this same logic gives

$$\frac{\delta \mathbf{Y}}{\delta \sigma_e} = \Sigma^{-1}(\mathbf{Y} - \mathbb{X}\beta) = (\sigma_\alpha^2 S_\alpha + \sigma_e^2 I)^{-1}(\mathbf{Y} - \mathbb{X}\beta).$$

A.3 Binomial Algorithm for Unknown n, Known p

As in Section 13.3.1, the generalized fiducial solution to the n unknown problem assigns mass to *sets* of integers rather than individual values. Our aim is to assign fiducial probabilities to all reasonable sets of n values in the sample space. Considering a *set* $\mathbf{s}$ of candidate n values and fixing the observed data $\mathbf{Y}$, we start by calculating the probability that we observe a set of uniforms values such that any $n^* \in \mathbf{s}$ could be the true n value. Notice that the set of viable uniform values for any superset of $\mathbf{s}$ must necessarily be a subset of the set of viable uniform values for $\mathbf{s}$. The astute reader will notice that this is the same as the notion of *commonality* found in Dempster-Schaffer calculus [30]. We start by calculating the following inversion of (13.9),

$$F_{n^*,p}(Y_i - 1) < U_i^* \leq F_{n^*,p}(Y_i), \quad \text{for } i = 1, \ldots, n. \tag{13.11}$$

Since $F_{n^*,p}(Y_i)$ is non-decreasing in n^*, we see that for any fixed $(U_1, \ldots, U_n^*)$ the set of n^* is an interval of consecutive integers. We will denote the set of all integer intervals $\mathbf{S} = \{\{i, i+1, \ldots, j\}, 1 \leq i \leq j\}$. The commonality of the set $\mathbf{s} \in \mathbf{S}$ will be

$$p(\mathbf{s}|\mathbf{Y}) = \prod_{i=1}^{m} \left[F_{\max\{\mathbf{s}\},p}(y_i) - F_{\min\{\mathbf{s}\},p}(y_i - 1)\right]^+.$$

Our algorithm begins by using this commonality to address the issue of no strict upper bound on the set of possible values for n^*. First, select a precision cutoff ϵ_1. Then, we define our set of candidate n^* values, which we will call $\mathbf{N}$, by way of the following algorithm: starting with the observed maximum of the data consider new values sequentially, comparing the commonality of the candidate n^* value, $p(\{n^*\}|\mathbf{Y})$, with commonalities observed so far. If the ratio of the commonality to the maximum commonality in $\mathbf{N}$ accepted so far

$$\frac{p(\{n^*\}|\mathbf{Y})}{\max_{k \in \mathbf{N}}\{p(\{k\}|\mathbf{Y})\}} > \epsilon_1,$$

add n^* to our set $\mathbf{N}$ and consider $n^* + 1$. Otherwise, halt the algorithm.

Next we approximate the GFD by restricting ourselves only to subsets of $\mathbf{N}$. Define $\hat{\mathbf{S}} = \{\mathbf{s} \in \mathbf{S}, \mathbf{s} \subset \mathbf{N}\}$. Using the analogue noted previously between Dempster-Shaffer calculus and the inversion of our DGA, we define the fiducial probability of a set $\mathbf{s}$ as the Dempster-Shaffer *mass* of $\mathbf{s}$ modified and renormalized so that $r_\mathbf{Y}(\emptyset) = 0$ [30, 43]. We calculate the the Dempster-Shaffer mass of $\mathbf{s}$ recursively by starting from $m(\mathbf{N}) = p(\mathbf{N}|\mathbf{Y})$ and then for $\mathbf{s} \in \hat{\mathbf{S}}$

$$m(\mathbf{s}) = \left[p(\mathbf{s}|\mathbf{Y}) - \left(\sum_{\mathbf{r} \in \hat{\mathbf{S}}:\mathbf{s} \subset \mathbf{r}} m(r)\right)\right]^+.$$

Then, after the probability of all sets are defined in this way, we renormalize them so that they add to 1. The fiducial probability for each element of $\mathbf{s} \in \hat{\mathbf{S}}$ is

$$r_\mathbf{Y}(\mathbf{s}) = \frac{m(\mathbf{s})}{1 - m(\emptyset)}.$$

A.4 Binomial Algorithm for Unknown n, Unknown p

We begin by making an important observation about the distribution of the predicted value $\hat{p}$. For random U_i^*, fixed $\mathbf{Y}$ and n, we have that

$$\hat{p}_i^{\text{upper}} \sim Beta(Y_i + 1, n - Y_i), \quad \hat{p}_i^{\text{lower}} \sim Beta(Y_i, n - Y_i + 1).$$

Writing out the distribution on the bounds of our fiducial sets for p in this way allows use of the following well-known result:

Lemma 13.1. Let $\mu := np$. Then

$$\hat{\mu}_i^{\text{upper}} = n\hat{p}_i^{\text{upper}} \xrightarrow{\mathcal{D}} Gamma(Y_i + 1, 1) \qquad \hat{\mu}_i^{\text{lower}} = n\hat{p}_i^{\text{lower}} \xrightarrow{\mathcal{D}} Gamma(Y_i, 1),$$

as $n \to \infty$.

Proof. See, for instance, [13]. □

Our aim is to develop a method that can simulate draws from the GFD by calculating sets with elements of the form $\hat{n} \times (\hat{\mu}^{\text{lower}}, \hat{\mu}^{\text{upper}}) \in \mathbb{N} \times \mathcal{B}(\Phi_+)$. A simplistic approach to generating values from the GFD would be to repeatedly re-sample the set of random variables $\mathbf{U}$ on the space $[0,1]^m$ and in turn recalculate the paired sets that satisfy (13.9). There are two issues that arise for this approach. First, there is no upper bound on the set of possible n values, and second, it is possible that a majority of the uniform sets $\mathbf{U}$ drawn completely randomly will yield no solution to (13.9). To address these issues we develop a stopping criterion for searching the sample space in the n direction and we develop a more computationally efficient way to randomly select uniform values $\mathbf{U}$.

We use the following protocol for choosing a range of potential n values. While there are many ways to choose a first candidate $\hat{n}$, we use the estimator suggested by [5]. From this first candidate we check each subsequent n by increment of one and calculate $(\hat{\mu}^{\text{lower}}, \hat{\mu}^{\text{upper}})$. We develop a stopping rule for $\hat{n}$ based on the asymptotic result in Lemma 13.1. Given a precision parameter $\epsilon_2 > 0$, we stop looking for further $\hat{n}$ values whenever

$$|\hat{\mu}_i^{\text{upper}} - H^{-1}(1 - U_i)_{Y_i+1,1}| < \epsilon_2 \quad \text{and} \quad |\hat{\mu}_i^{\text{lower}} - H^{-1}(1 - U_i)_{Y_i,1}| < \epsilon_2,$$

for all $i \in \{1, \ldots, m\}$, where $H_{\alpha,\beta}^{-1}$ is the quantile function of a $Gamma(\alpha, \beta)$ random variable. We stop searching for values of n that work as soon as we are sufficiently close to the limiting Gamma distribution, which does not depend on n. This stopping criterion leverages our assumption that our target data is binomial, and not Poisson, to reduce the space of possible n values to a finite set.

We implement an Gibbs sampler, sampling from the uniform distribution $\mathbf{U}$ randomly selected in such a way that there is a new solution set that is non-empty, i.e., satisfying Equation (13.10), or equivalently

$$G_{n-Y_i,Y_i+1}(1-p) \geq U_i > G_{n-Y_i+1,Y_i}(1-p), \quad i = 1, \ldots, m.$$

In particular, each value U_i is resampled iteratively by randomly selecting a new value from its conditional distribution given all the other Us: a uniform distribution on

$$\left(\min\{G_{n-Y_i+1,Y_i}\left(1 - \frac{\tilde{\mu}_{\tilde{n},j}^{\text{lower}}}{\tilde{n}}\right) : j \neq i\}, \max\{G_{n-Y_i,Y_i+1}\left(1 - \frac{\tilde{\mu}_{\tilde{n},j}^{\text{lower}}}{\tilde{n}}\right) : j \neq i\}\right],$$

where $\tilde{n}$, $\tilde{\mu}_{\tilde{n},j}^{\text{lower}}$, and $\tilde{\mu}_{\tilde{n},j}^{\text{upper}}$ for $j = \{1, \ldots, m\} \backslash \{i\}$ are the solution set that we would get if we were to remove the ith observation from our data. To speed up the computation we actually resample U_is in batches, so that all U_is corresponding to the same observed value of Y_i are sampled together.

Our investigations of the above method of resampling the uniform values $\mathbf{U}$ have revealed that there is a need for steps that would result large changes to n and μ. To allow for this, we add two Metropolis-Hastings steps at the end of each Gibbs sampler scan, one in the n direction and one in the μ direction.

The MH step in the μ direction was implemented using the following algorithm. Assume that you already have finished the above protocol for choosing a new set of uniform values $\mathbf{U}$ and that has a corresponding solution set

$$\bigcup_{\hat{n}} \{\hat{n}\} \times (\max\{\hat{\mu}_{1,\hat{n}}^{\text{lower}}, \dots, \hat{\mu}_{m,\hat{n}}^{\text{lower}}\}, \min\{\hat{\mu}_{1,\hat{n}}^{\text{upper}}, \dots, \hat{\mu}_{m,\hat{n}}^{\text{upper}}\}].$$

Randomly select a single value μ^* from the set of $\hat{\mu}$ values associated with the smallest $\hat{n}$

$$(\max\{\hat{\mu}_{1,\min\{\hat{n}\}}^{\text{lower}}, \dots, \hat{\mu}_{m,\min\{\hat{n}\}}^{\text{lower}}\}, \min\{\hat{\mu}_{1,\min\{\hat{n}\}}^{\text{upper}}, \dots, \hat{\mu}_{m,\min\{\hat{n}\}}^{\text{upper}}\}].$$

Next, define a new value, $\mu^\star$, that is drawn from the proposal distribution $N(\mu^*, \sigma^2)$ for some predetermined σ. Calculate a set of uniform values, $\mathbf{U}^\star$, such that $\mu^\star$ is contained in the solution set. Let $\dot{n}$, $\{\dot{\mu}_{\min\{\dot{n}\},j}^{\text{lower}}\}_{j=1}^m$, and $\{\dot{\mu}_{\min\{\dot{n}\},j}^{\text{upper}}\}_{j=1}^m$ be the values in the solution set created by using Equation (13.10) with $\mathbf{U}^\star$. The acceptance ratio for these new $\mathbf{U}^\star$ will then be

$$\frac{(\dot{\mu}_{\min\{\dot{n}\},j}^{\text{lower}} - \dot{\mu}_{\min\{\dot{n}\},j}^{\text{upper}}) \prod_{i=1}^m \left(G_{n-Y_i,Y_i+1}\left(1 - \frac{\hat{\mu}_{\hat{n},i}^{\text{lower}}}{\hat{n}}\right) - G_{n-Y_i+1,Y_i}\left(1 - \frac{\hat{\mu}_{\hat{n},i}^{\text{lower}}}{\hat{n}}\right)\right)}{(\hat{\mu}_{\min\{\hat{n}\},j}^{\text{lower}} - \hat{\mu}_{\min\{\hat{n}\},j}^{\text{upper}}) \prod_{i=1}^m \left(G_{n-Y_i,Y_i+1}\left(1 - \frac{\dot{\mu}_{\dot{n},i}^{\text{lower}}}{\dot{n}}\right) - G_{n-Y_i+1,Y_i}\left(1 - \frac{\dot{\mu}_{\dot{n},i}^{\text{lower}}}{\dot{n}}\right)\right)},$$

The MH step in the n direction was implemented using the following algorithm. Like the MH step in the μ direction, start by randomly selecting a single μ^* value from set of $\hat{\mu}$ values associated with the smallest $\hat{n}$, $\mu^* \in (\hat{\mu}_{\min\{\hat{n}\},j}^{\text{lower}}, \hat{\mu}_{\min\{\hat{n}\},j}^{\text{upper}})$. Then, let $n^\star := \min\{\hat{n}\} + 1 - 2X$, where $X \sim \text{Bernoulli}(1/2)$, be your new proposed n value. Next, find a set of uniform values $\mathbf{U}^\star$ such that $n^\star$ and $\mu^\star$ are both in the solution set. Let $\dot{n}$, $\{\dot{\mu}_{\min\{\dot{n}\},j}^{\text{lower}}\}_{j=1}^m$, and $\{\dot{\mu}_{\min\{\dot{n}\},j}^{\text{upper}}\}_{j=1}^m$ be the values in the solution set created by using Equation (13.10) with $\mathbf{U}^\star$. The acceptance ratio of $\mathbf{U}^\star$ has the same form as the acceptance ratio for the MH step in the μ direction.

The specific details of the algorithm can be read from our implementation (in R) on GitHub [https://github.com/sirmurphalot/IntroductionGFI]. On this page we have also posted a full pseudocode version of the algorithm for the curious reader.

Bibliography

[1] Ahrens, H. and Pincus, R., 1981. On two measures of unbalancedness in a one-way model and their relation to efficiency. *Biometrical Journal, 23*(3), pp. 227–235.

[2] Berger, J. O., Sun, D. and Song, C., 2020. Bayesian analysis of the covariance matrix of a multivariate normal distribution with a new class of priors. *Annals of Statistics, 48*(4), pp. 2381–2403.

[3] Berger, J. O., Sun, D. and Song, C., 2020. An objective prior for hyperparameters in normal hierarchical models. *Journal of Multivariate Analysis, 178*, pp. 1–13.

[4] Cisewski, J. and Hannig, J., 2008. Generalized fiducial inference for normal linear mixed models. *The Annals of Statistics, 40*, pp. 2102–2127.

[5] DasGupta, A. and Rubin, H., 2004. Estimation of binomial parameters when both n, p are unknown. *Journal of Statistical Planning and Inference, 130*(1–2), pp. 391–404.

[6] Dempster, A. P., 2008. The dempster–shafer calculus for statisticians. *International Journal of Approximate Reasoning, 48*(2), pp. 365–377. In Memory of Philippe Smets (1938–2005).

[7] Lidong, E., Hannig, J. and Iyer, H., 2008. Fiducial intervals for variance components in an unbalanced two-component normal mixed linear model. *Journal of the American Statistical Association, 103*(482), pp. 854–865.

[8] Edlefsen, P. T., Liu, C. and Dempster, A. P., 2009. Estimating limits from poisson counting data using dempster-shafer analysis. *Annals of Applied Statistics, 3*, pp. 764–790.

[9] Eves, H., 1996. *Elementary Matrix Theory* (1st ed.), pp. 265–267. Meneola, NY: Allyn and Bacon, Inc.

[10] Fazel, M., Hindi, H. and Boyd, S. P., 2003. Log-det heuristic for matrix rank minimization with applications to hankel and euclidean distance matrices. In *Proceedings of the 2003 American Control Conference, 2003*, volume 3, pp. 2156–2162. Danvers, MA.

[11] Fisher, R. A., 1935. The fiducial argument... *Annals of Eugenics, 6*(4), pp. 391–398. https://doi.org/10.1111/j.1469-1809.1935.tb02120.x.

[12] Grafarend, E. W., Krumm, F. W. and Schwarze, V. S. editors, 2003. *A Metric for Covariance Matrices*, pp. 299–309. Berlin, Heidelberg: Springer.

[13] Gut, A., 2005. *Probability: A Graduate Course.* New York: Springer.

[14] Hannig, J., Abdel-Karmin, A. and Iyer, H., 2003. Simultaneous fiducial generalized confidence intervals for ratios of means of lognormal distributions. *Austrian Journal of Statistics, 35*, pp. 261–269.

[15] Hannig, J., Iyer, L. E. H., and Wang, C.-M., 2007. Fiducial approach to uncertainty asessment: Accounting for error due to instrument resolution. *Metrologia, 44*(6), p. 476.

[16] Hannig, J. and Lee, T. C. M., 2009. Generalized fiducial inference for wavelet regression. *Biometrika, 96*, pp. 847–860.

[17] Hannig, J., Wang, C. M. and Iyer, H. K., 2013. Uncertainty calculation for the ratio of dependent measurements. *Metrologia, 4*, pp. 177–186.

[18] Hannig, J., 2009. On Generalized Fiducial Inference. *Statistica Sinica, 19*(2), pp. 491–544.

[19] Hannig, J., Iyer, H., Lai, R. C. S., and Lee, T. C. M., 2016. Generalized fiducial inference: A review and new results. *Journal of the American Statistical Association, 111*(515), pp. 1346–1361.

[20] Henderson, H. V. and Searle, S. R., 1979. Vec and vech operators for matrices, with some uses in jacobians and multivariate statistics. *The Canadian Journal of Statistics / La Revue Canadienne de Statistique, 7*(1), pp. 65–81.

[21] Horn, R. A. and Johnson, C. R., 2012. *Matrix Analysis* (2nd ed.). New York, NY: Cambridge University Press.

[22] Iyer, H. K., Wang, J. C. M. and Mathew, T., 2004. Models and confidence intervals for true values in interlaboratory trials. *Journal of the American Statistical Association, 99*, pp. 1060–1071.

[23] Konno, Y., 1995. Estimation of a normal covariance matrix with incomplete data under stein's loss. *Journal of Multivariate Analysis, 52*(2), pp. 308–324.

[24] Liu, Y. and Hannig, J., 2016. Generalized fiducial inference for binary logistic item response models. *Psychometrica, 81*, pp. 290–324.

[25] McNally, R. J., Iyer, H. K. and Mathew, T., 2003. Tests for individual and population bioequivalence based on generalized p-values. *Statistics in Medicine, 22*, pp. 31–53.

[26] Neupert, S. D., Growney, C. M., Zhu, X., Sorensen, J. K., Smith, E. L. and Hannig, J., 2021. Bff: Bayesian, fiducial, and frequentist analysis of cognitive engagement among cognitively impaired older adults. *Entropy, 23*, p. 428. https://doi.org/10.3390/e23040428.

[27] O'Dorney, E., 2014. Minimizing the cayley transform of an orthogonal matrix by multiplying by signature matrices. *Linear Algebra and Its Applications, 448*, pp. 97–103.

[28] Perron, F., 1992. Minimax estimators of a covariance matrix. *Journal of Multivariate Analysis, 43*(1), pp. 16–28.

[29] Schweder, T. and Hjort, N. L., 2016. *Confidence, Likelihood, Probability*, volume 41. New York, NY: Cambridge University Press.

[30] Shafer, G., 1976. *A Mathematical Theory of Evidence*. Princeton: Princeton University Press.

[31] Shi, W. J., Hannig, J., Lai, R. C. S. and Lee, T. C. M., 2021. Covariance estimation via fiducial inference. *Statistical Theory and Related Fields, 5*(4), pp. 316–331. https://doi.org/10.1080/24754269.2021.1877950.

[32] Stan Development Team, 2020. RStan: the R interface to Stan. R package version 2.21.2. https://cran.r-project.org/web/packages/rstan/vignettes/rstan.html.

[33] Wadsworth, G. P., 1960. *Introduction to Probability and Random Variables* (1st ed.), p. 52. New York: McGraw-Hill.

[34] Wandler, D. V. and Hannig, J., 2006. *Construction of Fiducial Confidence Intervals for the Mixture of Cauchy and Normal Distributions*. Master's Thesis, Department of Statistics, Colorado State University.

[35] Wandler, D. V. and Hannig, J., 2011. Fiducial inference on the maximum mean of a multivariate normal distribution. *Journal of Multivariate Analysis, 102*, pp. 87–104.

[36] Wandler, D. V. and Hannig, J., 2012. A fiducial approach to multiple comparisons. *Journal of Statistical Planning and Inference, 142*, pp. 878–895.

[37] Wandler, D. V. and Hannig, J., 2012. Generalized fiducial confidence intervals for extremes. *Extremes, 15*, pp. 67–87.

[38] Wang, C. M. and Iyer, H. K., 2006. Propagation of uncertainties in measurements using generalized inference. *Metrologia, 42*, pp. 145–153.

[39] Wang, C. M. and Iyer, H. K., 2006. Uncertainty analysis for vector measurands using fiducial inference. *Metrologia, 43*, pp. 486–494.

[40] Wang, J. C., Hannig, J. and Iyer, H. K., 2012. Pivotal methods in the propagation of distributions. *Metrologia, 49*, pp. 382–389.

[41] Williams, J. P. and Hannig, J., 2019. Nonpenalized variable selection in high-dimensional linear model settings via generalized fiducial inference. *The Annals of Statistics, 47*(3), pp. 1723–1753.

[42] Williams, J.P., Xie, Y. and Hannig, J., 2023. The EAS approach for graphical selection consistency in vector autoregression models. *Can J Statistics*, 51, pp. 674–703. https://doi.org/10.1002/cjs.11726

[43] Yager, R. R. and Liu, L., 2010. *Classic Works of the Dempster-Shafer Theory of Belief Functions* (1st ed.). New York, NY: Springer Publishing Company, Incorporated.

[44] Yang, R. and Berger, J. O., 1994. Estimation of a covariance matrix using the reference prior. *The Annals of Statistics, 22*, pp. 1195–1211.

14

Dempster-Shafer Theory for Statistical Inference

Ruobin Gong

Rutgers University

14.1 Introduction

The Dempster-Shafer (DS) theory is an uncertainty reasoning framework for drawing probabilistic assessments on assertions concerning unknown states of the world. The theory was conceived by Arthur Dempster as a generalization to the Fisherian fiducial inference and Bayesian inference [13, 15, 14, 17, 16, 18, 19]. It was further developed by Glenn Shafer [58, 60, 61, 62] and became widely known as the Dempster-Shafer theory. Shafer's book, *A Theory of Mathematical Evidence*, stands to be the landmark reference on the subject. Since its inception, the DS theory advised successful applications in various subfields of quantitative science and engineering, in particular signal processing, expert systems, and artificial intelligence. It excelled in areas and problems that demanded a rigorous treatment of uncertainty quantification. Despite its origin in statistics, the DS theory remained a relatively arcane subject in that literature, in part due to a lack of computational feasibility for pragmatic problems of statistical inference. In recent years, however, the DS theory has attracted renewed interest from the *BFF* (Bayesian, Fiducial, and Frequentist) community of statisticians.

A notable feature of the DS theory is the use of *belief functions* as the elementary vocabulary for encoding, computing, and expressing uncertain knowledge. The axioms for belief function generalize Kolmogorov's axioms for probability functions by relaxing additivity to *super-additivity.* The relaxation enables a richer expression of the degree of evidential support within a probabilistic model and marks the theory's distinguishing feature from the frequentist and Bayesian inferential paradigms.

Let Ω be a state space and $H \subseteq \Omega$ be an *assertion* concerning the state of the world. H may be a scientific hypothesis about an unobservable quantity or a range of interesting values for a future observation. Ordinary uncertain inference regarding H may be expressed by a probability $P(H)$. The value of $P(H)$ may convey meanings such as how likely H is under an assumed sampling model or the degree of evidence or belief held by the analyst toward H. The law of total probability dictates that $P(H) + P(H^c) = 1$, or that the uncertain inferences for H and against H share the probability of unity.

In contrast, inference from a DS model regarding H is summarized by a probability triple $(\mathtt{p}, \mathtt{q}, \mathtt{r})$, with $\mathtt{p} + \mathtt{q} + \mathtt{r} = 1$, to express the analyst's evidential support "for", "against", and "don't know" about the assertion H. The values of $(\mathtt{p}, \mathtt{q}, \mathtt{r})$ are derived from the belief function $\underline{P}$ that characterizes the joint DS model that aggregates all available statistical

 DOI: 10.1201/9780429341731-14

evidence using the *Extended Calculus of Probability*, which Section 14.3 introduces in detail. A large p probability expresses strong evidential support toward the assertion H. A large q probability expresses strong support toward its negation H^c. The r component represents the weight of evidence in support of a third, indeterminate logical state. A large r probability suggests that the model, encompassing the data as well as the modeler's knowledge and assumptions, is structurally deficient in making a precise pronouncement on the partition of interest $\{H, H^c\}$. Such an expression of quantified structural knowledge deficiency makes the $(\mathsf{p}, \mathsf{q}, \mathsf{r})$ probabilities distinct from ordinary probabilities.

In a problem of scientific inference, the agent may not make a definitive determination about an arbitrary question. She can have many reasons to withhold judgment, for example, when there is insufficient information to make that determination. The DS theory puts forth an uncertainty reasoning framework that enables the assessment of the indeterminate state via the "don't know" probability r. The explicit recognition of the "don't know" extends the range of probabilistic analysis available. The probability triple employed by the DS framework draws a stochastic parallel to the *three-valued logic*. Inductive inference paradigms that support three-valued decisions have been advocated in the philosophy of science (e.g. [45]), and the DS theory reflects these ideas as a quantitative framework.

In statistical inference applications, there are many situations in which there is insufficient evidence to support the construction of a fully specified probability model. For example, the data may be partially missing or coarse [40]; the model structure may be only partially identifiable [49, 27]; the analyst chooses to employ a family of priors or partial priors due to robustness considerations [4, 5]. These scenarios all encompass some form of low-resolution probabilistic information that is difficult to capture using a single probability specification. However, they can be naturally captured using the language of *imprecise probabilities* (IP), to which belief function is a useful special case.

This chapter discusses the application of DS theory to statistical inference. Section 14.2 introduces the axioms of belief function and its induction via the multi-valued mapping, and motivates its use within DS theory for non-invertible functional models. Section 14.3 describes the Dempster-Shafer *Extended Calculus of Probability* (DS-ECP), including the specification of marginal evidence, its aggregation, and the summary of post-data inference on the margins of interest. Section 14.4 uses an example of weakened simultaneous inference to illustrate the steps of DS-ECP. Section 14.5 discusses the design of Monte Carlo algorithms for DS computation, including recent advances in drawing categorical inference according to the *structure of the second kind* [13] using an efficient Gibbs sampler. Section 14.6 ends with a discussion about a set of open questions and debates, with pointers to further reading.

Before we delve in, a remark on the notation convention for the remainder of this chapter. The sans serif lowercase letters (such as x and u) denote unknown variables that may be assigned either a value or a probability distribution. Serif lowercase letters (such as x and u) are known variables, such as observed data or simulation realizations. Bold lowercase letters may be used to emphasize multi-dimensionality when needed. Boldface uppercase Roman and Greek letters (such as $\mathbf{X}$, $\mathbf{U}$, and $\mathbf{\Theta}$) are sets.

14.2 Belief Function via the Multi-Valued Mapping

Belief functions are the vocabulary of the DS theory. The axioms of belief function, stated below, generalize the familiar axioms of probability functions by supplanting additivity with super-additivity.

Definition 14.1 (axioms of belief function) *Let Ω be a state space and $\mathscr{P}(\Omega)$ the collection of subsets of Ω. A set function $\underline{P}: \mathscr{P}(\Omega) \to [0,1]$ is a belief function if*

1. $\underline{P}(\emptyset) = 0$,
2. $\underline{P}(\Omega) = 1$, *and*
3. *For every k and every denumerable collection $\{A_i, i = 1, \ldots\}$ of subsets of Ω,*

$$\underline{P}\left(\cup_{i=1}^{k} A_i\right) \geq \sum_{I \neq \emptyset; I \subset \{1,\ldots,k\}} (-1)^{|I|+1} \underline{P}\left(\cap_{i \in I} A_i\right). \qquad \text{(super-additivity)}$$

The conjugate function to $\underline{P}$, defined as $\overline{P}(A) = 1 - \underline{P}(A^c)$ for all $A \subseteq \Omega$, is its corresponding plausibility function.

As a special case, belief functions may arise as lower bounds to convex and closed sets of distributions. We give an example often encountered in robust Bayesian analysis. Let π_0 be a fixed distribution on the unit interval $[0,1]$. For some $\epsilon \in [0,1]$, the *ϵ-contamination class* is the set of distributions

$$\Pi_\epsilon = \{\pi \in \mathscr{P} : \pi = (1-\epsilon)\pi_0 + \epsilon\pi_1, \ \pi_1 \in \mathscr{P}\},$$

where $\mathscr{P}$ is the set of all probability distributions on $[0,1]$. In other words, the ϵ-contamination class is formed by mixing a given distribution π_0 with another arbitrary distribution defined on the same space. Then, the set function $\underline{P}: \mathscr{B}([0,1]) \to [0,1]$:

$$\underline{P}(A) = \inf_{\pi \in \Pi_\epsilon} \pi(A)$$

is a belief function. For every set A, $\underline{P}(A)$ is the minimum probability assigned to A by members of the class Π_ϵ. Specifically, $\underline{P}(A) = 0$ if A is the empty set, $\underline{P}(A) = 1$ if A is the full set $[0,1]$, and $\underline{P}(A) = (1-\epsilon)\pi_0(A)$ for all other measurable set A. One can verify that $\underline{P}$ is not additive: for example, $\underline{P}([0,0.5]) + \underline{P}((0.5,1]) = 1 - \epsilon \leq \underline{P}([0,1])$. However, $\underline{P}$ is super-additive and satisfies the axioms of belief function.

As a generalization to the familiar probability function, the belief function is no more than another mathematical object. However, it plays a central role in the inferential framework of the DS theory. To understand why belief functions are needed to support the working mechanism of the DS theory, a quick tour of the theory's intellectual origin is in order.

14.2.1 Non-Invertible Functional Models

Many statisticians active in foundations research during the mid-twentieth century were intrigued by R. A. Fisher's fiducial inference [9, 29, 38, 68, 46, 34, 33, 32, 6, 2, 35]. Arthur Dempster was no exception. To Dempster, fiducial inference was appealing because it dispensed with the need for a Bayesian prior yet still delivered uncertain inference in the form of probability distributions by *continuing to regard* the originally posited auxiliary randomness within the generative model as valid [11]. On the other hand, it was recognized that the Fisherian theory suffered from a plethora of logical issues and limitations [12]. In the words of Jimmie Savage, fiducial inference is "a bold attempt to make the Bayesian omelet without breaking the Bayesian eggs" [51, p. 578].

Fisher's fiducial inference can be understood as a type of functional model [7]. The starting point of a functional model is a deterministic equation, associating the observable data x with the unknown parameter θ:

$$\mathsf{x} = G(\theta, \mathsf{u}). \tag{14.1}$$

The equation involves an auxiliary random variable u, whose probabilistic distribution is assumed fully known. Upon observing the data, the equation is *inverted* while setting $\mathsf{x} = x$:

$$\theta = G^{-1}(\mathsf{u}; x) . \tag{14.2}$$

Implicitly assumed is that there exist at least some *compatible* $\{\mathsf{x}, \mathsf{u}\}$ pairs that satisfy (14.1) for at least one $\theta \in \Theta$, for which the inversion (14.2) is meaningful. In Equation (14.2), the data x is observed and fixed, whereas the auxiliary u is never observed and is assumed to maintain the same law of randomness as before the observation of x. Thus, the inversion of G results in a random quantity that characterizes the post-data inference for θ. Furthermore, if every $u \in \mathbf{U}$ is compatible with each given $x \in \mathbf{X}$, and if G in Equation (14.1) is strictly increasing in u for every θ, the inversion in Equation (14.2) produces a density-bearing random variable, which embodies the fiducial inference for θ given x in the Fisherian sense [30].

Dempster set off to address a class of fiducial problems that were not addressed under the Fisherian fiducial framework. It was the class of problems that resulted in *non-invertible* functional models [10], in the sense that for every $x \in \mathbf{X}$ and every compatible $u \in \mathbf{U}$, the solution to (14.2) is not unique. A functional model is noninvertible if, in some part of the $\mathbf{U}$-domain, the granularity of the functional equation $G(\cdot, u)$ is coarser in its image than that of its Θ-domain. This may happen, for example, if x is discrete but is equipped with a parametric model with a continuous parameter space, such as in the case of count and categorical data.

Upon observing the data, the inverted functional equation $G^{-1}(\cdot; x) : \mathbf{U} \to 2^{\Theta}$ forms a *multi-valued mapping* in the parameter space Θ, defined as follows.

Definition 14.2 (multi-valued mapping [14]) *Let* $(\mathbf{U}, \Omega)$ *be a pair of spaces, and* Γ *a* multi-valued mapping *such that* $\Gamma(u) \subset \Omega$ *for all* $u \in \mathbf{U}$. *For all* $A \subset \Omega$, *denote*

$$\begin{aligned} A_* &= \{u \in \mathbf{U} : \Gamma(u) \subseteq A\}, \\ A^* &= \{u \in \mathbf{U} : \Gamma(u) \cap A \neq \emptyset\}, \end{aligned}$$

and $\mathbf{U}^* \equiv \Omega^* = \Omega_*$ *is the* domain *of* Γ. Γ *is said to be* strongly measurable *if for every* $A \in \mathcal{B}(\Omega)$, $\Gamma^{-1}(A) \in \mathcal{B}(\mathbf{U})$.

In fiducial inference, the auxiliary variable u in Equation (14.1) carries a known probability measure γ on the $\mathbf{U}$ space, which would be deemed invariant before and after observing the data. Fiducial inference relies on the inversion G^{-1}, here represented by the map Γ, to allocate probability from the $\mathbf{U}$ space to the Θ space. If the map Γ were single-valued, γ would induce a pushforward measure on the Ω space by transferring the probability measure γ from the $\mathbf{U}$ space. In general, a multi-valued map is less "precise" than a single-valued one, in that the resulting allocation of probability on Ω through Γ may no longer be uniquely characterized by a single probability function.[1] What results from this allocation is a belief function on Ω, and we will say it is *induced* by the multi-valued mapping.

Definition 14.3 (belief function, multi-valued mapping-induced) *Following Definition 14.2, define*

$$\underline{P}(A) = \frac{\gamma(A_*)}{\gamma(\Omega^*)}, \qquad \overline{P}(A) = \frac{\gamma(A^*)}{\gamma(\Omega^*)}, \tag{14.3}$$

provided that $\gamma(\Omega^*) \neq 0$. *Then,* $\underline{P}$ *is a belief function, with* $\overline{P}$ *its conjugate plausibility function.*

The four-tuple $(\mathbf{U}, \mathcal{B}(\mathbf{U}), \gamma, \Gamma)$, consisting of an ordinary probability space plus the map Γ, is called the *source* of the belief function $\underline{P}$ [14, 69]. The source characterizes a *random set* in the Ω space in the following sense. For every u, $\Gamma(u)$ is a set in Ω termed the *focal set*. Serving as an index of the focal set, u is governed by a probability law specified by the triple $(\mathbf{U}, \mathcal{B}(\mathbf{U}), \gamma)$. Replacing u with the sans serif u to emphasize its probability law, the focal set $\Gamma(\mathsf{u})$ can be seen as governed by that law too. The probability that $\Gamma(\mathsf{u})$ belongs to any $A \in \Omega$, where both the set membership and the probability are defined by the source, is precisely $\underline{P}(A)$ or the belief function assessed at A.

When Γ takes a denumerable number of distinct set values, a helpful characterization of the source is the *mass function*, also called the *basic probability assignment*:

$$m(A) = \frac{\gamma(\{u \in \mathbf{U} : \Gamma(u) = A\})}{\gamma(\Omega^*)}. \tag{14.4}$$

The mass function corresponds uniquely to the belief function by the relation

$$\underline{P}(A) = \sum_{B \subseteq A} m(B), \tag{14.5}$$

and conversely by the *Möbius inversion* [58]. Another characterization of the belief function is the *commonality function*, defined as $Q(A) = \{\gamma(\Omega^*)\}^{-1} \gamma(\{u \in \mathbf{U} : A \subseteq \Gamma(u)\})$, which is useful in belief function computation [67, 43, 26].

The source plays a key role in the definition of *independence* among the marginal evidence of a DS model. For a collection of evidence to be eligible for combination via Dempster's rule, their full independence is required. Section 14.3.2 elaborates on this.

In the special case that Γ is a single-valued map, the source reduces to an ordinary random variable, and the associated belief function reduces to a countably additive probability function. In this case, the belief function is said to be *Bayesian.* The DS theory was first introduced as a generalization to Bayesian inference [13, 16], leveraging multi-valued maps to produce non-additive inferential summaries. A canonical example is Thomas Bayes's 1763 "probe example" concerning Binomial inference [3], for which the choice of the prior is controversial. Dempster's solution removes the prior, as well as the need to specify one [22]. At the same time, the DS theory differs substantially from the Bayesian and robust Bayesian approaches, both in its formulation of marginal evidence and the interpretation it attaches to the probabilistic inference. In particular, the $(\mathbf{p}, \mathbf{q}, \mathbf{r})$ probabilities produced by a DS model should not be interpreted as betting odds. We review additional related literature in Section 14.6.

14.2.2 DS Inference for Categorical Distributions: An Illustration

This section demonstrates with an example the key aspects of the DS inferential framework, including the structural equation, the multi-valued mapping obtained from its inversion, as well as the resulting belief function. We will consider the classic DS model for inference of the categorical distribution population parameter.

Denote the K-dimensional simplex as Δ, and for $\theta = (\theta_1, \ldots, \theta_K) \in \Theta \equiv \Delta$, let $\Delta_k(\theta)$ be the sub-simplex sharing all but the kth vertex with Δ, replacing it with θ. The sets $\{\Delta_k(\theta)\}_{k\in[K]}$ form a partition of Δ in the sense that their union is Δ and they have non-intersecting interiors. Each set has volume $\mathrm{Vol}(\Delta_k(\theta)) = \theta_k$. Let $\mathsf{u} \sim \gamma$, the uniform distribution on Δ, and define x through the structural equation

$$\mathsf{x} = G(\theta, \mathsf{u}) = \sum_{k\in[K]} k\mathbb{1}(\mathsf{u} \in \Delta_k(\theta)). \tag{14.6}$$

In other words, x takes the value of the unique index k such that u belongs to the sub-simplex $\Delta_k(\theta)$, which happens with probability that equals the volume of $\Delta_k(\theta)$. Then, x follows the categorical distribution with parameter θ. This sampling mechanism, termed the *structure of the second kind* [13],[2] has the notable feature that it is invariant to permutations of category labels. A useful characterization of the sampling mechanism in terms of the entries of the auxiliary u and the parameter θ through $\Delta_k(\theta)$ is the following.

Lemma 14.1 (Lemma 5.2 of [13]) *For $k \in [K]$, $\theta \in \Delta$ and $u \in \Delta$, $u \in \Delta_k(\theta)$ if and only if $u_\ell/u_k \geq \theta_\ell/\theta_k$ for all $\ell \in [K]$.*

The left panel of Figure 14.1 depicts the generative model (14.6). The draw u as shown falls into the red sub-simplex (x = 1) if and only if $u_1/u_2 \geq \theta_1/\theta_2$ and $u_1/u_3 \geq \theta_1/\theta_3$ by Lemma 14.1. This characterization becomes important when we discuss the Monte Carlo computation for the structure of the second kind model in Section 14.5.

Upon observing a single data point, inverting the functional equation (14.6) results in the following multi-valued mapping. For every u,

$$G^{-1}(u;x) = \{\theta \in \Delta : u \in \Delta_x(\theta)\} \tag{14.7}$$

with domain $\mathbf{U}^* = \{u \in \mathbf{U} : G^{-1}(u;x) \subseteq \Delta\}$. Note that $\mathbf{U}^*$ is dependent on the observed x. Equation (14.7) is a subset in Δ. Specifically, it is a polytope with vertices that include the xth vertex of Δ and u. The polytope becomes a random polytope as the index u in Equation (14.7) realizes from the auxiliary variable $\mathsf{u} \sim \gamma$, the uniform distribution on Δ as originally assumed. An instantiation of this polytope is depicted in the right panel of Figure 14.1. The random polytope encapsulates the post-data DS inference for θ. For every assertion $H \subseteq \Delta$ concerning the value of θ, according to (14.3), the belief and plausibility functions induced by (14.7) amount to

$$\underline{P}(H) = \{\gamma(\mathbf{U}^*)\}^{-1} \int \mathbb{1}\left(G^{-1}(u;x) \subseteq H\right) \gamma(du), \tag{14.8}$$

$$\overline{P}(H) = \{\gamma(\mathbf{U}^*)\}^{-1} \int \mathbb{1}\left(G^{-1}(u;x) \cap H \neq \emptyset\right) \gamma(du). \tag{14.9}$$

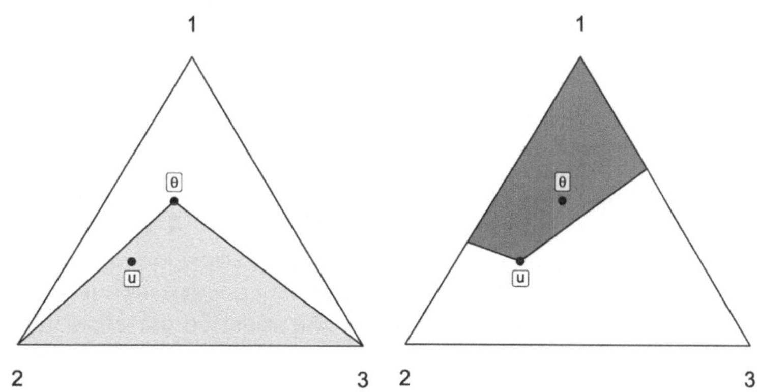

FIGURE 14.1
The *structure of the second kind* model with three categories and one observation. Left: relative to a given θ, the depicted draw of u falls into the light gray sub-simplex, resulting in x = 1 under the generative model (14.6). Right: given x = 1 and depicted draw of u, the inversion of the multi-valued map (14.7) results in the dark gray polytope, which is the set of all possible values of θ that could have generated this observation.

From here, the $(\mathtt{p}, \mathtt{q}, \mathtt{r})$ probability triple regarding the assertion H can be assessed according to (14.17), to be introduced in Section 14.3.3. The triple is interpreted as probabilities "for", "against", and "don't know" about the truth value of $\theta \in H$ given the observation x.

The categorical model described so far only utilizes a single observation. To model multiple independent observations, we must discuss how to aggregate multiple sources of independent marginal evidence under DS-ECP [20], the subject of the next section. We revisit the categorical model for multiple observations in Section 14.5, where a Monte Carlo method for computation is presented.

14.3 Dempster-Shafer Extended Calculus of Probability (DS-ECP)

Section 14.2 motivates the use of belief function in Dempster-Shafer theory as necessary vocabulary for the encoding and expression of uncertain evidence. As a framework of statistical inference, the Dempster-Shafer *Extended Calculus of Probability* (DS-ECP) [20, 21, 22] specifies a set of formal rules to perform probabilistic reasoning, including the specification, combination, and processing of marginal information encompassing observable data, model structure, as well as available (but optional) prior distributional information. The DS-ECP prescribes a stepwise recipe as follows:

1. Specify the model state space, variables, and independent marginal evidence;
2. Aggregate all specified marginal evidence via up-projection to the full state space, combine using *Dempster's rule of combination*, and down-project to margins of interest;
3. Summarize post-data inference in terms of $(\mathtt{p}, \mathtt{q}, \mathtt{r})$ probabilities.

The additional task of predicting future observations can be carried out after the last stage, by propagating inference on margins relating to future observations via another application of the second and the third steps. In what follows, we describe the three steps of DS-ECP in detail.

14.3.1 The State Space and the Specification of Marginal Evidence

A statistical model is a simplified description of the relationship between what we know or can observe, and what we wish to know but cannot observe. A suitable *state space* for a statistical model consists of all quantities which the scientist regards as potentially relevant to the problem at hand. The state space cannot be comprehensive but should encompass a scope broad enough to enable the encoding of the scientific narrative necessary for the inference task. In Savage's terms [52, Chapter 2], the state space is the *small world* with which the analyst chooses to concern themselves.

The *margins* of the state space are quantities involved in the inference task. The marginal quantities can be observable or unobservable. They can either be of direct scientific interest or serve intermediate roles to convey the model evidence. The usual constituents of a statistical model, including data (observed or missing) and parameters (of interest or nuisance), are margins of the state space. We restrict the scope of presentation in this chapter to state spaces that are Cartesian products (denoted by $\times$) of their margins, as they are the most common among statistical inference problems.

Model evidence can be expressed as statements involving margins of the state space. There are two types of evidence:

- *Deterministic evidence*: observed values of data, equality or inequality expressions that relate margins of the state space and reflect scientific knowledge or model assumptions, and hard subset restrictions on the state space;
- *Stochastic evidence*: distributional assumptions on state variables, or probabilistic functional model assumptions.

Like fiducial inference, DS-ECP uses auxiliary variables to express relational stochastic evidence. These are assertions that some margins of the state space relate to others in a way that can be captured by a probabilistic law. The auxiliary variables, introduced for the sole purpose of specifying stochastic evidence in functional relationships, are by construction not part of the state space. Their distributions are considered fully known and are not of interest.

As an example, the DS inference for a categorical observation described in Section 14.2 is performed on a state space $\Omega = [K] \times \Theta$. The evidence consists of:

i. The value of a single observable data point, $\mathsf{x} = x \in [K]$; and
ii. The functional equation that dictates the stochastic relationship between x and θ via G (14.6), where u is uniformly distributed on Δ.

If not one, but N data points were to be independently obtained given the same θ, the state space would be expanded to $\Omega = [K]^N \times \Theta$, and the collection of available evidence would include $(\mathsf{x}_1, \cdots, \mathsf{x}_N) = (x_1, \cdots, x_N) \in [K]^N$, as well as the N functional equations $\mathsf{x}_n = G(\theta, \mathsf{u}_n)$, where the auxiliary u_n's are i.i.d. uniform on Δ.

14.3.2 Evidence Aggregation: Projection and Combination

Statistical evidence in their raw form is nearly always *marginal*, in that they only concern a proper subset of margins of the full state space. In the categorical inference example, each independent realization of x_n from the same sampling model governed by the parameter θ constitutes marginal evidence concerning just x_n and θ. To combine all the information from $(\mathsf{x}_1, \cdots, \mathsf{x}_N)$ about θ requires aggregation of these pieces of marginal evidence. Aggregation in the DS framework is a three-step process: (1) *up-projecting* independent pieces of marginal evidence onto the full state space, (2) *combining* via *Dempster's rule of combination* to produce the joint DS model, and (3) *down-projecting* onto the margin of interest, resulting in the three-valued post-data inference for any assertions concerning that margin.

14.3.2.1 Up-Projection

To allow for the combination of different pieces of marginal evidence, they must be expressed in terms of the full state space Ω. To up-project a piece of marginal evidence is to *vacuously extend* it to the joint space, a finer frame of discernment compared to the one on which it was defined [58]. The up-projection maintains information in margins that deterministic or stochastic evidence are expressed while remaining maximally uncommitted on the rest of the margins. Each up-projected piece of evidence may be regarded as a multi-valued map onto the space Ω.

Definition 14.4 (Up-projection) *Suppose* Γ *is a multi-valued mapping defined on the pair of spaces* $(\mathbf{U}, \Omega^*)$, *and* $\Omega = \Omega^* \times \Omega^\circ$ *for some* Ω°. *The multi-valued mapping*

$$\Gamma^{\Omega^* \uparrow \Omega}(u) = \Gamma(u) \times \Omega^\circ, \quad \forall u \in \mathbf{U},$$

is the up-projection of Γ *from* Ω^* *to* Ω.

In the categorical inference example, the nth observation can be up-projected from a point in the marginal state space $[K]$ to the full state space $\Omega = [K]^N \times \Theta$, taking the form of a hyperplane that is

$$\{(\mathsf{x}_1, \ldots, \mathsf{x}_N, \theta) \in \Omega : \mathsf{x}_n = x_n\}. \tag{14.10}$$

The stochastic evidence derived from the inverted functional equations, $\theta = G^{-1}(\mathsf{u}_n, \mathsf{x}_n)$, can be expressed for each $n \in [N]$ as

$$\{(\mathsf{x}_1, \ldots, \mathsf{x}_N, \theta) \in \Omega : \theta \in \Delta_{\mathsf{x}_n}(\mathsf{u}_n)\} \tag{14.11}$$

due to (14.7). Notice that (14.10) is a deterministic subset of Ω, whereas (14.11) is a random subset whose randomness derives from the source variable u_n. These up-projected deterministic and random subsets of the state space Ω reflecting marginal evidence are called *marginal* DS models.

14.3.2.2 Dempster's Rule of Combination

The learning rule that dictates the aggregation of individual pieces of evidence and their transferal onto the target question is the pillar of reasoning for an inferential framework. In likelihood inference, the likelihood principle regulates the aggregation of information. In the Bayesian paradigm, this role is served by the Bayes rule. In DS-ECP, the learning rule is *Dempster's rule of combination*. It specifies the creation of a joint DS model via intersecting independent marginal DS models on the same space.

The application of Dempster's rule requires *independence* among all constituent pieces of marginal evidence. The concept of independence here is *cognitive* in nature [61]. It reflects a judgment by the analyst that all pieces of evidence are informative each on their own, present no interference with the informativeness of another, and simultaneously act as non-redundant inputs for the model. Mathematically, cognitive independence among a collection of evidence is equivalent to statistical independence among the sources of the constituent marginal DS models, defined below.

Definition 14.5 (Independent evidence) *A collection of* J *pieces of marginal evidence is said to be independent, if the probability spaces in their associated sources* $\{(\mathbf{U}_j, \mathcal{B}(\mathbf{U}_j), \gamma_j, \Gamma_j)\}_{j \in [J]}$ *are all statistically independent.*

Deterministic marginal DS models are associated with *constant* source variables. Since constants are statistically independent of other variables, deterministic evidence is considered independent whenever included in a collection, so long as they are not in *full conflict* (to be defined soon) with other evidence in the collection.

Definition 14.6 (Dempster's rule of combination) *Let* $\{(\mathbf{U}_j, \mathcal{B}(\mathbf{U}_j), \gamma_j, \Gamma_j)\}_{j \in [J]}$ *be a collection of independent marginal evidence, where* $\Gamma_j \in \Omega$ *for all* $j \in [J]$. *Writing* $\boldsymbol{u} = (u_1, \ldots, u_J)$, *and* $\otimes_{j=1}^J \gamma_j$ *the product measure of* $(\gamma_1, \ldots, \gamma_J)$, *the combined DS model is defined by the source* $(\mathbf{U}, \mathcal{B}(\mathbf{U}), \gamma, \Gamma)$ *where*

$$\Gamma(\boldsymbol{u}) = \cap_{j=1}^{J}\Gamma(u_j), \tag{14.12}$$

$$\mathbf{U} = \left\{\boldsymbol{u} \in \prod_{j=1}^{J}\mathbf{U}_j : \Gamma(\boldsymbol{u}) \neq \emptyset\right\}, \tag{14.13}$$

and

$$\gamma = \left(\int_{\mathbf{U}} \otimes_{j=1}^{J}\gamma_j(d\boldsymbol{u})\right)^{-1} \mathbb{1}_{\mathbf{U}} \otimes_{j=1}^{J} \gamma_j, \tag{14.14}$$

provided that $\mathbf{U} \neq \emptyset$.

In other words, the combined DS model is encapsulated by a new random set which is the *non-empty intersection* of the constituent independent random sets. The revised space $\mathbf{U}$ may be a proper subset of the product space $\prod_{j=1}^{J}\mathbf{U}_j$, eliminating incompatible $\boldsymbol{u}$ values that correspond to marginal focal sets with an empty intersection. The revised measure γ is well-defined whenever $\mathbf{U}$ has positive measure by the product measure $\otimes_{j=1}^{J}\gamma_j$. In the event $\mathbf{U}$ is nonempty but has measure zero, the revised measure (14.14) may be alternatively defined via regular conditional probability or limiting arguments. However, if $\mathbf{U}$ is the empty set, the collection of marginal evidence is said to be *in conflict* and is not suitable for combination via Dempster's rule.

Dempster's rule of combination and the concept of independence were first presented in [14, Section 5], using the language of sources that pertain to the constituent marginal evidence. Elsewhere in the literature (see e.g. [58]), the rule is also stated in equivalent terms using the belief functions associated with each piece of independent marginal evidence, denoted by $\underline{P}_j$. The combined belief function takes the simple form

$$\underline{P} := \oplus_{j=1}^{J}\underline{P}_j, \tag{14.15}$$

where the symbol $\oplus$, denoting the *orthogonal sum*, operates on the constituent $\underline{P}_j$'s in the way as spelled out in Definition 14.6. That is, $\underline{P}$ of (14.15) is the belief function associated with the combined DS model $(\mathbf{U}, \mathcal{B}(\mathbf{U}), \gamma, \Gamma)$ as defined therein.

14.3.2.3 Down-Projection

In an inference task, we are typically concerned with learning about margins of the state space, such as the parameter of interest, or an unknown future observation. To extract inference, the combined DS model is *down-projected* onto the margins of interest, reducing the dimensionality of the post-data inference summary.

Definition 14.7 (Down-projection) *Let* Γ *be a multi-valued mapping defined on the pair of spaces* $(\mathbf{U}, \Omega)$, *and* Ω^* *a space such that* $\Omega = \Omega^* \times \Omega^\circ$ *for some* Ω°. *The multi-valued mapping*

$$\Gamma^{\Omega\downarrow\Omega^*}(u) = \{\omega \in \Omega^* : \exists \nu \in \Omega^\circ \text{ such that } (\omega, \nu) \in \Gamma(u)\}, \quad \forall u \in \mathbf{U},$$

is the down-projection of Γ *from* Ω *to* Ω^*.

Down-projection is the counterpart operation to up-projection, in the sense that the former transforms a multi-valued mapping from the marginal space to the joint space, and the latter does the reverse. It is worth noting however, that while the up-projection of a marginal DS model ($\Omega^* \uparrow \Omega$) followed by a down-projection into the original margin ($\Omega \downarrow \Omega^*$) preserves the original marginal DS model, a down-projected DS model ($\Omega \downarrow \Omega^*$) followed by an up-projection into the original joint space ($\Omega^* \uparrow \Omega$) may not preserve the original joint DS model, unless the joint DS model accords to a multi-valued map that is cylindrical (i.e. with identical cross-sections) along Ω°, the dimensions over which the down-projection operation marginalizes.

14.3.3 Inference Summary

In DS-ECP, the post-data inference is summarized through the probability triple $(\mathtt{p}, \mathtt{q}, \mathtt{r})$ about an assertion concerning subsets of the state space. Let $(\mathbf{U}, \mathcal{B}(\mathbf{U}), \gamma, \Gamma)$ be the combined and down-projected DS model that embodies post-data inference on the margin of interest Ω^*. The belief function associated with this DS model is, for every $H \in \Omega^*$,

$$\underline{P}(H) = \{\gamma(\mathbf{U})\}^{-1} \int \mathbb{1}(\Gamma(u) \subseteq H) \gamma(du). \tag{14.16}$$

The $(\mathtt{p}, \mathtt{q}, \mathtt{r})$ probabilities about the assertion H are assessed as:

$$\mathtt{p}(H) = \underline{P}(H), \quad \mathtt{q}(H) = \underline{P}(H^c), \quad \mathtt{r}(H) = 1 - \mathtt{p}(H) - \mathtt{q}(H), \tag{14.17}$$

representing weights of evidence "for," "against," and "don't know" about H.

The $(\mathtt{p}, \mathtt{q}, \mathtt{r})$ probabilities may be calculated analytically if the integral in (14.16) is tractable. In practice, however, it is often the case that (14.16) needs to be approximated using numerical methods. Section 14.5 discusses the design considerations for Monte Carlo algorithms to perform DS post-data inference.

There are alternative ways to summarize inference embodied by a belief function. A commonly employed one is the *plausibility transform* [8], which can be regarded as the combination of the belief function with a Bayesian DS model following a uniform distribution. The transform results in a precise probability distribution in the margin of interest. While a distributional inference summary is simpler to convey, it forgoes the ability to express a non-zero "don't know" probability for events of interest, thus sacrificing much of the appeal of the DS framework.

14.4 Example: Weakened Simultaneous Inference with Measurement Errors

As motivated in Section 14.1, the use of a three-valued probabilistic assessment of evidence is particularly useful whenever the agent possesses only low-resolution information or wishes to weaken the input information and structural assumptions for robustness considerations. In this section, we showcase an analytical example of DS-ECP for drawing weakened simultaneous inference about multiple unknown means subject to measurement errors.

In a scientific experiment, a total of k unknown quantities, $\mathsf{m} = (\mathsf{m}_1, \ldots, \mathsf{m}_k)$, are each measured once by a summary statistic, $\mathsf{x} = (\mathsf{x}_1, \ldots, \mathsf{x}_k)$. The marginal reference distributions of the x_i's, typically governed by their respective m_i's, are well-understood, whereas information regarding their interdependence may not be readily available. Suppose that marginally for each i, after proper scaling and transformation

$$\mathsf{x}_i \sim N\left(\mathsf{m}_i, \mathsf{s}^2\right). \tag{14.18}$$

That is, each m_i is measured exactly once by x_i, with an additive measurement error $\mathsf{x}_i - \mathsf{m}_i$ that is normally distributed with mean zero and variance s^2. The goal is to make inferences about the uncertain values of m without additional prior information. In particular, the analyst does not know and is unwilling to assume any dependence structure (or the lack thereof) among the individual measurement errors. Accordingly, the statistical model to construct should be *weakened*, in the sense that it describes a probability law that governs only certain aspects of the measurement errors while leaving the remainder aspects unspecified. How can we specify a model to reflect such a state of partial knowledge?

One such weakened model we may employ is the *vacuous orientation* model, which regulates the probabilistic distribution of the sum of squares (their *configuration*) of the errors but is maximally uninformative of the direction of the errors (their *orientation*). We describe the model below. Suppose the sum of squares of the measurement errors follows a scaled Chi-squared distribution with k degrees of freedom. That is

$$(\mathsf{x}-\mathsf{m})^\top(\mathsf{x}-\mathsf{m})=\mathsf{s}^2\mathsf{u}_1, \tag{14.19}$$

where $\mathsf{u}_1 \sim \chi^2_k$. For every value of s^2, (14.19) is a logical consequence of (14.18), but is not a sufficient condition for it. Indeed, (14.18) encompasses the special case that all individual measurement errors are independently and identically distributed as $N(0,\mathsf{s}^2)$ random variables. It also encompasses an array of other cases, such as the case in which the errors are perfectly correlated, each marginally distributed as a scaled and signed χ_k random variable. The scaling factor s^2 may be given a prior distribution of choice:

$$\mathsf{s}^2=\mathsf{u}_2, \tag{14.20}$$

which we specify for concreteness that $\mathsf{u}_2 \sim \text{Inv-}\chi^2_\nu$, an inverse Chi-squared distribution with ν degrees of freedom, with $\mathsf{u}_2 \perp \mathsf{u}_1$. Together with the observed data

$$\mathsf{x}=x, \tag{14.21}$$

equations (14.19), (14.20) and (14.21) respectively specify three pieces of marginal evidence. Their up-projected marginal DS models accord with the following sources:

$$\begin{aligned}
(14.19):(\mathbf{U}_1,\mathcal{B}(\mathbf{U}_1),\gamma_1,\Gamma_1)\quad &\text{where}\quad \mathbf{U}_1=\mathbb{R},\ \gamma_1\sim\chi^2_k,\\
&\Gamma_1(u_1)\ =\left\{(\mathsf{x},\mathsf{m},\mathsf{s})\in\mathbb{R}^{2k}\times\mathbb{R}^+:(\mathsf{x}-\mathsf{m})^\top(\mathsf{x}-\mathsf{m})=\mathsf{s}^2u_1\right\},\\
(14.20):(\mathbf{U}_2,\mathcal{B}(\mathbf{U}_2),\gamma_2,\Gamma_2)\quad &\text{where}\quad \mathbf{U}_2=\mathbb{R},\ \gamma_2\sim\text{Inv-}\chi^2_\nu,\\
&\Gamma_2(u_2)=\left\{(\mathsf{x},\mathsf{m},\mathsf{s})\in\mathbb{R}^{2k}\times\mathbb{R}^+:\mathsf{s}^2=u_2\right\},\\
(14.21):(\mathbf{U}_3,\mathcal{B}(\mathbf{U}_3),\gamma_3,\Gamma_3)\quad &\text{where}\quad \Gamma_3(u_3)=\left\{(\mathsf{x},\mathsf{m},\mathsf{s})\in\mathbb{R}^{2k}\times\mathbb{R}^+:\mathsf{x}=x\right\},
\end{aligned}$$

where we abuse the notation "$\sim$" to mean that the measure γ accords with the said distribution. Note that since (14.21) is a deterministic piece of evidence, Γ_3 is a trivial function of u_3. After the combination of their respective sources via Dempster's rule, and down-projection to the m-margin, the DS model is characterized by $(\mathbf{U},\mathcal{B}(\mathbf{U}),\gamma,\Gamma)$, with

$$\mathbf{U}=\mathbf{U}_1\times\mathbf{U}_2=\mathbb{R}^2,\ \ \gamma=\gamma_1\times\gamma_2,\ \ \Gamma(u_1,u_2)=\left\{\mathsf{m}\in\mathbb{R}^k:(\mathsf{m}-x)^\top(\mathsf{m}-x)=u_1u_2\right\}.$$

As all marginal focal sets result in non-empty intersections in this model, all $(u_1,u_2)\in\mathbf{U}_1\times\mathbf{U}_2$ are compatible. The revised space is simply the product space: $\mathbf{U}=\mathbf{U}_1\times\mathbf{U}_2$, and no elimination is necessary. Moreover, since Γ depends on u_1,u_2 only through their product, the source of the resulting DS model is equivalently expressed by the following specification:

$$\mathbf{U}=\mathbb{R},\quad \gamma\sim(k/\nu)\,\mathsf{F}_{(k,\nu)},\quad \Gamma(u)=\left\{\mathsf{m}\in\mathbb{R}^k:(\mathsf{m}-x)^\top(\mathsf{m}-x)=u\right\}, \tag{14.22}$$

where $(k/\nu)\,\mathsf{F}_{(k,\nu)}$ is the scaled F-distribution with scaling factor k/ν and degrees of freedom k and ν. Indexed by their squared radius u, the multi-valued mapping Γ of (14.22) depicts concentric k-hyperspheres centered at the observed x. Figure 14.2 is a schematic illustration of the DS model when $k=2$.

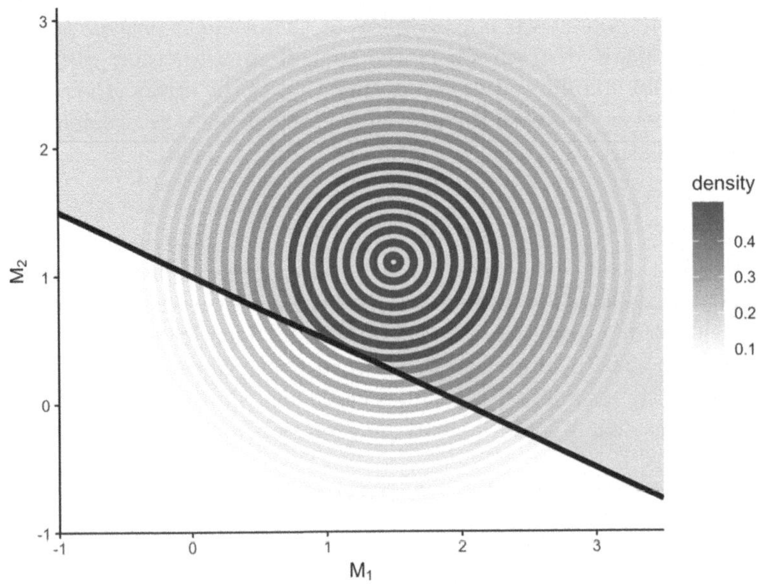

FIGURE 14.2
The vacuous orientation model when $k = 2$. Concentric hyperspheres are the mass-bearing random sets induced by the multi-valued mapping in Equation (14.22). The total probability mass of those contained in the shaded halfspace is the p probability for the linear equality assertion $H = \{\mathsf{m} \in \mathbb{R} : A\mathsf{m} \leq b\}$, where the observed data satisfies $Ax \leq b$.

To extract inference from the DS model in Equation (14.22) amounts to assessing the $(\mathtt{p}, \mathtt{q}, \mathtt{r})$ probabilities for assertions regarding the values of m. We first discuss the class of linear equality and inequality assertions, a commonly tested class of hypotheses about the unknown m. For assertions in the form of a consistent linear equality $H = \{\mathsf{m} \in \mathbb{R}^k : A\mathsf{m} = b\}$ where consistency means $H \neq \emptyset$, the post-data $(\mathtt{p}, \mathtt{q}, \mathtt{r})$ probabilities for H are

$$(0, F(t_x), 1 - F(t_x)). \tag{14.23}$$

Here $t_x = (b - Ax)^\top \left(AA^\top\right)^{-1} (b - Ax)$, and F is the cumulative probability function of γ in Equation (14.22), the scaled $\mathsf{F}_{(k,\nu)}$-distribution with scaling factor k/ν. For assertions in the form of a consistent linear inequality $H = \{\mathsf{m} \in \mathbb{R} : A\mathsf{m} \leq b\}$, the post-data $(\mathtt{p}, \mathtt{q}, \mathtt{r})$ probabilities for H are

$$\begin{cases} (F(t_x), 0, 1 - F(t_x)) & \text{if } Ax \leq b, \\ (0, F(t_x), 1 - F(t_x)) & \text{otherwise.} \end{cases} \tag{14.24}$$

Notice that the $\mathtt{p}$ probability for any linear equality assertion is always zero because it entails a sub-k dimensional hyperplane in $\mathbb{R}^k$, in which no mass-bearing random set (i.e. the concentric k-hyperspheres) can belong with positive probability. On the other hand, the $\mathtt{p}$ probability for a linear inequality assertion may be positive, and it is so precisely when the halfspace entailed by the inequality includes x, the center of the concentric random k-hyperspheres (see Figure 14.2). The $\mathtt{r}$ probability for both linear equality and inequality assertions takes the same value for the same A and b, reflecting the same degree of structural deficiency in the probabilistic evidence to supply a precise assessment.

The only type of assertions that has a zero "don't know" probability $\mathtt{r}$ in this model are of a quadratic form, corresponding to k-hyperspherical regions centered at the observed data. Write

$$C_{1-\alpha} = \left\{ \mathsf{m} \in \mathbb{R}^k : (\mathsf{m} - x)^\top (\mathsf{m} - x) \leq F_{1-\alpha}^{-1} \right\}, \quad \alpha \in [0, 1], \tag{14.25}$$

then the collection of $C_{1-\alpha}$ regions are concentric hyperspheres centered at x, having $(\mathbf{p}, \mathbf{q}, \mathbf{r})$ probabilities

$$(1 - \alpha, \alpha, 0) . \tag{14.26}$$

The fact that $\mathbf{r}(C_{1-\alpha}) = 0$ for all α means that the DS model produces *sharp* inference for the class of assertions $C_{1-\alpha}$. For this reason, they can be considered as the $100(1 - \alpha)\%$ post-data credible regions. Interested readers may refer to [36] which discusses the model in detail, including its calibration properties and a comparison with the Bonferroni family-wise error correction procedure.

14.5 Monte Carlo Computation for DS-ECP

Just like modern Bayesian inference, the computation task for DS-ECP models amounts to the efficient assessment of the post-data probabilities. The inference from a DS-ECP model is summarized by $(\mathbf{p}, \mathbf{q}, \mathbf{r})$ probabilities about assertions of interest. With the exception of a few conjugate or simple models (Section 14.3 being an example), most DS models would require nontrivial computation that would resort to Monte Carlo methods.

For a DS model characterized by the source $(\mathbf{U}, \mathcal{B}(\mathbf{U}), \gamma, \Gamma)$, its Monte Carlo estimation scheme obtains a collection of samples $\{u^{(i)}\}_{i \in [m]}$ with the target measure γ on $\mathbf{U}$. Then for any assertion H, the belief function in Equation (14.16) can be approximated via

$$\underline{P}(H) \doteq \frac{1}{m} \sum_{i=1}^{m} \mathbb{1} \left(\Gamma \left(u^{(i)} \right) \subseteq H \right), \tag{14.27}$$

based on which $(\mathbf{p}, \mathbf{q}, \mathbf{r})$ probabilities can be calculated. The consistency of the approximation is guaranteed by the law of large numbers under the usual conditions applicable for Monte Carlo methods. The central computational challenge is to efficiently sample the $u^{(i)}$'s from the target measure γ defined on $\mathbf{U}$. This task, however, can be nontrivial. The challenge is that the $\mathbf{U}$ space, a result of Dempster's combination, can be highly constrained, data-dependent, and often intractable. Simple rejection sampling, while conceptually appealing, can be hopelessly unviable.

Recently, [42] devised a Gibbs sampler to perform Monte Carlo estimation for the DS categorical model introduced in Section 14.2. It is the first algorithm to be able to perform efficient computation for the *structure of the second kind* model for a decently sized sample and problem dimension. Crucial to the feasibility of the Gibbs sampler is a characterization of the resulting DS model after combining marginal information from N independently observed data points.

We sketch the key idea behind the Gibbs sampler's feasibility by following the model setup introduced at the end of Section 14.3.1. For the nth categorical observation $\mathsf{x}_n = x_n$, the auxiliary variable associated with it, $\mathsf{u}_n = (\mathsf{u}_{n,1}, \ldots, \mathsf{u}_{n,K})$, follows the uniform distribution in the K-simplex. Also let $\mathcal{I}_k$ denote the set of indices corresponding to observations belonging to the kth category: $\{n \in [N] \colon \mathsf{x}_n = k\}$. For all $k, \ell \in [K]$, writing $\boldsymbol{u} = (u_1, \ldots u_N)$ and define

$$\eta_{k\to\ell}(\boldsymbol{u}) = \min_{n\in\mathcal{I}_k} \frac{u_{n,\ell}}{u_{n,k}}. \tag{14.28}$$

We have an important observation.

Theorem 14.1 [Proposition 3.1 of [42]] There exists $\theta \in \Delta$ satisfying $\theta_\ell/\theta_k \leq \eta_{k\to\ell}$ for all $k, \ell \in [K]$ if and only if the values $\{\eta_{k\to\ell}\}$, as functions of $\boldsymbol{u}$, satisfy

$$\eta_{j_1\to j_2}\eta_{j_2\to j_3}\cdots\eta_{j_L\to j_1} \geq 1, \tag{14.29}$$

for every $L \in [K]$ and every distinct set of indices $j_1, \ldots, j_L \in [K]$.

In other words, Theorem 14.1 gives a necessary and sufficient characterization of the revised space after Dempster's combination – i.e. $\mathbf{U}$ of (14.13) – on which the combined DS model for θ based on N independently sampled data points is defined. From here, a representation of the conditional distributions can be derived for the u_n's corresponding to observations in the kth category, given the u_n's corresponding to observations in all except the kth category; see Proposition 3.2 of [42]. The conditional distribution representation allows a Gibbs sampling scheme to loop over the K categories and obtain marginal updates to u at every iteration. Figure 14.3 depicts one configuration of feasible $\{\eta_{k\to\ell}\}$ values (dashed lines) for $K = 3$ on the left, and realized samples of the random polytope on the right.

The Gibbs algorithm effectively samples from the classic DS model summarizing post-data inference from a categorical distribution. Key to its feasibility is a characterization of the source resulting from Dempster's combination of model evidence, which enables the design of an efficient MCMC scheme to target it. However, research to date sheds little light on how to design a general computational solution for DS models. It seems that each model would still require a customized Monte Carlo scheme. The status quo of DS-ECP computation may be compared to the early days of non-conjugate Bayesian computation, though admittedly the former is still more challenging due to nontrivial specifications in terms of both probability and geometry. The richer the language of model expression, the fewer structural restrictions there are, the more tedious to satisfactorily summarize all information in it. The hope lies with future research to continue to explore simulation-based computational solutions for DS models, in particular for use in multi-dimensional real-valued spaces.

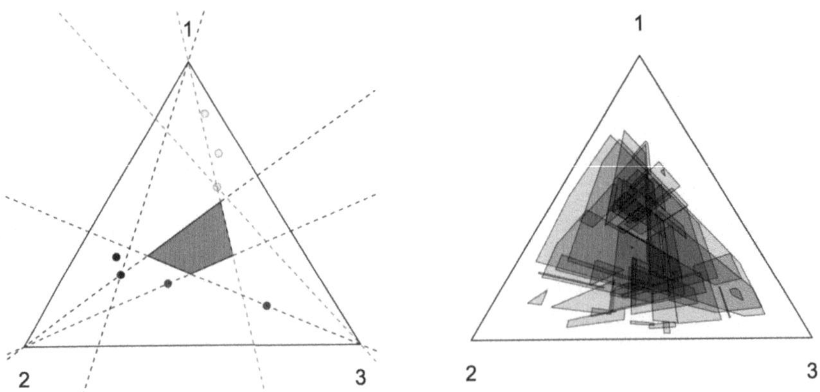

FIGURE 14.3
The Gibbs algorithm of [42] to sample from the structure of the second kind DS model with three categories. Left: dashed lines represent a set of feasible $\{\eta_{k\to l}\}$ values, for $k, l \in \{1, 2, 3\}$. The shaded area is one sample of the polytope. Right: an overlay of polytopes sampled by the Gibbs algorithm.

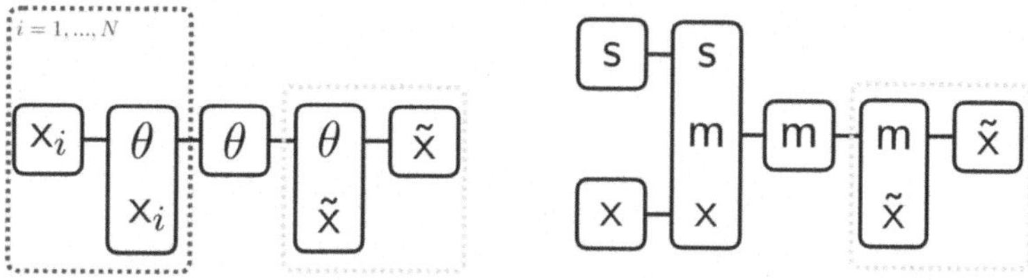

FIGURE 14.4
Join tree representations of the categorical inference model (left; Sections 14.2.2 and 14.5) and the vacuous orientation model (right; Section 14.4). Blue dashed square indicates independent repeated observations. Gray dashed squares indicate future observations and predictions (not discussed in the current chapter).

One potential way to reduce the computational burden for DS models may come from leveraging the model structure. The evidence contained in a full DS model can be captured with a join tree representation, in which every node in the tree represents one marginal DS model. As an illustration, Figure 14.4 depicts the join tree representations for the categorical inference model (Sections 14.2.2 and 14.5) and the vacuous orientation model (Section 14.4). In a join tree, evidence propagation is local and Markovian. One only needs to up-project, combine locally, and down-project to the adjacent node. Branches of the join tree whose information is already propagated can be successively "clipped off" and removed from the model. As a result, the computational task need not call for a joint space combination, which can be computationally resource-preserving if the state space is high-dimensional. Complex and sequential join tree structures, such as corresponding to a hidden Markov model expecting dynamic model input, can benefit from local computation greatly. More on graphical model computation for the DS framework, see [65, 44, 1].

14.6 Discussion and Further Reading

Many key elements of DS-ECP, including the use of non-additive probabilities and the combination of independent evidence encoded by them, are not part of the standard repertoire of statisticians. These ideas are far from radical, however, as traces of them can be found in the classical works that defined statistics. In *Ars Conjectandi*, Bernoulli distinguished the *pure* type of evidence from the *mixed* type, where the former is merely confirmatory whereas the latter can be both confirmatory and refutative, setting the stage for non-additive probabilistic accounting.[3] DS-ECP also shares elements with Don Fraser's structural inference [35], in particular an emphasis on the information contained in the model structure beyond distributions. The DS theory has also been a significant source of influence to the modern works on generalized fiducial inference (GFI; see [39]) and inferential models (IM; see [50]). Both subjects are discussed in detail elsewhere in this volume.

Due to the tremendous success the DS theory enjoyed in the engineering and computational sciences, the literature is vast and multi-disciplinary. Interested readers may refer to the 2008 edited collection, *Classic Works of the Dempster-Shafer Theory of Belief Functions* [71], as well as the International Journal of Approximate Reasoning (IJAR) 2016 special volume (No. 79), *Forty Years of Research on Dempster-Shafer Theory*. Thierry Denœux's editorial contains a categorized bibliography of 177 publications on the subject since the journal's inception in 1987 [23].

We conclude this chapter with a discussion of some open questions pertaining to the application of DS theory to statistical inference.

14.6.1 Dempster's Rule and the Independence Assumption

Individual belief functions represent the coarse and uncertain knowledge elicited from many experts. The rule that combines them reflects the modeler's judgment on the trustworthiness of these sources of evidence, as well as the logical independence among them. The DS-ECP employs Dempster's rule of combination to aggregate marginal evidence, but in certain applications, the modeler may not judge their sources of evidence in the same way as the rule requires. Motivated by the need to express alternative modeling judgments, combination rules for belief functions other than the Dempster's rule have been studied in the literature. Notable variations include the *discounting* of marginal belief functions with respect to degrees of reliability [58],[4] un-normalized combination as a different attribution of conflict [70], and the use of disjunction (rather than conjunction) to form consensus between marginal belief functions [24, 25]. [57] provides an excellent review.

Dempster's rule explicitly requires that the marginal DS models to be combined are independent from each other, in the sense defined in Section 14.3. In practice, however, the meaning of evidential independence is not always clear, nor does the modeler always have, or believe that they have, a collection of evidence that are truly independent.

Like other constituents of a model specification, the assumption of independence (or the lack thereof) is ultimately a subjective decision that the analyst must make. Shafer discussed the independence assumption at length in [63] and pointed out that the need for such an assessment is required in the Bayesian and likelihood inference paradigms as well. Indeed, the multiplication of the constituent likelihoods contributed by individual data points is based on the assumption that they are conditionally independent given the parameter. On the other hand, if dependence is suspect, the analyst would employ a joint likelihood that states explicitly the manners through which the multivariate data are dependent. [53] discussed the extension of capacity functionals to the multivariate case. [54, 55] explored the use of copulas to explicitly model the dependence of random sets via relating a joint belief function to its marginal belief functions.

14.6.2 The Choice of Structural Equation

As discussed in Section 14.3.1, stochastic marginal evidence within the DS framework is specified in the form of probabilistic functional models. Structural equations relate the observable data and the parameters of interest via the auxiliary random variables. The post-data inference resulting from the DS-ECP hinge on the inversion and aggregation of these stochastic functional relationships.

The structural equation, coupled with the probabilistic specification of the auxiliary variable, implies the sampling distribution of the observable data given the parameters. The reverse is not true. For the same data likelihood, there can exist multiple structural equations that correspond to it. Different structural equations have different inversions, define different marginal DS models, and result in different post-data inferences even if the implied sampling model for the data appears to be the same. What this means is that the DS framework, just like fiducial inference and other functional inference paradigms, does not obey the likelihood principle.

An example concerning a binomial experiment with n trials and θ the proportion parameter is as follows. Consider the two structural equations: (i) $\mathsf{x} = \sum_{i=1}^{n} \mathbf{1}(\mathsf{u}_i \leq \theta)$, where u_1 through u_n are independent uniform auxiliary random variables, and (ii) $\mathsf{x} = \sum_{i=0}^{n-1} \mathbf{1}(Q_{i+i,n-i}(\mathsf{u}) \leq \theta)$, where u is a uniform auxiliary random variable and $Q_{a,b}$ is the quantile function of the *Beta*(a, b) distribution. One may verify that both structural equations (i) and (ii) imply the same sampling distribution: $\mathsf{x} \sim Bin(n, \theta)$. However, their inversion upon observing $\mathsf{x} = x$ for any x result in different random sets.[5]

If the structural equations are a means to an end – that is, if they serve the sole purpose of inducing the desired sampling distribution on the observable data, the violation of the likelihood principle may be worrisome. But if the structural equation conveys substantive information about the generative process of the data beyond the sampling distribution, for example if it reflects a scientific theory which dictates how uncertain quantities connect together, the fact that the same data result in different inferences under different theories should not be surprising. In practice, the specification of the functional model may be advised by scientific motivations, as well as performance considerations for the resulting inference, such as the *validity* and *efficiency* criteria as put forth by the IM framework [50].

14.6.3 Dilation, Contraction, and Learning

Belief functions occupy a special corner within the world of imprecise probabilities. A belief function is a lower probability function, specifically a Choquet capacity of order ∞ [60]. In addition, every belief function characterizes a convex and closed set of probabilities, making them useful in describing sets of prior distributions in the robust Bayesian literature. The ϵ-contamination class of probabilities introduced in Section 14.2 is one such example.

Nevertheless, inference drawn from DS-ECP does not carry the same interpretation as that from a robust Bayesian model, even if both employ belief functions as the uncertainty descriptor. The robust Bayesian approach utilizes the generalized Bayes rule to perform model updating [28], which amounts to applying Bayes rule element-wise to every prior probability in a specified set. As an updating rule to sets of probabilities, the generalized Bayes rule prioritizes the avoidance of sure loss, whereas Dempster's rule doesn't. The distinction may be explained by that the former merely *focuses* the model on the conditioning evidence, whereas the latter *revises* the model in light of it [66]. In other words, updating via generalize Bayes rule respects the lower and upper probabilities as coherent expressions of personal belief, whereas updating via Dempster's rule emphasizes their nature as the logical encoding of aggregate probabilistic information.

We get a glimpse of this ideological difference through the seemingly puzzling behavior that the rules may incur. A notable phenomenon is *dilation*, studied systematically for the generalized Bayes rule [56]. Let Π be a convex and closed set of prior probabilities on Ω, with $\underline{P}$ its lower probability function. Let $\mathcal{B}$ be a partition of Ω, and A be an event. We say that $\mathcal{B}$ dilates A, if the lower and upper prior probability interval $[\underline{P}(A), \overline{P}(A)]$ is contained within the lower and upper posterior probability interval $[\underline{P}(A \mid B), \underline{P}(A \mid B)]$ for every $B \in \mathcal{B}$, where the conditional lower and upper probability function $\underline{P}(\cdot \mid \cdot)$ is induced by the updating rule, and the containment is strict for at least some B. [37] showed that Dempster's rule is less prone to dilation compared to generalized Bayes rule. However, it can incur *contraction* and *sure loss* whereas the generalized Bayes rule cannot.[6] Dempster's rule is also capable of revising a *vacuous* lower-upper prior interval (i.e. $[\underline{P}(A), \overline{P}(A)] = [0, 1]$) to a non-vacuous lower-upper posterior interval. It enables learning without the specification of a prior, granted that the application of Dempster's rule invokes assumptions of its own.

What we consider as paradoxical is ultimately a question about how we interpret the lower and upper probabilities, as well as how we expect them to evolve in light of new information. If one endorses the view that the lower and upper probability interval declares the likely values of a precise probability, similar to a confidence or a credible interval, dilation would seem odd because the expansion of the interval in all circumstances amounts to a "sure information loss" in some sense. But if the lower and upper probability interval reflects the range of betting odds within which the analyst abstains, dilation should not raise concern because the model may have focused on a piece of conditioning evidence with an indeterminate association with the event in question. The DS-ECP departs from the robust Bayesian tradition in that it enables the use of structurally deficient specifications of probabilistic models. It leverages marginal evidence in a rule-based manner to facilitate *learning*, that is, the refinement of uncertain information. Inference resulting from the DS-ECP should be interpreted with in mind the inevitable trade-off it strikes against the preservation of coherence.

Notes

1. Fisher recognized this kind of logical uncertainty as *deeper* than, and distinct from, the usual kind of uncertainty captured by known probabilities. He termed it *Uncertainty of Rank C* [31].
2. also termed the *simplicial* method [64]. In machine learning, the representation is also known as the "Gumbel-max trick" [48].
3. See Shafer [59] for an in-depth historical account.
4. Discounting is an important operation within the application of DS theory to expert systems and artificial intelligence; see e.g. [47].
5. See [42, Appendix A] for more details.
6. Also termed *constriction* [41], contraction happens when the lower and upper posterior probability intervals, $[\underline{P}(A \mid B), \underline{P}(A \mid B)]$, are contained in the lower and upper prior probability interval for every $B \in \mathcal{B}$. Similarly, *sure loss* happens when the lower and upper posterior probability intervals are fully displaced either to the left or to the right of the lower and upper prior probability interval, i.e. $\underline{P}(A \mid B) > \overline{P}(A)$ or $\overline{P}(A \mid B) < \underline{P}(A)$ for every $B \in \mathcal{B}$.

Bibliography

[1] Almond, R. G., 1995. *Graphical Belief Modeling.* Boca Raton, FL: CRC Press.

[2] Barnard, G. A., 1963. Some logical aspects of the fiducial argument. *Journal of the Royal Statistical Society: Series B (Methodological)*, *25*(1), pp. 111–114.

[3] Bayes, T., 1763. An essay towards solving a problem in the doctrine of chances. By the Late Rev. Mr. Bayes, F. R. S. Communicated by Mr. Price, in a Letter to John Canton, A. M. F. R. S. *Philosophical Transactions of the Royal Society of London*, *53*, pp. 370–418. https://doi.org/10.1098/rstl.1763.0053.

[4] Berger, J. O., 1990. Robust Bayesian analysis: Sensitivity to the prior. *Journal of Statistical Planning and Inference*, *25*(3), pp. 303–328.

[5] Berger, J. O., Moreno, E., Pericchi, L. R., Bayarri, M. J., Bernardo, J. M., Cano, J. A., De la Horra, J., Martín, J., Ríos-Insúa, D., Betrò, B. et al., 1994. An overview of robust Bayesian analysis. *Test*, *3*(1), pp. 5–124.

[6] Brillinger, D. R., 1962. Examples bearing on the definition of fiducial probability with a bibliography. *The Annals of Mathematical Statistics*, *33*(4), pp. 1349–1355.

[7] Bunke, H., 1975. Statistical inference: Fiducial and structural vs. likelihood. *Mathematische Operationsforschung und Statistik*, *6*(5), pp. 667–676.

[8] Cobb, B. R. and Shenoy, P. P., 2006. On the plausibility transformation method for translating belief function models to probability models. *International Journal of Approximate Reasoning*, *41*(3), pp. 314–330.

[9] Creasy, M. A., 1954. Limits for the ratio of means. *Journal of the Royal Statistical Society: Series B (Methodological)*, *16*(2), pp. 186–194.

[10] Dawid, P. A. and Stone, M., 1982. The functional-model basis of fiducial inference. *The Annals of Statistics*, *10*(4), pp. 1054–1067.

[11] Dempster, A. P., 1963. Further examples of inconsistencies in the fiducial argument. *The Annals of Mathematical Statistics*, *34*(3), pp. 884–891.

[12] Dempster, A. P., 1964. On the difficulties inherent in Fisher's fiducial argument. *Journal of the American Statistical Association*, *59*(305), pp. 56–66.

[13] Dempster, A. P., 1966. New methods for reasoning towards posterior distributions based on sample data. *The Annals of Mathematical Statistics*, *37*(2), pp. 355–374.

[14] Dempster, A. P., 1967. Upper and lower probabilities induced by a multivalued mapping. *The Annals of Mathematical Statistics*, *38*(2), pp. 325–339.

[15] Dempster, A. P., 1967. Upper and lower probability inferences based on a sample from a finite univariate population. *Biometrika*, *54*(3–4), pp. 515–528.

[16] Dempster, A. P., 1968. A generalization of Bayesian inference. *Journal of the Royal Statistical Society: Series B (Methodological)*, *30*(2), pp. 205–247.

[17] Dempster, A. P., 1968. Upper and lower probabilities generated by a random closed interval. *The Annals of Mathematical Statistics*, *39*(3), pp. 957–966.

[18] Dempster, A. P., 1969. Upper and lower probability inferences for families of hypotheses with monotone density ratios. *The Annals of Mathematical Statistics*, *40*(3), pp. 953–969.

[19] Dempster, A. P., 1972. A class of random convex polytopes. *The Annals of Mathematical Statistics*, *43*(1), pp. 260–272.

[20] Dempster, A. P., 2008. The Dempster-Shafer calculus for statisticians. *International Journal of Approximate Reasoning*, *48*(2), pp. 365–377.

[21] Dempster, A. P., 2014. Statistical inference from a Dempster-Shafer perspective. In *Past, Present, and Future of Statistical Science* (1st ed.), eds X. Lin, C. Genest, D. L. Banks, G. Molenberghs, D. W. Scott and J.-L. Wang, pp. 275–288. New York, NY: Chapman and Hall/CRC.

[22] Dempster, A. P., 2017. What Bayes did and what Bayes did not do. Fourth Bayesian, Fiducial and Frequentist Workshop, Harvard University, Cambridge, MA.

[23] Denœux, T., 2016. 40 years of Dempster-Shafer theory. *International Journal of Approximate Reasoning*, *79*(C), pp. 1–6.

[24] Dubois, D. and Prade, H., 1986. A set-theoretic view of belief functions logical operations and approximations by fuzzy sets. *International Journal of General System*, *12*(3), pp. 193–226.

[25] Dubois, D. and Prade, H., 1992. On the combination of evidence in various mathematical frameworks. In *Reliability Data Collection and Analysis*, eds J. Flamm and T. Luisi, pp. 213–241. Dordrecht, Netherlands.

[26] Edlefsen, P. T., Liu, C. and Dempster, A. P., 2009. Estimating limits from Poisson counting data using Dempster-Shafer analysis. *The Annals of Applied Statistics, 3*(2), pp. 764–790.

[27] Epstein, L. G., Kaido, H. and Seo, K., 2016. Robust confidence regions for incomplete models. *Econometrica, 84*(5), pp. 1799–1838.

[28] Fagin, R. and Halpern, J. Y., 1990. A new approach to updating beliefs. In *Proceedings of the Sixth Annual Conference on Uncertainty in Artificial Intelligence*, pp. 347–374. Cambridge, MA.

[29] Fieller, E. C., 1954. Some problems in interval estimation. *Journal of the Royal Statistical Society: Series B (Methodological), 16*(2), pp. 175–185.

[30] Fisher, R. A., 1930. Inverse probability. *Mathematical Proceedings of the Cambridge Philosophical Society, 26*(4), pp. 528–535.

[31] Fisher, R. A., 1957. The underworld of probability. *Sankhyā: The Indian Journal of Statistics (1933-1960), 18*(3/4), pp. 201–210.

[32] Fraser, D. A. S., 1962. On the consistency of the fiducial method. *Journal of the Royal Statistical Society: Series B (Methodological), 24*(2), pp. 425–434.

[33] Fraser, D. A. S., 1961. The fiducial method and invariance. *Biometrika, 48*(3/4), pp. 261–280.

[34] Fraser, D. A. S., 1961. On fiducial inference. *The Annals of Mathematical Statistics, 32*(3), pp. 661–676.

[35] Fraser, D. A. S., 1966. Structural probability and a generalization. *Biometrika, 53*(1–2), pp. 1–9.

[36] Gong, R., 2019. Simultaneous inference under the vacuous orientation assumption. *Proceedings of Machine Learning Research (ISIPTA'19), 103*, pp. 225–234.

[37] Gong, R. and Meng, X.-L., 2021. Judicious judgment meets unsettling updating: dilation, sure loss, and Simpson's paradox (with discussion). *Statistical Science, 36*(2), pp. 169–190.

[38] Grundy, P. M., 1956. Fiducial distributions and prior distributions: An example in which the former cannot be associated with the latter. *Journal of the Royal Statistical Society: Series B (Methodological), 18*(2), pp. 217–221.

[39] Hannig, J., Iyer, H., Lai, R. C. S. and Lee, T. C. M., 2016. Generalized fiducial inference: A review and new results. *Journal of the American Statistical Association, 111*(515), pp. 1346–1361.

[40] Heitjan, D. F. and Rubin, D. B., 1991. Ignorability and coarse data. *The Annals of Statistics, 19*(4), pp. 2244–2253.

[41] Herron, T., Seidenfeld, T. and Wasserman, L., 1994. The extent of dilation of sets of probabilities and the asymptotics of robust Bayesian inference. In *PSA: Proceedings of the Biennial Meeting of the Philosophy of Science Association*, volume 1994, pp. 250–259. Philosophy of Science Association.

[42] Jacob, P. E., Gong, R., Edlefsen, P. T. and Dempster, A. P., 2021. A Gibbs sampler for a class of random convex polytopes (with discussion). *Journal of the American Statistical Association*, *116*(535), pp. 1181–1192.

[43] Kennes, R., 1992. Computational aspects of the mobius transformation of graphs. *IEEE Transactions on Systems, Man, and Cybernetics*, *22*(2), pp. 201–223.

[44] Kong, C. T. A., 1987. *Multivariate Belief Functions and Graphical Models.* PhD thesis, Harvard University.

[45] Levi, I., 1983. *The Enterprise of Knowledge: An Essay on Knowledge, Credal Probability, and Chance.* Cambridge, MA: MIT Press.

[46] Lindley, D. V., 1958. Fiducial distributions and Bayes' theorem. *Journal of the Royal Statistical Society: Series B (Methodological)*, *20*(1), pp. 102–107.

[47] Lowrance, J. D., Garvey, T. D. and Strat, T. M., 1986. A framework for evidential-reasoning systems. In *Proceedings of the Fifth AAAI National Conference on Artificial Intelligence*, pp. 896–901. Philadelphia, Pennsylvania.

[48] Maddison, C. J., Tarlow, D. and Minka, T., 2014. A* sampling. *Advances in Neural Information Processing Systems*, *27*, pp. 3086–3094.

[49] Manski, C. F., 2003. *Partial Identification of Probability Distributions.* New York, NY: Springer Science & Business Media.

[50] Martin, R. and Liu, C., 2015. *Inferential Models: Reasoning with Uncertainty.* Boca Raton, FL: CRC Press.

[51] Savage, L. J., 1961. The foundations of statistics reconsidered. In *Proceedings of the Fourth Berkeley Symposium on Mathematical Statistics and Probability, Volume 1: Contributions to the Theory of Statistics.* The Regents of the University of California.

[52] Savage, L. J., 1972. *The Foundations of Statistics.* North Chelmsford, MA: Courier Corporation.

[53] Schmelzer, B., 2012. Characterizing joint distributions of random sets by multivariate capacities. *International Journal of Approximate Reasoning*, *53*(8), pp. 1228–1247.

[54] Schmelzer, B., 2015. Joint distributions of random sets and their relation to copulas. *International Journal of Approximate Reasoning*, *65*, pp. 59–69.

[55] Schmelzer, B., 2015. Sklar's theorem for minitive belief functions. *International Journal of Approximate Reasoning*, *63*, pp. 48–61.

[56] Seidenfeld, T. and Wasserman, L., 1993. Dilation for sets of probabilities. *The Annals of Statistics*, *21*(3), pp. 1139–1154.

[57] Sentz, K., Ferson, S. et al., 2002. *Combination of Evidence in Dempster-Shafer Theory*, volume 4015. United States, Web. https://doi.org/10.2172/800792.

[58] Shafer, G., 1976. *A Mathematical Theory of Evidence.* Princeton: Princeton University Press.

[59] Shafer, G., 1978. Non-additive probabilities in the work of Bernoulli and Lambert. *Archive for History of Exact Sciences*, *19*, pp. 309–370.

[60] Shafer, G., 1979. Allocations of probability. *The Annals of Probability, 7*(5), pp. 827–839.

[61] Shafer, G., 1981. Constructive probability. *Synthese, 48*(1), pp. 1–60.

[62] Shafer, G., 1982. Belief functions and parametric models. *Journal of the Royal Statistical Society: Series B (Methodological), 44*(3), pp. 322–339.

[63] Shafer, G., 1990. Perspectives on the theory and practice of belief functions. *International Journal of Approximate Reasoning, 4*(5–6), pp. 323–362.

[64] Shafer, G., 2021. Comment on "a Gibbs sampler for a class of random convex polytopes," by Pierre E. Jacob, Ruobin Gong, Paul T. Edlefsen, and Arthur P. Dempster. *Journal of the American Statistical Association, 116*(535), pp. 1196–1197.

[65] Shenoy, P. P. and Shafer, G., 1986. Propagating belief functions with local computations. *IEEE Expert, 1*(3), pp. 43–52.

[66] Smets, P., 1991. About updating. In *Proceedings of the Seventh Conference on Uncertainty in Artificial Intelligence*, pp. 378–385.

[67] Thoma, H. M., 1991. Belief function computations. *Conditional Logic in Expert Systems, 1*, pp. 269–308.

[68] Tukey, J. W., 1957. Some examples with fiducial relevance. *The Annals of Mathematical Statistics, 28*(3), pp. 687–695. https://doi.org/10.1214/aoms/1177706880.

[69] Wasserman, L. A., 1990. Belief functions and statistical inference. *Canadian Journal of Statistics, 18*(3), pp. 183–196.

[70] Yager, R. R., 1987. On the Dempster-Shafer framework and new combination rules. *Information Sciences, 41*(2), pp. 93–137.

[71] Yager, R. R. and Liu, L., 2008. *Classic Works of the Dempster-Shafer Theory of Belief Functions*, volume 219. New York, NY: Springer.

15

Slicing and Dicing a Path Through the Fiducial Forest

Joseph B. Lang
University of Iowa

15.1 Introduction

Of the three inference approaches, Bayesian, Frequentist, and Fiducial, the last is arguably the most controversial and least understood. R.A. Fisher (1930) introduced fiducial inference to reach his goal of deriving a prior-free post-data distribution *for*, or *of*, the unknown parameters. Paraphrasing L.J. Savage (1961), Fisher boldly attempted to "make the Bayesian omelet [compute a post-data distribution] without breaking the Bayesian eggs [without specifying a prior]." According to writers on its historical development (e.g. Dempster, 1964, Pedersen, 1978, Seidenfeld, 1992, Zabell, 1992, Efron, 1998, Hannig, 2009), the fiducial idea has had a checkered past, with its share of detractors and proponents. At times the fiducial approach betrayed glimpses of remarkable usefulness (see, for example, the recent work by Hannig and colleagues, and remarks in Efron, 1998), but at other times, owing to misunderstandings and missteps, it invited controversy.

Fisher's original fiducial conception, along with his one-parameter examples, sparked little controversy and was even credited by Neyman (1934) as the forerunner to his own confidence concept. In fact for these special examples, Fisher's fiducial probability statements could be viewed as shorthand for observed frequentist confidence statements. Ever the pragmatist, however, Fisher pushed his fiducial idea and applied it to more complicated, less directly amenable settings, like the multi-parameter, Behrens-Fisher setting. This push led to post-data distributions that no longer shared the observed frequentist properties of Neyman's confidence sets. Bartlett (1936) drove home this point, using Fisher's (1935) fiducial solution to the Behrens-Fisher problem as exhibit A. No longer moored to Neyman's confidence conception, Fisher's fiducial probabilities were in need of a separate rationale and interpretation. To Fisher, his "switching principle" fit the bill. It was this unmooring from observed frequentist confidence and the introduction of Fisher's switching principle that stirred up controversy and led to general confusion.

This chapter addresses this controversy and sets out to improve understanding through the introduction of a fiducial argument that does not invoke, nor rely upon, Fisher's switching principle or subsequent variants thereof (cf. Dempster, 1963, 1966, Fraser, 1966, Hannig, 2009, Hannig et al. 2016). The proposed slice-and-dice argument leads to straightforward interpretations of fiducial probabilities and distributions that are defensible using only an intuitive unremarkable realization, dicing proposition. Moreover, it is directly applicable for a broad class of multi-parameter models. To clarify the argument, the notation herein distinguishes between epistemic and aleatoric probability and between pre- and

DOI: 10.1201/9780429341731-15

post-data probabilities. It also avoids an asymmetric treatment of data and parameter values, a treatment that introduces an artificial and unhelpful dichotomy. To formulate a general theory of slice-and-dice fiducial inference, fundamental concepts, including slices and projections of fiducial sets, generalized pivots, and a taxonomy of relational pivotal models, are introduced. This chapter ends with several examples.

15.2 Probability Models and Inference Goal

A probability model $\mathscr{P}_Y$ for random vector Y is a collection of candidate distributions. A support of Y with respect to (wrt) this model is any set $su(Y)$ satisfying $P_Y(su(Y)) = 1, \forall P_Y \in \mathscr{P}_Y$. A minimal closed support of Y wrt model $\mathscr{P}_Y$ is defined as $min.su(Y) \equiv \{y : \exists P_Y \in \mathscr{P}_Y \ s.t. \ P_Y(N_\delta(y)) > 0, \forall \delta > 0\}$, where $N_\delta(y)$ is a δ-radius ball centered at y. A fiducial set $\mathcal{F}^Y$ for model $\mathscr{P}_Y$ is any set that is essentially equal to $min.su(Y)$, where A essentially equals B if $P_Y(A - B) = P_Y(B - A) = 0, \forall P_Y \in \mathscr{P}_Y$. In other words, **fiducial set** $\mathcal{F}^Y$ is the collection of possible values of Y under model $\mathscr{P}_Y$.

A probability model for the values in y_ω comprises two parts: (i) the assumption that $y_\omega \leftarrow Y$ and (ii) a model $\mathscr{P}_Y$ for Y. Part (i) states that the *a priori* uncertainty about the values in y_ω is characterized by viewing y_ω as a realization of the random vector Y, written $y_\omega \leftarrow Y$ or $y_\omega = Y(\omega)$, where ω is an outcome of the underlying random experiment.

In this chapter, the collection y_ω will have the form $y_\omega = (x_\omega, \theta_\omega)$ or $y_\omega = (x_\omega, \theta_\omega, u_\omega)$, where x_ω contains the observed data; θ_ω contains unobserved parameter values, which are the basis of targets of inferential interest; and u_ω contains other unobserved values that are not of direct inferential interest. More generally, as in prediction problems, the values in y_ω may include other unobserved values that are targets of inferential interest, but this chapter restricts attention to estimation problems where the inference targets are based only on θ_ω. Assuming that parameter value θ_ω is a realization of some random vector Θ should be non-controversial because degenerate distributions with unknown support points are allowed. For instance, if θ_ω is the 10^{10000} digit of π, we can model this as $\theta_\omega \leftarrow \Theta$, where $P(\Theta = \theta) = 1$, for $\theta \in \mathcal{F}^\Theta = \{0, 1, ..., 9\}$. This symmetric view of data and parameters aligns with the relevancy principle in Dempster (1963).

In closing this section, we state the **Inference Goal:** Use a model for $(x_\omega, \theta_\omega)$ or a model for $(x_\omega, \theta_\omega, u_\omega)$ along with the observed data x_ω to reduce uncertainty about the unobserved parameter values in θ_ω. To address this inference goal, we briefly address the following sub-goals: Goal (a) Decide how to measure uncertainty, see Section 15.3; Goal(b) Specify a reasonable model for $(x_\omega, \theta_\omega)$ or for $(x_\omega, \theta_\omega, u_\omega)$, see Sections 15.4 and 15.5; and finally Goal (c) Develop procedures that can be used to measure and communicate the reduced uncertainty, see the balance of this chapter.

15.3 Uncertainty and a Dicing Proposition

15.3.1 Two Types of Uncertainty

To reach the inference goal, we must first decide how to measure uncertainty. Herein, we focus on two types of uncertainty, epistemic and aleatoric, and their measures. Uncertainty stems from ignorance and hence is generally referred to as epistemic uncertainty. If the epistemic uncertainty stems from ignorance about the future outcome of a repeatable[1] process (i.e., a random experiment), then it is deemed aleatoric uncertainty (O'Hagan, 2004, Schweder, 2018). For example, there is epistemic uncertainty about the number of pips on the up-face of a hidden die. The epistemic uncertainty about the number of pips that will be on the up-face of a yet-to-be-rolled die could be deemed aleatoric–purely aleatoric if the die-rolling probability distribution were known. There is epistemic uncertainty about the 10^{1000th} digit of π and for most readers , there is epistemic uncertainty about the truth of the statement, "the English monarch in 1602 was female."

Epistemic uncertainty can be measured using an epistemic probability function $Pr(\cdot)$, which is a personalized function that maps statements to a value in the unit interval; the value $Pr(statement)$ is the epistemic probability that the *statement* is true. Aleatoric uncertainty can be measured using an aleatoric probability function $P(\cdot)$, which is a function that satisfies Kolmogorov's three axioms and maps sets from a σ-algebra to the unit interval. The aleatoric values $P(A)$ are tacitly taken to equal the epistemic $Pr(\textit{the process will result in an outcome in } A)$. Both epistemic and aleatoric probabilities are subjective or personal as they depend on the information available to the assigning agent about the statement or the repeatable process. For example, if Ed knows something about the die-rolling process that Sam does not, then what constitutes a replication of the experiment is different for Ed and Sam. The apparent objectivity of aleatoric probability stems from the [false] belief that the appropriateness of an aleatoric probability function can be empirically investigated by replicating the underlying process.

15.3.2 A Dicing Proposition

If $y_\omega \leftarrow Y$ and this is all that is known about y_ω, then we say that y_ω is an unremarkable realization of Y and write $y_\omega \overset{u.r.}{\leftarrow} Y$. If it is learned that $y_\omega \in B$, but nothing more, then y_ω is no longer an unremarkable realization of Y, but it is an unremarkable realization of $Y|(Y \in B)$, written $y_\omega \overset{u.r.}{\leftarrow} Y|(Y \in B)$. For instance, suppose that x_ω is the number of pips on the up-face of a die that is hidden under my hand. If you know that $x_\omega \leftarrow X \sim P(X = j) = \frac{1}{6}\mathbb{1}(j = 1, 2, \ldots, 6)$ and nothing more, then $x_\omega \overset{u.r.}{\leftarrow} X$. If you learn that x_ω is odd, then $x_\omega \overset{u.r.}{\leftarrow} X|(X \in \{1, 3, 5\})$. For this die-rolling example, the assignments $Pr(x_\omega = 3) \overset{u.r.}{=} P(X = 3) = 1/6$ and $Pr(x_\omega = 3|x_\omega \in \{1, 3, 5\}) \overset{u.r.}{=} P(X = 3|X \in \{1, 3, 5\}) = 1/3$ are intuitively appealing. This simple example motivates the following "dicing proposition," which is based on examples in Fraser (1961a, b, 1966), Dempster (1966:361), and Pedersen (1978:151). The proposition tells us that under an "unremarkable realization" assumption, epistemic probabilities can be assigned corresponding aleatoric probability values. The phrase, "unremarkable realization," is a nod to Fisher's (1958) description of a correct statement of probability, which includes the requirement that "no relevant subset can be recognized" (see Pedersen, 1978).

Dicing Proposition If it is known, assumed, or believed that $y_\omega \overset{u.r.}{\leftarrow} Y$, then a reasonable assignment is $Pr(y_\omega \in C) \overset{u.r.}{=} P(Y \in C)$. If it is learned that $y_\omega \in B$, then $y_\omega \overset{u.r.}{\leftarrow} Y|(Y \in B)$ and a reasonable assignment is $Pr(y_\omega \in C|y_\omega \in B) \overset{u.r.}{=} P(Y \in C|Y \in B)$. As a special case, suppose that (X, Θ) has a known distribution on $I\!R^2$. If $(x_\omega, \theta_\omega) \overset{u.r.}{\leftarrow} (X, \Theta)$ and the value x_ω is learned, then $(x_\omega, \theta_\omega) \overset{u.r.}{\leftarrow} (X, \Theta)|(X = x_\omega)$ and a reasonable assignment is $Pr(\theta_\omega \leq x_\omega|x_\omega) \overset{u.r.}{=} P(\Theta \leq x_\omega|X = x_\omega)$. Similarly, if instead of x_ω, the value θ_ω is learned, then a reasonable assignment is $Pr(\theta_\omega \leq x_\omega|\theta_\omega) \overset{u.r.}{=} P(\theta_\omega \leq X|\Theta = \theta_\omega)$.

15.4 A Brief Comparison of BFF Inference Approaches

The Bayesian, Frequentist, and Fiducial (BFF) inference approaches are related in that all three set out to assign values to conditional epistemic probabilities such as $Pr(\theta_\omega \in C(x_\omega)|$"*something*"$)$, where $C(\cdot)$ is a set-valued function of data x_ω. The Bayesian, Frequentist, and Fiducialist may even agree to use the same data-generating model (DGM), defined below. They do not, however, agree on the choice of "*something*" to condition on.

We begin with a definition of a data-generating model (DGM). A **DGM** has the form: $(x_\omega, \theta_\omega) \leftarrow (X, \Theta)$, where $\Theta \in \mathcal{F}^\Theta, X|(\Theta = \theta) \sim P_{X|\Theta}(\cdot|\theta)$, known for all $\theta \in \mathcal{F}^\Theta$. With a DGM, the distribution of Θ is left unspecified, save its support. With such a model, it is assumed pre-data that $(x_\omega, \theta_\omega) \overset{u.r.}{\leftarrow} (X, \Theta)$, which can be expressed in two distinct sequential forms:

$$\begin{array}{llll} \text{(i)} & \theta_\omega \overset{u.r.}{\leftarrow} \Theta \in \mathcal{F}^\Theta, & x_\omega \overset{u.r.}{\leftarrow} X|(\Theta = \theta_\omega) & \sim P_{X|\Theta}(\cdot|\theta_\omega), \\ \text{(ii)} & x_\omega \overset{u.r.}{\leftarrow} X \sim ?, & \theta_\omega \overset{u.r.}{\leftarrow} \Theta|(X = x_\omega) & \sim ?. \end{array} \tag{15.1}$$

The distributions of X and $\Theta|(X = x_\omega)$ are unknown because the distribution of Θ is unspecified. As examples, consider $(x_\omega, \theta_\omega) \leftarrow (X, \Theta)$, where...

$$\begin{array}{ll} \text{DGM1:} & \Theta \in \mathcal{F}^\Theta,\ X|(\Theta = \theta) \sim N(\theta, 1),\ \forall \theta \in \mathcal{F}^\Theta = I\!R \\ \text{DGM2:} & \Theta \in \mathcal{F}^\Theta,\ X|(\Theta = \theta) \sim IID(n)\ N_2(\theta),\ \forall \theta \equiv (\mu_1, \mu_2, \sigma_1^2, \sigma_2^2, \rho) \in \mathcal{F}^\Theta = \\ & I\!R^2 \times I\!R_+^2 \times (-1, 1),\ \text{an IID sample of } n \text{ from a bivariate Normal} \\ \text{DGM3:} & \Theta \in \mathcal{F}^\Theta,\ X|(\Theta = \theta) \sim IID(n)\ N(\mu, \sigma^2), \forall \theta \equiv (\mu, \sigma^2) \in \mathcal{F}^\Theta = I\!R \times I\!R_+. \end{array} \tag{15.2}$$

15.4.1 Bayesian Approach

The Bayesian focuses on line (ii) of (15.1), where it is observed that $x_\omega \overset{u.r.}{\leftarrow} X$ and $\theta_\omega \overset{u.r.}{\leftarrow} \Theta|(X = x_\omega)$. This, along with the dicing proposition, suggests a very reasonable inference approach based on assignments such as $Pr(\theta_\omega \in C(x_\omega)|x_\omega) \overset{u.r.}{=} P(\Theta \in C(x_\omega)|X = x_\omega)$. That is, the Bayesian conditions on the most obvious "*something*," namely, "*something*" = "*the observed data are* x_ω." Unfortunately, as hinted at by the question marks in (15.1), the Bayesian is confronted with a significant problem: the DGM does not lend itself to the computation of the distribution of $\Theta|(X = x_\omega)$ because the distribution of Θ is unspecified. As a practical, if controversial solution, the Bayesian replaces the non-restrictive "$\Theta \in \mathcal{F}^\Theta$" in the DGM with a more restrictive "prior" assumption along the lines of "$\Theta \sim P_\Theta$, where P_Θ is a known probability function." This revised, more restrictive Data-Target model can be specified as $(x_\omega, \theta_\omega) \leftarrow (X, \Theta)$, where $\Theta \sim P_\Theta,\ X|(\Theta = \theta) \sim P_{X|\Theta}(\cdot|\theta),\ \forall \theta \in \mathcal{F}^\Theta$. With this Data-Target Model, the distribution of $\Theta|(X = x_\omega)$ can be computed using a variant of Bayes' Rule (hence, the "Bayesian" nomenclature) and the Bayesian can go about assigning a value to the **post-data epistemic probability** $Pr(\theta_\omega \in C(x_\omega)|x_\omega)$. If, for instance, this probability were 0.95, the set $C(x_\omega)$ would be called a 95% Bayesian credible (or post-data probability) set for θ_ω. In other words, given that the observed data is x_ω, the epistemic probability is 0.95 that the unobserved parameter θ_ω is in the set $C(x_\omega)$. Usefully, the Bayesian can also assess the post-data epistemic plausibility of different candidate values for θ_ω. For example, intermediate values between 0 and 1 can be assigned to $Pr(\theta_\omega \leq t|x_\omega)$; specifically, $Pr(\theta_\omega \leq t|x_\omega) \overset{u.r.}{=} P(\Theta \leq t|X = x_\omega)$.

15.4.2 Frequentist Approach

The Frequentist registers some dissatisfaction with the Bayesian approach. In particular, the Frequentist is bothered by the fact that the Bayesian's post-data epistemic probabilities depend on the choice of Θ distribution in the Data-Target Model, and unfortunately, the observed data x_ω may have nothing much to say about the reasonableness of this choice. Ignoring debates about how to best choose a prior distribution for Θ or how sensitive post-data probabilities may be to different choices of Θ distribution, the Frequentist sets out to develop an inference approach that does not require the specification of the Θ distribution. Toward this end, the Frequentist focuses on line (i) of (15.1), where it is observed that $\theta_\omega \overset{u.r.}{\leftarrow} \Theta \in \mathcal{F}^\Theta$ and $x_\omega \overset{u.r.}{\leftarrow} X|(\Theta = \theta_\omega)$. This, along with the dicing proposition, suggests an inference approach based on assignments such as $Pr(\theta_\omega \in C(x_\omega)|\theta_\omega) \overset{u.r.}{=} P(\theta_\omega \in C(X)|\Theta = \theta_\omega)$. That is, the Frequentist conditions on "*something*" = "*the parameter value is* θ_ω." This is arguably a curious choice of "*something*" as it corresponds to a hypothetical probability that treats θ_ω as given and x_ω as unobserved, both of which are not true in reality.

Curiosities aside, the Frequentist is confronted with another, more significant problem: the aleatoric probability $P(\theta_\omega \in C(X)|\Theta = \theta_\omega)$ generally cannot be computed because θ_ω is unobserved. The Frequentist addresses this last problem by noting that although it cannot be computed, it is known that $P(\theta_\omega \in C(X)|\Theta = \theta_\omega)$ is a member of the set $\mathscr{P}_C \equiv \{P(\theta \in C(X)|\Theta = \theta) : \theta \in \mathcal{F}^\Theta\}$, which is computable using the DGM. Thus, the Frequentist finds that the **pre-data epistemic probability** $Pr(\theta_\omega \in C(x_\omega)|\theta_\omega)$ is a value in $\mathscr{P}_C$. The Frequentist's set-estimation goal is to find a set function $C(\cdot)$ that gives "small" $C(X)$ values and leads to an infimum of $\mathscr{P}_C$ that is large. If, for instance, the infimum were 0.95, then $Pr(\theta_\omega \in C(x_\omega)|\theta_\omega) \overset{u.r.}{=} P(\theta_\omega \in C(X)|\Theta = \theta_\omega) \geq 0.95$, and we would call $C(x_\omega)$ an observed frequentist, at least 95% confidence (or pre-data probability) set for θ_ω. This 95% confidence (or 95% pre-data probability) has a useful empirical (relative frequency) interpretation based on hypothetical replications of $X|(\Theta = \theta_\omega)$.

Unlike the Bayesian, the Frequentist's approach suffers an important drawback: the epistemic (pre- or post-data) plausibility of different candidate values for θ_ω cannot be reasonably assessed. For example, $Pr(\theta_\omega \leq t|\theta_\omega) \overset{u.r.}{=} P(\theta_\omega \leq t|\Theta = \theta_\omega) = \mathbb{1}(\theta_\omega \leq t) \in \{\mathbb{1}(\theta \leq t) : \theta \in \mathcal{F}^\Theta\}$ and this set generally equals $\{0, 1\}$. The Frequentist would come to the unhelpful conclusion, "I cannot compute a post-data probability, but the pre-data epistemic probability $Pr(\theta_\omega \leq t|\theta_\omega)$ is either 0 or 1."

15.4.3 Fiducial Approach

The Fiducialist registers dissatisfaction with both the Bayesian and frequentist approach. The Fiducialist part ways with the Frequentist, who conditions on "*something*" that leads to a hypothetical pre-data probability. Instead, the Fiducialist aligns with the Bayesian and uses post-data probabilities that condition on "*something*" akin to "*the observed data are* x_ω." Specifically, the Fiducialist assigns values to **post-data-slice epistemic probabilities** such as $Pr(\theta_\omega \in C(x_\omega)|\mathcal{F}(x_\omega))$, where $\mathcal{F}(x_\omega)$ is the "*post-data fiducial slice*," a subset of the model's original fiducial set (more on this below). The Fiducialist, like the Frequentist and unlike the Bayesian, avoids restricting the DGM to a Data-Target Model, which includes a "prior" distribution for Θ. Instead the Fiducialist introduces an augmented model for $(x_\omega, \theta_\omega, u_\omega)$, where u_ω is an unobservable realization of a pivot random variable that is related to X and Θ. This augmented model may induce or be compatible with the original DGM or it may be viewed as the starting point for inference without reference to any underlying DGM (Barnard, 1977, Dawid and Stone, 1982). Usefully, the Fiducialist, like the Bayesian and unlike the Frequentist, can measure the post-data epistemic plausibility of different candidate values for θ_ω. For example, intermediate values between 0 and 1 can be assigned to $Pr(\theta_\omega \leq t|\mathcal{F}(x_\omega))$. It has been said (cf. Xie and Singh, 2013) that the Fiducialist carries out inference about θ_ω using a prior-free post-data "distribution estimate," which is akin to the Bayesian's posterior distribution and is more informative than the Frequentist's point estimate combined with a single confidence set.

15.5 Relational Pivotal Models

The previous section defined a data-generating model (DGM) for $(x_\omega, \theta_\omega)$, which was used by the Frequentist, and a more restrictive Data-Target Model for $(x_\omega, \theta_\omega)$, which was used by the Bayesian. This section describes models for $(x_\omega, \theta_\omega, u_\omega)$, which can be used by the Fiducialist. Our descriptions of these models, most of which have appeared in the fiducial and related literature studies, are necessarily brief.

In the following, random variable U is said to be model independent of Θ, denoted $U \perp \Theta$, if $U|(\Theta = \theta) \sim U, \forall\theta \in \mathcal{F}^\Theta$. (Model independence implies independence, but the converse is not generally true.) A random variable U is conventionally defined to be a pivot for $k(\Theta)$, or for estimating $k(\theta_\omega)$, if $U \perp \Theta$ and $U = H(X, k(\Theta))$ has a known distribution. We will generalize this definition of a pivot in Section 15.8 after introducing relational models and fiducial sets.

Fisher, in his introduction to fiducial inference (1930), made use of a pivotal model (PM, see e.g. Dawid and Stone, 1982) for estimating $k(\theta_\omega)$, which can be defined as PM: $(x_\omega, \theta_\omega, u_\omega) \leftarrow (X, \Theta, U)$, where $\Theta \in \mathcal{F}^\Theta$, $U = H(X, k(\Theta))$, $U \perp \Theta$, $U \sim P_U$, (known). A Fisherian pivotal model (FPM) for estimating $k(\theta_\omega)$ is a PM that satisfies the additional restrictive conditions: (i) U has a continuous, univariate distribution; (ii) $k(\theta)$ is $I\!R^1$-valued; and (iii) for every $x \in \mathcal{F}^X$, the mapping $k \mapsto u = H(x, k)$ is a strictly monotonic, continuous bijection onto the space $\mathcal{F}^U$, which does not depend on x; equivalently, the set of possible values of U does not depend on the observed value of X. The condition FPM(iii) implies that the pivot $U = H(X, k(\Theta))$ is smoothly invertible of class 0 (Tukey, 1957). As examples of FPMs, consider $(x_\omega, \theta_\omega, u_\omega) \leftarrow (X, \Theta, U)$, where

$$
\begin{array}{ll}
\text{FPM1:} & \Theta \in I\!R^1,\ U = X - \Theta,\ \ U \perp \Theta,\ \ U \sim N(0,1). \\
\text{FPM2:} & \Theta \in \mathcal{F}^\Theta,\ U = F(t(X)|k(\Theta)),\ U \perp \Theta,\ U \sim U(0,1), \\
& \text{where } F \text{ is a continuous,univariate cdf of } t(X)|(\Theta = \theta), \text{ which depends} \\
& \text{only on } k(\theta). \text{ Moreover, } \ k \mapsto F(t(x)|k) \text{ is a strictly monotonic,} \\
& \text{differentiable map onto a set containing } (0,1) \text{ for each } x \in \mathcal{F}^X. \\
\text{FPM3:} & \Theta \in I\!R^1 \times I\!R^1_+,\ U = \frac{\overline{X} - \Theta_1}{S/\sqrt{n}},\ U \perp \Theta,\ U \sim t(n-1). \text{ Here, } \Theta_1 = k(\Theta).
\end{array}
\tag{15.3}
$$

Fisher's (e.g. 1930, 1935) original one-parameter fiducial examples were all FPMs. His very first example in 1930 had the form FPM2, where $t(X)$ was the sample correlation r based on data modeled as in DGM2 of (15.2). FPM3 appeared in Fisher (1935), where the statistics $\overline{X}$ and S^2 were sufficient statistics corresponding to DGM3.

On the one hand, a PM is sometimes related to an underlying DGM. For instance, the PM may induce a DGM, as FPM1 induces DGM1. Or the two may be compatible in that the PM specifies $H(X, k(\Theta)) = U, U \perp \Theta, U \sim P_U$ and the DGM implies that $H(X, k(\Theta))|(\Theta = \theta) \sim U \sim P_U$, as FPM2 and DGM2 are compatible when $t(X)$ is the sample correlation and as FPM3 and DGM3 are compatible. On the other hand, a PM need not be related to an underlying DGM and may not induce a DGM at all. In these cases, the PM, rather than a DGM, can be viewed as the starting point for statistical inference (Barnard, 1977).

A structural model (SM), as described in Fraser (1961a, b, 1966) and elsewhere (Bunke 1975, Dawid and Stone, 1982, Hannig 2009, etc.) is defined as SM: $(x_\omega, \theta_\omega, u_\omega) \leftarrow (X, \Theta, U)$, where $\Theta \in \mathcal{F}^\Theta$, $X = G(\Theta, U)$, $U \perp \Theta$, $U \sim P_U$ (known). An SM can be thought of as a generalization of a measurement model, where $X = \Theta + U$ and U is viewed as a collection of measurement error variables (Fraser, 1966). When G corresponds to a transformation group, Fraser (1961a, b, 1966) also refers to the structural model as a transformation model. Note that SMs and PMs are isomorphic only when the mapping $x \mapsto u = H(x, \theta)$ is a bijection (Fraser, 1982).

For our purposes, it is convenient to subsume PMs and SMs within the broader class of relational pivotal models (RPM). An RPM has the form $(x_\omega, \theta_\omega, u_\omega) \leftarrow (X, \Theta, U)$, where

$$\text{RPM:} \quad \Theta \in \mathcal{F}^{\Theta}, \ \ \mathcal{R}(X, \Theta, U), \ U \perp \Theta, \ U \sim P_U \text{ (known)}. \tag{15.4}$$

and $\mathcal{R}(x, \theta, u)$ states that triple (x, θ, u) satisfies the constraints imposed by relation $\mathcal{R}$. Technically, the "P" in RPM requires that the random variable U be a [weak or strong] pivot for some $k(\Theta)$ wrt relation $\mathcal{R}$. We defer the definition of a pivot wrt $\mathcal{R}$ to Section 15.8. Examples of relations include, $\mathcal{R} : u = H(x, k(\theta))$, $\mathcal{R} : x = G(\theta, u)$, $\mathcal{R} : ||x - G(\theta, u)|| \leq \epsilon$. Quite generally, RPMs can be used to model continuous or discrete X, multivariate X and or U, and multi-parameter Θ. Importantly, RPMs include relations used in contemporary generalized fiducial inference (GFI; cf. Hannig, 2009, Hannig et al., 2016).

It is not the goal of this paper to address non-uniqueness and inefficiency issues. Therefore, like Barnard (1977), we do not require that distributional assumptions about (Θ, U) in the RPM determine (uniquely or otherwise) a DGM for X. Instead, we view the RPM as the starting point for inference, thereby avoiding issues of non-uniqueness whereby multiple RPMs induce the same DGM, but lead to different fiducial distributions (e.g., Tukey, 1957). Relatedly, we do not require that "the whole of the information available" (Fisher, 1935) is used in the RPM; e.g., we make no assumption that RPM depends on X only through sufficient statistics corresponding to some underlying DGM. On this point, we align more with Neyman (1934, 1937, 1941) and his views of confidence procedures, than with Fisher. Finally, as with generalized fiducial inference (GFI), we do not require the existence of a transformation group structure nor do we assume that the RPM satisfies invertibility conditions, smooth or otherwise.

15.6 Arguments Based on Switching Principles

To this point, we have considered Goals (a) and (b) of Section 15.2. This brings us to Goal(c): Develop inference procedures that can be used to measure and communicate the reduced uncertainty. The current section describes historical and contemporary pivot-and-switch arguments that are based on a variety of existing switching principles. These arguments lead to prior-free post-data probabilities, including fiducial probabilities, that can be used to address Goal(c). Section 15.7 below introduces an alternative to these pivot-and-switch arguments.

15.6.1 Fisher's Fiducial Switching Principle

We begin by considering the Fisherian pivotal model (FPM), a very special RPM. With an FPM, there exists a unique value k, denoted $H^{-1}(x, u)$, satisfying $H(x, k) = u$. It follows that $P(k(\theta) \leq k_\beta(X) | \Theta = \theta) = 1 - \beta$, where for all x, $k_\beta(x) = H^{-1}(x, q(\beta; U))$ or $H^{-1}(x, q(1 - \beta; U))$, as the mapping $k \mapsto H(x, k)$ is decreasing or increasing. This led Fisher to his infamous–or "extra-logical" as Neyman euphemistically called it–switch maneuver,

$$1 - \beta = \begin{array}{c} P(k(\theta) \le k_\beta(X)|\Theta = \theta) \overset{sw}{=} P_{sw}(k(\theta_\omega) \le k_\beta(x_\omega)|x_\omega), \\ \text{where } k(\theta_\omega)|x_\omega \overset{sw}{\sim} H^{-1}(x_\omega, U). \end{array} \tag{15.5}$$

Note that the $1-\beta$ quantile of the $H^{-1}(x_\omega, U)$ distribution, namely $k_\beta(x_\omega)$, is an observed frequentist (OF) $1-\beta$ upper bound for $k(\theta_\omega)$. Hence, $P_{sw}(k(\theta_\omega) \le y|x_\omega) \overset{sw}{=} 1-\beta$ implies that $y = k^H(x_\omega)$, where $P(k(\theta) \le k^H(X)|\Theta = \theta) = 1-\beta$ for all $0 < \beta < 1$ and the fiducial distribution of $k(\theta_\omega)|x_\omega$ is called an OF fiducial distribution. In other words, in this FPM setting, the pivot-and-switch fiducial distribution $k(\theta_\omega)|x_\omega \overset{sw}{\sim} H^{-1}(x_\omega, U)$ has an OF property, which endows statements such as "$P_{sw}(k(\theta_\omega) \le y|x_\omega) = 0.95$" with two candidate interpretations: (i) The non-controversial OF interpretation, which takes the statement as shorthand for "$(-\infty, y]$ is an observed frequentist 0.95 confidence interval for $k(\theta_\omega)$" and (ii) the more controversial switching principle interpretation, which interprets $P_{sw}(k(\theta_\omega) \le k_\beta(x_\omega)|x_\omega) = 0.95$ as a post-data probability. To interpret this object as a probability and to motivate the reasonableness of its assigned value 0.95, Fisher needed some rationale for going from the pre-data probability assignment "$1-\beta = P(k(\theta) \le k_\beta(X)|\Theta = \theta)$" to the post-data probability assignment "$1-\beta = P_{sw}(k(\theta_\omega) \le k_\beta(x_\omega)|x_\omega)$." To Fisher, the obvious rationale was his **switching principle**, which simply asserts that the measure $P(k(\theta) \le k_\beta(X)|\Theta = \theta)$ of pre-data uncertainty about x_ω can be transferred without a change in value to the measure $P_{sw}(k(\theta_\omega) \le k_\beta(x_\omega)|x_\omega)$ of post-data uncertainty about $k(\theta_\omega)$. The subscript on P_{sw} was not used by Fisher. It is introduced here to highlight the fact that the switch leads to a post-data probability function that need not share any mathematical structure with the aleatoric probability function $P(\cdot|\theta)$; hence, Neyman's "extra-logical." Contrast this to the Bayesian pre- and post-data probabilities, which both derive from the joint probability function $P_{X,\Theta}$. The subscript sw was also added because it was not clear (Zabell, 1992) whether Fisher meant this statement to be read epistemically "$\overset{sw}{=} Pr(k(\theta_\omega) \le k_\beta(x_\omega)|x_\omega)$" or aleatorically "$\overset{sw}{=} P(k(\Theta) \le k_\beta(x_\omega)|X = x_\omega)$," or perhaps something altogether different. Later, Fisher leaned toward the aleatoric and added to his switching principle that the pre-data unknown θ_ω becomes the post-data random variable Θ. Deficient notation is at least partially to blame for the acceptance of this mysterious switch from fixed value to random variable: the generic symbol θ was, and continues to be, used to represent three distinct objects, θ_ω, Θ, and the dummy index in statements like $X|(\Theta = \theta) \sim P_{X|\Theta}(\cdot|\theta), \theta \in \mathcal{F}^\Theta$.

Fisher originally used language that aligned more with the non-controversial OF interpretation of his fiducial "probability" (Seidenfeld, 1992, Zabell, 1992). This was tantamount to treating the distribution function of $H^{-1}(x_\omega, U)$ as a "confidence distribution" (Cox, 1958, Efron, 1998, Schweder and Hjort, 2002, Xie and Singh, 2013). Over time, however, driven by his overarching goal to derive prior-free post-data distributions in more general settings, Fisher began using language that aligned with the switching-principle interpretation. Neyman initially chalked such language up to a simple case of *lapsus lingua*, but by 1937 or 1941 at the latest (Neyman, 1937, 1941), Neyman saw Fisher's fiducial approach, with its switching-principle interpretation, as completely distinct from his confidence conception. In contrast to his 1934 paper, Neyman's (1937) confidence interval opus avoided all mention of Fisher's fiducial work, much to Fisher's chagrin (Lehmann, 2011:82).

15.6.2 The Direct Switching Principle

As an alternative to fiducial probabilities, Dempster (1963) introduced direct probabilities, which were interpreted as aleatoric, prior-free post-data probabilities based on the $\Theta|(X = x_\omega)$ distribution. In his development of direct probabilities, Dempster (1963) introduced an alternative to Fisher's switching principle. Consider an RPM where the set of possible U

values when X takes value x is $\mathcal{F}^U(x) \equiv \{u : \mathcal{R}(x,\theta,u),\ u \in \mathcal{F}^U, \theta \in \mathcal{F}^\Theta\}$. The **direct sw′itching principle** can be stated as follows: If the set of possible U values does not depend on the observed value of X (i.e. $\mathcal{F}^U(x) = \mathcal{F}^U, \forall x \in \mathcal{F}^X$), then post-data measures of uncertainty about θ_ω based on $\Theta \in \mathcal{F}^\Theta$, $\mathcal{R}(X,\Theta,U)$, $U \perp \Theta$, $U \sim P_U$ are identical to those based on $\Theta \in \mathcal{F}^\Theta, \mathcal{R}(X,\Theta,U)$, $U \perp X$, $U \sim P_U$. That is, the roles of random variables X and Θ can be formally switched. Dempster noted a potential drawback to this principle: The assumptions $U \perp \Theta$ and $U \perp X$ cannot be simultaneously satisfied except in a special location setting with improper uniform Θ distribution (Lindley, 1958). He did, however, proceed with an example designed to make the "assumption less implausible than it at first appears."

In highlighting parallels between his direct probabilities and fiducial probabilities, Dempster (1963) noted that Fisher's switching principle is formally equivalent to the direct sw′itching principle. As an illustration, for invertible PMs the direct pivot-and-*sw′*itch approach can be formally applied as follows: Switch from pre-data $U = H(X, k(\Theta))$ to post-data $U = H(x_\omega, k(\Theta))$ and continue to assume that U has known, pre-data distribution P_U. This is tantamount to using the *sw′*itching principle to note that $H(x_\omega, k(\Theta))|(X = x_\omega) \sim U|(X = x_\omega) \overset{sw'}{\sim} U$ has a known distribution. Hence, by invertibility, the distribution of $k(\Theta)|(X = x_\omega)$ is known, and is identical to the distribution of $H^{-1}(x_\omega, U)$. For FPMs, this distribution is identical to Fisher's pivot-and-switch fiducial distribution, $k(\theta_\omega)|x_\omega \overset{sw}{\sim} H^{-1}(x_\omega, U)$. The same argument applies in the invertible SM setting.

15.6.3 The Structural Switching Principle

The direct *sw′*itching principle applies only when it can be assumed that $U \perp X$. The structural *sw″*itching principle applies more generally in settings where the pivot U is clearly not model independent of X ($U \not\perp X$). In particular, the principle can be invoked when such non-independence occurs because the relation $\mathcal{R}$ in the RPM implies that possible values of U depend on the observed value of X. As an example, suppose that $\Theta \geq 0$ and $X = \Theta + U$, where $U \sim N(0, 1)$. In this case, given $X = x$, the set of possible values of U is $\{u : u \leq x\}$ and hence $U|(X = x)$ has distribution that depends on x. The adjective "structural" is used because both Dempster (1966) and Fraser (1966) applied an epistemic version of the structural *sw″*itching principle in the derivation of their prior-free post-data distributions; Dempster referred to distributions based on structures of the first and second kind and Fraser referred to structural distributions. More recently, Hannig (2009) also employed this structural *sw″*itching principle in his derivation of generalized fiducial distributions.

Consider an RPM where the set of possible U values, when X takes value x, is $\mathcal{F}^U(x) \equiv \{u : \mathcal{R}(x,\theta,u),\ u \in \mathcal{F}^U, \theta \in \mathcal{F}^\Theta\}$. The **structural sw″itching principle** (aleatoric version) can be stated as follows: Post-data measures of uncertainty about θ_ω based on $\Theta \in \mathcal{F}^\Theta, \mathcal{R}(X,\Theta,U)$, $U \perp \Theta$, $U \sim P_U$ are identical to those based on $\Theta \in \mathcal{F}^\Theta, \mathcal{R}(X,\Theta,U)$, $U|(X = x_\omega) \sim U|(U \in \mathcal{F}^U(x_\omega))$, $U \sim P_U$, and this holds for any observed x_ω. The structural *sw″*itching principle is equivalent to the "non-informative" principle of Bunke (1975): Post-data, it is known that $\mathcal{F}^U(x_\omega)$ is a support of $U|(X = x_\omega)$. The non-informative principle asserts that this is the only thing that is known and hence it can be assumed that $U|(X = x_\omega) \sim U|(U \in \mathcal{F}^U(x_\omega))$. The direct *sw′*itching principle follows as a special case of the structural *sw″*itching principle: the direct principle applies when $\mathcal{F}^U(x) = \mathcal{F}^U$, $\forall x \in \mathcal{F}^X$, in which case the structural principle states that $U|(X = x) \overset{sw''}{\sim} U|(U \in \mathcal{F}^U)$, which is equivalent to the direct principle, $U|(X = x) \overset{sw'}{\sim} U$ because $P(U \in \mathcal{F}^U) = 1$.

As an illustration, consider an RPM with an invertible relation $X = G(\Theta, U)$, say $\Theta = G^{-1}(X, U)$. (The following argument can be easily modified for the PM setting with invertible relation $U = H(X, k(\Theta))$.) By the structural sw''itching principle, the aleatoric post-data distribution of Θ can be derived as follows: $\Theta|(X = x_\omega) \sim G^{-1}(X,U)|(X = x_\omega) \sim G^{-1}(x_\omega, U)|(X = x_\omega) \overset{sw''}{\sim} G^{-1}(x_\omega, U)|(U \in \mathcal{F}^U(x_\omega))$. If we take the post-data view that $\theta_\omega \overset{u.r.}{\leftarrow} \Theta|(X = x_\omega)$ and apply the dicing proposition, we can use this aleatoric post-data distribution to assign values to epistemic probabilities, e.g., $Pr(\theta_\omega \in C(x_\omega)|x_\omega) \overset{u.r.}{=} P(\Theta \in C(x_\omega)|X = x_\omega) \overset{sw''}{=} P(G^{-1}(x_\omega, U) \in C(x_\omega)|U \in \mathcal{F}^U(x_\omega))$.

15.7 An Alternative to Switching Principle Arguments

As an alternative to the pivot-and-switch arguments of the previous section, we introduce a qualitatively different slice-and-dice argument for deriving prior-free post-data distributions. This argument does not invoke, nor does it rely upon, a switching principle. Instead it is based only on the information in post-data "fiducial slices" and the dicing principle. We begin with definitions of fiducial sets and projections and slices thereof.

15.7.1 Fiducial Sets, Projections, and Slices

Consider an RPM for $(x_\omega, \theta_\omega, u_\omega)$, say $\Theta \in \Omega, \mathcal{R}(X, \Theta, U),\ U \perp \Theta, U \sim P_U$. The fiducial set corresponding to an RPM is the set of possible values of (X, Θ, U) under the model, which is assumed to be essentially equal to $min.su(X, \Theta, U)$, the minimal closed support of (X, Θ, U) (see Section 15.2). We can express this fiducial set as $\mathcal{F} \equiv \mathcal{F}^{X,\Theta,U} = \{(x, \theta, u) : \mathcal{R}(x, \theta, u), x \in I\!R^p, \theta \in \Omega, u \in min.su(U)\}$. Any value in $\mathcal{F}$ is said to be fiducial (true or faithful) to the model. Observing a value outside $\mathcal{F}$ would lead us to no longer trust or have faith in our model. Hence, the "fiducial" nomenclature (from Latin *fiducia* 'trust'). As a concrete example, consider a non-FPM variant of FPM1 of (15.3),

$$\text{RPM1:}\ \ \Theta \geq 0,\ \ X = \Theta + U,\ \ U \perp \Theta,\ \ U \sim N(0, 1). \tag{15.6}$$

Its fiducial set is $\mathcal{F} = \{(x, \theta, u) : x = \theta + u,\ x \in I\!R^1, \theta \geq 0, u \in I\!R^1\}$. The triple $(5, 2, 3)$ is fiducial, but the triples $(5, 2, 4)$ and $(5, -1, 6)$ are not. Fiducial sets, as the name implies, play a foundational role in fiducial inference. In fact, slices and projections of fiducial sets, as described next, are the primary objects of interest to the Fiducialist.

The projection of $\mathcal{F}$ onto the U space is defined as $\mathcal{F}^U = \{u : (x, \theta, u) \in \mathcal{F}\}$; other projections, such as $\mathcal{F}^\Theta, \mathcal{F}^X, \mathcal{F}^{\Theta,U}$, are defined similarly. It follows that an RPM and its fiducial set can be re-expressed as $\Theta \in \mathcal{F}^\Theta,\ \mathcal{R}(X, \Theta, U),\ U \perp \Theta,\ U \sim P_U$ and $\mathcal{F} = \{(x, \theta, u) : \mathcal{R}(x, \theta, u), x \in \mathcal{F}^X, \theta \in \mathcal{F}^\Theta, u \in \mathcal{F}^U\}$. A fiducial slice of $\mathcal{F}$ wrt $X = x_\omega$ is defined as $\mathcal{F}(x_\omega) \equiv \{(x_\omega, \theta, u) : (x_\omega, \theta, u) \in \mathcal{F}\}$ and can be interpreted as the possible (aka fiducial) values of (X, Θ, U), given that $(X = x_\omega)$ is observed. Finally, an example of a fiducial projected slice is $\mathcal{F}^U(x_\omega) = \{u : (x_\omega, \theta, u) \in \mathcal{F}\}$. This is the set of possible U values given that $X = x_\omega$ is observed.

For a general RPM, we have $\mathcal{F}(x_\omega) = \{(x_\omega, \theta, u) : \mathcal{R}(x_\omega, \theta, u), \theta \in \mathcal{F}^\Theta, u \in \mathcal{F}^U\}$; $\mathcal{F}^U = \{u : \mathcal{R}(x, \theta, u),\ x \in \mathcal{F}^X, \theta \in \mathcal{F}^\Theta, u \in \mathcal{F}^U\}$; $\mathcal{F}^U(x_\omega) = \{u : \mathcal{R}(x_\omega, \theta, u),\ \theta \in \mathcal{F}^\Theta, u \in \mathcal{F}^U\}$; and $\mathcal{F}^\Theta(x_\omega, u) = \{\theta : \mathcal{R}(x_\omega, \theta, u), \theta \in \mathcal{F}^\Theta\}$. For the special RPM1 of (15.6) with $x_\omega = 5$, $\mathcal{F}^U = \{u : x = \theta + u, x \in I\!R^1, \theta \geq 0, u \in I\!R^1\} = I\!R^1$; $\mathcal{F}^U(x_\omega) = \{u : 5 = \theta + u, \theta \geq 0, u \in I\!R^1\} = \{u : u \leq 5\}$; and $\mathcal{F}^\Theta(x_\omega, u) = \{\theta : 5 = \theta + u, \theta \geq 0\} = \{5 - u\}$ or $\emptyset$, as u is in $\mathcal{F}^U(x_\omega)$ or not.

Because x_ω represents the observed data, we will refer to $\mathcal{F}(x_\omega)$ as the "**post-data fiducial slice.**" Carrying the "fiducial" terminology a bit further, the observed data can be viewed as a "fiducial marker" (defn.: *anything accepted as a fixed basis of reference or comparison*) in statistical inference: If we initially know only that a triple, say $(x_\omega, \theta_\omega, u_\omega)$, falls in $\mathcal{F}$, so $(\theta_\omega, u_\omega)$ falls in $\mathcal{F}^{\Theta,U}$, then once the data x_ω is observed, the set of candidate values for the unobserved pair $(\theta_\omega, u_\omega)$ is reduced to the projected post-data fiducial slice $\mathcal{F}^{\Theta,U}(x_\omega)$. This is analogous to including a ruler, or line gauge, in a photograph to give a sense of scale, or more precisely, to reduce the set of candidate values for the sizes of the objects in the photograph. By the generic dictionary definition, the ruler, and by analogy, the observed data, play the role of a *fiducial marker.*

In closing this section, we give a set identity that will prove particularly useful for the slice-and-dice argument described below:

$$\mathcal{F}^{\Theta,U}(x_\omega) \stackrel{result}{=} \{(\theta,u):\ u \in \mathcal{F}^U(x_\omega),\ \theta \in \mathcal{F}^\Theta(x_\omega,u)\}. \tag{15.7}$$

15.7.2 The Slice-and-Dice Fiducial Argument

Suppose that pre-data the uncertainty can be modeled as $(x_\omega, \theta_\omega, u_\omega) \stackrel{u.r.}{\leftarrow} (X, \Theta, U)$, where $\Theta \in \mathcal{F}^\Theta$, $\mathcal{R}(X,\Theta,U)$, $U \perp \Theta$, $U \sim P_U$. This RPM has fiducial set $\mathcal{F} = \{(x,\theta,u) : \mathcal{R}(x,\theta,u),\ x \in \mathcal{F}^X,\ \theta \in \mathcal{F}^\Theta,\ u \in \mathcal{F}^U\}$. After x_ω is observed, the fiducial set of candidate values for $(x_\omega, \theta_\omega, u_\omega)$ reduces from $\mathcal{F}$ to the post-data fiducial slice, $\mathcal{F}(x_\omega) = \{(x_\omega,\theta,u) : \mathcal{R}(x_\omega,\theta,u),\ \theta \in \mathcal{F}^\Theta,\ u \in \mathcal{F}^U\}$. Therefore, post-data, we have learned that $(\theta_\omega, u_\omega)$ is in $\mathcal{F}^{\Theta,U}(x_\omega)$, which, by equation (15.7), can be expressed as $\mathcal{F}^{\Theta,U}(x_\omega) = \{(\theta,u) :\ u \in \mathcal{F}^U(x_\omega),\ \theta \in \mathcal{F}^\Theta(x_\omega,u)\}$. That is, post-data we know that $(\theta_\omega, u_\omega) \leftarrow (\Theta, U)$ and have learned that $u_\omega \in \mathcal{F}^U(x_\omega)$ and $\theta_\omega \in \mathcal{F}^\Theta(x_\omega, u_\omega)$. Let us break down what has been learned about u_ω and θ_ω individually.

Regarding what has been learned about u_ω, once we are given that $u_\omega \in \mathcal{F}^U(x_\omega)$, the additional information $\theta_\omega \in \mathcal{F}^\Theta(x_\omega, u_\omega)$ is redundant, because without specifying the distribution of Θ, the position of θ_ω in $\mathcal{F}^\Theta$ is completely unknown. Thus, the post-data uncertainty about u_ω can be expressed as $u_\omega \stackrel{u.r.}{\leftarrow} U|(U \in \mathcal{F}^U(x_\omega))$. As an example, suppose that $\Theta \in I\!R^1, X_i = \Theta + U_i, i = 1,...,n,\ U \perp \Theta,\ U \sim IID(n)\ N(0,1)$. Then $\mathcal{F}^U(x_\omega) = \{u \in I\!R^n : Au = Ax_\omega\}$, where $Az = (z_2 - z_1, ..., z_n - z_1)$ and $\mathcal{F}^\Theta(x_\omega, u_\omega) = \{x_{1\omega} - u_{1\omega}\}$. Having learned that $u_\omega \in \mathcal{F}^U(x_\omega)$, i.e. that $Au_\omega = Ax_\omega$, the additional information that $\theta_\omega \in \mathcal{F}^\Theta(x_\omega, u_\omega)$, i.e. that $u_{1\omega} = x_{1\omega} - \theta_\omega$, tells us nothing more about u_ω because the position of θ_ω in $I\!R^1$ is completely unknown.

Regarding θ_ω, given u_ω and $\theta_\omega \in \mathcal{F}^\Theta(x_\omega, u_\omega)$, the post-data uncertainty about θ_ω can be expressed as $\theta_\omega \stackrel{u.r.}{\leftarrow} \Theta|(U = u_\omega, \Theta \in \mathcal{F}^\Theta(x_\omega, u_\omega))$, and because the distribution of Θ is left unspecified, all that is known about this conditional distribution is that its support is $\mathcal{F}^\Theta(x_\omega, u_\omega)$. It follows that the post-data uncertainty about $(\theta_\omega, u_\omega)$ can be stated sequentially as $u_\omega \stackrel{u.r.}{\leftarrow} U|(U \in \mathcal{F}^U(x_\omega)),\ \ \theta_\omega \stackrel{u.r.}{\leftarrow} \Theta|(U = u_\omega, \Theta \in \mathcal{F}^\Theta(x_\omega, u_\omega)) \in \mathcal{F}^\Theta(x_\omega, u_\omega)$. And this is equivalent to $(\theta_\omega, u_\omega) \stackrel{u.r.}{\leftarrow} (\Theta_\omega, U_\omega)$, where newly introduced $(\Theta_\omega, U_\omega)$ has joint distribution satisfying,

$$\begin{aligned} U_\omega &\sim U|(U \in \mathcal{F}^U(x_\omega)) \\ \Theta_\omega|(U_\omega = u) &\in \mathcal{F}^\Theta(x_\omega, u), \qquad \forall u \in \mathcal{F}^U(x_\omega). \end{aligned} \tag{15.8}$$

Because post-$\mathcal{F}(x_\omega)$, $(\theta_\omega, u_\omega) \stackrel{u.r.}{\leftarrow} (\Theta_\omega, U_\omega)$, any random variable $(\Theta_\omega, U_\omega)$ satisfying (15.8) is called a post-$\mathcal{F}(x_\omega)$ **fiducial generator of** $(\theta_\omega, u_\omega)$. And the distribution of $(\Theta_\omega, U_\omega)$

is called a post-$\mathcal{F}(x_\omega)$ **fiducial distribution for estimating** $(\theta_\omega, u_\omega)$. Similarly, for any function $k(\cdot)$, because $k(\theta_\omega, u_\omega) \overset{u.r.}{\leftarrow} k(\Theta_\omega, U_\omega)$, the random variable $k(\Theta_\omega, U_\omega)$ is called a post-$\mathcal{F}(x_\omega)$ fiducial generator of $k(\theta_\omega, u_\omega)$, and the distribution of $k(\Theta_\omega, U_\omega)$ is a post-$\mathcal{F}(x_\omega)$ fiducial distribution for estimating $k(\theta_\omega, u_\omega)$. The slice-and-dice fiducial generator Θ_ω formally plays the same role as Hannig's (2009) pivot-and-*sw''*itch "generalized fiducial quantity $R_\xi(x)$." On the one hand, if the support $\mathcal{F}^\Theta(x_\omega, u)$ is a singleton set, with sole element denoted $\tau(x_\omega, u)$, then $(\Theta_\omega, U_\omega) = (\tau(x_\omega, U_\omega), U_\omega)$ is the unique fiducial generator of $(\theta_\omega, u_\omega)$ and the joint distribution of $(\Theta_\omega, U_\omega)$ is completely specified. On the other hand, when the support $\mathcal{F}^\Theta(x_\omega, u)$ is multi-valued, then there is no unique fiducial generator and the distribution of Θ_ω is incompletely specified. This "incomplete-specification" problem, slightly rephrased, was addressed in the generalized fiducial inference literature (cf. Hannig et al., 2009, 2013, 2016), where a pivot-and-sw''itch approach was taken. We will briefly address this problem from the slice-and-dice perspective after a taxonomy of RPMs is given.

Suppose that $(\Theta_\omega, U_\omega)$ has a known distribution that satisfies (15.8), so it is a fiducial generator of $(\theta_\omega, u_\omega)$. Because $(\theta_\omega, u_\omega) \overset{u.r.}{\leftarrow} (\Theta_\omega, U_\omega)$, we can assign values to post-$\mathcal{F}(x_\omega)$ epistemic probabilities using the dicing proposition. For example, $Pr(\theta_\omega \in C(x_\omega)|\mathcal{F}(x_\omega)) \overset{u.r.}{=} P(\Theta_\omega \in C(x_\omega))$. We are led to the following definition: A post-$\mathcal{F}(x_\omega)$ epistemic probability, such as $Pr(\theta_\omega \in C(x_\omega)|\mathcal{F}(x_\omega))$, that is assigned the aleatoric probability $P(\Theta_\omega \in C(x_\omega))$, where Θ_ω is a post-$\mathcal{F}(x_\omega)$ fiducial generator of θ_ω, is called a **fiducial probability**. For direct comparison, the Bayesian uses $Pr(\theta_\omega \in C(x_\omega)|x_\omega) \overset{u.r.}{=} P(\Theta \in C(x_\omega)|X = x_\omega)$; the Frequentist uses $Pr(\theta_\omega \in C(x_\omega)|\theta_\omega) \overset{u.r.}{=} P(\theta_\omega \in C(X)|\Theta = \theta_\omega) \in \mathscr{P}_C$, and the Fiducialist uses $Pr(\theta_\omega \in C(x_\omega)|\mathcal{F}(x_\omega)) \overset{u.r.}{=} P(\Theta_\omega \in C(x_\omega))$. There is nothing unusual about the Fiducialist's probability. All three approaches (BFF) use standard probability distributions ($\Theta|(X = x_\omega)$, $X|(\Theta = \theta)$, and Θ_ω, respectively) for epistemic probability assignments.

Finally, any set $C(x_\omega)$ that satisfies $P(k(\Theta_\omega) \in C(x_\omega)) = 1 - \alpha$ is a **post-$\mathcal{F}(x_\omega)$ fiducial $1-\alpha$ credible set for** $k(\theta_\omega)$. When $k(\theta_\omega) \in I\!R^1$, a simple-to-compute fiducial $1-\alpha$ credible interval for $k(\theta_\omega)$ is

$$C(x_\omega) = [q(\alpha_{1\omega}; k(\Theta_\omega)),\ q(1 - \alpha_{2\omega}; k(\Theta_\omega))], \tag{15.9}$$

where $(\alpha_{1\omega}, \alpha_{2\omega})$ is that member of $\{(\alpha/2, \alpha/2), (0, \alpha), (\alpha, 0)\}$ that gives the shortest interval. Allowing the α's to depend on x_ω is permissible because fiducial credible sets, like Bayesian credible sets, have post-data epistemic interpretations, not pre-data frequentist interpretations.

In summary, the **slice-and-dice argument** is based on the post-data-fiducial-slice, unremarkable realization results in equation (15.8), along with an application of the dicing proposition. Specifically, the post-data-slice uncertainty about the unobserved values is modeled as $(\theta_\omega, u_\omega) \overset{u.r.}{\leftarrow} (\Theta_\omega, U_\omega)$, and by the dicing proposition, post-$\mathcal{F}(x_\omega)$ probability assignments such as $Pr(\theta_\omega \in C(x_\omega)|\mathcal{F}(x_\omega)) \overset{u.r.}{=} P(\Theta_\omega \in C(x_\omega))$ are made. To emphasize the derivation approach, a fiducial distribution for estimating θ_ω, i.e. the distribution of Θ_ω, will also be called a **slice-and-dice fiducial distribution *for estimating*** θ_ω. This stands in contrast to Fisher's pivot-and-*sw*itch fiducial distribution *of* θ_ω; the pivot-and-*sw'*itch direct probability distribution of $\Theta|(X = x_\omega)$ (Dempster, 1963); the pivot-and-*sw''*itch structural distribution for θ_ω (Dempster, 1966, Fraser, 1966); and the pivot-and-*sw''*itch generalized fiducial distribution for θ_ω (Hannig et al., 2009, 2013, 2016). For FPMs, these distributions are all formally, but not conceptually, identical to that of fiducial generator Θ_ω. And for generalized fiducial inference RPMs, the distribution of the generalized fiducial quantity (Hannig, 2009) is formally, but not conceptually, identical to that of Θ_ω.

15.8 A Taxonomy for Relational Pivotal Models

This section uses the fiducial objects from the slice-and-dice argument to develop a taxonomy for relational pivotal models (RPMs). The purpose of this taxonomy is to give us a language to better communicate the qualitative differences and similarities across RPMs. Importantly, the taxonomy will give us a clearer view of the issues that may or may not arise in any given problem. Consider the general RPM with relation $\mathcal{R}$ and fiducial set $\mathcal{F} = \{(x, \theta, u) : \mathcal{R}(x, \theta, u), x \in \mathcal{F}^X, \theta \in \mathcal{F}^\Theta, u \in \mathcal{F}^U\}$.

We begin with a definition of a pivot wrt relation $\mathcal{R}$, thus completing our technical definition of an RPM of Section 15.5. A variable U is a [strong] **pivot for $k(\Theta)$ wrt relation $\mathcal{R}$** if (i) $U \perp \Theta$, (ii) U has known distribution, and (iii) for each $x \in \mathcal{F}^X$, $|\mathcal{F}^U(x)| > 1$ and if $\mathcal{R}(x, \theta', u')$, $\mathcal{R}(x, \theta'', u'')$, and $u' \neq u''$, then $k(\theta') \neq k(\theta'')$. The condition $|\mathcal{F}^U(x)| > 1$ simply states that $\mathcal{F}^U(x) \equiv \{u \in \mathcal{F}^U : \mathcal{R}(x, \theta, u), \theta \in \mathcal{F}^\Theta\}$ has at least two members; that is, there exists fiducial triples (x, θ', u') and (x, θ'', u'') with $u' \neq u''$. A variable U is a **weak pivot for $k(\Theta)$ wrt relation $\mathcal{R}$** if (i) $U \perp \Theta$, (ii) U has known distribution, and (iii) for each $x \in \mathcal{F}^X$, $k(\mathcal{F}^\Theta(x)) \neq k(\mathcal{F}^\Theta(x, u))$, for some $u \in \mathcal{F}^U$.

In other words, if U is a [strong] pivot for $k(\Theta)$, then for fixed data, different values of U lead to different values of $k(\theta)$, and if U is a weak pivot for $k(\Theta)$, then for fixed data, different values of U lead to different sets of possible values of $k(\Theta)$. It can be argued that (a) for a model with relation $\mathcal{R} : U = H(X, \Theta)$, U is a [strong] pivot for $k(\Theta)$ wrt $\mathcal{R}$ if and only if $H(X, \Theta)$ is a pivot for $k(\Theta)$, using the conventional definition, and hence this definition of a pivot indeed generalizes the conventional definition of a pivot. (b) if U is a [strong] pivot then it is also a weak pivot; and (c) If U is a weak or strong pivot for $k(\Theta)$, then it is correspondingly a weak or strong pivot for Θ.

In the following taxonomic definitions, the phrase, "at $X = x$" could be replaced by "over X values" or dropped altogether if the simple property holds for all $x \in \mathcal{F}^X$. An RPM is **simple wrt** U at $X = x$ if $\mathcal{F}^U(x) = \mathcal{F}^U$. In other words, when an RPM is simple wrt U, observing x_ω places no additional fiducial constraints on u_ω; i.e. pre- and post-data possible U values, namely $\mathcal{F}^U$ and $\mathcal{F}^U(x_\omega)$, are identical. An RPM is **simple wrt Θ given** U at $X = x$, if $\mathcal{F}^\Theta(x, u)$ is a singleton set for every $u \in \mathcal{F}^U(x)$. More generally, **an RPM is simple wrt** $k(\Theta)$ **given** U, at $X = x$, if $k(\mathcal{F}^\Theta(x, u))$ is a singleton set for every $u \in \mathcal{F}^U(x)$. When an RPM is simple wrt Θ [wrt $k(\Theta)$] given U, the relation $\mathcal{R}(x_\omega, \theta, u)$ has a single solution in θ [$k(\theta)$] for any given $u \in \mathcal{F}^U(x_\omega)$, which implies that the distribution of $(\Theta_\omega, U_\omega)$ in (15.8) is completely specified. If the RPM is not simple wrt Θ [wrt $k(\Theta)$] given U, the fiducial generator distribution is incompletely specified. This "incomplete-specification" problem is addressed below.

An RPM is **doubly simple [wrt** (U, Θ)] at $X = x$, if it is simple wrt U and simple wrt Θ given U, at $X = x$. More generally, an RPM is **doubly simple wrt** $(U, k(\Theta))$ at $X = x$, if it is simple wrt U and simple wrt $k(\Theta)$ given U, at $X = x$. A Fisherian Pivotal Model (FPM) is an example of a doubly simple RPM. Finally, an RPM is **ample wrt** U at $X = x$, if $P(U \in \mathcal{F}^U(x)) > 0$. An RPM that is not ample is a **skinny** RPM. Notice that if an RPM is simple wrt U then it is ample, but the converse is false. Because the fiducial generator U_ω has conditional distribution $U|(U \in \mathcal{F}^U(x_\omega))$, Borel's paradox tells us that a skinny RPM can lead to non-uniqueness problems when the 0-probability event $(U \in \mathcal{F}^U(x_\omega))$ does not correspond to an obvious choice of conditioning σ-algebra. As an example, with the event $(U_1 = U_2)$, it is not clear whether $\sigma(U_1 - U_2)$ or $\sigma(U_1/U_2)$ should be used. (cf. Proschan and Presnell, 1998). This "skinny" problem is addressed below.

15.9 Examples

This section illustrates the slice-and-dice approach to fiducial inference and highlights the utility of the RPM taxonomy. The first subsection gives several historical examples and the second subsection focuses on examples from contemporary generalized fiducial inference.

15.9.1 Historical Examples

Example 15.1 The RPM1 of (15.6), which has fiducial set $\mathcal{F} = \{(x, \theta, u): \ x = \theta + u, \ \theta \geq 0, \ u \in I\!R\}$, is not simple wrt U at $X = x_\omega$, because $\mathcal{F}^U(x) = \{u : u \leq x_\omega\} \neq \mathcal{F}^U = I\!R$. It is, however, ample wrt U and simple wrt Θ given U because $\mathcal{F}^\Theta(x_\omega, u) = \{\tau(x_\omega, u) \equiv x_\omega - u\}$, a singleton, for every $u \in \mathcal{F}^U(x_\omega)$, and this is true for any $x_\omega \in \mathcal{F}^X$. It follows that $\Theta_\omega = x_\omega - U_\omega \sim x_\omega - U|(U \leq x_\omega)$ is a fiducial generator of θ_ω. Hence, the fiducial distribution for estimating θ_ω, i.e., the distribution of Θ_ω, is a truncated Normal $Y|(Y \geq 0)$, where $Y \sim N(x_\omega, 1)$. Note that owing to the constraint $\Theta \geq 0$, this RPM1 is not a Fisherian PM. Hence, the quantiles of Θ_ω cannot be interpreted as observed frequentist confidence bounds. In particular, the 95% fiducial credible interval for θ_ω does not have a frequentist 95% confidence interpretation. Interestingly, however, the frequentist coverage of the fiducial credible interval of form (15.9) is around 0.975 at $\theta_\omega = 0$, dips to around 0.93 at $\theta_\omega = 3$, and then levels off at 0.95 for larger θ_ω. □

Example 15.2 Consider FPM1 of (15.3), which is RPM1 with "$\Theta \geq 0$" replaced by "$\Theta \in I\!R$." This FPM1 is doubly simple; i.e. simple (and hence ample) wrt U and simple wrt Θ given U. In this case, the fiducial generator of θ_ω is $\Theta_\omega = x_\omega - U \sim N(x_\omega, 1)$. For this Fisherian PM, the quantiles of Θ_ω can be interpreted as observed frequentist confidence bounds and $[q(.025; \Theta_\omega), q(.975; \Theta_\omega)]$ is an observed 95% frequentist confidence interval. □

Example 15.3 Consider Fisher's (1930) original example, FPM2 of (15.3): $\Theta \in \mathcal{F}^\Theta$, $U = F(t(X)|k(\Theta))$, $U \perp \Theta$, $U \sim U(0,1)$, where the cdf F satisfies the conditions listed in (15.3), including $F(t|k)$ strictly monotone in k. Under these conditions, the relevant fiducial sets are $\mathcal{F}^U(x_\omega) = (0,1) = \mathcal{F}^U$, $\mathcal{F}^\Theta(x_\omega, u) = \{\theta \in \mathcal{F}^\Theta : \ u = F(t(x_\omega)|k(\theta))\}$, and $k(\mathcal{F}^\Theta(x_\omega, u)) = \{\kappa(x_\omega, u)\}$, where $\kappa(x_\omega, u)$ is the unique value satisfying $F(t(x_\omega)|\kappa(x_\omega, u)) = u$. It follows that FPM2 is doubly simple wrt $(U, k(\Theta))$. Let Θ_ω satisfy $\Theta_\omega|(U_\omega = u) \in \mathcal{F}^\Theta(x_\omega, u), \forall u \in \mathcal{F}^U(x_\omega)$. Then Θ_ω is a fiducial generator of θ_ω and hence $k(\Theta_\omega)$ is a fiducial generator of $k(\theta_\omega)$. But $k(\Theta_\omega)|(U_\omega = u) \in k(\mathcal{F}^\Theta(x_\omega, u)) = \{\kappa(x_\omega, u)\}$, so $k(\Theta_\omega) \stackrel{wp\ 1}{=} \kappa(x_\omega, U_\omega)$. When $F(t|k)$ is strictly decreasing (the increasing argument is similar), the cdf of $k(\Theta_\omega)$ can be computed as, $F_{k(\Theta_\omega)}(y) = P(\kappa(x_\omega, U_\omega) \leq y) = P(F(t(x_\omega)|\kappa(x_\omega, U_\omega)) \geq F(t(x_\omega)|y)) = P(U_\omega \geq F(t(x_\omega)|y)) = 1 - F(t(x_\omega)|y)$, because $U_\omega \sim U \sim U(0,1)$. Assuming $F(t|k)$ is differentiable in k, we also have that the probability density function of $k(\Theta_\omega)$ is $f_{k(\Theta_\omega)}(y) = -\frac{\partial F(t(x_\omega)|y)}{\partial y}$, which is the same result Fisher (1930) arrived at using his pivot-and-switch argument. In general, the FPM2 fiducial pdf has the form $f_{k(\Theta_\omega)}(y) = |\frac{\partial F(t(x_\omega)|y)}{\partial y}|$. □

Example 15.4 Consider RPM2: $\Theta \in I\!R^1$, $X_i = \Theta + U_i$, $i = 1, \ldots, n$, $U \perp \Theta$, $U \sim IID(n)\ N(0,1)$. RPM2 is not simple nor is it ample wrt U, but it is simple wrt Θ given U. In particular, the relevant fiducial sets are $\mathcal{F}^U(x_\omega) = \{u : \ Au = Ax_\omega\}$, where $Az = (z_2 - z_1, \ldots, z_n - z_1)$ and $\mathcal{F}^\Theta(x_\omega, u) = \{x_{1\omega} - u_1\}, \forall u \in \mathcal{F}^U(x_\omega)$. The fiducial generator of θ_ω is $\Theta_\omega = x_{1\omega} - U_{\omega 1} \sim x_{1\omega} - U_1|(AU = Ax_\omega)$. By multivariate normal conditional distribution properties, we have that $\Theta_\omega \sim N(\overline{x}_\omega, 1/n)$. □

Example 15.5 Note that RPM2 induces DGM: $X|(\Theta = \theta) \sim IID(n)\ N(\theta, 1),\ \forall\theta \in I\!R^1$, for which $\overline{X}$ is sufficient. The sufficient DGM is $\overline{X}|(\Theta = \theta) \sim N(\theta, 1/n)$, which is induced by RPM_{2s}: $\Theta \in I\!R^1,\ \overline{X} = \Theta + U,\ U \perp \Theta,\ U \sim N(0, 1/n)$. Now RPM_{2s} is doubly simple, in fact it is an FPM, with $\mathcal{F}^U(\overline{x}_\omega) = \mathcal{F}^U = I\!R^1$ and $\mathcal{F}^\Theta(\overline{x}_\omega, u) = \{\overline{x}_\omega - u\}$. Hence, the post-$\mathcal{F}(\overline{x}_\omega)$ fiducial generator of θ_ω is $\Theta_\omega = \overline{x}_\omega - U \sim N(\overline{x}_\omega, 1/n)$, which is identical to the post-$\mathcal{F}(x_\omega)$ fiducial generator of Example 15.4. □

Example 15.6 Consider FPM3 of (15.3): $\Theta = (\Theta_1, \Theta_2) \in \mathcal{F}^\Theta \equiv I\!R \times I\!R_+,\ U = \frac{\overline{X}-\Theta_1}{\sqrt{S^2/n}}$, $U \perp \Theta$, and $U \sim t(n-1)$. FPM3 was considered by Fisher (1935). Here we note that FPM3 is simple, and hence ample, wrt U and simple wrt $k(\Theta) \equiv \Theta_1$ given U, but it is not simple wrt Θ given U. In particular, letting $d_\omega = (\overline{x}_\omega, s^2_\omega)$, note that $\mathcal{F}^\Theta(d_\omega, u) = \{\theta : \theta \in I\!R \times I\!R_+,\ u = \frac{\overline{x}_\omega - \theta_1}{\sqrt{s^2_\omega/n}}\} = \{\theta : \theta_1 = \overline{x}_\omega - u\sqrt{s^2_\omega/n},\ \theta_2 > 0\}$ is not a singleton set, so FPM3 is not simple wrt Θ given U, but $k(\mathcal{F}^\Theta(d_\omega, u)) = \{\overline{x}_\omega - u\sqrt{s^2_\omega/n}\}$, so it is simple wrt $k(\Theta)$ given U. In summary, FPM3 is doubly simple wrt $(U, k(\Theta))$, but not wrt (U, Θ). It follows by the slice-and-dice argument that the post-$\mathcal{F}(d_\omega)$ fiducial generator of $k(\theta_\omega) \equiv \theta_{1\omega}$ is $k(\Theta_\omega) \equiv \Theta_{\omega 1} = \overline{x}_\omega - U\sqrt{s^2_\omega/n}$, where $U \sim t(n-1)$. Notice that the $1-\alpha$ quantile of $\Theta_{\omega 1}$, which is $\overline{x}_\omega + q(1-\alpha; t(n-1))\sqrt{s^2_\omega/n}$, is an observed frequentist $1-\alpha$ upper bound for $\theta_{1\omega}$. This should not be surprising given that FPM3 is an FPM for estimating $\theta_{1\omega}$ and FPM fiducial distributions have observed frequency properties. □

15.9.2 Generalized Fiducial Inference Examples

The slice-and-dice argument and RPM taxonomy are illustrated in the context of Generalized Fiducial Inference (GFI; Hannig, 2009, Hannig et al., 2016). GFI employs RPMs with relations of the general structural form $X = G(\Theta, U)$, where for broadened applicability, very few restrictions are imposed. There is a cost to this flexibility. In the GFI setting, the RPM may not be simple or ample wrt U and it may not be simple wrt Θ given U. The issue of non-simple-wrt-U was addressed by Hannig et al. (2009, 2013, 2016) using a structural pivot-and-sw''itch argument. Herein, this issue is resolved because of the fiducial generating distribution in (15.8) conditions on the projected post-data slice $\mathcal{F}^U(x_\omega)$. The not-ample ("skinny") problem and the not-simple-wrt-Θ-given-U ("incomplete-specification") problem were also addressed by Hannig et al. using a pivot-and-sw$''$itch argument. It is instructive to view these problems and candidate solutions through the lens of the slice-and-dice argument.

We begin with the skinny problem. As noted earlier, an RPM that is skinny (not ample) wrt U presents a non-uniqueness problem because the distribution of $U_\omega \sim U|(U \in \mathcal{F}^U(x_\omega))$ depends on the choice of conditioning σ-algebra associated with the 0-probability event $(U \in \mathcal{F}^U(x_\omega))$. Using a structural pivot-and-sw''itch argument, and under mild regularity conditions, Hannig et al. presented a clever solution to the skinny problem, one that essentially automates the choice of a very reasonable σ-algebra.

We present a slice-and-dice solution to the skinny problem that buttresses the GFI pivot-and-sw$''$itch solution. Any GFI RPM can be re-expressed as $\text{RPM}^{(0)} : \Theta \in \mathcal{F}^\Theta,\ ||X - G(\Theta, U)|| = 0,\ U \perp \Theta,\ U \sim P_U$, where $||\cdot||$ is a norm. Consider the coarsened RPM, $\text{RPM}^{(\epsilon)} : \Theta \in \mathcal{F}^\Theta, ||X - G(\Theta, U)|| \le \epsilon, U \perp \Theta, U \sim P_U$, where it is assumed that $||x - G(\theta, u)||$ has a minimum over $\mathcal{F}^\Theta$. For convenience, also assume that the minimizer in $\mathcal{F}^\Theta$ is unique. This coarsening leads to a model with fiducial set satisfying $\mathcal{F}^{(\epsilon)} \supseteq \mathcal{F}^{(0)}$ and corresponds to "fattening" the conditioning event (Hannig et al., 2016). The projected fiducial set $\mathcal{F}^{(\epsilon)U}(x_\omega) = \{u \in \mathcal{F}^U : ||x_\omega - G(\theta, u)|| \le \epsilon, \theta \in \mathcal{F}^\Theta\}$ can be expressed as

$\mathcal{F}^{(\epsilon)U}(x_\omega) = \{u \in \mathcal{F}^U : \min_{\theta \in \mathcal{F}^\Theta} ||x_\omega - G(\theta, u)|| \leq \epsilon\}$, which for $\epsilon > 0$ leads to an ample RPM. Unfortunately, ample RPM$^{(\epsilon)}$ is not simple wrt Θ given U because $\mathcal{F}^{(\epsilon)\Theta}(x_\omega, u) = \{\theta : ||x_\omega - G(\theta, u)|| \leq \epsilon,\ \theta \in \mathcal{F}^\Theta\}$ is multi-valued. Fortunately, by the slice-and-dice argument, any $(\Theta_\omega, U_\omega)$ satisfying (15.8), with $\mathcal{F}^{(\epsilon)U}(x_\omega)$ and $\mathcal{F}^{(\epsilon)\Theta}(x_\omega, u)$, is a fiducial generator of $(\theta_\omega, u_\omega)$. One intuitive choice that works is $U_\omega^{(\epsilon)} \sim U|(U \in \mathcal{F}^{(\epsilon)U}(x_\omega)) \sim U|(\min_{\theta \in \mathcal{F}^\Theta} ||x_\omega - G(\theta, U)|| \leq \epsilon)$ and $\Theta_\omega^{(\epsilon)} \sim \text{argmin}_{\theta \in \mathcal{F}^\Theta} ||x_\omega - G(\theta, U_\omega^{(\epsilon)})||$. To see that this choice works, note that $\Theta_\omega^{(\epsilon)}|(U_\omega^{(\epsilon)} = u) \overset{wp1}{=} \text{argmin}_{\theta \in \mathcal{F}^\Theta} ||x_\omega - G(\theta, u)|| \in \{\theta : ||x_\omega - G(\theta, u)|| \leq \epsilon,\ \theta \in \mathcal{F}^\Theta\} = \mathcal{F}^{(\epsilon)\Theta}(x_\omega, u)$, for all $u \in \mathcal{F}^{(\epsilon)U}(x_\omega)$. In summary, the slice-and-dice argument leads to the post-$\mathcal{F}^{(\epsilon)}(x_\omega)$ fiducial generator of θ_ω,

$$\Theta_\omega^{(\epsilon)} \sim \text{argmin}_{\theta \in \mathcal{F}^\Theta} ||x_\omega - G(\theta, U)||\ |\ (\min_{\theta \in \mathcal{F}^\Theta} ||x_\omega - G(\theta, U)|| \leq \epsilon). \tag{15.10}$$

By the dicing proposition, because $\theta_\omega \overset{u.r.}{\leftarrow} \Theta_\omega^{(\epsilon)}$, we can make post-$\mathcal{F}^{(\epsilon)}(x_\omega)$ assignments such as $Pr(\theta_\omega \in C(x_\omega)|\mathcal{F}^{(\epsilon)}(x_\omega)) \overset{u.r.}{=} P(\Theta_\omega^{(\epsilon)} \in C(x_\omega))$. The expression on the right-hand side of equation (15.10) is analogous to that in Hannig et al. (2016, equation (2)), where a pivot-and-sw"itch argument was used. If desired, and under regularity conditions, one could also make post-$\mathcal{F}^{(0)}(x_\omega)$ assignments. Specifically, when the weak limit of $\Theta_\omega^{(\epsilon)}$ exists as $\epsilon \to 0$, call it $\Theta_\omega^{(0)}$, the assignment $Pr(\theta_\omega \in C(x_\omega)|\mathcal{F}^{(0)}(x_\omega)) \overset{u.r.}{=} P(\Theta_\omega^{(0)} \in C(x_\omega))$ is reasonable. Hannig's Theorem 1 (2016) can be used to give sufficient conditions for the existence of this weak limit and it can be used to find the probability density function of limiting fiducial generator $\Theta_\omega^{(0)}$. We illustrate the slice-and-dice solution to the skinny problem in the next example.

Example 15.7. Examples 15.4 and 15.5 above gave two approaches to carrying out fiducial inference about θ_ω based on RPM2. They both led to a fiducial generator of θ_ω with $N(\overline{x}_\omega, 1/n)$ distribution. Here we give a third approach based on a coarsened model. Because RPM2 is skinny, consider the coarsened RPM2$^{(\epsilon)}$: $\Theta \in I\!R^1$, $||X - \Theta 1_n - U|| \leq \epsilon$, $U \perp \Theta$, $U \sim\ IID(n)\ N(0, 1)$. RPM2$^{(\epsilon)}$ is ample wrt U, but neither simple wrt U nor wrt Θ given U. The relevant fiducial sets are $\mathcal{F}^{(\epsilon)U}(x_\omega) = \{u \in I\!R^n : \min_{\theta \in I\!R^1} ||x_\omega - \theta 1_n - u|| \leq \epsilon\} = \{u \in I\!R^n : ||u - \overline{u}1_n - (x_\omega - \overline{x}_\omega 1_n)|| \leq \epsilon\}$ and $\mathcal{F}^{(\epsilon)\Theta}(x_\omega, u) = \{\theta \in \{\theta \in I\!R^1 : ||x_\omega - \theta 1_n - u|| \leq \epsilon\}$. By the slice-and-dice argument, if $\tau(x_\omega, u) \in \mathcal{F}^{(\epsilon)\Theta}(x_\omega, u), \forall u \in\in \mathcal{F}^{(\epsilon)U}(x_\omega)$, then $\Theta_\omega^{(\epsilon)} = \tau(x_\omega, U_\omega)$ is a fiducial generator of θ_ω. One candidate is $\tau(x_\omega, u) = \text{argmin}_{\theta \in I\!R^1} ||x_\omega - \theta 1_n - u|| = \overline{x}_\omega - \overline{u}$. The corresponding fiducial generator of θ_ω is:

$$\Theta_\omega^{(\epsilon)} \quad = \tau(x_\omega, U_\omega) \quad \sim \quad \overline{x}_\omega - \overline{U} \quad | \ (||U - \overline{U}1_n - (x_\omega - \overline{x}_\omega 1_n)|| \leq \epsilon),$$

which is precisely that in (15.10). It is an attractive choice because $\overline{U}$ is independent of $(U - \overline{U}1_n)$, and hence $\Theta_\omega^{(\epsilon)} \sim N(\overline{x}_\omega, 1/n)$, *for every* $\epsilon \geq 0$. To be clear, any fiducial generator of θ_ω under model RPM2$^{(\epsilon)}$is legitimate for assigning values to $Pr(\theta_\omega \in C(x_\omega)|\mathcal{F}^{(\epsilon)}(x_\omega))$, which is an epistemic probability that conditions on an ϵ-coarsened post-data fiducial slice. It is just that in this setting there exists a fiducial generator that works for every $\epsilon \geq 0$, namely $\tau(x_\omega, U) = \overline{x}_\omega - \overline{U}$. This makes for a good choice because the assigned epistemic probabilities satisfy $Pr(\theta_\omega \in C(x_\omega)|\mathcal{F}^{(\epsilon)}(x_\omega)) = Pr(\theta_\omega \in C(x_\omega)|\mathcal{F}^{(0)}(x_\omega))$ for every $\epsilon \geq 0$, and the right-hand epistemic probability conditions on all the information in the uncoarsened post-data fiducial slice. □

We turn our attention to the incomplete-specification problem. Recall that if the RPM were simple wrt Θ given U, then $\mathcal{F}^\Theta(x_\omega, u)$ is a singleton set, and hence $(\Theta_\omega, U_\omega) \overset{wp\ 1}{=}$

$(\tau(x_\omega, U_\omega), U_\omega)$, where $\tau(x_\omega, u)$ is the sole member of $\mathcal{F}^\Theta(x_\omega, u)$. That is, the fiducial generator has a completely specified distribution. If, however, the RPM is not simple wrt Θ given U, then the joint distribution of $(\Theta_\omega, U_\omega)$ is not completely specified in (15.8) because it is only known that the distribution of $\Theta_\omega|(U_\omega = u)$ has support $\mathcal{F}^\Theta(x_\omega, u)$.

There are two main approaches to addressing the incomplete-specification problem. The first, **probability-bounds approach** leaves the distribution of $\Theta_\omega|(U_\omega = u)$ unspecified and employs lower and/or upper bounds on probabilities derived from the incompletely specified distribution of $(\Theta_\omega, U_\omega)$. For example, for scalar $k(\Theta)$, it is straightforward to see that for any distribution of $\Theta_\omega|(U_\omega = u)$ with support $\mathcal{F}^\Theta(x_\omega, u)$,

$$\begin{aligned} &P(\sup k(\mathcal{F}^\Theta(x_\omega, U_\omega)) \le b) - P(\inf k(\mathcal{F}^\Theta(x_\omega, U_\omega)) < a) \\ &\quad \le \; P(a \le k(\Theta_\omega) \le b) \\ &\quad \le \; P(\inf k(\mathcal{F}^\Theta(x_\omega, U_\omega)) \le b) - P(\sup k(\mathcal{F}^\Theta(x_\omega, U_\omega)) < a). \end{aligned} \tag{15.11}$$

This approach is related to the *inferential models, belief function* approach of Martin et al. (2010, 2013), which in turn is based on the Dempster-Shafer theory of lower and upper probabilities (Dempster, 1966, 1967, 2008, Shafer, 1982).

A second, quite practical approach to solving the incomplete-specification problem is the **partial-assignment approach**, whereby a distribution of $\Theta_\omega|(U_\omega = u)$ over its multi-valued support $\mathcal{F}^\Theta(x_\omega, u)$ is *assigned* by the analyst (see, e.g., Hannig, 2009, Hannig et al., 2016, for a related description). This is tantamount to choosing a particular fiducial generator $(\Theta_\omega, U_\omega)$ that satisfies (15.8). In fact, this was the approach taken in Example 15.7, where the "argmin" candidate was chosen. The "partial" is used as a reminder that the joint distribution of $(\Theta_\omega, U_\omega)$ is only partially assigned because U_ω has known distribution. This partial-assignment approach simplifies inferences, but the choice of distribution introduces a bit of subjectivity into the fiducial analysis. Fortunately, there are results (cf. Hannig, 2009, 2013) suggesting that fiducial analyses are often robust, e.g. asymptotically invariant, to the choice of assigned distribution over $\mathcal{F}^\Theta(x_\omega, u)$; essentially, the fiducial slices $\mathcal{F}^\Theta(x_\omega, u)$ often get smaller asymptotically as the sample size grows or the coarsening parameter ϵ shrinks. Moreover, Hannig (2009, 2013) notes that a high-variance distribution over any multi-valued $\mathcal{F}^\Theta(x_\omega, u)$ often leads to fiducial intervals with good frequentist coverage properties, even in small-sample settings. As an illustrative example, if $\mathcal{F}^\Theta(x_\omega, u)$ were an interval, a reasonable approach would be to set $\Theta_\omega^{(R)} = \tau(x_\omega, U_\omega, R)$, where $\tau(x_\omega, u, R) \equiv R \sup \mathcal{F}^\Theta(x_\omega, u) + (1 - R) \inf \mathcal{F}^\Theta(x_\omega, u)$ and $R|(U_\omega = u) \sim$ Beta(a, a) for some small a, say $a = 10^{-1000}$; this leads to a high-variance distribution for $\Theta_\omega^{(R)}|(U_\omega = u)$ over $\mathcal{F}^\Theta(x_\omega, u)$. That $\Theta_\omega^{(R)} \equiv \tau(x_\omega, U_\omega, R)$ is indeed a fiducial generator of θ_ω follows by the slice-and-dice argument upon noting that $\Theta_\omega^{(R)}|(U_\omega = u) \sim \tau(x_\omega, u, R)$ and $\tau(x_\omega, u, r) \in \mathcal{F}^\Theta(x_\omega, u)$, $\forall u \in \mathcal{F}^U(x_\omega)$ and $0 < r < 1$. If, in addition, $\mathcal{F}^\Theta(x_\omega, u)$ were closed, so the infimum and supremum are in the fiducial interval, then a valid fiducial generator could be based on choosing $a = 0$, so $R|(U_\omega = u) \sim \text{Beta}(0, 0) \overset{\text{def}}{\sim} \text{binom}(1, 0.5)$. This choice gives the highest possible variance distribution with support over $\mathcal{F}^\Theta(x_\omega, u)$ by Popoviciu's inequality (1935).

Example 15.8 (Two Independent Binomials) This example illustrates the two approaches to solving the incomplete-specification problem and highlights the utility of the slice-and-dice argument. Let $x_\omega = (x_{1\omega}, x_{2\omega})$ be the number of successes for two independent samples, where the success probabilities are $\theta_\omega = (\theta_{1\omega}, \theta_{2\omega})$. More specifically, consider the DGM: $(x_\omega, \theta_\omega) \leftarrow (X, \Theta)$, where $\Theta \in \mathcal{F}^\Theta \equiv [0, 1]^2$, $X_i|(\Theta = \theta)$ *indep* $\sim$ *binom*(n_i, θ_i), $i = 1, 2; \forall \theta \in \mathcal{F}^\Theta$. This DGM is induced by the RPM: $(x_\omega, \theta_\omega, u_\omega) \leftarrow (X, \Theta, U)$, where $\Theta \in \mathcal{F}^\Theta$, $X_i = \sum_{j=1}^{n_i} \mathbb{1}(U_{ij} \le \Theta_i), i = 1, 2$, $U \perp \Theta$, $U \sim$ *IID*$(n_1 + n_2)$ $U(0, 1)$.

The estimands of interest are $k_1(\theta_\omega) = \theta_{1\omega}$, $k_2(\theta_\omega) = \theta_{2\omega}$, $k_3(\theta_\omega) = \theta_{2\omega} - \theta_{1\omega}$, and $k_4(\theta_\omega) = \theta_{2\omega}/\theta_{1\omega}$.

For convenience, drop the ω subscript on the observed data and use $x \equiv x_\omega$. Define $U_{i(k)}$ as the kth order statistic for sample $U_i \equiv (U_{i1}, ..., U_{in_i})$. The relevant fiducial slices are $\mathcal{F}^U(x) = \{u \in [0,1]^{n_1+n_2} : x_i = \sum_{j=1}^{n_i} \mathbb{1}(u_{ij} \leq \theta_i),\ i = 1,2,\ \theta \in [0,1]^2\} = [0,1]^{n_1+n_2} = \mathcal{F}^U$ and $\mathcal{F}^\Theta(x,u) = \{\theta \in [0,1]^2 :\ x_i = \sum_{j=1}^{n_i} \mathbb{1}(u_{ij} \leq \theta_i),\ i = 1,2\} = \{\theta : u_{i(x_i)} \leq \theta_i < u_{i(x_i+1)}, i = 1,2\}$, where $u_{i(0)} \equiv 0$, $u_{i(n_i+1)} \equiv 1$, and "$< u_{i(n_i+1)}$" is taken to mean "≤ 1." It follows that this RPM is simple (and hence ample) wrt U, so the fiducial generator $U_\omega \sim U$. The RPM, however, is not simple wrt Θ given U. Nor is it simple wrt $k_h(\Theta)$ given U for any of the estimand functions $k_h, h = 1,2,3,4$. Thus, there is an "incomplete-specification" problem for all of the fiducial generators, Θ_ω and $k_h(\Theta_\omega)$, $h = 1,2,3,4$.

By the slice-and-dice argument, any $\Theta_\omega = \tau(x,U)$, where $\tau(x,u) \in \mathcal{F}^\Theta(x,u)$, $\forall u \in \mathcal{F}^U(x)$, is a fiducial generator of θ_ω, which in turn implies that $k_h(\Theta_\omega)$ is a fiducial generator of $k_h(\theta_\omega)$. Our goal is to compute fiducial credible intervals for the estimands of interest. Krishnamoorthy et al. (2017) tacitly used pivot-and-switch arguments based on Stevens (1950) to compute fiducial credible intervals in this two-independent-binomial setting.

We begin with the "probability bounds" solution to the incomplete-specification problem. Define $\inf_h \equiv \inf(k_h(\mathcal{F}^\Theta(x,U)))$ and $\sup_h \equiv \sup(k_h(\mathcal{F}^\Theta(x,U)))$. (There is no inadvertent omission of the ω subscript on U, because $U_\omega \sim U$ in this simple-wrt-U setting.) Because $\Theta_\omega \in \mathcal{F}^\Theta(x,U)$ and $k_h(\Theta_\omega) \in k_h(\mathcal{F}^\Theta(x,U))$ for every candidate fiducial generator, it follows that $P(\sup_h \leq b) - P(\inf_h < a) \leq P(a \leq k_h(\Theta_\omega) \leq b)$. Therefore, $[q(\alpha_{1\omega}; \inf_h), q(1 - \alpha_{2\omega}; \sup_h)]$, where α's are defined as in (15.9), is an at least $1-\alpha$ fiducial credible interval for $k_h(\theta_\omega)$. Conveniently, the infimums and supremums have simple explicit forms, e.g., $\inf_1 = U_{1(x_1)}, \sup_1 = U_{1(x_1+1)}$, $\inf_3 = U_{2(x_2)} - U_{1(x_1+1)}$, $\sup_3 = U_{2(x_2+1)} - U_{1(x_1)}$, $\inf_4 = U_{2(x_2)}/U_{1(x_1+1)}$, and $\sup_4 = U_{2(x_2+1)}/U_{1(x_1)}$. Thus, the at least $1-\alpha$ fiducial credible intervals are simple to compute or approximate via simulation.

Note that the fixed-α intervals $[q(\alpha_1; \inf_h), q(1-\alpha_2; \sup_h)]$, where $\alpha_1 + \alpha_2 = \alpha$ and α_j's do not vary with the data, are also at least $1-\alpha$ fiducial credible intervals. Of course, these fixed-α intervals are no narrower and may be wider than the varying-α intervals of (15.9), so the latter may be preferred. As an interesting historical aside, the fixed-α at least $1-\alpha$ fiducial credible interval for $\theta_{1\omega}$, namely, $[q(\alpha_1; \inf_1), q(1-\alpha_2; \sup_1)] = [q(\alpha_1; U_{1(x_1)}), q(1-\alpha_2; U_{1(x_1+1)})] = [q(\alpha_1; Beta(x_1, n_1 - x_1 + 1)), q(1-\alpha_2; Beta(x_1+1, n_1 - x_1))]$, is identical to the observed frequentist Clopper-Pearson $(1-\alpha)100\%$ confidence interval for $\theta_{1\omega}$. This identity gives retrospective support for Clopper and Pearson's (1934) choice of title, "The use of *confidence or fiducial* limits illustrated in the case of the binomial," with emphasis added.

The "partial assignment" solution to the incomplete-specification problem is also straightforward. In their pivot-and-switch approach, Krishnamoorthy et al. (2017) addressed this problem using "approximate fiducial variables," which could also be classified as partial assignment fiducial generators. The following high-variance partial assignment fiducial generators, which align with recommendations in Hannig (2009, 2013), are different from those in Krishnamoorthy et al.: Define $\Theta_\omega^{(R)} \equiv \tau(x,U,R) = (R_1 \sup_1 + (1-R_1)\inf_1,\ R_2 \sup_2 + (1-R_2)\inf_2)$, where $R_i|(U = u)$ *indep* $\sim$ *binom*$(1, 0.5)$. It follows that $\tau(x,u,r) \in \mathcal{F}^\Theta(x,u)$, $\forall u \in \mathcal{F}^U(x)$, and $r \in \{0,1\}^2$. Hence, by the slice-and-dice argument, $\Theta_\omega^{(R)}$ is a fiducial generator of θ_ω, where $\Theta_\omega^{(R)}|(U = u)$ has high variance over the rectangular region $\mathcal{F}^\Theta(x,u)$. By the same reasoning, $k_h(\Theta_\omega^{(R)})$ is a fiducial generator of $k_h(\theta_\omega)$, where $k_h(\Theta_\omega^{(R)})|(U = u)$ has high variance over the region $k_h(\mathcal{F}^\Theta(x,u))$. Thus, $[q(\alpha_{1\omega}; k_h(\Theta_\omega^{(R)})),\ q(1-\alpha_{2\omega};\ k_h(\Theta_\omega^{(R)}))]$, where the α's are defined as in (15.9), is a $1-\alpha$ fiducial credible interval for $k_h(\theta_\omega)$. □

15.10 Summary

Regardless of what one may think of a fiducial inference built upon a foundation supported by switching principles, the mounting empirical evidence (e.g., see the impressive body of work by Hannig and colleagues, cf. Hannig et al., 2016) suggests that fiducial solutions are often attractive alternatives to frequentist and Bayesian solutions. It is precisely for this reason that we ventured to inspect the fiducial foundation through a different lens.

To be clear, this chapter does not argue that the fiducial approach leads to post-data uncertainty measures that can, or should, align with frequentist or Bayesian interpretations. Rather, this chapter sets out to avoid the more controversial switching principles in the development of reasonable interpretations of fiducial distributions and probabilities that are untethered from either of the other two interpretations. As an example, as noted in the introduction, it was quickly realized that Fisher's fiducial solution to the Behrens-Fisher problem did not directly lend itself to frequentist interpretations. This unmooring from frequentism forced Fisher to rely upon a controversial switching principle to interpret his fiducial measures. This chapter's slice-and-dice approach leads to the same Behrens-Fisher fiducial solution as Fisher's, but the interpretation of post-data uncertainty avoids any reliance on any of the more controversial switching principles described herein.

Simply stated, this chapter's modest goal was not to replace the existing fiducial foundation, but rather to buttress it with a slice-and-dice argument that does not rely on switching principles. By buttressing the foundation, we aimed to further legitimize the fiducial approach. We clarified the approach by explicitly distinguishing between concepts such as epistemic and aleatoric probability, pre- and post-data probability, and realizations and random variables. To avoid an unhelpful dichotomoy, data and parameters were treated symmetrically. By developing a taxonomy of relational pivotal models and by clearly defining fiducial objects, including fiducial generators, distributions, and probabilities, we hope to have chipped away at some of the controversy and misunderstandings associated with fiducial inference. Perhaps with a fortified and clearer logical framework in place, future researchers will tread more confidently into the fiducial forest.

Note

1. We avoid discussions of what it means to "repeat" a process, and whether that is even possible.

Bibliography

Barnard, G. A., 1977. Pivotal inference and the Bayesian controversy. *Bulletin of the International Statistical Institute, 47*, pp. 543–551.

Bartlett, M. S., 1936. The information available in small samples. *Proceedings of the Cambridge Philosophical Society, 32*, pp. 560–566.

Bunke, H., 1975. Statistical inference: Fiducial and structural vs. likelihood. *Mathematische Operationsforschung und Statistik, 6*(5), pp. 667–676.

Clopper, C. J. and Pearson, E. S., 1934. The use of confidence or fiducial limits illustrated in the case of the binomial. *Biometrika, 26*(4), pp. 404–413.

Cox, D. R., 1958. Some problems connected with statistical inference. *The Annals of Mathematical Statistics, 29*(2), pp. 357–372.

Dawid, A. P. and Stone, M., 1982. The functional-model basis of fiducial inference. *The Annals of Statistics, 10*(4), pp. 1054–1067.

Dempster, A. P., 1963. On direct probabilities. *Journal of the Royal Statistical Society, Series B, 25*(1), pp. 100–110.

Dempster, A. P., 1964. On the difficulties inherent in Fisher's fiducial argument. *Journal of the American Statistical Association, 59*(305), pp. 56–66.

Dempster, A. P., 1966. New methods for reasoning towards posterior distributions based on sample data. *The Annals of Mathematical Statistics, 37*(2), pp. 355–374.

Dempster, A. P., 1967. Upper and lower probabilities induced by a multivalued mapping. *Annals of Mathematical Statistics, 38*, pp. 325–339.

Dempster, A. P., 2008. The Dempster-Shafer calculus for statisticians. *International Journal of Approximate Reasoning, 48*, pp. 365–377.

Efron, B., 1998. R.A. Fisher in the 21st century. *Statistical Science, 13*, pp. 95–122.

Fisher, R. A., 1930. Inverse probability. *Proceedings of the Cambridge Philosophical Society, xxvi*, pp. 528–535.

Fisher, R. A., 1935. The fiducial argument in statistical inference. *Annals of Eugenics, VI*, pp. 91–98.

Fisher, R. A., 1958. The nature of probability. *The Centennial Review, 2*, pp. 261–274.

Fraser, D. A. S., 1961a. On fiducial inference. *The Annals of Mathematical Statistics, 32*, pp. 661–676.

Fraser, D. A. S., 1961b. The fiducial method and invariance. *Biometrika, 48*, pp. 261–280.

Fraser, D. A. S., 1966. Structural probability and a generalization. *Biometrika, 53*, pp. 1–9.

Fraser, D. A. S., 1982. Discussion: The functional-model basis of fiducial inference. *The Annals of Statistics, 10*(4), pp. 1070–1073.

Hannig, J., 2009. On generalized fiducial inference. *Statistica Sinica, 19*, pp. 491–544.

Hannig, J., 2013. Generalized fiducial inference via discretization. *Statistica Sinica, 23*, pp. 489–514.

Hannig, J., Iyer, H., Lai, R. C. and Lee, T. C., 2016. Generalized fiducial inference: A review and new results. *Journal of the American Statistical Association, 111*(515), pp. 1346–1361.

Krishnamoorthy, K., Lee, M. and Zhang, D., 2017. Closed-form fiducial confidence intervals for some functions of independent binomial parameters with comparisons. *Statistical Methods in Medical Research, 26*(1), pp. 43–63.

Lehmann, E. L., 2011. *Fisher, Neyman, and the Creation of Classical Statistics.* New York, Springer Science & Business Media.

Lindley, D. V., 1958. Fiducial distributions and Bayes' theorem. *Journal of the Royal Statistical Society, B, 20*, pp. 102–107.

Martin, R., Zhang, J. and Liu, C., 2010. Dempster-Shafer theory and statistical inference with weak beliefs. *Statistical Science, 25*(1), pp. 72–87.

Martin, R. and Liu, C., 2013. Inferential models: A framework for prior-free posterior probabilistic inference. *Journal of the American Statistical Association, 108*(501), pp. 301–313.

Neyman, J., 1934. On the two different aspects of the representative method: The method of stratified sampling and the method of purposive selection. *Journal of the Royal Statistical Society, 97*(4), pp. 558–625.

Neyman, J., 1937. Outline of a theory of statistical estimation based on the classical theory of probability. *Philosophical Transactions of the Royal Society of London. Series A, Mathematical and Physical Sciences, 236*(767), pp. 333–380.

Neyman, J., 1941. Fiducial argument and the theory of confidence intervals. *Biometrika, 32*(2), pp. 128–150.

O'Hagan, A., 2004. Dicing with the unknown. *Significance, 1*, pp. 132–133.

Pedersen, J. G., 1978. Fiducial inference. *International Statistical Review, 46*(2), pp. 147–170.

Popoviciu, T., 1935. Sur les équations algébriques ayant toutes leurs racines réelles. *Mathematica* (Cluj) *9*, pp. 129–145.

Proschan, M. A. and Presnell, B., 1998. Expect the unexpected from conditional expectation. *The American Statistician 52*(3), pp. 248–252.

Savage, L. J., 1961. The foundations of statistics reconsidered. In *Proceedings of the Fourth Berkeley Symposium on Mathematical Statistics and Probability, Volume 1: Contributions to the Theory of Statistics.* The Regents of the University of California.

Schweder, T. and Hjort, N. L., 2002. Confidence and likelihood. *Scandinavian Journal of Statistics, 29*, pp. 309–332.

Schweder, T., 2018. Confidence is epistemic probability for empirical science. *Journal of Statistical Planning and Inference, 195*, pp. 116–125.

Seidenfeld, T., 1992. RA Fisher's fiducial argument and Bayes' theorem. *Statistical Science, 7*(3), pp. 358–368.

Shafer, G., 1982. Belief functions and parametric models (with discussion). *Journal of the Royal Statistical Society, Series B, 44*, pp. 322–352.

Stevens, W. L., 1950. Fiducial limits of the parameter of a discontinuous distribution. *Biometrika, 37*(1/2), pp. 117–129.

Tukey, J. W., 1957. Some examples with fiducial relevance. *The Annals of Mathematical Statistics, 28*, pp. 687–695.

Xie, M. and Singh, K., 2013. Confidence distribution, the frequentist distribution estimator of a parameter: A review. *International Statistical Review, 81*, pp. 3–39.

Zabell, S. L., 1992. RA Fisher and the fiducial argument. *Statistical Science, 7*(3), pp. 369–387.

16

Inferential Models and Possibility Measures

Chuanhai Liu
Purdue University

Ryan Martin
North Carolina State University

The inferential model (IM) framework produces data-dependent, non-additive degrees of belief about the unknown parameter that is provably valid. The validity property guarantees, among other things, that inference procedures derived from the IM control frequentist error rates at the nominal level. A technical complication is that IMs are built on a relatively unfamiliar theory of random sets. Here we develop an alternative—and practically equivalent—formulation, based on a theory of *possibility measures*, which is simpler in many respects. This new perspective also sheds light on the relationship between IMs and Fisher's fiducial inference, as well as on the construction of optimal IMs.

16.1 Introduction

Broadly speaking, statistics aims to quantify uncertainty about relevant unknowns based on observed data and, therefore, plays a fundamental role in science. More precisely, the goal of statistical inference, as we see it, is to take inputs—including data, posited statistical model (if any), and prior information (if any)—and return output in the form of meaningful numerical degrees of belief for relevant hypotheses concerning the unknowns. But despite its importance, there is still no consensus in the statistical community about how the inputs should be turned into output, what mathematical form the output should take, and what statistical properties they ought to satisfy. Reading between the lines, these questions make up what Efron (2013) called the "most important unresolved problem in statistical inference." The *BFF* or *Bayes, Fiducial, and Frequentist* group (e.g., Meng, 2017) was created, circa 2014, largely to foster research efforts that could help resolve this unresolved problem. While the depth and breadth of the group's contributions have been remarkable, the diversity of perspectives in this volume reveals that we are no closer to a resolution than we were in 2014. In light of the replication crisis in science (e.g., Camerer et al., 2018), the general confusion about and lack of trust in statistics (e.g. McShane et al., 2019; Trafimow and Marks, 2015), and new competition from applied mathematics, computer science, and engineering for resources and opportunities, resolving these problems is more important now than ever.

 DOI: 10.1201/9780429341731-16

Given that a theory of statistical inference is intended to be a foundation on which methods will be developed for solving scientific problems, it is essential to consider the totality of applications in order to ensure a base level of reliability or replicability. As Reid and Cox (2015) put it

> *it is unacceptable if a procedure...of representing uncertain knowledge would, if used repeatedly, give systematically misleading conclusions.*

Central to the development of our *inferential model* (IM) framework is the principle that belief assignments must be calibrated, or *valid* in the sense of (16.3) or (16.4) below. Our validity condition equips the belief assignments with a necessary external qualification that determines an objective scale for interpretation. A startling realization was that validity, in the strong sense that we advocate, cannot be achieved by a framework whose belief assignments satisfy the mathematical properties of additive probabilities. Therefore, if one strictly adheres to validity, then a departure from virtually all of the existing schools of thought is required, and the IM framework—first described in the sequence of papers by Martin and Liu (2013, 2015b,c) and then in our monograph (Martin and Liu, 2015a)—provides a guide into the world of statistical inference based on non-additive or imprecise probabilities.

After recognizing the apparently fundamental role that imprecise probability plays in statistical inference, we spent the last year digging into the imprecise probability literature in hopes of better understanding this important new connection. The present paper, the first of several resulting from these recent efforts, offers an alternative perspective on and construction of IMs. An essential and distinguishing feature of the IM framework is its use of random sets to quantify uncertainty about the unobserved value of a certain auxiliary variable. While some may find the use of random sets intuitively appealing, others will surely be bewildered. Therefore, we offer here an alternative, more direct, and arguably simpler construction that bypasses consideration of random sets. The key to this simplification is the fact that our recommended random sets are *nested*, a characteristic that provides valuable additional structure compared to random sets which are not nested. Roughly speaking, nested random sets are equivalent to *possibility measures* (e.g., Dubois and Prade, 1988; Zadeh, 1978), and the latter can be characterized by ordinary functions—what we call *possibility contours*—rather than non-additive set functions. In addition to their relative simplicity, possibility measures have a surprisingly close and inherent connection with the validity property we seek in the context of statistical inference. This stems from a beautiful characterization of a possibility measure's credal set in terms of probabilities assigned to possibility contour's level sets. Consequently, possibility measures are fundamental to statistical inference, so drawing this connection provides some important insights beyond the simplified IM construction.

The remainder of this chapter is organized as follows. After setting up the notation, the need for non-additivity, and the basic IM construction in Section 16.2, we introduce the necessary concepts and terminology from possibility theory in Section 16.3.1. Then Section 16.3.2 develops the new possibility measure-based construction, various implications are discussed in Section 16.3.3, and its equivalence to the original IM formulation is established in Section 16.3.4. Two illustrative examples are presented in Section 16.4 with an emphasis on both the possibility measure interpretation and the key dimension reduction steps originally presented in Martin and Liu (2015b,c). Finally, Section 16.5 gives some concluding remarks and poses some open questions.

16.2 Background

16.2.1 Probabilistic Inference and False Confidence

Statisticians, engineers, and others choose to quantify their uncertainties in terms of probabilities for various reasons. And centered around virtually every one of those individual reasons, a framework for probabilistic inference has been developed. So what makes the IM framework unique? Like many, we were originally troubled by the Bayesians' need for a prior distribution and we started down that now well-worn path, but we eventually realized the problem went much deeper. Specifically, all existing frameworks for probabilistic inference fail to provide certain guarantees that we believe are crucial to the interpretation of those probabilities. We start with a brief description of this observation.

To set the scene, let Y denote the observable data, with statistical model $Y \sim \mathsf{P}_{Y|\theta}$, where $\mathsf{P}_{Y|\theta}$ is a probability distribution for Y, supported on (a sigma-algebra containing subsets of) the space $\mathbb{Y}$, depending on a parameter $\theta \in \Theta$. Now let Q_y be a data-dependent probability measure defined on Θ—a Bayes or empirical Bayes posterior (Berger, 1985; Ghosh et al., 2006), a fiducial or generalized fiducial distribution (Fisher, 1935; Hannig et al., 2016; Zabell, 1992), a confidence distribution (Schweder and Hjort, 2016; Xie and Singh, 2013), or something else—based on which inferences will be drawn. That is, the truthfulness of assertions $A \subseteq \Theta$ concerning the unknown θ will be assessed based on the magnitude of $Q_y(A)$. Therefore, in the spirit of replicability, a minimal requirement is that Q_y assigning high probability to a false assertion should be a rare event. More precisely, if $\theta \notin A$, then $Q_Y(A)$, as a function of $Y \sim \mathsf{P}_{Y|\theta}$, should not tend to be large. Since Q_y does not return "real probabilities" (Fraser, 2014, p. 249), if the aforementioned requirement is not met, and inferences are systematically misleading, then the Q_y-probabilities are not meaningful in any sense. Balch et al. (2019) showed that every data-dependent probability measure is afflicted with *false confidence*, i.e., there always exists false assertions that tend to be assigned high probability; see Section 16.4.2 below for an example, and Martin (2019) for more discussion. Of course, not all assertions are afflicted with false confidence, and those that are afflicted might not be "practical." But since there are examples where practically relevant assertions are afflicted with false confidence, including the satellite collision application in Balch et al. (2019), and no assurance that risks of "systematically misleading conclusions" are under control, this must be taken seriously.

16.2.2 Basic Inferential Models

The only way to manage false confidence is to abandon the use of probability for uncertainty quantification in the context of statistical inference. But if not probability, then what? Our proposal in Martin and Liu (2013, 2015a) was to construct a non-additive/imprecise probability, in particular, a special type of belief function (e.g., Dempster, 1967, 1968; Kohlas and Monney, 1995; Shafer, 1976) derived from the distribution of a (nested) random set (Molchanov, 2005; Nguyen, 2006). The proposed construction was broken down into three steps: *associate* (A), *predict* (P), and *combine* (C).

A-step. Associate the observable data $Y \in \mathbb{Y}$ and unknown parameter $\theta \in \Theta$ to an unobservable auxiliary variable $U \in \mathbb{U}$, with known distribution $U \sim \mathsf{P}_U$, via the relation

$$a(Y, \theta, U) = 0, \tag{16.1}$$

where the mapping a is known. This basically boils down to an algorithm for simulating data $Y \sim \mathsf{P}_{Y|\theta}$ on a computer.

P-step. Predict the unobserved value of U, associated with (Y, θ) in (16.1), with a random set $\mathcal{S} \subseteq \mathbb{U}$, with distribution $\mathsf{P}_{\mathcal{S}}$; more details on this are below.

C-step. Combine the observed data $Y = y$, the association (16.1), and the random set $\mathcal{S} \sim \mathsf{P}_{\mathcal{S}}$ into a new data-dependent random set $\Theta_y(\mathcal{S}) \subseteq \Theta$, given by

$$\Theta_y(\mathcal{S}) = \bigcup_{u \in \mathcal{S}} \{\vartheta : a(y, \vartheta, u) = 0\}.$$

The inferential output is determined by the distribution of $\Theta_y(\mathcal{S})$, as a function of $\mathcal{S} \sim \mathsf{P}_{\mathcal{S}}$.

There are a number of relevant summaries of the distribution of $\Theta_y(\mathcal{S})$. In particular, the *belief function* or lower probability is

$$\underline{\Pi}_y(A) = \mathsf{P}_{\mathcal{S}}\{\Theta_y(\mathcal{S}) \subseteq A\}, \quad A \subseteq \Theta,$$

and the *plausibility function* or upper probability is

$$\overline{\Pi}_y(A) = \mathsf{P}_{\mathcal{S}}\{\Theta_y(\mathcal{S}) \cap A \neq \varnothing\} = 1 - \underline{\Pi}_y(A^c), \quad A \subseteq \Theta.$$

Note, first, that the belief and plausibility functions are non-additive, e.g., $\underline{\Pi}_y(A) + \underline{\Pi}_y(A^c) \leq 1$, from which it follows that $\underline{\Pi}_y(A) \leq \overline{\Pi}_y(A)$, hence the interpretation as lower and upper probabilities. More details on the imprecise probability structure of the inferential output will be given in Section 16.3. Second, if there is positive $\mathsf{P}_{\mathcal{S}}$-probability that $\Theta_y(\mathcal{S})$ is empty, then the above are replaced by conditional probabilities given "$\Theta_y(\mathcal{S}) \neq \varnothing$." Non-emptiness can fail when there are non-trivial constraints on the parameter space (Ermini Leaf and Liu, 2012), but it is most common when the auxiliary variable dimension can be reduced; see Section 16.4.

While the user's choice of random set $\mathcal{S} \sim \mathsf{P}_{\mathcal{S}}$ is quite flexible, a good choice is critical to the interpretation and properties of the IM output. The key is to make a connection between the distribution of the random set $\mathcal{S}$ and that of the unobservable auxiliary variable U. Toward this, define the random set's *hitting probability* as

$$\gamma(u) = \mathsf{P}_{\mathcal{S}}(\mathcal{S} \ni u), \quad u \in \mathbb{U}.$$

Clearly, this depends on $\mathsf{P}_{\mathcal{S}}$, and we make a link to P_U by requiring that

$$\gamma(U) \geq_{\text{st}} \mathsf{Unif}(0,1) \text{ as a function of } U \sim \mathsf{P}_U. \tag{16.2}$$

The relation $\geq_{\text{st}}$ is read "stochastically no smaller than" and, in this case, means that the distribution function of $\gamma(U)$ is on or below that of $\mathsf{Unif}(0,1)$. This condition is mild and easy to arrange, and we discuss this further below.

With this, it can be shown that the IM output is *valid* in the sense of the following proposition. In other words, validity guarantees that assigning large belief to false assertions—or small plausibility to true assertions—is a rare event. Consequently, unlike any probability-based framework (see Section 16.2.1), IMs are not afflicted by false confidence.

Proposition 16.1 *If the random set $\mathcal{S} \sim \mathsf{P}_{\mathcal{S}}$ satisfies (16.2), then the IM output is valid in the sense that*

$$\sup_{\theta \notin A} \mathsf{P}_{Y|\theta}\{\underline{\Pi}_Y(A) \geq 1 - \alpha\} \leq \alpha, \quad \textit{for all } A \subseteq \Theta \textit{ and all } \alpha \in [0,1]. \tag{16.3}$$

Since this holds for all A, the duality between $\underline{\Pi}$ *and* $\overline{\Pi}$ *implies the following:*

$$\sup_{\theta \in A} \mathsf{P}_{Y|\theta}\{\overline{\Pi}_Y(A) \leq \alpha\} \leq \alpha, \quad \textit{for all } A \subseteq \Theta \textit{ and all } \alpha \in [0,1]. \tag{16.4}$$

The validity theorem establishes a meaningful interpretation for the IM belief and plausibility output. Beyond that, it provides desirable frequentist guarantees for decision procedures derived from the same IM output.

Proposition 16.2 *Let* $(\underline{\Pi}_Y, \overline{\Pi}_Y)$ *be valid IM output as described above.*

- *Consider testing* $H_0 : \theta \in A$, *for any* $A \subseteq \Theta$. *Then, for any* $\alpha \in (0,1)$, *the test that rejects* H_0 *if and only if* $\overline{\Pi}_Y(A) \leq \alpha$ *has Type I error probability bounded by* α.
- *For any* $\alpha \in (0,1)$, *the* $100(1-\alpha)\%$ *plausibility region* $\{\vartheta \in \Theta : \overline{\Pi}_y(\{\vartheta\}) > \alpha\}$ *has coverage probability lower-bounded by* $1-\alpha$.

There is no shortage of, e.g., tests that can control Type I errors for a given class of hypotheses. But if the class of hypotheses changes, then the test procedure changes too. The IM-based tests, on the other hand, provide Type I error control for all hypotheses. In addition to tests and confidence regions, valid predictions can also be achieved (Martin and Lingham, 2016). Finally, aside from producing valid solutions for a wide range of statistical problems, characterization results (e.g., Cella and Martin, 2020; Martin, 2021; Martin and Liu, 2014) show that, roughly, for any valid test or confidence region, there exists a valid IM whose corresponding test or confidence region is at least as efficient.

16.3 New IM Construction

16.3.1 Possibility Measures

Toward a more direct IM construction, one that apparently skips the random set specification, we provide the necessary background on a special imprecise probability model, namely, *possibility measures*. Key references include Dubois and Prade (1988), De Cooman (1997), Dubois (2006), and the chapter by Destercke and Dubois (2014) in the introductory volume Augustin et al. (2014). Possibility measures also have close connections to Shafer's consonant belief functions and to fuzzy sets (e.g., Zadeh, 1978).

A possibility measure on a space $\mathbb{U}$ is determined by a function $\pi : \mathbb{U} \to [0,1]$, which we call the *possibility contour*, that satisfies $\sup_u \pi(u) = 1$. This defines a *possibility measure* $\overline{\Pi}$, a set function defined on the power set of $\mathbb{U}$, according to the rule

$$\overline{\Pi}(K) = \sup_{u \in K} \pi(u), \quad K \subseteq \mathbb{U}, \tag{16.5}$$

where supremum over the empty set is defined to be 0. Clearly, $\overline{\Pi}$ satisfies $\overline{\Pi}(\varnothing) = 0$ and $\overline{\Pi}(\mathbb{U}) = 1$ by definition, just like a probability measure, but it is not additive. Indeed, possibility measures are supremum-preserving (e.g., De Cooman and Aeyels, 1999), i.e.,

$$\overline{\Pi}\big(\textstyle\bigcup_\lambda K_\lambda\big) = \sup_\lambda \overline{\Pi}(K_\lambda), \quad \text{for any collection } K_\lambda \subseteq \mathbb{U}.$$

Restricting to countable collections and using the fact that sums of non-negative numbers are bigger than suprema,it follows that $\overline{\Pi}$ is subadditive. A possibility measure $\overline{\Pi}$ has a dual

$\underline{\Pi}$, called a *necessity measure*, given by $\underline{\Pi}(K) = 1 - \overline{\Pi}(K^c)$, and it defines a coherent lower prevision (Troffaes and de Cooman, 2014; Walley, 1991).

An interesting class of possibility measure examples—in fact, the only class relevant to use here, see (16.9) below—are those determined by a pair (P, h), where P is probability distribution and h is real-valued function, both on the same space $\mathbb{U}$. Then

$$\pi(u) = \mathsf{P}\{h(U) < h(u)\}, \quad u \in \mathbb{U}, \tag{16.6}$$

is a possibility contour, and the corresponding $\overline{\Pi}$ is defined via optimization in (16.5).

Of all the imprecise probability models, possibility measures are among the most restrictive—in particular, necessity and possibility measures are special cases of Shafer's belief and plausibility functions. Here "restrictive" is a positive quality because it implies a level of simplicity that more general models do not have. This makes it possible to answer questions such as how close a possibility measure is to a probability measure, etc. Toward this, like any other imprecise probability model, a possibility measure $\overline{\Pi}$ determines a *credal set*, or set of compatible probability measures,

$$\mathscr{C}(\overline{\Pi}) = \{\mathsf{P} \in \text{prob}(\mathbb{U}) : \mathsf{P}(K) \leq \overline{\Pi}(K) \text{ for all measurable } K\}, \tag{16.7}$$

where prob($\mathbb{U}$) denotes the set of all probability measures on (a specified sigma-algebra of subsets of) $\mathbb{U}$. There are interesting characterizations of the property "$\mathsf{P} \in \mathscr{C}(\overline{\Pi})$" in terms of the P-probability assigned to the so-called α-cuts of $\overline{\Pi}$. In particular, if π is the possibility contour, then the α-cut of $\overline{\Pi}$ is defined as

$$C_{\overline{\Pi}}^{\alpha} = \{u : \pi(u) \geq \alpha\}, \quad \alpha \in [0, 1].$$

These are simply the upper-level sets of π. Then Dubois et al. (2004) and, in a more general context, Couso et al. (2001) established that

$$\mathsf{P} \in \mathscr{C}(\overline{\Pi}) \iff \mathsf{P}(C_{\overline{\Pi}}^{\alpha}) \geq 1 - \alpha \quad \text{for all } \alpha \in [0, 1]. \tag{16.8}$$

Questions about "how imprecise" a given possibility measure $\overline{\Pi}$ is can then be addressed by looking at how diverse is the set of P that satisfies the condition on the right-hand side of (16.8). That is, if very different P are such that $\mathsf{P}(C_{\overline{\Pi}}^{\alpha}) \geq 1 - \alpha$ for all α, then the credal set $\mathscr{C}(\overline{\Pi})$ is large and $\overline{\Pi}$ is rather imprecise.

Conversely, if there is a specific probability measure P of interest, then (16.8) can be used to answer questions about "how precise" a possibility measure can be and remain compatible with the given P. This latter question is particularly relevant to us. The idea is to first define a measure of a possibility measure's precision or *specificity*, and then solve the corresponding maximization problem. This is the *maximum specificity principle* (e.g., Dubois and Prade, 1986). Intuitively, contours π being pointwise smaller gives the possibility measure more specificity, so the goal is to make this contour function small while maintaining the P-probability assignments to the corresponding α-cuts. If P has a unimodal density function f, then the goal is to let π match the shape of f as closely as possible. Dubois et al. (2004) showed that the maximum specificity possibility distribution compatible with P has contour

$$\pi_{\mathsf{P}}(u) = \mathsf{P}\{f(U) < f(u)\}. \tag{16.9}$$

Note how simple the solution to this complex optimization problem is; see, also, (16.6). Computation of π_{P} is also relatively straightforward: it can be done exactly for nice distributions P, numerically (e.g., via quadrature) for complex P in low dimensions, and via Monte Carlo in higher dimensions. The most important property is a curious consequence of the probability assignments to α-cuts, namely,

$$\text{if } U \sim \mathsf{P}, \text{ then } \pi_{\mathsf{P}}(U) \geq_{\text{st}} \mathsf{Unif}(0,1). \tag{16.10}$$

For continuous data problems, if there are no sets of positive P-measure where f is constant, then the above stochastic inequality is an equality. We say this property is "curious" because it suggests an inherent connection between possibility measures and the validity property (16.2) described above.

16.3.2 IMs from Possibility Measures

For a given baseline association (16.1), let $\mathbb{U}_y(\theta) = \{u : a(y,\theta,u) = 0\}$ for a given (y,θ). In continuous-data problems, this will be a singleton, which we denote as $u_{y,\theta}$. We do not require that the association equation can be solved for θ, as a function of (y,u), as would be needed to derive, say, a fiducial or structural distribution. However, being able to reduce the dimension so that a solution for θ can be found is important, for efficiency purposes. Details about dimension reduction are discussed in Martin and Liu (2015b,c), and we summarize this in the context of two examples in Section 16.4.

Our key assumption here is that $\bigcup_\theta \mathbb{U}_y(\theta) = \mathbb{U}$ for all y. In other words, this means that there are no constraints on the possible u values induced by any observation y. This assumption might fail, for example, in problems that involve non-trivial constraints on the parameter space (Ermini Leaf and Liu, 2012). In any case, if there are u values that can be ruled out based on an observation $Y = y$, then that information would be known to the data analyst and should be used to sharpen his/her inference. Here we are simply assuming that no such side information is available.

Returning to the IM construction, start with the same baseline association as in (16.1) and auxiliary variable distribution P_U, but now consider these new P- and C-steps.

P-step. Model the post-data uncertainty about the unobserved value of U by a possibility measure compatible with P_U. Reasonable choices include those with contours like in (16.6), in particular, the maximum specificity contour

$$\pi(u) = \mathsf{P}_U\{f(U) < f(u)\}, \quad u \in \mathbb{U}, \tag{16.11}$$

with f the density corresponding to P_U.

C-step. Combine the data $Y = y$, connection to θ and the auxiliary variable in association (16.1), with the above possibility measure to get a *posterior possibility contour*

$$\pi_y(\vartheta) = \sup_{u \in \mathbb{U}_y(\vartheta)} \pi(u), \quad \vartheta \in \Theta.$$

From the posterior possibility contour π_y, we can extend it to a posterior possibility measure defined on all subsets of Θ according to the optimization rule above:

$$\overline{\Pi}_y(A) = \sup_{\vartheta \in A} \pi_y(\vartheta), \quad A \subseteq \Theta.$$

Define the corresponding posterior necessity measure as $\underline{\Pi}_y(A) = 1 - \overline{\Pi}_y(A^c)$. This construction holds even for multivariate parameters, so the above formula can be used for marginal inference on any relevant feature $\phi = \phi(\theta)$ of the full parameter.

There are some parallels between this and Fisher's fiducial inference; see Remark 16.1 below. But there are some obvious differences too. One important difference is that a strong validity result like Proposition 16.1 above can easily be proved.

Proposition 16.3 *The posterior possibility measure $\overline{\Pi}_Y$ is valid in the sense of* (16.4), *i.e.,*

$$\sup_{\theta \in A} \mathsf{P}_{Y|\theta}\{\overline{\Pi}_Y(A) \leq \alpha\} \leq \alpha, \quad \text{for all } A \subseteq \Theta \text{ and all } \alpha \in [0,1].$$

Similarly, a claim analogous to (16.3) *holds for the posterior necessity measure $\underline{\Pi}_Y$.*

Proof. First, we have $\overline{\Pi}_Y(A) \geq \pi_Y(\theta)$ by monotonicity. Second, $\pi_Y(\theta)$ equals $\pi(u_{Y,\theta})$, which, as a function of $Y \sim \mathsf{P}_{Y|\theta}$, has the same distribution as $\pi(U)$. Finally, the latter random variable is stochastically no smaller than $\mathsf{Unif}(0,1)$ by (16.10). □

The possibility contour based on the maximum specificity criterion is persuasive, but it is not the only option. In fact, there are some cases where this is not an option at all. For example, if $\mathsf{P}_U = \mathsf{Unif}(0,1)$, which often appears in applications, then the density is constant—not unimodal—so the above version of the maximum specificity construction cannot be used. Fortunately, Dubois et al. (2004) show that the most precise possibility measure compatible with a symmetric probability distribution having bounded support is the so-called *triangular* possibility measure. For symmetric distributions on $[0,1]$, the triangular possibility contour is given by

$$\pi(u) = 1 - |2u - 1|, \quad u \in [0,1]. \tag{16.12}$$

It is easy to check that, with $\mathsf{P}_U = \mathsf{Unif}(0,1)$, this contour satisfies the important distributional property (16.10). Other choices as in (16.6) are possible, but the benefit of using (16.11) or (16.12) is that these are principled choices, motivated by optimality.

For a quick illustration, consider a single observation Y from a Cauchy distribution with location parameter θ. Then the association is $Y - \theta - U = 0$, where $U \sim \mathsf{P}_U$, with P_U a standard Cauchy distribution, having density

$$f(u) \propto (1 + u^2)^{-1}, \quad u \in \mathbb{R}.$$

An easy derivation shows that the "optimal" possibility measure has contour

$$\pi(u) = \mathsf{P}_U\{(1+U^2)^{-1} < (1+u^2)^{-1}\} = 2\{1 - F(|u|)\},$$

where F is the cumulative distribution function corresponding to f. Since there is a unique solution $u_{y,\vartheta} = y - \vartheta$ for u, the posterior possibility contour π_y is simple:

$$\pi_y(\vartheta) = 2\{1 - F(|y - \vartheta|)\}.$$

The expression on the right-hand side is familiar to those who work with IMs; it is also related to the so-called *confidence curve* (Birnbaum, 1961) that often appears in the confidence distribution literature. A plot of this function is displayed in Figure 16.1, for the case with $y = 0$. For example, the upper α-level set, with $\alpha = 0.05$, say, returns a 95% confidence interval for θ. If one was interested in the hypothesis $A = (10, \infty)$, then it is clear from the contour's monotonicity that $\overline{\Pi}_y(A) = \pi_y(10)$ and $\underline{\Pi}_y(A) = 0$.

16.3.3 Technical Remarks

Fiducial inference, Dempster's extension, and IMs all share one thing in common: their use of the auxiliary variable with known distribution P_U. The difference is that the former use P_U directly to carry out inference while the latter adjusts P_U either by introducing a random set (Section 16.2.2) or a possibility measure (Section 16.3.2). Here we make three remarks to compare and contrast fiducial and IMs when the latter are formulated via possibility

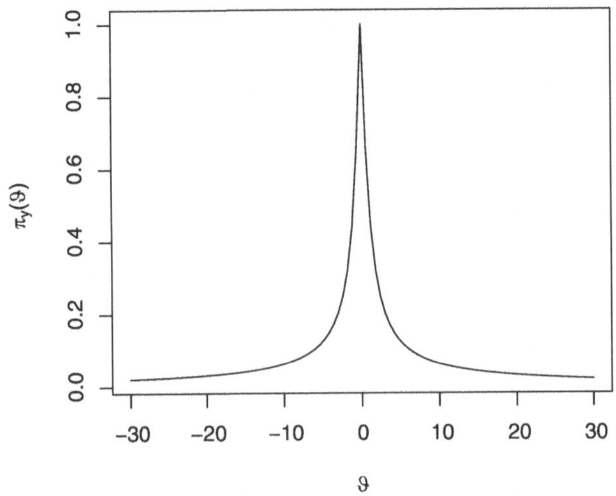

FIGURE 16.1
Possibility contour $\vartheta \mapsto \pi_y(\vartheta)$ for the Cauchy example with $y = 0$.

measures. Since the available IM optimality theory in Martin and Liu (2013, Section 16.4) relies on a connection to fiducial, the possibility measure formulation has some implications there too; see Remark 16.3.

Remark 16.1 Various connections between this new IM construction and fiducial inference can be made. Assume here that (y, θ, u) are all of the same dimension, so that we can solve the association equation for $u = u_{y,\theta}$ and $\theta = \theta_{y,u}$; see Martin and Liu (2015b) and Section 16.4.1 below for how this can be arranged. Then the fiducial distribution for θ, given $Y = y$, can be understood as the distribution for $\theta_{y,U}$, as a function of $U \sim \mathsf{P}_U$, induced by probability calculus; this is the fiducial argument based on the *continue to regard* step in Dempster (1963) or the *switching principle* in Hannig et al. (2016). What we described in Section 16.3.2 is similar in spirit, but we replace "$U \sim \mathsf{P}_U$" by a compatible possibility measure $\overline{\Pi}$ on $\mathbb{U}$ and then propagate using possibility calculus. An interesting question is if one could first obtain the fiducial distribution for θ using probability calculus—which, of course, is not valid—and then construct a compatible possibility measure from it to achieve validity. Specifically, this latter approach would proceed by constructing the fiducial probability density function $\vartheta \mapsto f(u_{y,\vartheta})J_y(\vartheta)$, where $J_y(\theta) = |\partial u_{y,\theta}/\partial\theta|$ is the Jacobian (Hannig et al., 2016, Theorem 1), and then a compatible possibility measure with contour

$$\vartheta \mapsto \mathsf{P}_U\{f(U)J_y(\theta_{y,U}) < f(u_{y,\vartheta})J_y(\vartheta)\}.$$

The only way this can equal $\pi_y(\vartheta) = \mathsf{P}_U\{f(U) < f(u_{y,\vartheta})\}$ is if the Jacobian term is constant, that is, if $u_{y,\vartheta}$ is linear in ϑ or, equivalently, if θ is a location parameter. Therefore, in general, our proposal is *not* equivalent to first finding the fiducial distribution for θ and then constructing a compatible possibility measure.

Remark 16.2 Our strategy for finding $\overline{\Pi}$ is *global* in the sense that it focuses on a measure of compatibility that takes into consideration the probability/possibility assignments to all (measurable) subsets of $\mathbb{U}$; see (16.7). Consider an alternative *local* strategy that focuses on a specified subclass $\mathscr{U}$ of subsets of $\mathbb{U}$ and a weaker notion of compatibility between a probability P and a possibility $\overline{\Pi}$, namely,

$$P(K) \le \overline{\Pi}(K) \text{ for all } K \in \mathscr{U}.$$

We have in mind that $\mathscr{U}$ could be determined by a corresponding subclass of assertions $A \subseteq \Theta$ about θ deemed to be of special relevance, while all the other assertions are dismissed as irrelevant. In other words, this local strategy boils down to loosening the requirement that validity apply to *all assertions* about θ, which puts the reliability of inferences in jeopardy by the false confidence theorem. It turns out that the default-prior Bayes or fiducial solutions drop out as special cases of the above construction *only* under this risky local strategy. To see this, consider a simple case where (y, θ, u) are all scalars, and let $\theta_{y,u}$ denote the θ-solution to the association equation $a(y, \theta, u) = 0$. Suppose that the only assertions about θ of interest are half-lines in the class $\mathscr{A} = \{[\vartheta, \infty) : \vartheta \in \mathbb{R}\}$. If $u \mapsto \theta_{y,u}$ is monotone increasing, the above collection of half-lines on the parameter space corresponds to half-lines

$$\mathscr{U} = \{(-\infty, u] : u \in \mathbb{R}\}$$

on the auxiliary variable space. (If $u \mapsto \theta_{y,u}$ is monotone decreasing, then the orientation of the intervals in $\mathscr{U}$ reverses.) If the focus is exclusively on half-lines in the u-space, then the "best" compatible possibility measure is exactly the probability distribution, i.e., $\overline{\Pi} = \mathsf{P}_U$. With this choice, it is easy to see that the posterior possibility satisfies

$$\overline{\Pi}_y\big([\vartheta, \infty)\big) = \mathsf{P}_U(\theta_{y,U} \geq \vartheta).$$

The right-hand side is exactly Fisher's fiducial probability of the assertion "$\theta \geq \vartheta$" and, at least in transformation models, this would agree with the default-prior Bayesian posterior probability and Fraser's structural probability. But the false confidence theorem applies to these probabilities, so the corresponding inference is not valid in the strong sense of Section 16.2. Thus, we recover the familiar result that fiducial inference is valid only in a limited sense, i.e., only for half-line assertions.

Remark 16.3 In the original IM framework, summarized in Section 16.2.2, an important open question concerns *optimality*, i.e., what is the "best" random set $\mathcal{S} \sim \mathsf{P}_\mathcal{S}$? The specific optimality results presented in Martin and Liu (2013) are not fully satisfactory, but one of their preliminary results provides some valuable intuition. Roughly, their Proposition 16.1 says that an IM cannot be more efficient than the primitive fiducial posterior from Remark 16.1 above. Of course, the latter is not valid, but the key take-away is that, for efficiency's sake, the IM's adjustment to achieve validity—via random sets or possibility measures—should be minimal. Depending on the perspective, this boils down to choosing $\mathcal{S}$ as small as possible, in some sense, relative to the constraint (16.2), or choosing $\overline{\Pi}$ as close to P_U as possible subject to the compatibility constraint. An advantage of the new possibility measure-based approach is that the latter objective is apparently easier to formulate and solve mathematically, leading to $\overline{\Pi}$ defined by the contour (16.11). Incidentally, in cases where P_U has a unimodal density f, our conjecture of what the "smallest possible $\mathcal{S}$ subject to (16.2)" would be is a random upper level set of f, i.e.,

$$\mathcal{S} = \{u \in \mathbb{U} : f(u) \geq f(\tilde{U})\}, \quad \tilde{U} \sim \mathsf{P}_U.$$

It is easy to check that this random set's hitting probability $u \mapsto \mathsf{P}_\mathcal{S}(\mathcal{S} \ni u)$ is exactly the possibility contour in (16.11). Therefore, the new possibility measure perspective provides some rigorous justification for our intuition-based random set recommendations. Unfortunately, this possibility measure perspective does not provide a completely satisfactory answer to the question of optimality; see Section 16.5.

16.3.4 Equivalence of the Two Constructions?

The distinguishing feature is that the original construction, as discussed in Section 16.2, makes use of random sets while the new construction uses possibility measures. A natural question is if the two are equivalent, at least in some sense.

The fundamental admissibility result in Martin and Liu (2013, Theorem 3) implies that the only random sets to be considered in the IM framework are closed and *nested*; here, $\mathcal{S}$ is nested if, for any pair S and S' in the support of $\mathsf{P}_{\mathcal{S}}$, either $S \subseteq S'$ or $S \supseteq S'$. Examples of closed nested random sets include those with form like in the displayed equation of Remark 16.3. It is well known that hitting probabilities for nested random sets are possibility contour functions (e.g., De Cooman and Aeyels, 2000), so it is no surprise that, if we start from a nested random set $\mathcal{S}$, then we arrive at a possibility measure. What about the other direction? Suppose we start, as in Section 16.3.2, with a possibility measure $\overline{\Pi}$ on $\mathbb{U}$, with contour π. Is there a corresponding closed and nested random set whose hitting probability agrees with π? Questions about the connections between different models are important, so there is guidance in the imprecise probability literature. Indeed, Miranda et al. (2004, Theorem 4.4) show that, in all practically relevant settings (e.g., $\mathbb{U}$ a Euclidean space), every possibility measure corresponds to a closed nested random set. Therefore, in our setting, when we directly introduce a possibility measure $\overline{\Pi}$ on $\mathbb{U}$, with contour π, we have implicitly defined a nested random set $\mathcal{S} \sim \mathsf{P}_{\mathcal{S}}$ such that $\pi(u) = \mathsf{P}_{\mathcal{S}}(\mathcal{S} \ni u)$ for all u. From this random set $\mathcal{S}$, the corresponding IM from Section 16.2.2 would have a plausibility function which, for any $A \subseteq \Theta$, satisfies

$$
\begin{aligned}
\overline{\Pi}_y(A) &= \mathsf{P}_{\mathcal{S}}\{\Theta_y(\mathcal{S}) \cap A \neq \varnothing\} \\
&= \mathsf{P}_{\mathcal{S}}\{\mathcal{S} \ni u_{y,\vartheta} \text{ for some } u \in \mathbb{U}_y(\vartheta) \text{ and some } \vartheta \in A\} \\
&= \sup_{\vartheta \in A} \sup_{u \in \mathbb{U}_y(\vartheta)} \mathsf{P}_{\mathcal{S}}(\mathcal{S} \ni u) \\
&= \sup_{\vartheta \in A} \sup_{u \in \mathbb{U}_y(\vartheta)} \pi(u),
\end{aligned}
$$

where the third equality follows from $\mathcal{S}$ being nested, and the fourth by the above connection between π and $\mathcal{S}$. We immediately recognize the right-hand side as $\overline{\Pi}_y(A)$, the possibility measure in Section 16.3.2, which establishes the equivalence.

Altogether, there is no loss (or gain) of generality in working with possibility measures instead of nested random sets. However, we believe this alternative possibility-based construction is valuable for several reasons. First, the use of random sets is a distinguishing feature of the IM framework, which leads to the important validity property, but this is admittedly complicated and unfamiliar to practitioners, so a formulation based on ordinary functions might be more easily accessible. Second, since possibility measures are among the simplest of the imprecise probability models, drawing a connection between these and IMs can lead to some new insights or results. Third, as discussed in Remark 16.3, the existing theory for possibility measures gives rigorous confirmation of our intuition about a default multipurpose random set that is optimal in a certain sense.

Pushing this equivalence inquiry further leads to some interesting open questions. If $\mathcal{S}$ is nested, then it is easy to verify that the corresponding random set $\Theta_y(\mathcal{S})$ on the parameter space is nested too, so the corresponding hitting probability is a possibility contour. But, as before, the opposite direction is more interesting. That is, suppose we directly define the posterior possibility measure $\overline{\Pi}_y$, without starting with $\overline{\Pi}$ on $\mathbb{U}$ and propagating to Θ. Then the results of Miranda et al. (2004) imply the existence of a closed, nested, and data-dependent random set $\mathcal{T}_y \sim \mathsf{P}_{\mathcal{T}_y}$ on Θ that satisfies

$$\pi_y(\vartheta) = \mathsf{P}_{\mathcal{T}_y}(\mathcal{T}_y \ni \vartheta), \quad \vartheta \in \Theta.$$

Then the question is if there exists a nested random set $\mathcal{S}$ on $\mathbb{U}$ such that $\Theta_y(\mathcal{S}) = \mathcal{T}_y$? Based on how the map from $\mathcal{S}$ to $\Theta_y(\mathcal{S})$ is defined, the only reasonable candidate $\mathcal{S}$ would be of the form

$$\mathcal{S} = \mathcal{S}_y = \bigcup_{\vartheta \in \mathcal{T}_y} \{u_{y,\vartheta}\},$$

which, in general, depends on data y. One can imagine certain cases where the dependence on y would naturally drop out, for example, in transformation models, but are there others? And how to interpret a data-dependent random set? Fraser has discussed data-dependent priors in Bayesian analysis (e.g., Fraser et al., 2010, 2016; Fraser, 2011), and there could be a connection. There is also a question about if and, if so, then under what conditions, would a data-dependent random set lead to valid inference? One situation where validity can be achieved with data-dependent random sets is in Ermini Leaf and Liu (2012); see, also, Cella and Martin (2019). There is also a yet-to-be-fully-understood connection between the parameter-dependent random sets in Martin and Liu (2015b), Martin (2021), etc., and their data-dependent counterparts.

16.4 Examples

As indicated above, steps to reduce the dimension of the auxiliary variable space are crucial to making the IM solution (both valid and) efficient. These dimension-reduction techniques are described in detail in Martin and Liu (2015b,c). Rather than give a general survey, we have opted to describe these methods in the context of two examples. These will also serve as illustrations of the new possibility measure-based approach.

16.4.1 Curved Normal

Let Y consist of n observations, independent and identically distributed according to $\mathsf{N}(\theta, \theta^2)$, with $\theta \in \Theta = \mathbb{R}$. Since θ controls both the mean and variance, this normal model is said to be *curved*. This curvature makes the problem non-regular and generally more challenging than the version with separate mean and variance parameters.

We can immediately reduce the problem down to one that involves minimal sufficient statistics. With a slight abuse of notation, let Y_1 and Y_2 denote the mean and standard deviation, respectively, of the original n-sample, and write the association (16.1) as

$$Y_1 = \theta + |\theta| U_1 \quad \text{and} \quad Y_2 = |\theta| U_2,$$

with P_U the joint distribution of $U_1 \sim \mathsf{N}(0, n^{-1})$ and $U_2 \sim \{(n-1)^{-1}\mathsf{Chisq}(n-1)\}^{1/2}$, independent. Since θ is a scalar while $U = (U_1, U_2)$ is two-dimensional, there is an opportunity to reduce dimension further.

For the moment, assume that $\text{sign}(\theta) \in \{-1, +1\}$ is *known*; this is not an unrealistic assumption, but we will discuss below how it can easily be removed. To reduce the dimension, our strategy is to first find a feature of U whose value is actually *observed*, and then condition on the value of that observed feature. Aside from problems with a group transformation structure, where certain invariant statistics can be identified, there are very few general methods for finding observed features of U. One suggested in Martin and Liu (2015b) is based on solving a partial differential equation. That is, we seek a function η such that $\eta(u_{y,\theta})$ is not sensitive to changes in θ, in other words, such that $\partial\eta(u_{y,\theta})/\partial\theta = 0$. By the chain rule, this amounts to finding η that satisfies

$$\frac{\partial\eta(u)}{\partial u}\bigg|_{u=u_{y,\theta}} \frac{\partial u_{y,\theta}}{\partial\theta} = 0.$$

In the present case, since $u = u_{y,\theta} = |\theta|^{-1}(y_1 - \theta,\, y_2)^\top$, and

$$\frac{\partial u}{\partial\theta} = -\frac{\text{sign}(\theta)}{|\theta|}\begin{pmatrix} u_1 + \text{sign}(\theta) \\ u_2 \end{pmatrix},$$

where sign(θ) is assumed known, it is easy to check that

$$\eta(u) = u_2^{-1}\{\text{sign}(\theta) + u_1\}$$

solves the differential equation. Therefore, with known sign(θ), the quantity $\eta(U)$ is observed, and its value in the sample is

$$\eta(u_{y,\theta}) = (y_2/|\theta|)^{-1}\{\text{sign}(\theta) + (y_1 - \theta)/|\theta|\} = y_1/y_2.$$

If we let $H(y) = y_1/y_2$, then we have a reduced association

$$Y_1 = \theta + Y_2\tau(U) \quad \text{and} \quad \eta(U) = H(Y),$$

where $\tau(u) = u_1/u_2$ is a scalar. Then the proposal is to build an IM based on the conditional distribution of $V = \tau(U)$, given $\eta(U) = H(y)$. If we write h for the observed value of $H(Y)$, then that conditional distribution for V has density f_h that satisfies

$$\log f_h(v) = \log|\tfrac{1}{h-v}| - \tfrac{n}{2}(\tfrac{v}{h-v})^2 + (n-2)\log\tfrac{\text{sign}(\theta)}{h-v} - \tfrac{n-1}{2}(\tfrac{1}{h-v})^2 + \text{constant},$$

for all v such that $\text{sign}(\theta)/(h - v) > 0$. From here, construct a (conditional) IM for θ by defining the f_h-dependent possibility contour

$$\pi_h(v) = \mathsf{P}_{\tau(U)|\eta(U)=h}\{f_h(\tau(U)) < f_h(v)\},$$

and then the corresponding posterior version:

$$\pi_{y|h}(\vartheta) = \pi_h(v_{y,\vartheta}),$$

where $v_{y,\vartheta} = (y_1 - \theta)/y_2$. It follows from Theorem 1 in Martin and Liu (2015b) that the corresponding (conditional) IM is valid.

For illustration, suppose data of size $n = 10$ are sampled from the curved normal distribution with $\theta = 2$. The sample mean and standard deviation are $y_1 = 1.86$ and $y_2 = 2.12$, respectively, so that $h = H(y) = 0.88$. A plot of the posterior possibility contour is shown in Figure 16.2. The horizontal cut—and corresponding vertical fences—at $\alpha = 0.05$ determines a 95% confidence interval for θ based on these data. To assess the IM's validity, we do a small simulation study, where we repeat the above experiment 1000 times. For each data set, we produce 95% confidence intervals based on the above IM and also a (higher-order accurate) generalized fiducial interval. For the IM intervals, the estimated coverage probability and mean length are 0.956 and 1.82, respectively; for the fiducial intervals, the estimated coverage probability and mean length are 0.932 and 1.45, respectively. On the one hand, the fiducial intervals are a little shorter, but they tend to under-cover. On the other hand, the IM intervals are guaranteed to achieve the nominal coverage thanks to validity.

As promised, it is possible to remove the assumption that sign(θ) is known. This requires a so-called *local* conditional IM in Martin and Liu (2015b), allowing the differential equation's solution to depend on a local parameter, and then the local parameter-dependent IMs get suitably glued together. Here the dependence on the local parameter is very weak—only depends on the sign—so this is relatively easy to manage. But since assuming sign(θ) to be known is very mild, we opt for the simpler solution here.

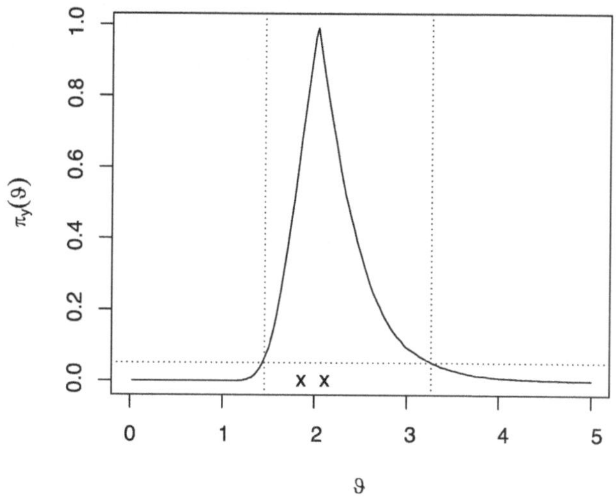

FIGURE 16.2
Possibility contour $\vartheta \mapsto \pi_{y|h}(\vartheta)$ for the curved normal example with $n = 10$, $\theta = 2$, and sample mean and standard deviation marked with X.

16.4.2 Exponential Errors-in-Variables

Suppose we have a pair of independent observations $Y = (Y_1, Y_2)$, where

$$Y_i = \theta_i + U_i, \quad i = 1, 2.$$

The distribution P_U of $U = (U_1, U_2)$ is known and here we consider the case where $U_i \sim \mathsf{Exp}(\lambda_i)$, independent, but where the rate parameters, λ_1 and λ_2, need not be equal. As an example, if there are n_i independent and identically distributed exponentials shifted by θ_i, then Y_i represents the minimum of those and $\lambda_i = n_i$; this is what would emerge from the conditioning argument described in Section 16.4.1 above. For joint inference on $\theta = (\theta_1, \theta_2)$, there is a density f for U and it is straightforward to follow the maximum specificity principle and construct a joint posterior possibility contour for θ as described above. Similarly, for a location parameter problem like this one, a flat prior is the appropriate default choice, and one can immediately derive a Bayesian posterior distribution; the fiducial and structural distributions are the same.

Things are less straightforward, however, if we add some additional structure. Imagine that the support lower bound θ_1 for Y_1 is believed to be a linear function of θ_2, and the target is to estimate that function, but only a noisy measurement Y_2 of θ_2 is available. This boils down to a simple version of the *measurement error* problem, wherein the true "covariate" θ_2 can only be measured with error. From this point of view, it is natural to rewrite the problem as

$$Y_1 = \phi\xi + U_1 \quad \text{and} \quad Y_2 = \xi + U_2,$$

where $\phi > 0$ is the interest parameter and $\xi > 0$ is a nuisance parameter. Then the goal is marginal inference on ϕ in the presence of unknown ξ (cf. Creasy, 1954; Fieller, 1954).

Plugging the second equation in the above display into the first, we arrive at

$$Y_1 - \phi Y_2 = U_1 - \phi U_2 \quad \text{and} \quad Y_2 = \xi + U_2.$$

Notice that the first equation is free of ξ. For such cases, the general theory in Martin and Liu (2015c) says that exact marginal inference can be achieved by simply ignoring the second ξ-dependent equation. Not only does this eliminate the nuisance parameter, but it creates an opportunity for dimension reduction. Since the distribution of $U_1 - \phi U_2$ is known, namely, *asymmetric Laplace*, we can employ a probability integral transformation. That is, $U_1 - \psi U_2$ has the same distribution as $G_{\phi}^{-1}(V)$, where $V \sim \mathsf{Unif}(0,1)$ and $G_\psi = F_{\lambda_1, \phi\lambda_2}$ is the asymmetric Laplace distribution function parametrized as

$$F_{r_1,r_2}(x) = \begin{cases} 1 - \frac{r_2}{r_1+r_2} e^{-r_1 x} & \text{if } x \geq 0 \\ \frac{r_1}{r_1+r_2} e^{r_2 x}, & \text{if } x < 0. \end{cases}$$

If we use the triangular possibility contour (16.12) for V, then the corresponding marginal posterior possibility contour for ψ is

$$\pi_y(\varphi) = 1 - \left|2G_\varphi(y_1 - \varphi y_2) - 1\right|, \quad \varphi > 0.$$

It follows from Theorem 2 in Martin and Liu (2015c) that the corresponding IM is valid, which implies, for example, that confidence regions derived from this posterior possibility contour achieve the nominal coverage probability.

For illustration, consider the case where $\lambda_1 = \lambda_2 = 5$, $\theta_1 = 1$, and $\theta_2 = 0.1$; this implies $\xi = 0.1$ and $\phi = 10$. Figure 16.3(a) plots the posterior possibility contour. Note that the peak of this plot is close to the true $\phi = 10$, but that the curve does not vanish in the right tail. This is because the data cannot rule out arbitrarily large values of ϕ, hence a 95% IM confidence interval would be unbounded. This unboundedness is actually necessary to achieve the nominal coverage in this problem; see Gleser and Hwang (1987). To compare the IM with a Bayes solution, we can find the marginal posterior for $\phi = \theta_1/\theta_2$ based on a default flat prior for (θ_1, θ_2), which we denote by Q_y. Consider an assertion $A = \{\phi \leq 9\}$, which happens to be false in this example. We compare the distribution of belief assignments to A based on the Bayes and IM posteriors. In particular, we compare the two distribution functions

$$\alpha \mapsto \mathsf{P}_{Y|\theta}\{Q_Y(A) \leq \alpha\} \quad \text{and} \quad \alpha \mapsto \mathsf{P}_{Y|\theta}\{\underline{\Pi}_Y(A) \leq \alpha\}.$$

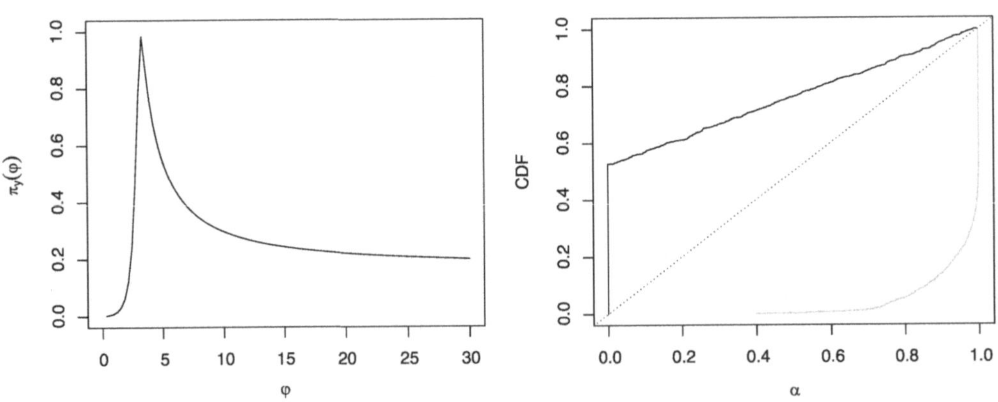

FIGURE 16.3
Left: plot of the posterior possibility contour $\varphi \mapsto \pi_y(\varphi)$ in the exponential error-in-variables example, with true value $\phi = 10$ and $(y_1, y_2) = (1.40, 0.50)$. Right: plots of the distribution function $\alpha \mapsto \mathsf{P}_{Y|\theta}\{\star \leq \alpha\}$, for $\star$ equal to $Q_Y(A)$ (gray) and $\underline{\Pi}_Y(A)$ (black), where $A = \{\phi \leq 9\}$ is a false assertion. (a) Posterior possibility contour. (b) Distribution function of $Q_Y(A)$ and $\underline{\Pi}_Y(A)$.

On the belief/lower probability/necessity scale, validity corresponds to this distribution function being on or above the diagonal line. As expected, we find the IM curve above the diagonal line, while the Bayesian curve is well below. Being so far below the diagonal line in this case is rather problematic because it means the Bayesian posterior probabilities assigned to the false assertion are *almost always large*. In light of this false confidence, it is clear that the Bayesian inference is not reliable.

16.5 Conclusion

Here we have presented a new perspective on and construction of IMs based on possibility measures. The chief advantage of this formulation is its relative simplicity, i.e., a user need not directly consider the specification of and computation with a random set. And this benefit comes without sacrificing on the essential validity property. In fact, certain characterizations suggest that validity is somehow inherent in possibility measures, giving them a fundamental role in statistical inference. Beyond validity, the existing possibility theory provides some guidance in terms of efficiency and optimality, as discussed in Remark 16.3, and the dimension-reduction strategies for improved efficiency fit in seamlessly. There is certainly more work to be done, but we hope that the reader can see the IM framework's potential as a solution to Efron's "most important unresolved problem."

We conclude by mentioning a few ideas and open problems.

- As mentioned in Remark 16.3, the possibility measure connection does not completely resolve the optimality question. A reason is that the maximum specificity-based possibility measure apparently depends on the particular choice of (a, U), association and auxiliary variable. Intuitively, an "optimal" IM should depend only on the assumed statistical model, $Y \sim \mathsf{P}_{Y|\theta}$. Toward a better understanding of efficiency and optimality, it would be interesting to investigate the IM's sensitivity to different choices of (a, U) when using the maximum specificity-based possibility measure.

- When $\theta = (\phi, \xi)$ and ξ is a nuisance parameter,in cases where ξ cannot be completely eliminated, Martin and Liu (2015c) work with a ξ-dependent auxiliary variable and then suitably fatten the corresponding random set to achieve valid marginal inference on ϕ. In the context of the present paper, an alternative strategy for this would be to find the most precise $\overline{\Pi}$ that is compatible with *all* of the ξ-dependent auxiliary variable distributions. Is the latter compatibility task easier or more efficient than the random set fattening?

- In statistics, we often have data *and* some vaguely specified structural assumptions, such as sparsity or smoothness. These separate pieces of information are typically combined by introducing either a penalty or a prior distribution. An important topic in imprecise probability is combining relevant information from different sources into an overall assessment of uncertainty. If validity dictates that we work with imprecise probability, then we should investigate the properties of these alternative combination rules in modern, high-dimensional statistical applications involving structural assumptions.

Bibliography

Augustin, T., Coolen, F. P. A., de Cooman, G. and Troffaes, M. C. M. editors, 2014. *Introduction to Imprecise Probabilities*. Wiley Series in Probability and Statistics. Chichester: John Wiley & Sons, Ltd.

Balch, M. S., Martin, R. and Ferson, S. 2019. Satellite conjunction analysis and the false confidence theorem. *Proceedings of the Royal Society A*, *475*(2227), p. 20180565.

Berger, J. O., 1985. *Statistical Decision Theory and Bayesian Analysis* (2nd ed.). New York: Springer-Verlag.

Birnbaum, A., 1961. Confidence curves: An omnibus technique for estimation and testing statistical hypotheses. *Journal of the American Statistical Association*, *56*, pp. 246–249.

Camerer, C. F., Dreber, A., Holzmeister, F., Ho, T.-H., Huber, J., Johannesson, M., Kirchler, M., Nave, G., Nosek, B. A., Pfeiffer, T., Altmejd, A., Buttrick, N., Chan, T., Chen, Y., Forsell, E., Gampa, A., Heikensten, E., Hummer, L., Imai, T., Isaksson, S., Manfredi, D., Rose, J., Wagenmakers, E.-J. and Wu, H., 2018. Evaluating the replicability of social science experiments in *nature* and *science* between 2010 and 2015. *Nature Human Behaviour*, *2*(9), pp. 637–644.

Cella, L. and Martin, R., 03–06 July 2019. Incorporating expert opinion in an inferential model while maintaining validity. In *Proceedings of the Eleventh International Symposium on Imprecise Probabilities: Theories and Applications*, volume 103 of *Proceedings of Machine Learning Research*, eds J. De Bock, C. P. de Campos, G. de Cooman, E. Quaeghebeur and G. Wheeler, pp. 68–77. Thagaste, Ghent, Belgium: PMLR.

Cella, L. and Martin, R., 2020. Validity, consonant plausibility measures, and conformal prediction. *International Journal of Approximate Reasoning*, *141*, pp. 110–130.

Couso, I., Montes, S. and Gil, P., 2001. The necessity of the strong α-cuts of a fuzzy set. *International Journal of Uncertainty, Fuzziness and Knowledge-Based Systems*, *9*(2), pp. 249–262.

Creasy, M. A., 1954. Symposium on interval estimation: Limits for the ratio of means. *Journal of the Royal Statistical Society, Series B*, *16*, pp. 186–194.

De Cooman, G., 1997. Possibility theory. I. The measure- and integral-theoretic groundwork. *International Journal of General Systems*, *25*(4), pp. 291–323.

De Cooman, G. and Aeyels, D., 1999. Supremum preserving upper probabilities. *Information Science*, *118*(1–4), pp. 173–212.

De Cooman, G. and Aeyels, D., 2000. A random set description of a possibility measure and its natural extension. *IEEE Transactions on Systems, Man, and Cybernetics*, *30*, pp. 124–131.

Dempster, A. P., 1963. Further examples of inconsistencies in the fiducial argument. *Annals of Mathematical Statistics*, *34*, pp. 884–891.

Dempster, A. P., 1967. Upper and lower probabilities induced by a multivalued mapping. *Annals of Mathematical Statistics*, *38*, pp. 325–339.

Dempster, A. P., 1968. Upper and lower probabilities generated by a random closed interval. *Annals of Mathematical Statistics*, *39*, pp. 957–966.

Destercke, S. and Dubois, D., 2014. Special cases. In *Introduction to Imprecise Probabilities*, Wiley Series in Probability and Statistics, eds Thomas Augustin, Frank Coolen, Gert de Cooman, and Matthias Troffaes, pp. 79–92. Chichester: Wiley.

Dubois, D., 2006. Possibility theory and statistical reasoning. *Computational Statistics & Data Analysis*, *51*(1), pp. 47–69.

Dubois, D. and Prade, H., 1986. The principle of minimum specificity as a basis for evidential reasoning. In *Proceedings of International Conference on Information Processing and Management of Uncertainty in Knowledge-Based Systems*, pp. 75–84. Springer.

Dubois, D. and Prade, H., 1988. *Possibility Theory.* New York: Plenum Press.

Dubois, D., Foulloy, L., Mauris, G. and Prade, H., 2004. Probability-possibility transformations, triangular fuzzy sets, and probabilistic inequalities. *Reliable Computer*, *10*(4), pp. 273–297.

Efron, B., 2013. Discussion: "Confidence distribution, the frequentist distribution estimator of a parameter: A review" [mr3047496]. *International Statistical Review*, *81*(1), pp. 41–42.

Ermini Leaf, D. and Liu, C., 2012. Inference about constrained parameters using the elastic belief method. *International Journal of Approximate Reasoning*, *53*(5), pp. 709–727.

Fieller, E. C., 1954. Symposium on interval estimation: Some problems in interval estimation. *Journal of the Royal Statistical Society: Series B*, *16*, pp. 175–185.

Fisher, R. A., 1935. The fiducial argument in statistical inference. *Annals of Eugenics*, *6*, pp. 391–398.

Fraser, D. A. S., 2011. Is Bayes posterior just quick and dirty confidence? *Statistical Science*, *26*(3), pp. 299–316.

Fraser, D. A. S., 2014. Why does statistics have two theories? In *Past, Present, and Future of Statistical Science*, Chapter 22, eds X. Lin, C. Genest, D. L. Banks, G. Molenberghs, D. W. Scott and J.-L. Wang, pp. 237–252. New York, NY: Chapman & Hall/CRC Press.

Fraser, D. A. S., Reid, N., Marras, E. and Yi, G. Y., 2010. Default priors for Bayesian and frequentist inference. *Journal of the Royal Statistical Society, Series B Statistical Methodology*, *72*(5), pp. 631–654. ISSN 1369-7412.

Fraser, D. A. S., Bédard, M., Wong, A., Lin, W. and Fraser, A. M., 2016. Bayes, reproducibility and the quest for truth. *Statistical Science*, *31*(4), pp. 578–590.

Ghosh, J. K., Delampady, M. and Samanta, T., 2006. *An Introduction to Bayesian Analysis.* New York: Springer.

Gleser, L. J. and Hwang, J. T., 1987. The nonexistence of $100(1-\alpha)\%$ confidence sets of finite expected diameter in errors-in-variables and related models. *Annals of Statistics*, *15*(4), pp. 1351–1362.

Hannig, J., Iyer, H., Lai, R. C. S. and Lee, T. C. M., 2016. Generalized fiducial inference: A review and new results. *Journal of the American Statistical Association*, *111*(515), pp. 1346–1361.

Kohlas, J. and Monney, P.-A., 1995. *A Mathematical Theory of Hints*, volume 425 of *Lecture Notes in Economics and Mathematical Systems.* Berlin: Springer-Verlag.

Martin, R., 2019. False confidence, non-additive beliefs, and valid statistical inference. *International Journal of Approximate Reasoning, 113*, pp. 39–73.

Martin, R., 2021. An imprecise-probabilistic characterization of frequentist statistical inference. https://researchers.one/articles/21.01.00002.

Martin, R. and Lingham, R. T., 2016. Prior-free probabilistic prediction of future observations. *Technometrics, 58*(2), pp. 225–235.

Martin, R. and Liu, C., 2013. Inferential models: A framework for prior-free posterior probabilistic inference. *Journal of the American Statistical Association, 108*(501), pp. 301–313.

Martin, R. and Liu, C., 2014. A note on p-values interpreted as plausibilities. *Statistica Sinica, 24*(4), pp. 1703–1716.

Martin, R. and Liu, C., 2015a. *Inferential Models: Reasoning with Uncertainty*, volume 147 of *Monographs on Statistics and Applied Probability*. Boca Raton, FL: CRC Press.

Martin, R. and Liu, C., 2015b. Conditional inferential models: Combining information for prior-free probabilistic inference. *Journal of the Royal Statistical Society, Series B Statistical Methodology, 77*(1), pp. 195–217.

Martin, R. and Liu, C., 2015c. Marginal inferential models: Prior-free probabilistic inference on interest parameters. *Journal of the American Statistical Association, 110*(512), pp. 1621–1631.

McShane, B. B., Gal, D., Gelman, A., Robert, C., and Tackett, J. L., 2019. Abandon statistical significance. *American Statistical, 73*(suppl. 1), pp. 235–245.

Meng, X.-L., June/July 2017. X-L Files: Bayesian, fiducial and frequentist: BFF4EVER. *IMS Bulletin, 46*(4), p. 17.

Miranda, E., Couso, I. and Gil, P., 2004. A random set characterization of possibility measures. *Information Science, 168*(1–4), pp. 51–75.

Molchanov, I., 2005. *Theory of Random Sets.* Probability and Its Applications (New York). London: Springer-Verlag London Ltd.

Nguyen, H. T., 2006. *An Introduction to Random Sets.* Boca Raton, FL: Chapman & Hall/CRC.

Reid, N. and Cox, D. R., 2015. On some principles of statistical inference. *International Statistical Review, 83*(2), pp. 293–308.

Schweder, T. and Hjort, N. L., 2016. *Confidence, Likelihood, Probability*, volume 41 of *Cambridge Series in Statistical and Probabilistic Mathematics.* New York: Cambridge University Press.

Shafer, G., 1976. *A Mathematical Theory of Evidence.* Princeton, NJ: Princeton University Press.

Trafimow, D. and Marks, M., 2015. Editorial. *Basic and Applied Social Psychology, 37*(1), pp. 1–2.

Troffaes, M. C. M. and de Cooman, G., 2014. *Lower Previsions.* Wiley Series in Probability and Statistics. Chichester: John Wiley & Sons, Ltd.

Walley, P., 1991. *Statistical Reasoning with Imprecise Probabilities*, volume 42 of *Monographs on Statistics and Applied Probability*. London: Chapman & Hall Ltd.

Xie, M. and Singh, K., 2013. Confidence distribution, the frequentist distribution estimator of a parameter: A review. *International Statistical Review, 81*(1), pp. 3–39.

Zabell, S. L., 1992. R. A. Fisher and the fiducial argument. *Statistical Science, 7*(3), pp. 369–387.

Zadeh, L. A., 1978. Fuzzy sets as a basis for a theory of possibility. *Fuzzy Sets and Systems, 1*(1), pp. 3–28.

17

Conformal Predictive Distributions: An Approach to Nonparametric Fiducial Prediction

Vladimir Vovk

Royal Holloway, University of London

17.1 Introduction

The most developed approach to nonparametric fiducial prediction is via conformal predictive distributions, and these are the main subject of this chapter. We start in Section 17.2 from reviewing parts of traditional fiducial inference that are relevant to conformal predictive distributions. At the end of the section, we discuss key desiderata for fiducial prediction, validity and efficiency.

Fiducial prediction in the nonparametric setting was at least implicit in Fisher's work and has been greatly developed in the work of Dempster, Hill, and Coolen. It will be the topic of Section 17.3, where we describe what we call the Dempster–Hill procedure and embed it into fiducial prediction by expressing it in terms of a randomized pivot.

What makes (supervised) machine learning a powerful practical approach to various kinds of prediction problems is that it deals with observations that are pairs (x, y), where x is a potentially complicated object (such as a movie, or all available information about a patient) and y is an associated label (such as a movie's sales figures). In nonparametric fiducial prediction, as developed in statistics, the xs are absent and the observations are just the ys. Conformal predictive distributions add the objects x to the Dempster–Hill picture, which only involves the labels y. This is discussed in Section 17.4.

Conformal predictive distributions can be derived only under serious restrictions. Getting rid of these restrictions extends their application area and logically leads to the more general area of conformal prediction, which is the topic of Section 17.5. In that section, we will also discuss a generalization of conformal prediction to general repetitive structures (including many of the models considered by Fisher in his parametric fiducial inference) and its application to online methods of hypothesis testing.

We will not make notational distinction between random variables and their possible values, which should always be clear from context. The phrase "uniformly distributed on the interval $[0, 1]$" (applied to a random variable) will sometimes be abbreviated to "uniformly distributed" or even "uniform".

 DOI: 10.1201/9780429341731-17

17.2 Parametric Fiducial Prediction

Fisher's fiducial inference, at least in his publications, can be divided into two parts (not strictly disjoint). One part concerns inference about the parameters of the unknown true distribution generating the data. This part is controversial and is often regarded as Fisher's greatest blunder [12]. Now it has been revived under the name of confidence distributions (see, e.g., [38] and [54]), but it is still common to deemphasize connections with Fisher's ideas, to stay away from controversy. This chapter is about the other part of fiducial inference, namely inference about future observations, or fiducial prediction. This part remains feasible even without parametric assumptions.

Fiducial prediction is much less prominent in Fisher's oeuvre, and it is much less controversial. This section reviews Fisher's work and some other closely related work in this area.

Remark 17.2.1. In this chapter, we only consider the simple case of predicting one future observation, although Fisher and later authors are sometimes interested in predicting several future observations (which may blur the difference between fiducial prediction and fiducial inference about parameters, since in many interesting cases parameters can be recovered as limiting values of some function of k future observations as $k \to \infty$).

17.2.1 Fisher's Fiducial Prediction

Fisher's two main publications that discuss fiducial prediction are his 1935 paper [15] and his 1956 book [19]. One of the examples in his 1935 paper (Section II of reference [15]) is particularly simple, treating the Gaussian IID model (in this chapter, the adjective "IID" signals being related to a sequence of independent and identically distributed random variables, but rarely simply stands for "independent and identically distributed"). Consider an IID sequence of observations $y_1, y_2, \dots$, each observation distributed as $y_i \sim N(\mu, \sigma^2)$ with unknown mean μ and variance σ^2. After observing IID $y_1, \dots, y_n$ (our *past data* y^{past}) we can compute

$$\bar{y} := \frac{1}{n}\sum_{i=1}^{n} y_i, \qquad s^2 := \frac{1}{n-1}\sum_{i=1}^{n}(y_i - \bar{y})^2.$$

Then

$$t := \sqrt{\frac{n}{n+1}}\frac{y_{n+1} - \bar{y}}{s}, \tag{17.1}$$

where y_{n+1} is the next observation, has Student's t-distribution with $n-1$ degrees of freedom; such a statistic with an invariant (independent of the parameters) distribution is known as a *pivot.*

Once we have a pivot, we can project its distribution onto the next observation. For example, if $\epsilon \in (0,1)$ is our target probability of error, we may predict that the pivot will be in the interval $(F^t_{n-1}(\epsilon/2), F^t_{n-1}(1-\epsilon/2))$, where F^t_{n-1} is the t-distribution function with $n-1$ degrees of freedom. This gives us the prediction

$$y_{n+1} \in \left(\bar{y} + sF^t_{n-1}(\epsilon/2), \bar{y} + sF^t_{n-1}(1-\epsilon/2)\right) \tag{17.2}$$

for the next observation. This set prediction is valid in the sense that it is correct with probability $1-\epsilon$.

The general scheme of fiducial prediction in Fisher's work is that we combine the past observations y^{past} and future observation y to obtain a pivot U,

$$U := Q(y^{\text{past}}, y).$$

The distribution of U is independent of the parameters, and without loss of generality, we can assume that U is uniformly distributed (on $[0,1]$), at least when it is continuous: if it is not uniformly distributed already, replace U by $F_U(U)$, where F_U is U's distribution function.

The existence of, and using, a pivot is a hallmark of fiducial inference. Fisher's approach was invariably *direct*, in that he saw the pivot immediately, without any auxiliary tools. Some kind of normalization was used to get rid of the parameters. For example, in the case of (17.1), we get rid of the location parameter μ by subtracting $\bar{y}$ from the future observation, and we get rid of the scale parameter σ by dividing the difference by s. In future sections, where we consider nonparametric situations, pivots will be derived indirectly via conditioning on sufficient statistics.

If y ranges over $\mathbb{R}$ and the uniformly distributed $Q(y^{\text{past}}, y)$ is increasing in y, $y \mapsto Q(y^{\text{past}}, y)$ will often be a distribution function, and we may call it the *fiducial (predictive) distribution*. In Fisher's example (17.1), the pivot

$$t = \sqrt{\frac{n}{n+1}}\frac{y-\bar{y}}{s}$$

becomes the fiducial (predictive) distribution (function)

$$Q(y^{\text{past}}, y) := F^t_{n-1}\left(\sqrt{\frac{n}{n+1}}\frac{y-\bar{y}}{s}\right). \tag{17.3}$$

Therefore, a fiducial predictive distribution can be defined simply as a uniform pivot, provided the pivot *is* a distribution function (i.e., is increasing from 0 to 1 over $(-\infty, \infty)$). Notice that no explicit "fiducial inversion" is needed in this exposition (unless we want prediction sets).

Fisher imposed some further restrictions in both fiducial prediction and non-predictive fiducial inference, emphasizing both continuity (in many publications) and monotonicity (in 1962 in letters to Barnard and Sprott [4, pp. 44, 218–219], for parametric fiducial inference). In a non-predictive context, he writes in his last letter to Barnard in 1962 (Bennett [4, p. 44] and Barnard [3]):

> A pivotal quantity is a function of parameters and statistics, the distribution of which is independent of all parameters. To be of any use in deducing probability statements about parameters, let me add
>
> a. it involves only one parameter,
>
> b. the statistics involved are jointly exhaustive for that parameter,
>
> c. it varies monotonically with that parameter.

The publications where Fisher prominently mentions *continuity* includes his 1956 book [19] ("the observations should not be discontinuous").

Even if the pivot is increasing in the future observation, there is still a possibility that it will fail to be a distribution function. Namely, there is no guarantee that $Q(y^{\text{past}}, -\infty) = 0$ and $Q(y^{\text{past}}, \infty) = 1$ is satisfied for all y^{past}. For example, $Q(y^{\text{past}}, -\infty) > 0$ means that there is a positive mass at $-\infty$. Using the pivot for computing prediction sets is more flexible.

17.2.2 Validity and Efficiency of Fiducial Prediction

Probabilistic forecasting has become very popular in many application areas, such as economics and weather forecasting (see, e.g., [21, Subsection 1.1]). The most fundamental property of validity is probabilistic calibration, promoted by, among others, Philip Dawid [9] and Tilmann Gneiting (see, e.g., [21, Definition 3(b)]). In general, a random distribution function Q is *probabilistically calibrated* if the random variable

$$Q(y^{\text{past}}, y-) + \tau\left(Q(y^{\text{past}}, y) - Q(y^{\text{past}}, y-)\right)$$

has a uniform distribution on $[0, 1]$ provided τ is a uniform random variable that is independent of the past and future observations y^{past} and y. In Fisher's continuous case, probabilistic calibration simply means the uniform distribution of $Q(y^{\text{past}}, y)$ and, therefore, is achieved automatically. In fact, this simplified understanding of probabilistic calibration will be sufficient in the whole of this chapter, even when we move on to the Dempster–Hill procedure and conformal predictive distributions.

Probabilistic calibration may be the most fundamental notion of validity, but there are several other requirements of this kind, such as marginal calibration [21, Definition 3(a)]. The strongest requirement of validity would be that $Q(y^{\text{past}}, y)$ have the uniform distribution given the σ-algebra generated by the past observations y^{past}. Such strong validity is attainable only with the full knowledge of the stochastic mechanism generating the data (such as under Bayesian assumptions). But we might hope to achieve the uniformity of the distribution of $Q(y^{\text{past}}, y)$ given a smaller σ-algebra $\mathcal{F}$. An interesting example of such conditional probabilistic calibration is due to Peter McCullagh [34, 35]. Our model is that of linear regression

$$y_i = \beta \cdot x_i + \sigma\xi_i, \tag{17.4}$$

where x_i are fixed vectors in $\mathbb{R}^p$, $\beta \in \mathbb{R}^p$ is a vector of parameters, $\sigma > 0$ is another parameter, and ξ_i are IID noise random variables with a known distribution P (which does not have to be Gaussian). Let $\mathcal{F}$ be the σ-algebra of events invariant under the transformations $(y_1, y_2, \dots) \mapsto (a \cdot x_1 + by_1, a \cdot x_2 + by_2, \dots)$, $a \in \mathbb{R}^p$ being a vector and $b > 0$ a positive number. Then the fiducial predictive distribution constructed by McCullagh is probabilistically calibrated given $\mathcal{F}$. Intuitively, the σ-algebra $\mathcal{F}$ corresponds to forgetting just $p + 1$ numbers, where, remember, $p + 1$ is the total number of parameters.

Fisher was motivated by a search for fiducial statements that "may claim unique validity" [36, footnote in Fisher's comment]. He was not satisfied with fiducial statements that were merely true (or valid); he wanted them to be the whole truth.

This was the reason for his introduction of various restrictions, such as the use of exhaustive statistics, as in his letter to Barnard. The insistence on unique validity, leading to serious difficulties and numerous paradoxes, was perhaps the main reason for the rejection of fiducial inference by many statisticians. In various extensions of fiducial inference (such as by Hannig [23]), the requirement of uniqueness has been abandoned. To compare various valid procedures, we then need additional desiderata. An appealing statement of the overall goal of probabilistic forecasting is due to Gneiting and his co-authors (see, e.g., [21, Section 1.2]):

> Probabilistic forecasting has the general goal of maximizing the sharpness of predictive distributions, subject to calibration.

The idea is that validity is the requirement of agreement between the predictive distributions and the actual observations (they should tell the truth), whereas the sharpness characterizes the predictive distributions only (measures how concentrated they are). More generally, instead of the sharpness we can also discuss the efficiency of predictive distributions (the truth should be informative, even if it is not the whole truth), without insisting that the efficiency does not depend on the actual observations. This leads to Martin and Liu's [32, Section 3.3] *efficiency principle*:

> Subject to the validity constraint, probabilistic inference should be made as efficient as possible.

17.3 Dempster–Hill Procedure

This section makes a first step toward nonparametric prediction. In the nonparametric setting, Fisher's requirement of continuity becomes unnecessary if we are allowed to randomize (at least a little).

17.3.1 Fisher's Nonparametric Fiducial Inference

There are only hints of nonparametric fiducial prediction in Fisher's work [40], but numerous authors trace their ideas in this area to Fisher: see, e.g., Dempster 1963 [10], Lane and Sudderth 1984 [28], Hill 1992 [26], and Coolen 1998 [8].

However, Fisher definitely introduced nonparametric fiducial inference for parameter values. In [17], he traced the idea back to Student. In the case of two observations y_1 and y_2 from $N(\mu, 1)$, the probability that $\mu < y_{(1)}$ is $1/4$, the probability that $\mu \in (y_{(1)}, y_{(2)})$ is $1/2$, and the probability that $\mu > y_{(2)}$ is $1/4$, where $y_{(1)} := \min(y_1, y_2)$ and $y_{(2)} := \max(y_1, y_2)$ are the order statistics. In that paper, Fisher extended this statement to an arbitrary sample size n and to the pth quantile μ_p (dropping the Gaussian assumption). Another extension, stated in [17] and in his 1948 paper [18, Subsection 4.7], is to simultaneous inference about several quantiles μ_p. In view of Remark 17.2.1, this may be regarded as a kind of prediction, namely predicting some features of infinitely many future observations.

17.3.2 Dempster–Hill Procedure

Genuine nonparametric fiducial prediction may be said to have been started by Jeffreys in his 1932 paper [27]. He discussed a very special case, predicting a third observation (but it was sufficient for his goal of justifying his noninformative prior for the parameter $1/\sigma$ of the Gaussian family $N(\mu, \sigma^2)$). Assuming (implicitly) IID observations $y_1, y_2, \dots$ from a continuous distribution, what is the probability that $y_3 \in (y_1, y_2)$? According to Jeffreys, the answer is easily seen to be one-third. Indeed, all orders of observations in $\{y_1, y_2, y_3\}$ (assumed different) have the same probability, and the middle observation (not the largest and not the smallest) will be y_3 with probability $1/3$.

Fisher [14] did not accept Jeffreys's argument, let alone accept it as fiducial. One reason might be that Jeffreys did not use the fiducial language (this was done by Seidenfeld in 1995 [39]) and instead couched it in improper Bayesian terms, which Fisher did not like. But another important reason may be that Jeffreys's argument was blatantly discontinuous. We may consider the rank of y_3 as pivot, but we can't even say that it is approximately continuous (which will often be the case for the general Dempster–Hill procedure), since there are only three observations in Jeffreys's picture. In his later paper [16, p. 51] (1935), Fisher did introduce the device of randomization to turn "a discontinuous distribution, leading to statements of fiducial inequality, into a continuous distribution, capable of yielding exact fiducial statements", but he never reconsidered his rejection of Jeffreys's argument.

Jeffreys's procedure is a special case of what I will call the *Dempster–Hill procedure*; the latter extends the former to the case of any finite number of observations and any interval between adjacent observations. Dempster [10, (5.7)] derives it by modifying Fisher's fiducial argument into what he calls a direct probability argument. Dempster makes it look as if his method is known, but he refers to Wilks [53, Chapter 11], who describes prediction intervals rather than predictive distributions. Hill [24, 25] refers to the Dempster–Hill procedure as A_n; this is the statement that, given the data $y_1, \dots, y_n$, the probability that the next observation y falls in $(y_{(i)}, y_{(i+1)})$ is $1/(n+1)$, for each $i = 0, \dots, n$; by definition, $y_{(0)} := -\infty$, and $y_{(n+1)} := \infty$.

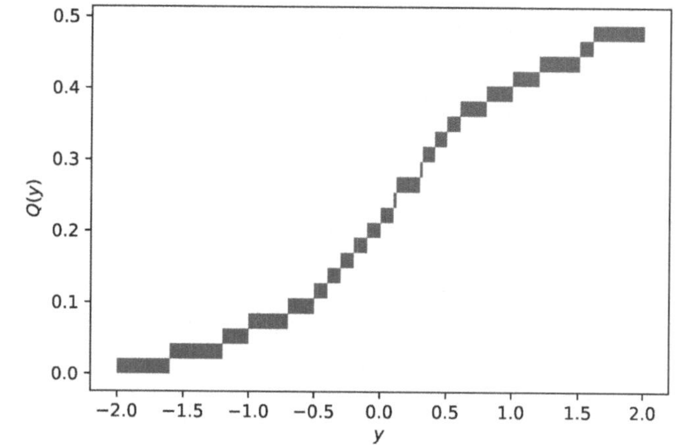

FIGURE 17.1
Sample predictive distribution.

As already mentioned, both Dempster and Hill trace their ideas back to Fisher. In his 1992 paper [26] Hill writes:

> Note that for all three of these authors [Student, Fisher, Dempster] the justification for A_n seems to be purely intuitive. Thus none give anything vaguely representing a "proof" for A_n. . . .

Hill [25] also referred to his procedure as *Bayesian nonparametric predictive inference*. This was abbreviated to *nonparametric predictive inference* (or *NPI*) by Frank Coolen [1, 2], which makes it very wide; in fact, this whole chapter belongs to the area of nonparametric predictive inference.

To embed the Dempster–Hill procedure into fiducial prediction, let us define formally a continuous pivot. As usual, we can achieve continuity by using randomization: if τ is distributed uniformly on $[0, 1]$ and is independent of the observations, the *Dempster–Hill pivot*

$$Q(y^{\text{past}}, y, \tau) := \frac{|\{i \mid y_i < y\}| + \tau + \tau\,|\{i \mid y_i = y\}|}{n+1} \tag{17.5}$$

(i ranging over $\{1, \ldots, n\}$, where n is the number of past observations) is distributed uniformly on $[0, 1]$. The last addend in the numerator in (17.5) takes care of possible ties, but even in the continuous case, we need the penultimate addend to achieve the uniformity of the distribution. For large n, the distribution will be approximately uniform even if we ignore τ (say, set $\tau := 0$ or $\tau := 1$).

As function of y, $Q(y^{\text{past}}, y, \tau)$ as defined by (17.5) can be considered to be a predictive distribution; for a large number n of past observations, it is an approximate distribution function. An example of such a predictive distribution is given in Figure 17.1. The horizontal axis is y, and the vertical axis gives the interval of the values $Q(y^{\text{past}}, y, \tau)$ for all $\tau \in [0, 1]$, where the past observations y^{past} are fixed.

17.3.3 Validity and Efficiency of the Dempster–Hill Procedure

As other fiducial procedures, the Dempster–Hill procedure is automatically valid in that the pivot (17.5) is distributed uniformly on $[0, 1]$. As for its efficiency, we do not really need it in this section for the purpose of comparison, since the Dempster–Hill procedure is uniquely determined (efficiency will become important in the next section, where this procedure will be greatly generalized). Still the efficiency of the Dempster–Hill procedure in the sense of the closeness of the empirical distribution function to the true data-generating distribution function has been extensively studied under the rubric of empirical processes (see, e.g., [42]).

17.4 Conformal Predictive Distributions

As mentioned in Section 17.1, a serious limitation of the Dempster–Hill procedure is that it does not cover typical problems of machine learning, such as regression and classification, where the observations are pairs (x, y) and not just the labels y; the task is to predict the unknown label y of a test object x. The procedure has been criticized on this account; e.g., Genest and Kalbfleisch [20] say in their response to Berliner and Hill [5]: "To be truly useful, however, the methods need extension to regression models with unknown regression parameters".

In most of the rest of this chapter, our statistical model will be the one standard in machine learning and much of nonparametric statistics: the observations are assumed to be IID (and nothing more is assumed). We will refer to it as the *IID model* (or *hypothesis of randomness*). There are several related ways, treated in Subsections 17.4.1 and 17.4.2, to construct a *conformal pivot*, a random variable that is distributed uniformly on $[0, 1]$ (under the IID model) and can be used as a predictive distribution (*conformal predictive distribution*).

While the Dempster–Hill distributions are uniquely determined, there is a huge variety of conformal predictive distributions. For example, outputs of many prediction algorithms in machine learning and statistics can be turned (usually in more than one way) into predictive distributions.

In this section, we consider the case of regression, where the labels y are real numbers, $y \in \mathbb{R}$; the objects x are elements of a measurable space (often Euclidean space $\mathbb{R}^p$).

17.4.1 Full Conformal Predictive Distributions

A *conformity measure* is a measurable function A mapping finite sequences of observations $(z_1, \dots, z_l)$ to *conformity scores* $(\alpha_1, \dots, \alpha_l)$ (sequences of real numbers of the same length) that is equivariant: for any l and any permutation π of $\{1, \dots, l\}$,

$$A(z_1, \dots, z_l) = (\alpha_1, \dots, \alpha_l) \Longrightarrow A\left(z_{\pi(1)}, \dots, z_{\pi(l)}\right) = \left(\alpha_{\pi(1)}, \dots, \alpha_{\pi(l)}\right). \tag{17.6}$$

The origin of the terminology is that α_i measures how well z_i conforms to the other observations among $z_1, \dots, z_l$. In many cases, we can build A on top of some *underlying* algorithm. A simple example is where the conformity scores are defined by

$$\alpha_i := y_i - \hat{y}_i, \tag{17.7}$$

where $\hat{y}_i$ is the prediction for the label of x_i produced by a prediction algorithm trained on $z_1, \dots, z_l$. Another example, that might better agree with the terminology, is

$$\alpha_i := -\left|y_i - \hat{y}_i\right|. \tag{17.8}$$

The *conformal pivot* determined by a conformity measure A is

$$Q(z^{\text{past}}, (x, y), \tau) := \frac{\left|\{i \mid \alpha_i^y < \alpha^y\}\right| + \tau + \tau\left|\{i \mid \alpha_i^y = \alpha^y\}\right|}{n+1}, \tag{17.9}$$

where z^{past} is a sequence $z_1, \dots, z_n$ of observations of length n (our *training sequence*), $i = 1, \dots, n$,

$$(\alpha_1^y, \dots, \alpha_n^y, \alpha^y) := A(z_1, \dots, z_n, (x, y)), \tag{17.10}$$

and $\tau \in [0, 1]$. For a proof that the distribution of $Q(z^{\text{past}}, (x, y), \tau)$ is uniform on $[0, 1]$, provided τ is distributed uniformly on $[0, 1]$ and independent of the observations, see, e.g., [48, Proposition 2.4].

Notice that the direct implementation of (17.9) involves heavy computations, even if we limit ourselves to a finite grid of values for the label y (which can be done rigorously in some cases [6]). If we have several test objects x for which we would like to have predictive distributions, we need to recompute all the $n+1$ conformity scores (17.10) for each possible value of y and for each test object x.

We say that (17.9) is a *conformal predictive system* if:

- $Q(z^{\text{past}}, (x, y), \tau)$ is an increasing function of $y \in \mathbb{R}$ (by definition it is an increasing and linear function of τ);
- for each training sequence z^{past} and each test object x,

$$\lim_{y \to -\infty} Q(z^{\text{past}}, (x, y), 0) = 0 \tag{17.11}$$

$$\lim_{y \to \infty} Q(z^{\text{past}}, (x, y), 1) = 1. \tag{17.12}$$

In this case, we will say that $Q(z^{\text{past}}, (x, y), \tau)$ as function of y is a *conformal predictive distribution.*

With a careful choice of the conformity measure A the conformal pivot will be an efficiently computable conformal predictive system. A trivial example is where A strips a sequence of observations of the objects:

$$A\left((x_1, y_1), \dots, (x_l, y_l)\right) := (y_1, \dots, y_l)$$

(or, alternatively, we have no objects and A is the identity function). In this case, we obtain the Dempster–Hill pivot (17.5).

Our first nontrivial example is the *Least Squares Predictive Machine* (LSPM). This is the case of the conformity measure (17.7), where $\hat{y}_i$ is the Least Squares (LS) estimate of the label of ith object, which is assumed to be a vector in $\mathbb{R}^p$. The answer to the question of monotonicity of the conformal pivot in y depends on the details of the definition of the LS estimate [52, Section 3]:

- If $\hat{y}_i$ is the LS estimate based on all of $z_1, \dots, z_l$ (so that α_i is the "full residual"), monotonicity can be violated (albeit only in pathological cases of a high-leverage test object x).
- If $\hat{y}_i$ is the LS estimate based on $z_1, \dots, z_l$ with z_i removed (so that α_i is the "deleted residual"), monotonicity can be violated (albeit only in pathological cases of high-leverage training objects in z^{past}).
- But in an intermediate situation (of a "studentized residual" α_i), monotonicity always holds.

Therefore, only the studentized LSPM is a conformal predictive system [52, Proposition 5]. Its predictions (conformal predictive distributions) can be computed in time $O(n^2 + kn)$, where n is the length of the training sequence and k is the length of the test sequence (preprocessing takes time $O(n^2)$ and then processing each test object takes time $O(n)$); the dimension p is assumed to be fixed.

Fisher's idea of using exhaustive statistics to obtain uniquely valid predictive distributions does not work at all in the case of conformal predictive systems. Even when constructing a conformity measure from an underlying algorithm, we have a plenty of choices, and arguably some underlying algorithms available in machine learning may involve an element of intelligence (such as artificial neural networks). The variety of conformal predictive systems is real, all of them are valid (in the sense of being probabilistically calibrated), and we need some notion of efficiency to distinguish between them. Fisher's whole truth [36, footnote in Fisher's comment] is usually not attainable.

A useful notion of efficiency for a conformal predictive system is how close the predictive distributions that it outputs are to the true predictive distributions. One way to answer this question is to impose strong assumptions on the data-generating distribution. In the case of the LSPM, it is natural to assume the model (17.4), where ξ_i are IID standard Gaussian random variables, in which case we will refer to it as the *Gauss linear model*. Under this parametric assumption, we can construct nearly optimal, or *oracular*, predictive distributions, and it turns out that, under natural regularity conditions, the conformal predictive distributions output by the LSPM and the oracular predictive distributions approach each other at the usual rate $O(n^{-1/2})$: see Theorems 2–4 of reference [52].

Why are such efficiency results useful? Can't we just use the oracular predictive distributions? This would be risky in situations where we are willing to accept the nonparametric IID model but reluctant to accept the parametric Gauss linear model. We have validity under the IID model, so the conformal predictive distributions will not be misleading (at worst they will be useless). But if we are lucky and the Gauss linear model also holds, we will quickly approach the true predictive distributions.

The LSPM provides an example of the procedure of *conformalization*. We take a point prediction procedure, Least Squares, that is optimal, in some sense, under the Gauss linear model. Then we pass it through the "conformal machine" (17.9)–(17.10) to obtain guaranteed validity under the IID model.

A disadvantage of the LSPM is that its underlying prediction algorithm, Least Squares, is linear, and therefore, the LSPM is likely to be efficient only when the true relation between the labels and objects is linear. A method of extending linear methods to nonlinear situations that is standard in machine learning is to apply a feature mapping to the objects, which corresponds to replacing the dot product with another kernel in the object space. The LSPM can be "kernelized" in this way while maintaining its computational efficiency [49]. Pre-processing a training sequence of length n takes, asymptotically, the same time as inverting an $n \times n$ matrix ($O(n^3)$ with the schoolbook method) and, after that, processing a test object takes time $O(n^2)$.

If we are only interested in a weak form of asymptotic efficiency of conformal predictive distributions, their universal consistency, there is no need to adopt narrow statistical models such as the Gauss linear model, and it holds under the IID model for a suitable conformity measure (Theorem 28 of reference [43]). Universally consistent conformal predictive systems can be built on top of many classical universally consistent algorithms, such as nearest neighbours, and their construction adapts standard arguments for universal consistency in classification and regression [41, 11, 22].

An advantage of predictive distributions over other, less complete, forms of prediction is that in decision-making problems predictive distributions can be combined with a utility function and the expected utility maximization principle to obtain optimal decisions. Using the predictive distributions produced by a universally consistent predictive system leads, under natural conditions, to decisions whose regret tends to 0 in probability under the IID model (Theorem 3 of reference [46]).

17.4.2 Split-Conformal and Cross Conformal Predictive Distributions

As mentioned earlier, conformal predictive distributions, also called *full* conformal predictive distributions, are difficult to compute apart from a small number of conformity measures that are particularly mathematically tractable. It may also be difficult to check that a full conformal pivot is an increasing function of the label. Both problems simplify drastically if we have enough data.

A *split-conformity measure* is a measurable function A that maps an observation z and a sequence of observations $z_1, \dots, z_l$ into a *conformity score* $A(z; z_1, \dots, z_l) \in \mathbb{R}$; intuitively, the conformity score shows how similar z is to the elements of the sequence $z_1, \dots, z_l$. An example is again given by (17.7), which in our current notation becomes

$$A((x, y); z_1, \dots, z_l) := y - \hat{y}, \tag{17.13}$$

where $\hat{y}$ is the prediction for the label of x produced by a prediction algorithm trained on $z_1, \dots, z_l$.

Let us divide the training sequence $z^{\text{past}} = (z_1, \dots, z_n)$ into two parts:

- the *training sequence proper*, $z_1, \dots, z_m$, of length m,
- and the *calibration sequence*, $z_{m+1}, \dots, z_n$, of length $n - m$.

The *split-conformal pivot* for a test object x is

$$Q(z^{\text{past}}, (x, y), \tau) := \frac{|\{i \mid \alpha_i < \alpha\}| + \tau + \tau\,|\{i \mid \alpha_i = \alpha\}|}{n - m + 1}, \tag{17.14}$$

where i ranges over $m + 1, \dots, n$ and

$$\alpha_i := A(z_i; z_1, \dots, z_m), \quad \alpha := A((x, y); z_1, \dots, z_m). \tag{17.15}$$

It is easy to construct computationally efficient split-conformal pivots. For example, if the split-conformity measure is (17.13), there is no need to retrain the underlying prediction algorithm for each calibration and test object.

It is also easy to ensure that the split-conformal pivot $Q(z^{\text{past}}, (x, y), \tau)$ is an increasing function of y: it is enough to require that $A((x, y); \dots)$ is increasing in y. If, in addition, the convex closure of $A((x, \mathbb{R}); \dots)$ does not depend on x, the split-conformal pivot will satisfy (17.11)–(17.12), in which case it is called a *split-conformal predictive system* and its output is called a *split-conformal predictive distribution.* For details, see (Section 3 of reference [50]).

Split-conformal predictive systems are computationally efficient but may lose predictive efficiency as compared with full conformal predictive systems, which use the full training sequence as both training sequence proper and calibration sequence. A natural way out is to divide the training sequence into a number of folds (as in cross-validation), use each fold in turn as calibration sequence, and combine the corresponding split-conformal predictive distributions. The resulting *cross-conformal predictive distributions* (Section 4 of reference [50]) lose guaranteed validity but are well-calibrated in practice (in the absence of excessive randomization in the underlying algorithm [31, 50]). Under a suitable choice of a split-conformity measure, split-conformal and cross-conformal predictive distributions are universally consistent [50, Theorem 7.2].

One limitation of the conformity measure (17.7) and the split-conformity measure (17.13) is that they implicitly assume homoscedasticity. As a result, the predictive distributions for all test objects are, essentially, horizontal translates of each other (even though the shape of the predictive distribution can be very adaptive, as is the case for the kernelized LSPM [49] with a universal kernel). We can generalize, e.g., (17.13) to

$$A((x, y); z_1, \dots, z_l) := \frac{y - \hat{y}}{s},$$

where $s > 0$ is an estimate, based on $z_1, \dots, z_l$ and x, of the accuracy of the prediction $\hat{y}$ for the label of x. This will still produce essentially the same shape of the predictive distributions, and we will just add a scale parameter.

The split-conformal (and by extension cross-conformal) method allows a much more radical solution. As split-conformity measure, we can take

$$A((x, y); z_1, \dots, z_m) := F(y),$$

where F is a predictive distribution function, not required to satisfy any properties of validity under the IID model, for the label of x computed from $z_1, \dots, z_m$ as training sequence. Examples of possible F are the Nadaraya–Watson procedure [37], random forest regression, procedures in the TensorFlow probability module, and various Gaussian processes. See [51] for experimental results. This gives another example of conformalization; the application of the split-conformal machine (17.14)–(17.15) ensures the validity under the IID model.

17.5 Conformal Prediction

The requirement of monotonicity of conformal pivots in the label is restrictive and dropping it greatly extends the application area of the conformal method. For example, it then extends immediately to the case of classification, where the label is only allowed to take values in a finite set (which does not even have to be ordered). Instead of probability distributions, we only have p-values, (17.9) or (17.14) (they are valid p-values since they are distributed uniformly on $[0, 1]$). In order for p-values to become probabilities (to form a distribution function, at least approximate) we need to impose strict discipline, first of all monotonicity, but even mere p-values have many important uses.

The step from conformal predictive distributions to p-values is somewhat similar to the steps from fiducial distributions for parameter values to Neyman's confidence intervals [36] and from fiducial predictive distributions to prediction intervals (e.g., from (17.3) to (17.2)). The theory simplifies and generalizes. We lose something but enough survives. Our goal is still to design prediction algorithms that are automatically valid and, under the constraint of validity, are as efficient as possible, in various formal and informal senses.

17.5.1 Examples of Conformal Predictors

We will refer to the system of p-values (17.9) (without further requirements such as monotonicity) as *conformal predictor*. An example is the conformal predictor (*1-nearest neighbour conformal predictor* [48, Section 3.1]) determined by the dimensionless conformity measure

$$\alpha_i := \frac{\min_{j \in \{1,\dots,l\}: y_j \neq y_i} \|x_j - x_i\|}{\min_{j \in \{1,\dots,l\} \setminus \{i\}: y_j = y_i} \|x_j - x_i\|}$$

(using the same notation as in (17.6), with $z_i = (x_i, y_i)$ and with the objects x_i being vectors in a Euclidean space).

Similarly, we can define a *split-conformal predictor* as the system of p-values (17.14), without further requirements. An example is the split conformal predictor determined by the split-conformity measure $A((x, y); z_1, \dots, z_l) := P(\{y\})$, where P is the predictive distribution for the label of the object x output by a Bayesian algorithm trained on $z_1, \dots, z_l$ as training set [48, Section 4.2]. An advantage of this split-conformal predictor as compared with the underlying Bayesian algorithm is that the former is valid (in the sense of leading to uniformly distributed p-values) under the nonparametric IID model rather than a fully specified Bayesian model.

In the rest of this section, we will concentrate on full conformal prediction, although many ideas can be extended to split-conformal and cross-conformal prediction.

17.5.2 Validity of Conformal Prediction

We know that conformal predictors are valid in the sense of (17.9) being distributed uniformly on $[0, 1]$. This allows us to compute prediction sets with a guaranteed probability of error; e.g., defining the prediction set as the set of all labels leading to a p-value greater than a fixed significance level $\epsilon \in (0, 1)$ ensures the probability of error ϵ. (In order to obtain a bounded prediction interval in the context of regression, we should use, e.g., (17.8) rather than (17.7).)

This property of validity can be strengthened in the online prediction protocol. In this protocol, the observations $z_n = (x_n, y_n)$, $n = 1, 2, \dots$, arrive sequentially, and at the nth step, we predict the label y_{n+1} of x_{n+1} given x_{n+1} itself and the previous observations $z^{\text{past}} := (z_1, \dots, z_n)$. It turns out that the p-values $p_n := Q(z^{\text{past}}, (x_{n+1}, y_{n+1}), \tau_n)$ are independent provided the uniform random numbers τ_n are independent between themselves and of the observations [48, Proposition 2.4]. Therefore, by the law of large numbers, the guaranteed probability of error for the prediction set will be reflected in the frequency of errors.

17.5.3 Efficiency of Conformal Prediction and Training Conformal Predictors

The question of measuring the efficiency of conformal predictors is difficult, and during the short history of conformal prediction several unsuitable measures have been used. In the case of classification, a natural desideratum for a criterion of efficiency is that the true conditional probability of the label y of an object x should be an ideal conformity measure as evaluated by that criterion. This desideratum is formalized in Section 4 of reference [47] and criteria of efficiency that satisfy it are called *probabilistic.*

An example of a probabilistic criterion of efficiency is evaluating the quality of a conformal predictor by the average p-value computed for a test object x with postulated label y, with x ranging over the test sequence and y ranging over all possible labels. This criterion does not depend on the true labels of the test objects, and so agrees with the notion of sharpness in Gneiting et al.'s version of the efficiency principle. The rationale behind it is that we would like the p-values for labels different from the true one to be as small as possible (the p-value for the true label is not under our control; because of the validity, it is distributed uniformly on $[0, 1]$).

The disadvantage of this criterion of efficiency is that the average contains a lot of noise created by the true label. Another criterion of efficiency ("observed fuzziness") is defined in the same way except that (x, y) ranges over all pairs where x is a test object and y is a label different from x's true label. A disadvantage is that this criterion does depend on the true labels of the test objects, but the average now involves less noise; in the ideal case, it will be close to zero.

The original goal of conformal prediction was to complement the predictions output by state-of-the-art algorithms of mainstream machine learning by provably valid measures of their accuracy and reliability [48]. Recently, first steps have been made in designing conformal predictors *ab initio* [7]; we can train conformal predictors directly by minimizing the observed fuzziness on a calibration sequence (or in the framework of cross-validation).

17.5.4 Conformal Martingales and Online Hypothesis Testing

An interesting application of the validity of conformal predictors, including the independence of p-values in the online protocol, is to test the hypothesis of randomness online [44]. Suppose an IID sequence of observations at some point ceases to be IID. How can we detect the change and, for example, raise an alarm as soon as possible after the change occurs? (A special case is where the hypothesis of randomness never holds, which corresponds to testing randomness.)

Existing methods can deal with the case where the pre-change distribution is known. In combination with conformal prediction, we can apply these methods to the uniformly distributed and independent p-values. This gives us procedures for change detection in the nonparametric situation where the pre-change distribution is only known to belong to the IID model.

17.5.5 Repetitive Structures

Conformal prediction, as discussed so far, produces predictions that are valid under the IID model. However, it can be generalized in a straightforward manner to arbitrary *repetitive structures*, as introduced by Per Martin-Löf [33] and further developed by Lauritzen [29].

The IID model is an important repetitive structure, but another example is provided by the Gaussian IID model [48, Section 8.5]. Interestingly, Fisher's fiducial distribution (17.3) coincides with the conformal predictive distribution for the Gaussian IID model and trivial conformity measure (the identity function). This implies, in particular, that Fisher's fiducial distribution (17.3) satisfies a stronger property of validity than probabilistic calibration; we can also say that the random variables $Q(y^{\text{past}}, y_n)$ are independent under the Gaussian IID model (as defined at the beginning of Subsection 17.2.1) in the online prediction protocol. Therefore, the prediction intervals (17.2) fail to cover the true future observations with probability ϵ independently.

Yet another example of repetitive structure is provided by graphical models [13]. The testing methods described in Subsection 17.5.4 are applicable to any repetitive structure, including graphical models.

17.6 Conclusion

The key message of this chapter is that there are ways to extend ideas of fiducial prediction to nonparametric settings, including those useful in regression problems. The notion of validity known as probabilistic calibration is attained automatically, but there are other notions of validity that deserve to be explored in modern approaches to fiducial prediction. Efficiency is never automatic and is an extensive and largely unexplored area of research.

This includes developing suitable criteria of efficiency for predictive systems, fiducial and conformal. Such criteria can then be used both for evaluating predictive systems for practical purposes and for theoretical studies, such as complementing the known existence results [43] for universally consistent conformal predictive distributions by results about the attainable rates of convergence for those distributions, perhaps along the lines of known results (see, e.g., [30]) in conformal regression.

Acknowledgments

This chapter is essentially a written version of my talk at the Sixth BFF meeting in 2019 (29 April 2019, Duke University, Durham, NC). Out of the six BFF meetings so far I have attended two (Fourth in 2017 and Sixth in 2019), and both deeply affected this chapter. I am grateful to the organizers of and participants in both meetings. The work on this chapter has been partially supported by Amazon, Astra Zeneca, and Stena Line.

Bibliography

[1] Augustin, T. and Coolen, F. P. A., 2004. Nonparametric predictive inference and interval probability. *Journal of Statistical Planning and Inference, 124*, pp. 251–272.

[2] Augustin, T., Walter, G. and Coolen, F. P. A., 2014. Statistical inference. In *Introduction to Imprecise Probabilities*, Chapter 7, eds T. Augustin, F. P. A. Coolen, G. de Cooman and M. C. M. Troffaes, pp. 135–189. Chichester: Wiley.

[3] Barnard, G. A., 1963. Fisher's contributions to mathematical statistics. *Journal of the Royal Statistical Society A, 126*, pp. 162–166.

[4] Bennett, J. H. editor, 1990. *Statistical Inference and Analysis: Selected Correspondence of R. A. Fisher.* Oxford: Clarendon Press.

[5] Berliner, L. M. and Hill, B. M., 1988. Bayesian nonparametric survival analysis. *Journal of the American Statistical Association, 83*, pp. 772–779.

[6] Chen, W., Chun, K.-J. and Barber, R. F., 2018. Discretized conformal prediction for efficient distribution-free inference. *Stat, 7*, e173.

[7] Colombo, N. and Vovk, V., 2020. Training conformal predictors. *Proceedings of Machine Learning Research, 128*, pp. 55–64. COPA.

[8] Coolen, F. P. A., 1998. Low structure imprecise predictive inference for Bayes' problem. *Statistics and Probability Letters, 36*, pp. 349–357.

[9] Dawid, A. P., 1984. Statistical theory: The prequential approach (with discussion). *Journal of the Royal Statistical Society A, 147*, pp. 278–292.

[10] Dempster, A. P., 1963. On direct probabilities. *Journal of the Royal Statistical Society B, 25*, pp. 100–110.

[11] Devroye, L., Györfi, L. and Lugosi, G., 1996. *A Probabilistic Theory of Pattern Recognition*. New York: Springer.

[12] Efron, B., 1998. R. A. Fisher in the 21st century. *Statistical Science, 13*, pp. 95–122.

[13] Fedorova, V., Gammerman, A., Nouretdinov, I. and Vovk, V., 2015. Hypergraphical conformal predictors. *International Journal on Artificial Intelligence Tools, 24*(6), p. 1560003. COPA 2013 Special Issue.

[14] Fisher, R. A., 1933. The concepts of inverse probability and fiducial probability referring to unknown parameters. *Proceedings of the Royal Society of London A, 139*, pp. 343–348.

[15] Fisher, R. A., 1935. The fiducial argument in statistical inference. *Annals of Eugenics, 6*, pp. 391–398.

[16] Fisher, R. A., 1935. The logic of inductive inference. *Journal of the Royal Statistical Society, 98*, pp. 39–54.

[17] Fisher, R. A., 1939. "Student". *Annals of Eugenics, 9*, pp. 1–9.

[18] Fisher, R. A., 1948. Conclusions fiduciaires. *Annales de l'Institut Henry Poincaré, 10*, pp. 191–213.

[19] Fisher, R. A., 1973. *Statistical Methods and Scientific Inference* (3rd ed.). New York: Hafner. First edition: 1956.

[20] Genest, C. and Kalbfleisch, J., 1988. Bayesian nonparametric survival analysis: comment. *Journal of the American Statistical Association, 83*, pp. 780–781. Comment on [5].

[21] Gneiting, T. and Katzfuss, M., 2014. Probabilistic forecasting. *Annual Review of Statistics and Its Application, 1*, pp. 125–151.

[22] Györfi, L., Kohler, M., Krzyżak, A. and Walk, H., 2002. *A Distribution-Free Theory of Nonparametric Regression*. New York: Springer.

[23] Hannig, J., 2009. On generalized fiducial inference. *Statistica Sinica, 19*, pp. 491–544.

[24] Hill, B. M., 1968. Posterior distribution of percentiles: Bayes' theorem for sampling from a population. *Journal of the American Statistical Association, 63*, pp. 677–691.

[25] Hill, B. M., 1988. De Finetti's theorem, induction, and $A_{(n)}$ or Bayesian nonparametric predictive inference (with discussion). In *Bayesian Statistics 3*, eds D. V. Lindley, J. M. Bernardo, M. H. DeGroot and A. F. M. Smith, pp. 211–241. Oxford: Oxford University Press.

[26] Hill, B. M., 1992. Bayesian nonparametric prediction and statistical inference. In *Bayesian Analysis in Statistics and Econometrics*, volume 75 of Lecture Notes in Statistics, Chapter 4, eds P. K. Goel and N. S. Iyengar, pp. 43–94. New York: Springer.

[27] Jeffreys, H., 1932. On the theory of errors and least squares. *Proceedings of the Royal Society of London A, 138*, pp. 48–55.

[28] Lane, D. A. and Sudderth, W. D., 1984. Coherent predictive inference. *Sankhyā A, 46*, pp. 166–185.

[29] Lauritzen, S. L., 1988. *Extremal Families and Systems of Sufficient Statistics*, volume 49 of Lecture Notes in Statistics. New York: Springer.

[30] Lei, J. and Wasserman, L., 2014. Distribution-free prediction bands for non-parametric regression. *Journal of the Royal Statistical Society B*, *76*, pp. 71–96.

[31] Linusson, H., Norinder, U., Boström, H., Johansson, U. and Löfström, T., 2017. On the calibration of aggregated conformal predictors. *Proceedings of Machine Learning Research*, *60*, pp. 154–173. COPA 2017.

[32] Martin, R. and Liu, C., 2016. *Inferential Models: Reasoning with Uncertainty*. Boca Raton, FL: CRC Press.

[33] Martin-Löf, P., 1974. Repetitive structures and the relation between canonical and microcanonical distributions in statistics and statistical mechanics. In *Proceedings of Conference on Foundational Questions in Statistical Inference*, eds O. Barndorff-Nielsen, P. Blæsild and G. Schou, pp. 271–294. Aarhus.

[34] McCullagh, P., 2004. Fiducial prediction. Manuscript. https://www.stat.uchicago.edu/~pmcc/reports/fiducial.pdf

[35] McCullagh, P., Vovk, V., Nouretdinov, I., Devetyarov, D. and Gammerman, A., 2009. Conditional prediction intervals for linear regression. In *Proceedings of the Eighth International Conference on Machine Learning and Applications (ICMLA 2009)*, pp. 131–138. http://www.stat.uchicago.edu/~pmcc/reports/predict.pdf.

[36] Neyman, J., 1934. On the two different aspects of the representative method: the method of stratified sampling and the method of purposive selection (with discussion). *Journal of the Royal Statistical Society*, *97*, pp. 558–625. Fisher's comment: 614–619.

[37] Rosenblatt, M., 1969. Conditional probability density and regression estimators. In *Multivariate Analysis II*, eds P. R. Krishnaiah, pp. 25–31. New York: Academic Press.

[38] Schweder, T. and Hjort, N. L., 2016. *Confidence, Likelihood, Probability: Statistical Inference with Confidence Distributions*. Cambridge: Cambridge University Press.

[39] Seidenfeld, T., 1995. Jeffreys, Fisher, and Keynes: predicting the third observation, given the first two. In *New Perspectives on Keynes*, eds A. F. Cottrell and M. S. Lawlor, pp. 39–52. Durham: Duke University Press.

[40] Seidenfeld, T., 2017. Personal communication. Fourth BFF meeting, May 2017.

[41] Stone, C. J., 1977. Consistent nonparametric regression (with discussion). *Annals of Statistics*, *5*, pp. 595–645.

[42] Vaart, A. W. van der and Wellner, J. A., 1996., *Weak Convergence and Empirical Processes: With Applications to Statistics*. New York: Springer.

[43] Vovk, V., 2019. Universally consistent conformal predictive distributions. *Proceedings of Machine Learning Research*, *105*, pp. 105–122. COPA 2019, to appear in *Pattern Recognition*.

[44] Vovk, V., 2021. Testing randomness online. *Statistical Science*, *36*, pp. 595–611.

[45] Vovk, V., 2022. Universal predictive systems. *Pattern Recognition*, 126, 108536.

[46] Vovk, V. and Bendtsen, C., 2018. Conformal predictive decision making. *Proceedings of Machine Learning Research, 91*, pp. 52–62. COPA 2018.

[47] Vovk, V., Fedorova, V., Nouretdinov, I. and Gammerman, A., 2017. Criteria of efficiency for set-valued classification. *Annals of Mathematics and Artificial Intelligence, 81*, pp. 21–46. COPA 2016 Special Issue.

[48] Vovk, V., Gammerman, A. and Shafer, G., 2005. *Algorithmic Learning in a Random World.* New York: Springer.

[49] Vovk, V., Nouretdinov, I., Manokhin, V. and Gammerman, A., 2018. Conformal predictive distributions with kernels. In *Braverman's Readings in Machine Learning: Key Ideas from Inception to Current State*, volume 11100, eds L. Rozonoer, B. Mirkin and I. Muchnik, pp. 103–121. Cham, Switzerland: Springer.

[50] Vovk, V., Petej, I., Nouretdinov, I., Manokhin, V. and Gammerman, A., 2020. Computationally efficient versions of conformal predictive distributions. *Neurocomputing, 397*, pp. 292–308. COPA 2018 Special Issue.

[51] Vovk, V., Petej, I., Toccaceli, P. Gammerman, A., Ahlberg, E. and Carlsson, L., 2020. Conformal calibration. *Proceedings of Machine Learning Research, 128*, 84–99. COPA 2020.

[52] Vovk, V., Shen, J., Manokhin, V. and Xie, M., 2019. Nonparametric predictive distributions based on conformal prediction. *Machine Learning, 108*, pp. 445–474. COPA 2017 Special Issue.

[53] Wilks, S. S., 1962. *Mathematical Statistics.* New York: Wiley.

[54] Xie, M. and Singh, K., 2013. Confidence distribution, the frequentist distribution estimator of a parameter: A review. *International Statistical Review, 81*, pp. 3–39.

18

Fiducial Inference and Decision Theory

Gunnar Taraldsen and Bo Henry Lindqvist
Norwegian University of Science and Technology

The majority of the statisticians concluded many decades ago that fiducial inference was nonsensical to them. Hannig et al. (2016) and others have, however, contributed to a renewed interest and focus. Fiducial inference is like Bayesian analysis, but without requiring a prior. The prior information is replaced by assuming a particular data generating equation. Berger (1985) explains that Bayesian analysis and statistical decision theory are in harmony. Taraldsen and Lindqvist (2013) show that fiducial theory and statistical decision theory also play well together. The purpose of this chapter is to explain and exemplify this together with recent mathematical results.

18.1 Introduction

The fiducial density $\pi(\rho \,|\, r)$ of the correlation ρ given the empirical correlation r is

$$\pi(\rho \,|\, r) = \frac{(1-r^2)^{\frac{\nu-1}{2}} \cdot (1-\rho^2)^{\frac{\nu-2}{2}} \cdot (1-r\rho)^{\frac{1-2\nu}{2}}}{\sqrt{2}B(\nu+\frac{1}{2},\frac{1}{2})} \cdot F(\frac{3}{2},-\frac{1}{2};\nu+\frac{1}{2};\frac{1+r\rho}{2}) \tag{18.1}$$

where B is the Beta function, F is the Gaussian hypergeometric function, and $n = \nu - 1$ is the sample size (Taraldsen, 2021a). The fiducial is also given by Rao's formula

$$\pi(\rho \,|\, r) = \frac{(1-r^2)^{\frac{\nu-1}{2}} \cdot (1-\rho^2)^{\frac{\nu-2}{2}}}{\pi(\nu-2)!} \partial_{\rho r}^{\nu-2} \left\{ \frac{\theta - \frac{1}{2}\sin 2\theta}{\sin^3\theta} \right\} \tag{18.2}$$

where $\cos\theta = -\rho r$ and $0 < \theta < \pi$. Rao's formula (18.2) is stated as the very last formula in the classical book *Statistical Methods and Scientific Inference* by Fisher (1956, equation (234)). The fiducial for the correlation is also the very first example of a fiducial distribution presented by Fisher (1930).

It is no longer possible to ask Fisher in person, but from his writings it seems he would still claim:

The fiducial is the answer!

We agree. The problem Fisher gave us is then to explain and characterize relevant questions. This is a very open-ended problem, and it still needs more investigation. The original claim of the naturalness of the fiducial by Fisher (1930), and the many questions and more specific claims that followed have inspired the development of classical and Bayesian practical, theoretical, and mathematical statistics. It still does.

DOI: 10.1201/9780429341731-18

Recent mathematical results that give links between different modes of inference are presented in Section 18.4 without any particular interpretation. Section 18.5 includes a brief discussion of related work and comments on open problems. Much of the text is devoted to exemplifying the interplay between fiducial arguments and decision theory as described by Taraldsen and Lindqvist (2013). Section 18.2 presents decision theory illustrated with a simple example and Section 18.3 contains more examples. In the rest of this Introduction a direct interpretation of the fiducial distribution is explained.

Consider the following two equivalent equations

$$r = r(\rho, u) \iff \rho = \rho(r, u) \tag{18.3}$$

The first equation gives the uncertainty of the empirical correlation r for a known correlation ρ from the uncertainty in a Monte Carlo variable u. It is called a data generating equation since computer-generated samples from u's distribution gives samples from the distribution of r when ρ is known. A data generating equation $r = r(\rho, u)$ follows from a data generating equation for the binormal. It is not unique, but $\pi(\rho \,|\, r)$ is unique.

The second equality in equation (18.3) is equivalent with the first equality. It gives the uncertainty of the correlation ρ for a known empirical correlation r from the uncertainty in u. It is called a parameter generating equation since computer-generated samples from u's distribution give samples from the distribution of ρ when r is known. The corresponding density of ρ when r is known is given by the two alternative expressions in equation (18.1) and equation (18.2) respectively. The fiducial distribution is hence simply the result of propagation of the uncertainty as defined by the uncertainty in u and the generating equations (18.3).

The interpretation of the fiducial in a pure fiducial perspective is as the state of knowledge of the parameter ρ given the data and the parameter generating model. This determines then also the state of knowledge of any derived parameter $\lambda = \lambda(\rho)$. This interpretation is exactly as for a Bayesian prior and for a Bayesian posterior, but it is to be used for cases without a prior. Taraldsen and Lindqvist (2019, pp. 523–524) explain and discuss this point in more detail. It is also this interpretation which is stated directly by Fisher (1956, pp. 54–55):

> *By contrast, the fiducial argument uses the observations only to change the logical status of the parameter from one in which nothing is known of it, and no probability statement about it can be made, to the status of a random variable having a well-defined distribution.*

Consider an example where the result of an experiment is given by four points with (x, y) coordinates $(773, 727)$, $(777, 735)$, $(284, 286)$, and $(519, 573)$ (Taraldsen, 2021a). There are reasons a priori for assuming a linear relationship. This is further supported by Figure 18.1, and a high value for the coefficient of determination $R^2 = 97.00\%$. The R^2 equals the square of the empirical correlation $r = 98.49\%$. This is an example of linear regression used extensively in applied sciences.

A natural question is: What about uncertainty? In fiducial inference, the uncertainty is given by the fiducial distribution. The state of knowledge regarding the correlation is given by the fiducial density shown in Figure 18.2. The uncertainty of any derived parameter $\lambda(\rho)$ follows by propagation of uncertainty.

What about decision theory? In this case, a natural question would be to find a point or interval esimate $\hat{\rho}$ for ρ. Decision theory gives one possible approach. A possible loss function for a point estimate is the absolute loss

$$l = |\hat{\rho} - \rho| \tag{18.4}$$

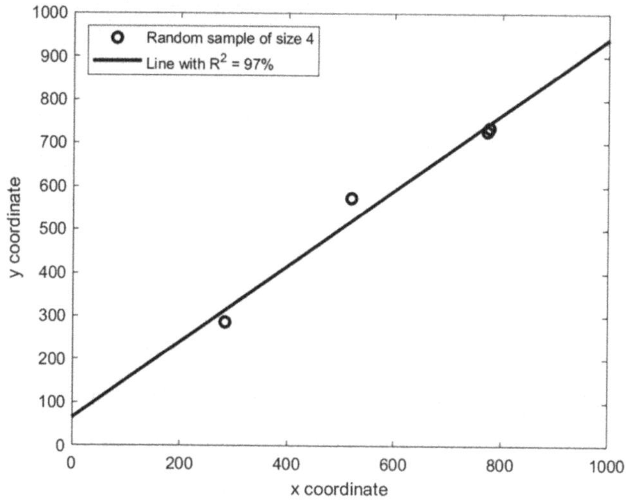

FIGURE 18.1
A regression line corresponding to an example by Fisher (1930, p. 434).

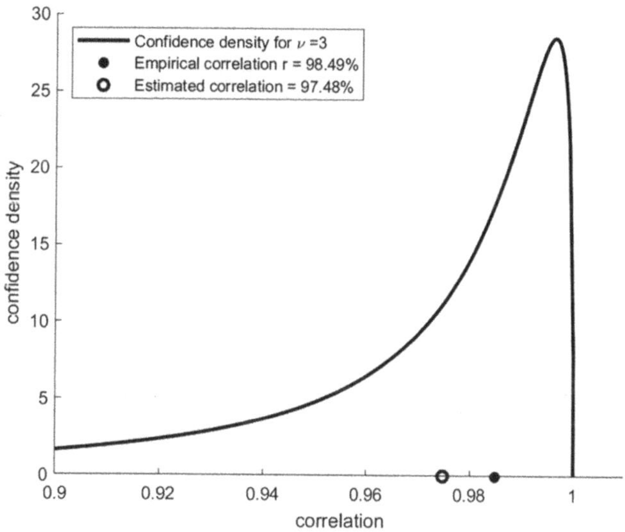

FIGURE 18.2
The fiducial density for correlation in the Fisher (1930, p. 434) example.

The fiducial risk

$$\int l(\hat{\rho}, \rho)\pi(\rho \,|\, r)\, d\rho = \int |\hat{\rho} - \rho|\, \pi(\rho \,|\, r)\, d\rho \tag{18.5}$$

is minimized by the median $\hat{\rho} = 97.48\%$. This is different from the ordinary estimate given directly by $r = 98.49\%$.

The median is the optimal estimate given the particular loss in equation (18.4). In general, the loss depends on the particular application of the analysis. The absolute loss has the advantage of giving an estimate which is invariant under one-on-one transformations of the parameter.

18.2 Fiducial Decision Theory

Given data y it is required to find an *action* $x = x(y)$ that is optimal. Optimality can be formalized starting with the specification of a *loss*

$$l = l(\gamma, x) \tag{18.6}$$

corresponding to a parameter γ and the action x. An example for a real statistic is given by squared error loss

$$l = (\gamma - x)^2 \tag{18.7}$$

The estimate x can be a test, a set estimate, a distribution estimate, or more complicated objects, and then other loss functions are natural.

It will be assumed in this section that a fiducial distribution for γ is given by the distribution of a random quantity Γ^y. An example was given in the Introduction and further examples will be given. A more general class of fiducial distributions are given by Definition 18.2 in Section 18.4. This class of fiducial distributions includes all the examples in the following. The random quantity $\Gamma^y = \gamma(U^a, y)$ is obtained by solving a data generating equation. The distribution of U^a is the updated state of knowledge of u given the data y. The value $a = a(y)$ is the observed value of a maximal invariant statistic $A = a(Y)$.

The distribution of the loss $l(\Gamma^y, x(y))$ for each y determines if an action x is good. More generally, the joint distribution of $l(\Gamma^y, x_A(y))$ and $l(\Gamma^y, x_B(y))$ for each y can be used in a comparison of two actions. The analysis can sometimes be simplified by only considering a simpler statistic given by a single number. The *fiducial risk*, if it exists, is defined to be the expected loss

$$r = \mathrm{E}l(\Gamma^y, x) \tag{18.8}$$

The fiducial risk of an action $x = x(y)$ is a statistic $r = r(y)$. An action x^* is the best fiducial action if its fiducial risk r^* is smaller than or equal to any other risk.

A best fiducial action $x = x(y)$ can in good cases be obtained directly by minimizing the risk. As a solution of an optimization problem, the best fiducial action is also called an optimal fiducial action. The prototypical example is given by

$$x(y) = \mathrm{E}\Gamma^y \tag{18.9}$$

which minimizes the risk from the squared error loss in equation (18.7). More generally, the optimization problem is much simplified both practically and theoretically if $x \mapsto l(\gamma, x)$ is convex for all γ.

An example is given by the maximum likelihood estimator y of the scale $\theta > 0$ from a random sample of size n from a gamma distribution with known shape $\alpha > 0$. The maximum likelihood estimator y of θ equals the empirical mean of the sample and the law of u is a gamma distribution with shape $n\alpha$ and scale $(n\alpha)^{-1}$. The model generating equation is $y = \theta u$ is given simply by multiplication. Taraldsen and Lindqvist (2015) investigate this and some related models in more detail.

Fiducial inference is in this simple case given by solving $y = \theta u$ with the result

$$\theta = yu^{-1} \tag{18.10}$$

The fiducial distribution is hence an inverse gamma distribution scaled by the data y. Exact simulation from this fiducial, and also the fiducial for any focus parameter $\gamma = \gamma(\theta)$, is obtained directly from the generator u.

In this case, the maximal invariant is trivial (a constant), and the state of knowledge of u is unchanged by the observation. Formally, in this case, $U^a \sim U$. The three examples in Section 18.3 demonstrate the more complicated case where the state of knowledge of u is updated by the data and $U^a \not\sim U$.

For the loss in equation (18.7) for the gamma scale model the fiducial optimal action is

$$x = \mathrm{E}\Theta^y = \left(\frac{n\alpha}{n\alpha - 1}\right) y \tag{18.11}$$

assuming $n\alpha > 1$. It can be observed that this estimate is different from the MLE y, but for large $n\alpha$ the difference is small.

Existence and calculation of a fiducial optimal action is hence similar to the problem of finding an optimal Bayesian posterior action. Berger (1985) and Eaton (1989) provide many examples and theories related to calculating optimal Bayesian actions. In addition to estimation, this includes hypothesis testing and randomized actions.

The next aim is, as in textbooks on Bayesian decision theory, to provide a link to frequentist decision theory. The distribution of $l(\gamma, X)$ determines if the action $X = x(Y)$ is good. A data generating equation can be used to investigate this for each model θ. This gives a method for comparison of two actions X_A and X_B. Instead of a comparison of the distribution of X_A and X_B one should instead compare the distribution of the losses $l(\gamma, X_A)$ and $l(\gamma, X_B)$. It can be observed that a data generating model allows simulation from the joint law of the actions, and also from the joint law of the losses.

As for the fiducial risk, it is convenient, if possible, to simplify the analysis down to a single number. The *risk* is defined as the expected loss

$$\rho = \mathrm{E}l(\gamma, X) \tag{18.12}$$

The risk is a function of the model θ since $\gamma = \gamma(\theta)$ and the distribution of X depends on θ. The risk of a statistic X is hence a parameter. The action X^* is *uniformly optimal* if the risk ρ^* is smaller than or equal to the risk ρ of any other action. The statement $\rho^* \leq \rho$ means that $\rho^*(\theta) \leq \rho(\theta)$ for all models θ. Existence and uniqueness of a uniformly optimal action are challenging since the unknown is the function $x(y)$, and the bound is to hold uniformly for all models θ.

A much simpler problem is to find a fiducial optimal action as defined by minimizing the fiducial risk. A candidate action is hence given by a fiducial optimal action. This turns out to be also uniformly optimal given some additional restrictions as proved by Taraldsen and Lindqvist (2013). The estimators are required to be equivariant and the loss is assumed to be invariant. A precise statement is given in Theorem 18.5 in Section 18.4.

The MLE y and the optimal x in equation (18.11) for the gamma scale model are both equivariant estimates in the sense that $x(ay) = ax(y)$. The loss in equation (18.7) is not invariant, but the losses

$$l = \theta^{-2}(\theta - x)^2 \tag{18.13}$$

$$l = [\ln(\theta) - \ln(x)]^2 \tag{18.14}$$

are both invariant for the gamma scale model. These losses obey $l(a\theta, ax) = l(\theta, x)$.

Uniformly optimal actions corresponding to the losses in equation (18.13) and equation (18.14) are respectively given by equation (18.11) and

$$x = \exp(\mathrm{E} \ln \Theta^y) = \frac{y}{\alpha} \exp\left[\ln(n\alpha) - \psi(n\alpha)\right] \tag{18.15}$$

where ψ is the digamma function. The action defined by equation (18.11) is also the equivariant estimate that uniformly minimizes the squared error risk. The result in equation (18.15) is derived by Taraldsen and Lindqvist (2015, p. 3759). It gives a natural estimator for the problem since the loss equals the squared Fisher information distance (Taraldsen and Lindqvist, 2013, p. 333).

18.3 Examples

Three examples are considered in the following three subsections.

18.3.1 The Original Fiducial

Fisher (1930) introduced the concept of a fiducial distribution. Fisher's first example is the fiducial density $\pi(\rho\,|\,r)$ for the correlation ρ of the binormal distribution. It is given by

$$\pi(\rho\,|\,r) = -\partial_\rho F(r\,|\,\rho), \tag{18.16}$$

where $F(r\,|\,\rho)$ is the cumulative distribution function for the empirical correlation r of a random sample of size n from the binormal distribution.

Taraldsen (2021a) reconsidered this problem and derived an exact explicit formula for the fiducial density $\pi(\rho\,|\,r)$ as stated in equation (18.1). This theory is here reconsidered and generalized. The purpose is to demonstrate fiducial inference for a nontrivial problem. It should be noted that the likelihood is not used in the arguments. The argument is valid also for models that are not defined by a likelihood.

Example 18.1 (Random points in the plane) *Assume that the joint law of* $(u_1, v_1, \ldots, u_n, v_n)$ *is known. A data generating model for* n *random points in the plane is defined by*

$$\begin{bmatrix} x_i \\ y_i \end{bmatrix} = \begin{bmatrix} \mu_x \\ \mu_y \end{bmatrix} + \begin{bmatrix} \sigma_x & 0 \\ \rho\sigma_y & \sqrt{1-\rho^2}\sigma_y \end{bmatrix} \cdot \begin{bmatrix} u_i \\ v_i \end{bmatrix} \tag{18.17}$$

and $\theta = (\mu_x, \mu_y, \sigma_x, \sigma_y, \rho) \in \Omega_\Theta \subset \mathbb{R}^2 \times [0,\infty)^2 \times (-1, 1)$.

A general definition of a data generating equation is given in Definition 18.1 in Section 18.4. The data generating model in equation (18.17) is equivalent with generation of x-coordinates by $x_i = \mu_x + \sigma_x u_i$ followed by generation of y-coordinates by the linear regression $y_i = ax_i + b + \sigma v_i$. Conventional statistical inference takes the resulting probability distribution for the data as a starting point for analysis. Fiducial inference includes the given data generating equation as a part of the model formulation. This has consequences in the analysis, and for the resulting fiducial.

The slope $a = \rho\sigma_y/\sigma_x$, the intercept $b = \mu_y - \rho\mu_x\sigma_y/\sigma_x$, and the conditional standard deviation $\sigma = \sqrt{1-\rho^2}\sigma_y$ give the alternative parameterization $(\mu_x, \sigma_x, a, b, \sigma)$. The original parameterization $(\mu_x, \mu_y, \sigma_x, \sigma_y, \rho)$ corresponds to location, scale, and correlation ρ if it is assumed that $\mathrm{E}(U_iV_i) = \mathrm{E}(U_i) = \mathrm{E}(V_i) = 0$ and $\mathrm{Var}(U_i) = \mathrm{Var}(V_i) = 1$.

Different Monte Carlo distributions and different choices for Ω_Θ give different data generating models. All possibilities in Definition 18.1 can be illustrated. The case $n = 1$ and σ_x, σ_y, ρ known gives a conventional, simple, pivotal data generating model. The case $n > 1$ and σ_x, σ_y, ρ known can give a data generating model which is not solvable. The case $n > 2$ and the largest possible model space can give a data generating model which is not solvable. All combinations of known and unknown components for the two alternative parametrizations give a rich arsenal of different problems for analysis.

A model corresponding to regression of x on y gives an alternative data generating model. Fiducial inference for the two models gives different fiducial distributions. This is in contrast with conventional statistical inference. Conventional statistical inference gives identical conclusions for the two models since the probability distribution for the data is identical if it is assumed that $(u_1, u_1, \ldots, u_n, v_n)$ is a random sample from the $\mathrm{N}(0,1)$.

If the focus is on the correlation coefficient ρ, then the analysis can be simplified. The empirical correlation r of the random points from equation (18.17) has a probability distribution that only depends on ρ. This case was analyzed by Elfving (1947, Equation 57) for the binormal case. Motivated by this we consider a more general case.

Example 18.2 (The Elfving Equation) *Let the Monte Carlo variable be a known law for $m \in [0,\infty)^2 \times \mathbb{R}$, let the parameter be the correlation $\rho \in (-1,1)$, and let the statistic be the empirical correlation $r \in (-1,1)$. The equation*

$$\sqrt{m_1}\frac{\rho}{\sqrt{1-\rho^2}} - \sqrt{m_2}\frac{r}{\sqrt{1-r^2}} = m_3 \tag{18.18}$$

can be solved to give a unique data generating model $r = r(m,\rho)$ and a unique parameter generating model $\rho = \rho(m,r)$.

Theorem 18.1 (Confidence for correlation) *The fiducial distribution from the parameter generating model $\rho = \rho(m,r)$ from equation (18.18) is an exact confidence distribution. It equals the Bayes posterior for ρ from the model in equation (18.17) with either of the priors*

$$\pi(\mu_x, \mu_y, \sigma_x, \sigma_y, \rho) = \frac{1}{\sigma_y^2(1-\rho^2)} \tag{18.19}$$

and also from the prior density

$$\pi(\mu_x, \mu_y, \sigma_x, \sigma_y, \rho) = \frac{1}{\sigma_x^2(1-\rho^2)} \tag{18.20}$$

where the law of m is determined by the conditional law of $(u_1, v_1, \ldots, u_n, v_n)$ given the maximal invariant a for the group action in equation (18.17). If $(u_1, v_1, \ldots, u_n, v_n)$ is a random sample from $N(0,1)$, then $m_1 \sim \chi^2(\nu), m_2 \sim \chi^2(\nu-1), m_3 \sim N(0,1)$ are independent with $\nu = n-1 > 1$, and the fiducial density $\pi(\rho \,|\, r)$ is given by equation (18.1). The one-sided confidence intervals from $\pi(\rho \,|\, r)$ are then uniformly most accurate invariant with respect to the location-scale groups on the two coordinates.

Proof *The last part is proved by Taraldsen (2021a). The first claim follows from Theorem 18.2. Equality of the fiducial and the posterior follows from Theorem 18.3 applied to the conditional model given a. It is important to note that the fiducial is not only given by r, but it depends also on the maximal invariant a. The Basu theorem ensures equality in the normal case since in this case the r is a part of a complete sufficient statistic so r and a are independent. In this case the fiducial depends only on r as given by $\pi(\rho \,|\, r)$.* □

18.3.2 The Scaled Uniform Model

Consider the data generating equation

$$y = \theta u \tag{18.21}$$

where $\theta > 0$ and the generator $u = (u_1, u_2)$ is given by the largest and smallest observation from a random sample of size n from the uniform distribution on the interval $(1-k, 1+k)$. The corresponding statistical model was recently discussed by Mandel (2020) and before that by Galili and Meilijson (2016). The existence of optimal estimators was left open by their analysis. The minimal sufficient statistic y is not complete, but unique optimal equivariant decision rules exist. This was proved by Taraldsen (2020a) and summarized by Taraldsen (2020b).

The data generating equation does not have a measurable selection solution for all data y and all generators u. It is an over determined system with two equations for the single unknown parameter. This is a general problem in fiducial inference, and many different solutions have been suggested. One approach is to replace the original data with a minimal sufficient statistic as for the scaled gamma model. The problem was there conveniently formulated directly with the minimal sufficient statistic. This does not always solve the problem as exemplified here with the *scaled uniform model.* The minimal sufficient is two-dimensional, but the model space is one-dimensional.

A second approach is to expand the parameter space so that solutions exist, and then condition the fiducial down to the original space. In the present example, a fiducial is obtained naturally for a parameter (θ_1, θ_2). The original model is obtained by restriction to the curve where $(\theta_1, \theta_2) = (\theta, \theta)$. A final fiducial can be obtained by conditioning on $C(\Theta) = \Theta_1 - \Theta_2 = 0$ or by $C(\Theta) = \Theta_1/\Theta_2 = 1$, or by an infinity of other possibilities. The result will depend on the choice of the condition $C(\Theta)$, and the function C is then part of the model assumptions. This approach is introduced and discussed in more detail by Taraldsen and Lindqvist (2017) and by Taraldsen and Lindqvist (2018). A related approach is used by Cisewski and Hannig (2012).

A third approach, which is related to the previous, is to condition on the generator U so that a solution for the parameter exists. This is the approach explained after Definition 18.2 and indicated previously. One possibility is to rely on a particular choice of an ancillary statistic $A = a(Y)$. Assume that a is invariant in the sense of $a(\theta u)$ being a function of only u. This can be used to replace the original model by a conditional fiducial model given the condition

$$a(\theta^* U) = a \tag{18.22}$$

where $a = a(y)$. The law of U^a defined by this condition is the updated state of knowledge about u given by the data y. The choice of θ^* in equation (18.22) does not influence the fiducial since $a(\theta U)$ does not depend on θ. A fiducial is then defined as the distribution of $\Theta^y = \theta(U^a, y)$.

A natural choice for the ancillary would be to choose a maximal invariant: Any other invariant is then, by definition, a function of this invariant. Conditions for existence and uniqueness (up till 1-1 correspondences) of a maximal invariant is left for future investigations. Basu (1964) demonstrates that the choice of an ancillary can be problematic, but Cox (1971) gives a possible resolution for some particular cases. Conditioning on an ancillary to obtain a fiducial is also discussed briefly by Hannig et al. (2016, p. 1350). A fully satisfactory solution exists, however, in the group case, as explained by Taraldsen and Lindqvist (2013).

For the scaled uniform model, a unique maximal invariant is given by

$$a(U) = U_2/U_1 \tag{18.23}$$

The fiducial is given by the distribution of $U_1 \,|(a(U) = y_2/y_1)$. Calculus gives that it is Pareto with truncation interval $[(1-k)/a, (1+k)]$ and index $-n$. The fiducial is then the corresponding distribution of $\Theta = y_1/U_1$. It is Pareto with index n and truncation interval $[\hat{\theta}_{\text{ML}}, \hat{\theta}_{\text{MU}}]$, so

$$\Theta^y \sim \pi(\theta \mid y) \propto (\hat{\theta}_{\mathrm{ML}} \leq \theta \leq \hat{\theta}_{\mathrm{MU}}) \cdot \theta^{-n-1} \tag{18.24}$$

where $\hat{\theta}_{\mathrm{ML}} = y_1/(1+k)$ and $\hat{\theta}_{\mathrm{MU}} = y_2/(1-k)$ (Taraldsen, 2020a).

The conditional model just described fulfills all the assumptions of Theorem 18.5. Uniformly optimal actions for the conditional model corresponding to the losses in equation (18.13) and equation (18.14) are respectively

$$x = \frac{\mathrm{E}([\Theta^y]^{-1})}{\mathrm{E}([\Theta^y]^{-2})} = \frac{\alpha}{\alpha - 1} \cdot \frac{1 - {b_*}^{1-\alpha}}{1 - {b_*}^{-\alpha}} \cdot \hat{\theta}_{\mathrm{ML}} \tag{18.25}$$

$$x = \exp \mathrm{E} \ln \Theta^y = \exp\left\{\frac{1}{n} - \frac{\ln b_*}{1 - {b_*}^n}\right\} \cdot \hat{\theta}_{\mathrm{ML}} \tag{18.26}$$

with $b_* = \hat{\theta}_{\mathrm{MU}}/\hat{\theta}_{\mathrm{ML}}$ and $\alpha = n + 2$. Both actions are uniformly optimal.

The action defined by equation (18.25) is also the action that minimizes the expected squared error in the class of equivariant estimators. The action defined by equation (18.26) is, however, possibly preferable for some investigations since it minimizes also the loss corresponding to the squared Fisher information distance (Taraldsen and Lindqvist, 2013, p. 333).

18.3.3 Optimal Linear Prediction

Consider a data generating equation

$$y = \theta + u \tag{18.27}$$

where y and u belong to a Hilbert space, and θ is restricted to be in a closed subspace Ω_Θ. A known distribution for u defines then a conventional fiducial model. The prototypical example is given by $\theta = X\beta$ where β are unknown regression coefficients, X is the design matrix, and Ω_Θ is the range of X. The case where the dimension of β is larger than the dimension of y has received considerable attention. The analysis that follows gives an optimal equivariant action for a prediction problem.

The real or complex Hilbert space may be infinite dimensional as discussed by Taraldsen and Lindqvist (2013, p. 331) for the case where Ω_Θ equals the Hilbert space. In this sense, this exemplifies a non-parametric fiducial model. The case with Ω_Θ being a strict subspace can be treated by conditioning on a maximal invariant as explained next. The analysis here is an extension of the analysis given by Taraldsen and Lindqvist (2013, p. 331). It exemplifies in particular that the fiducial can be non-unique, but that a unique optimal decision can be derived anyway.

Let p be the orthogonal projection on Ω_Θ. It follows that $a = a(y) = (1-p)(y)$ is a maximal invariant. This is the projection on the orthogonal complement of Ω_Θ. The law of U^a is determined by conditioning U on

$$a(U) = a \tag{18.28}$$

where $a = a(y)$ is the observed value of the maximal invariant. The law of U^a is the updated state of knowledge of u given by the observation.

The fiducial is simply the law of

$$\Theta^y = y - U^a \tag{18.29}$$

In the case of i.i.d. Gaussian U it follows that $U^a = a + pU$, but this does not hold in general.

Consider the parameter $\gamma = A\theta$ where $A : \Omega_\Theta \to \Omega_\Gamma$ is a bounded linear operator. This includes the case $A = I$, but also the case when Ω_Γ is one dimensional. The action of θ on Ω_Γ is given by

$$\theta(\gamma) = A\theta + \gamma \tag{18.30}$$

The loss

$$l = |\hat{\gamma} - \gamma|^2 \tag{18.31}$$

is invariant. A fiducial optimal estimate of $\gamma = A\theta$ is

$$\hat{\gamma} = \mathrm{E}\Gamma^y = \mathrm{E}(A\Theta^y) \tag{18.32}$$

Theorem 18.5 in Section 18.4 gives that this is the uniformly optimal equivariant estimate. This is an equivariant estimator, $\gamma(\theta + y) = A\theta + \gamma(y)$, since

$$\Theta^{\theta+y} = \theta + \Theta^y \tag{18.33}$$

Let x_* be given and consider the prediction of a real $Y_* = x_*\beta + U_*$. Assume that $\mathrm{E}U_* = 0$, and that the distribution of u and u_* are independent. Assume furthermore that there exist an A such that $AX = x_*$, and if not let A be the least squares solution found by projection. The previous argument gives an optimal estimate of $\gamma = AX\beta$. If $AX = x_*$, then this is an optimal estimate of the best linear prediction $x_*\beta$ of Y_*.

Many other cases can be considered and treated smoothly along the indicated path. One case is given by replacing u by σu where σ is also unknown. It should also be noted that the case $\theta = X\beta$ can give examples of a non-unique fiducial distribution for β, but they all give the same unique optimal estimator $\hat{\gamma}$. Fraser (1979) demonstrates fiducial inference for many other possible cases.

18.4 General Mathematical Results and Definitions

The purpose of this section is to briefly summarize mathematical definitions and theorems providing links between fiducial inference, Bayesian inference, and confidence distributions. Any *function* used in the following is assumed to be measurable. Any *set* is likewise assumed to be a *measurable space* (Rudin, 1987, p. 8). The qualifier *measurable* is henceforth usually dropped. A statistic $x = x(y) \in \Omega_X$ is a (measurable!) function of the data $y \in \Omega_Y$. A particular sample space Ω_X of a particular statistic X is sometimes called the action space in the context of decision theory to differentiate this from all other statistics. An action is a statistic.

The initial assumption for conventional statistical analysis is a specification of an indexed family $\{\mathrm{P}_Y^\theta \mid \theta \in \Omega_\Theta\}$ of probability distributions on the data space Ω_Y. It is assumed that $Y \sim \mathrm{P}_Y$, but the model $\theta \in \Omega_\Theta$ is unknown. A parameter $\gamma = \gamma(\theta) \in \Omega_\Gamma$ is a function of the model θ. Both the parameter space Ω_Γ and the model space Ω_Θ are measurable spaces as explained above. A particular parameter γ is sometimes called the focus parameter to differentiate this from all other parameters. In any particular problem all of the above spaces with its structure must be specified in a mathematical analysis of the problem.

Fiducial inference, as formulated mathematically by Taraldsen and Lindqvist (2013, 2015, 2019), and Taraldsen (2021a), is inference where a given data generating model is part of the problem formulation. Fiducial inference is then different from both Bayesian and frequentist inference since neither is based on the concept of a data generating model.

Definition 18.1 (**Generating Models**) *A data generating model for a statistic x is given by a function*

$$x = x(\gamma, u) \tag{18.34}$$

where $\gamma = \gamma(\theta)$ is a parameter. The law of the Monte Carlo variable u is given by a family of probability distribution $\{\mathrm{P}_U^\theta \mid \theta \in \Omega_\Theta\}$ indexed by the model parameter $\theta \in \Omega_\Theta$. A data generating model is conventional if the law P_U of the Monte Carlo variable is known. The data generating model is solvable if it can be solved to give a parameter generating model

$$\gamma = \gamma(x, u) \tag{18.35}$$

A data generating model is simple if it is solvable with a unique solution. The model is pivotal if it is conventional and can be uniquely solved to give a pivotal generating model

$$u = u(x, \gamma) \tag{18.36}$$

It has been explained and exemplified how a fiducial distribution can be determined by a data generating model. It will always be defined as the distribution of a parameter generating model. The simplest case is given by a solvable model.

Definition 18.2 (**Fiducial**) *Assume a data generating model $x = x(\gamma, u)$ is a solvable and that $\gamma = \gamma(x, u)$ is a corresponding parameter generating model. A fiducial distribution for γ is then defined to be the distribution of $\gamma(x, U)$ for fixed x.*

Definition 18.2 gives a unique fiducial from any simple conventional data generating model. Different choices of measurable selection solutions of equation (18.34) give different fiducials if the model is only solvable. If the data generating model is solvable conditionally given a maximal invariant $a(X) = a$, then a fiducial distribution for γ is defined to be the distribution of $\gamma(x, U^a)$ for fixed x.

There are many different data generating models for a given statistical model. In general, as demonstrated, the corresponding fiducial depends on the data generating model. For the case of a real statistic, exemplified by the empirical correlation r, the fiducial is, however, uniquely determined by the statistical model under mild assumptions.

Theorem 18.2 (**Uniqueness**) *The fiducial distribution from of a real valued, strictly monotonic, and simple data generating model is uniquely determined by the sampling distribution of the statistic. If, additionally, the sampling distribution of the statistic is continuous, then the fiducial distribution is an exact confidence distribution.*

Proof *This is Theorem 18.1 proved by Taraldsen and Lindqvist (2018, p. 143). The idea is that inversion of the cumulative distribution function gives a data generating model which is both simple and pivotal and a calculation shows that the fiducial from this coincides with the initial model.* □

As explained by Taraldsen and Lindqvist (2013, p. 329) it is known that the fiducial coincides with the posterior from a right Haar prior for a group model. A more general result was proven by Taraldsen and Lindqvist (2015, p. 3756, Theorem 2.1).

Theorem 18.3 (Bayes) *The fiducial from a simple data generating model* $t = t(\theta, u)$ *is the Bayes posterior from a* σ*-finite prior* P_Θ *if the distribution of* $t(\theta, u)$ *does not depend on* u *when* $\theta \sim \mathrm{P}_\Theta$.

Proof *The idea of the proof by Taraldsen and Lindqvist (2015, p. 3765) is given by establishing a joint density for* $(t(\Theta, U), U)$ *using the Fubini theorem together with the general change-of-variables theorem from measure theory. This is different from the more common change-of-variables theorem which requires differentiability. It should also be noted that there is no assumption of existence of a dominating measure. The proof holds in the generality as stated. It follows also that the posterior is proper.* □

The assumption of Theorem 18.3 is fulfilled when $t = t(\theta, u) = \theta u$ is given by group multiplication and P_Θ is chosen as the right Haar prior. The priors in Theorem 18.1 correspond to the right Haar priors corresponding to the choice of a Cholesky decomposition in lower- or upper-triangular matrices. Explicit expressions for right Haar priors are derived and discussed by Eaton (1989) in the more general context of group invariance in statistics. It is noteworthy that the fiducial solution does not need the expressions for the priors. The fiducial provides hence a simplified approach for Bayesian analysis for this case.

Definition 18.3 (Confidence) *A confidence distribution* C *for a parameter* γ *is a distribution estimator with*

$$\mathrm{C}(A_p) = p \tag{18.37}$$

for a family of confidence sets A_p *with level* p *for all levels* $0 < p < 1$. *If equation (18.37) holds only for all* $p \in I \subset]0, 1[$, *then* C *is a confidence distribution with level set* I.

Theorem 18.4 (Pivotal) *The fiducial from a conventional simple pivotal data generating model* $t = t(\gamma, u)$ *is an exact confidence distribution if* P_U *is continuous.*

Proof *Let* $\mathrm{P}(U \in B) = p$ *for an event* $A \subset \Omega_U$. *Define* $A(t) = \{\gamma \,|\, u = u(\gamma, t) \in B\}$. *It follows that* $\mathrm{P}(\gamma(t, U) \in A(t)) = p$ *so equation (18.37) is fulfilled. Furthermore,* $\mathrm{P}(\gamma \in A(T)) = p$, *so* A *is an exact confidence set with level* p. *The set of possible levels* I *is determined by* P_U, *but* $I =]0, 1[$ *when the Monte Carlo variable is continuous.* □

Taraldsen and Lindqvist (2013, p. 334) observe that data generating models can be used to derive confidence distributions. The assumed existence in Definition 18.3 of a family $\{A_p \,|\, p \in I\}$ of confidence sets is important, but the family is not unique. Definition 18.3 is discussed in more technical detail by Taraldsen (2021b). The construction of suitable confidence sets associated with a confidence distribution is an important problem. This is discussed by Liu et al. (2021) using the concept of a depth-CD.

It is seen from the previous that a fiducial can be a Bayes posterior and it can also be a confidence distribution. The fiducial can also determine optimal decision rules.

Theorem 18.5 *The risk of an equivariant rule in a group invariant problem is determined by a fiducial distribution if the group acts transitively on the model parameter space.*

Proof *This is Theorem 18.1 in Taraldsen and Lindqvist (2013).* □

It should be noted that a fiducial distribution always exists in the previous statement, but uniqueness of a fiducial distribution is not assumed. Optimal inference procedures are determined by the fiducial distribution regardless of the choice of a measurable selection for the determination of a fiducial distribution. This is exemplified by the linear prediction example by cases where the null space of the design matrix is nonzero.

18.5 Discussion

Fiducial inference is generally speaking not a well-defined concept. Different authors give different definitions. The adjective "fiducial" comes from the Latin *fiducia* for faith and means something taken as standard of reference or something founded on faith or trust. The term "fiducial" has been used in the previous also as a noun instead of the longer "fiducial distribution". This is in harmony with the tradition of referring to the "posterior" used as a noun instead of using the longer "posterior distribution".

Pedersen (1978, p. 147) concluded that the fiducial argument has had very limited success and that it was essentially dead. There was then hence little confidence in fiducial inference among most statisticians. This is also true today. Fiducial inference is viewed by many statisticians today as a somewhat obscure topic. Different versions are presented in textbooks and reference works, and most often with critical remarks. Casella and Berger (2002, p. 291), Kendall and Stuart (1961, p. 134), and Stuart et al. (1999, p. 440) present versions based directly on the likelihood, but Cox and Hinkley (1974, p. 246), Sprott (2000, p. 77), and Barnard (1995) present versions based on pivotal quantities.

Fiducial inference, as presented here, is a generalization of the version based on pivotal quantities. It is in particular not restricted to models defined by a likelihood. It is based on a data generating model. A data generating model is a most convenient starting point for inference since data y can be simulated by simulating u on a computer for each given model θ. A data generating model gives also a data generating model for any statistic by $x = x(y(u, \theta)) = x(u, \theta)$, and its properties can be determined by simulation. Properties of the data generating model $x = x(u, \theta)$ of the statistic x decides if the statistic x is good for inference. This is one advantage of a data generating model as compared with a statistical model given by a specification of the conditional law of the data directly.

Eaton (1989) demonstrates the importance of symmetry considerations in statistics. Symmetry is also important for fiducial inference as discussed in considerable detail by Fraser (1968). A more complete review of fiducial inference, including its history, is beyond the scope of this text, but Seidenfeld (1979), Dawid and Wang (1993), Lehmann and Romano (2005, p. 175), Veronese and Melilli (2015), Hannig et al. (2016), Taraldsen and Lindqvist (2018), and Dawid (2020) can be consulted for discussion and further references.

Fraser (1968, p. 50) refers to the generator variable u as the *error variable* and the data generating equation as the *structural equation*, but then in a more restrictive setting where the model θ is given by a 1-1 transformation of the data space $\Omega_Y = \Omega_U$. The data generating model is then $y = \theta u$ where θ belongs to a group and is acting on the data space $\Omega_Y = \Omega_U$. The notation $y = \theta u$ is also used by Dawid and Stone (1982) for the more general case of a jointly measurable mapping of the *model* parameter $\theta \in \Omega_\Theta$ and the *generator* variable $u \in \Omega_U$ to the *data* $y \in \Omega_Y$.

Consider again the model given by equation (18.17). The case with $\mu_x, \mu_y, \sigma_x, \sigma_y$ known is particularly challenging. It gives a model which is not solvable even after conditioning on a maximal invariant. The resulting statistical model for the data given by assuming that the Monte Carlo variables are a random sample from the standard normal was used by Basu (1964, p. 11) to illustrate the difficulty with some of the arguments by Fisher (1935). The x coordinates is an ancillary statistic, and the conditional model for the y coordinates can be solved. Interchanging the roles of the coordinates gives a different solution. Neither solutions are in fact good from a frequentist point of view. Helland-Moe (2021) considers many alternative solutions. The best solution is obtained by considering the larger two-parameter model $\theta = (\sigma, \rho)$ with $\mu_x = \mu_y = 0$, $\sigma_x = \sigma_y = \sigma$. The resulting fiducial is in particular better than the fiducial given by equation (18.1), so the information $\sigma_x = \sigma_y$ does give improved inference for the correlation. An alternative data generating model was used to get this improvement. It remains a challenge to obtain an improved frequentist solution using the additional information given by $\sigma = 1$.

Bibliography

Barnard, G. A., 1995. Pivotal models and the fiducial argument. *International Statistical Review / Revue Internationale de Statistique*, *63*(3), pp. 309–323.

Basu, D., 1964. Recovery of ancillary information. *Sankhyā: The Indian Journal of Statistics, Series A (1961–2002)*, *26*(1), pp. 3–16.

Berger, J. O., 1985. *Statistical Decision Theory and Bayesian Analysis.* New York: Springer.

Casella, G. and Berger, R. L., 2002. *Statistical Inference* (2nd ed.). USA: Duxbury, Thomson Learning.

Cisewski, J. and Hannig, J., 2012, August. Generalized fiducial inference for normal linear mixed models. *Annals of Statistics*, *40*(4), pp. 2102–2127.

Cox, D. R., 1971. The choice between alternative ancillary statistics. *Journal of the Royal Statistical Society: Series B (Methodological)*, *33*(2), pp. 251–255.

Cox, D. R. and Hinkley, D. V., 1974. *Theoretical Statistics.* Boca Raton: Chapman and Hall.

Dawid, A. P. and Stone, M., 1982. The functional-model basis of fiducial inference (with discussion). *The Annals of Statistics*, *10*(4), pp. 1054–1074.

Dawid, A. P. and Wang, J., 1993. Fiducial prediction and semi-Bayesian inference. *The Annals of Statistics*, *21*(3), pp. 1119–1138.

Dawid, P., 2020. Fiducial inference then and now. arXiv:2012.10689 [math, stat].

Eaton, M. L., 1989. Group invariance applications in statistics. *Regional Conference Series in Probability and Statistics*, *1*, pp. i–133.

Elfving, G., 1947. A simple method of deducing certain distributions connected with multivariate sampling. *Scandinavian Actuarial Journal*, *1947*(1), pp. 56–74.

Fisher, R. A., 1930. Inverse probability. *Proceedings of the Cambridge Philosophical Society*, *26*, pp. 528–535.

Fisher, R. A., 1935. The logic of inductive inference. *Journal of the Royal Statistical Society*, *98*(1), pp. 39–82.

Fisher, R. A., 1956. *Statistical Methods and Scientific Inference.* Collier-Macmillan Canada Ltd., Toronto, Ontario: Hafner Press (revised and enlarged 1973 edition).

Fraser, D. A. S., 1968. *The Structure of Inference.* New York: John Wiley & Sons, Inc.

Fraser, D. A. S., 1979. *Inference and Linear Models.* New York: McGraw-Hill.

Galili, T. and Meilijson, I., 2016. An example of an improvable rao–blackwell improvement, inefficient maximum likelihood estimator, and unbiased generalized bayes estimator. *The American Statistician*, *70*(1), pp. 108–113.

Hannig, J., Iyer, H., Lai, R. C. S. and Lee, T. C. M., 2016. Generalized fiducial inference: A review and new results. *Journal of the American Statistical Association*, *111*(515), pp. 1346–1361.

Helland-Moe, O., 2021. *Objective Inference for Correlation.* Master's thesis, NTNU.

Kendall, M. G. and Stuart, A., 1961. *The Advanced Theory of Statistics. Volume 2: Inference and Relationship.* New York: Hafner Publishing Company.

Lehmann, E. L. and Romano, J. P., 2005. *Testing Statistical Hypotheses.* New York: Springer.

Liu, D., Liu, R. Y. and Xie, M.-G., 2021. Nonparametric fusion learning for multiparameters: Synthesize inferences from diverse sources using data depth and confidence distribution. *Journal of the American Statistical Association*, *117*(540), pp. 2086–2104. https://doi.org/10.1080/01621459.2021.1902817.

Mandel, M., 2020. The scaled uniform model revisited. *The American Statistician*, *74*(1), pp. 98–100.

Pedersen, J. G., 1978. Fiducial inference. *International Statistical Review/Revue Internationale de Statistique*, *46*(2), pp. 147–170.

Rudin, W., 1987. *Real and Complex Analysis.* New York: McGraw-Hill.

Seidenfeld, T., 1979. *Philosophical Problems of Statistical Inference: Learning from R.A. Fisher.* Theory and Decision Library. Dordrecht: Springer.

Sprott, D. A., 2000. *Statistical Inference in Science.* Springer Series in Statistics. New York: Springer-Verlag.

Stuart, A., Ord, K. and Arnold, S., 1999. *Kendall's Advanced Theory of Statistics, Classical Inference and the Linear Model* (6th ed.), Volume 2A. London: Wiley.

Taraldsen, G., 2020a. Fiducial symmetry in action. arXiv:2004.11878.

Taraldsen, G., 2020b. Micha Mandel (2020), "The Scaled Uniform Model Revisited," The American Statistician, 74:1, 98–100: Comment. *The American Statistician*, *74*(1), p. 315. https://doi.org/10.1080/00031305.2020.1769727.

Taraldsen, G., 2021. Joint Confidence Distributions. Technical Report. https://doi.org/10.13140/RG.2.2.33079.85920.

Taraldsen, G., 2023. The confidence density for correlation. *Sankhya A*, *85*, pp. 600–616. https://doi.org/10.1007/s13171-021-00267-y.

Taraldsen, G. and Lindqvist, B. H., 2013. Fiducial theory and optimal inference. *Annals of Statistics*, *41*(1), pp. 323–341.

Taraldsen, G. and Lindqvist, B. H., 2015. Fiducial and posterior sampling. *Communications in Statistics: Theory and Methods*, *44*(17), pp. 3754–3767.

Taraldsen, G. and Lindqvist, B. H., 2017. Fiducial on a string. arXiv:1706.03805 [stat].

Taraldsen, G. and Lindqvist, B. H., 2018. Conditional fiducial models. *Journal of Statistical Planning and Inference*, *195*, pp. 141–152.

Taraldsen, G. and Lindqvist, B. H., 2019. Discussion of 'Nonparametric generalized fiducial inference for survival functions under censoring'. *Biometrika*, *106*(3), pp. 523–526.

Veronese, P. and Melilli, E., 2015. Fiducial and confidence distributions for real exponential families. *Scandinavian Journal of Statistics*, *42*(2), pp. 471–484.

Index

Note: Bold page numbers refer to tables; *italic* page numbers refer to figures.

ABC *see* approximate Bayesian computing (ABC)
absolute loss 382, 383
adjusted profile h-likelihood (APHL) method 177–178
admissible 18, 77, 134, 183
adolescent BFFer 161
  managing nuisance constructs 161–162
  trio of BFF-learning (and practicing) stories 162–163
aleatoric
  probability 323, 325, 327, 330, 334, 341
  uncertainty 325
aleatory probability 38
ancillary statistic 46, 70, 84, 96, 213, 215, 216, 228–233, 393
APHL method *see* adjusted profile h-likelihood (APHL) method
approximate Bayesian computing (ABC) 118, 121–124, *123,* **124**
artificial
  sample 118
  randomness 117, 118, 120, 122
asymptotic theory, higher-order 49
auxiliary 43, 50, 99, 163, 302–303, 305, 307, 313, 316, 345–347, 350, 351, 353, 355, 359, 366

backpropagation technique 189
Barndorff-Nielsen's formula 215
basic inferential models 346–348
basic probability assignment 304
batting average 32n5
Bayes
  factor 133, 134, 136–139, 143–144, 149
  information criterion 153
Bayes algorithms, variational 190
Bayes and Frequentist (BF) paradigm 238, *240*
Bayes Billiard 124, 125
Bayes' Billiard Table Experiment 124, 125
Bayes estimation approach 21
Bayes factors 133
  fractional 143–144
Bayesian algorithm 121
Bayesian analysis 133
Bayesian and frequentist perspectives 241–244
Bayesian approach 122, 323
  to data integration 242
  in multi-parameter problems 40
Bayesian decision theory 385
Bayesian, frequentist, and fiducial (BFF) inferences 106, 326
  duality of parameters in 124–125
Bayesian, frequentist, and fiducial (BFF) paradigms
  using CD framework to unify and combine inference across 114–116, **115**
Bayesian, frequentist, and fiducial (BFF) trinity 163–166
  contrasting frequentist and Bayesian imaginaries *164,* 164–165
  imaginary for BFF 163–164
  imagine the fiducial imaginary 165–166
Bayesian, frequentist, and fiducial (BFF) union: randomness matching and parameter duality 116–118
Bayesian, frequentist, and fiducial (BFF) workshops 35
Bayesian fusion learning 242
Bayesian inference 3, 35, 36, 40, 41, 58, 108, 121–124, 162, 264, 304
Bayesian methods 47
Bayesian model 20, 23, 171, 317
Bayesian model uncertainty 138
  assigning model prior probabilities 139
  model averaging 140–142
  notation for model uncertainty 138–139
Bayesian model uncertainty, issues with 152
  approximations and asymptotics 153

Bayesian model uncertainty, issues with (*Cont.*)
    computation and moving in model space 152
Bayesian motivator: teacher's taste of two-sample
    nuanced (or nuisance) question 168
    seriously missed educational opportunity 168–169
Bayesian neural network (BNN) 191
    MCMC algorithms for 192
    variational inference 191
Bayesian paradigm 257
Bayesian paths 177
Bayesian setting 38
Bayesian solutions, informal objective 61–63
Bayesian stochastic inversion algorithm 121
Bayesian tests 134
Bayesian *vs.* frequentist divide 264
Bayes inference, empirical 48
Bayes meta-estimator 242
Bayes' rule 21, 326
Bayes' theorem 35, 106, 276
Behrens-Fisher problem 169
belief function 47, 300–306,
belief function via the multi-valued mapping 301–302
    non-invertible functional models 302–304
Benjamini and Hochberg 22, 23, *24*, 25
Bernoulli
    distribution 100
    law of large numbers 265
    model 100
Bernoullian inference via game theory 267–269
Bernstein-von Mises theorem 40, 120–124
Best Linear Unbiased Estimator (BLUE) 241, 243
beyond all-or-nothing bets 265–266
BIC 153
binomial
    distribution 99, 277, 284–292
    model 124
binomial algorithm 278
    for unknown n, known p 294
    for unknown n, unknown p 295–296
binomial distribution 99
    n known, p unknown 284–285
    n unknown, p known 285–288, *286*, *287*
    n unknown, p unknown 288–291, *291*, **292**
BLUE *see* Best Linear Unbiased Estimator (BLUE)
BNN *see* Bayesian neural network (BNN)
bootstrap
    distribution 106, 108, 111, 115
    sample 108, 118, 119, 125
bootstrap central limit theorem 118
bootstrap estimators 118–119

Cartesian products 306
Cauchy
    distribution 163, 351
    method 182
Cayley transform 279
CD *see* confidence distribution (CD)
children obesity dataset (OBICE study) 150
Chi-squared distribution 311
choice of the π, background in 142
    fractional Bayes factors 143–144
    mixture g-priors in linear model 142–143, **143**
CNNs *see* convolutional neural networks (CNNs)
coefficient of variation, 90, 92–93
    normal 90
    normal model conditioned on 92–93
communication-efficient methods 247
composite testing 133
conditional
    density 36, 43, 216, 217, 230
    inference 215–218
    principle 216
conditional consistency 92–93
conditional inference 5, 59
conditional inference, approximate
    ancillary and sufficient directions 215–217
    computation of V 217–218
conditioning difficulties 93–94
confidence curve 109, 110, 112, *113*, *290*
confidence distribution (CD) 35, 37, 38, 44–45
    construction with different inferential methods 111–113, *113*
    definitions 107–111, *108*
    epistemic nature of 48
    framework to unify and combine inference across BFF paradigms 114–116, **115**

fusion learning 243–244
inference 48
random variables 118–119
conformal
predictive distributions 364–377
predictive inference 369
conformal martingales 376
conformal prediction 373
efficiency of 375–376
validity of 375
conformal predictive distributions 364, 370
cross 373–374
full 370–372
conformal predictors
examples of 374–375
training 375–376
constriction 318n6
continuous non-partitionable model 97
convolutional neural networks (CNNs) 192
Cournot's principle 265
covariance matrices estimation 278

data-dependent priors 42
data generating algorithm (DGA) 276
data generating equation 382, 388, 389
data generating model (DGM) 326, 328, 391
data harmonization *see* model fusion
data integration 238, 239, 251
Bayesian approach to 242
in settings 251
dataset specific parameters 239
data-target model 327
Da Vinci Code for h-likelihood 176–177
decision theory 382
deep learning (DL) 188
Dempster—Hill 364, 367–371
Dempster—Hill distributions 370
full conformal predictive distributions 370–372
Dempster–Hill procedure 368–370
validity and efficiency of 369–370
Dempster-Shafer (DS) inference for categorical distributions 304–306, *305*
Dempster-Shafer (DS) theory 47, 277, 285, 300–318
of inference 98
Dempster-Shafer Extended Calculus of Probability (DS-ECP) 306
Dempster's rule of combination 308–309
down-projection 309
evidence aggregation: projection and combination 307
inference summary 310
Monte Carlo computation for 313–315, *314*, *315*
state space and the specification of marginal evidence 306–307
up-projection 307–308
Dempster's rule and independence assumption 316
dependent data sources, integration of 254–255
DGA *see* data generating algorithm (DGA)
DGM *see* data-generating model (DGM)
dicing proposition 325
difficulties
choice of group 91
conditional consistency 92–93
conditioning and partitioning 93–94
marginalization inconsistency 91–92
dimensionality reduction 190
direct switching principle 330–331
Dirichlet 67, 75, 76
distribution estimator 107–111
distribution for parameters
belief functions and inferential models 47
comparisons 47–50
confidence distributions 44–45
generalized fiducial 42–44
historical sketch 35–38
nature of probability 38–39
objective Bayes 40–42
significance functions 45–47
divide-and-conquer strategies 238
DL *see* deep learning (DL)
DS-ECP *see* Dempster-Shafer Extended Calculus of Probability (DS-ECP)
duality of parameters 124–125

Efron's Statistical Triangle 5
electronic health record data 251
Elfving equation 387
empirical Bayes 8
deconvolution 27
f-modeling 26–29
g-modeling 26–29
inference 48
empirical expected posterior prior, for non-nested models 148–149

empirical probability 38
empirical uncertainty 125, 126
ensemble methods 29
epistemic
probability 38–40, 48, 325–327
uncertainty 325
epistemic nature of confidence distribution 48
epistemic probability, objective 39
EPPrior 145, 146
expected posterior priors 144
choice of base model 149–151
general definition 144–145
general testing 152
studying robustness by varying the training sample size 151–152
expected posterior priors (intrinsic priors)
for linear model 146–148
for nested models 145–146
exponential errors-in-variables 357–359
exponential family 28, 46, 210, 214–216, 218–231
exponential family inferences 219
general exponential family 221–223
linear exponential family 220–221
one-parameter case 219–220
renormalisation 223–224, *224*

false confidence 346
false discovery
proportion 22
rates 22–26, *23*, *24*, *26*
Fiducial
distribution 35, 37, 38, 43–45, 47–50, 83–98, 384, 387, 390–393
inference 83–102, 106–127, 276–296
prediction 364–377
probability 37, 39, 89, 285, 294, 352, 353
fiducial activity and sunspot occurrence *7*
fiducial approach 327–328
fiducial decision theory 384–386
fiducial distribution 114, 115, 384
generalized 48
fiducial inference 4, 5, 276, 381, 393
Fisher's 365
generalized 119–121
fiducialistic encounter: BFFer's taste of trinity 175
Da Vinci Code for h-likelihood 176–177
fall in Fisher's pit 177–179
following the frequentist and Bayesian paths 177
multi-perspective prelude: integration/optimization 175
fiducial prediction
Fisher 365–366
genuine nonparametric 368
in nonparametric setting 364
parametric 365–367
validity and efficiency of 366–367
fiducial sampling 119–121
Fisherian pivotal model (FPM) 85, 329
Fisherian theory 302
Fisher information 40, 42
Fisher information matrix 42, 64, 66, 176, 211
Fisher's fiducial approach or method 44, 119
distribution 37, 376
equation 45
inference 365
prediction 365–366
switching principle 329–330
Fisher's Gamma model 14
Fisher's method 15
Fisher's nonparametric fiducial inference 368
Fisher's original fiducial argument 84–85
Fisher's original fiducial conception 323
f -modeling 26–29
FPM *see* Fisherian pivotal model (FPM)
fractional Bayes factors 143–144
frequentism 37, 38, 50, 58–79, 341
frequentist 118–119
approach 326–327
fusion learning 241–242
inference 1
frequentist matching
invariance priors 69–71
matching priors 67–68
multiparameters 68–69
priors 60
single parameter case 68
frequentist performance *60*, 60–61
full conformal predictive distributions 370–372
functional model 88
general non-invertible 99–100
monotonic 89
non-simple 95–96
simple functional model (SFM) 88–89
simple non-invertible 98–99

fusion learning 239
Bayesian 242
confidence distribution 243–244
frequentist 241–242
fusion of heterogeneous parameters 255–256
fusion procedure 240

game theory 272–273
Bernoullian inference via 267–269
Gamma MLE 9
Gaussian processes 190
Gauss-Markov estimation paradigm 241
generalized fiducial 42–44
distribution 48, 276, 280, **281**, *281*, 285
inference 118, 119–121, 337–340
general likelihood
basic approximation 225
generalising V 226–227
general model fusion analysis 239
general non-invertible functional model 99–100
generative model 190, 302, 305, *305*
genuine nonparametric fiducial prediction 368
Geometric Langevin Monte Carlo 192
GFD *see* generalized fiducial distribution (GFD)
GFI *see* generalized fiducial inference (GFI)
Gibbs algorithm 314
g-modeling 26–29, *29*
Google's NGram Viewer 1
g-priors 142
graphical processing units (GPUs) 188
group models, non-transitive 95
group-structured models 87
simple group model 87–88
structural models 88
group structure, normal model with 87

Hald regression 140, 141, **141**
heterogeneous parameters 250–251
hierarchical normal models 77–79
hierarchical reference approach 76–77
high-dimensional 43, 190, 244, 246–248
high-dimensional data sources, integration of 246–248
higher-order asymptotic theory 49
Hilbert space 389
homogeneity assumption, on model parameters 244
homogeneous parameters, integration of 248–251
hyper-g intrinsic prior (IHG) **151**

identifiability 191–194, 196, 197
IID model 372
IM *see* inferential model (IM)
imprecise probability 39, 47, 269, 270, 273n3, 301, 345–349
improper priors, justifying posteriors from 62–63
improvisation: Bob tests alice 266–267
independence assumption, Dempster's rule and 316
independent data sources
fusion of heterogeneous parameters with 255–256
integration of 245–248
integration of homogeneous parameters 250–251
independent data sources, fusion of parameters with 251
fusion of parameters 252–254
random effects approach 252
individual-level information, integration of 245–246
inferential method 106, 111–115, 117, 125
inferential model (IM) 47, 49
framework 344, 345
theory of 50
inferential model (IM) construction
equivalence of two constructions 354–355
possibility measures 348–351
technical remarks 351–353, *352*
informal objective Bayesian solutions 61
justifying posteriors from improper priors 62–63
truncation of parameter space 61
vague proper priors 61–62
INLA 153
integration of dependent data sources 254–255
integration of independent data sources 245
integration of high-dimensional data sources 246–248
integration of individual- and summary-level information 245–246
intrinsic hyper-g prior (IHG prior) 147

intrinsic prior
derivation of 147–148
properties of 147
for testing a precise hypothesis 136–137
inverse probability 63–64
inversion method 119, 121

James–Stein Estimator 15–22, **16, 18, 19**
rule 18, 22
theorem 16, 18, 19
Jeffreys-rule Prior 64–65
multi-parameter case 66–67
single parameter case 65–66

Kullback-Liebler (KL)
distance 191
divergence 71, 191

Laplace approximation 49, 153, 178, 231, 232
Laplace's method 64
latent factor models through neural networks
anchor points 194–195
combining parameter constraints with anchor points 195
general model 192–193
identifiability 193–194
least squares predictive machine (LSPM) 371, 372
left Haar prior 70
likelihood as pivot 215
Lindsey's method 25
linear dimensionality reduction methods 19
linear model
expected posterior priors (intrinsic priors) for 146–148
mixture g-priors in 142–143
linear prediction, optimal 389–390
local asymptotic normality 211
location model 65, 94–95
location-scale model 96
lower triangular group, Wishart model with 87–88
LSPM *see* least squares predictive machine (LSPM)
Lugannani-Rice formula 220

Mann–Whitney test 112
many-normal-means problem 42
marginal
density 10, 12, 20–22, 25–27, 219, 231
inference 210, 350, 357, 359
likelihood 60, *60*, 133, 145, 152, 153
marginalization
consistency 89–91
inconsistency 91–92
of probabilities 38
paradox 41, 91
Markov Chain Monte Carlo (MCMC) 41, 169, 278
algorithm 280
algorithm for Bayesian neural network (BNN) 192
methods 192
sampling 278
mathematical enlightenment
rare statistically surprising conjecture 179–180
Cauchy's rescue to hotelling 180–182
measure-based approach
curved normal 355–356
exponential errors-in-variables 357–359
measure theory 272–273
median probability model 141–142
meta-analysis 115, *117*, 240–242
missing species problem 10–15, **11**, *12*, **13, 15**
mixed models Jacobian quantity, derivation of 293–294
mixture g-priors, in linear model 142–143
MLE 127
MLPs *see* multilayer perceptrons (MLPs)
Mobius inversion 304
model averaging 140–142
basics 140
median probability model 141–142
prediction 141
model fusion 238, 239
methods 238
model parameters, homogeneity assumption on 244
model uncertainty, notation for 138–139
monotonic functional model 89
monotonic model 98–99
Monte Carlo computation for DS-ECP 313–315, *314*, *315*
Monte-Carlo randomness 126
multilayer perceptrons (MLPs) 192

multinomial
distribution 27, 66, 76–77
model 67
multi-parameter problems, Bayesian approach in 40
multiple continuous parameters 73–74
multivariate Jacobian quantity, derivation of 292–293
multivariate normal distribution
generalization of one way random effects model 282–284
multivariate normal data 278–282

negative binomial 12, 148
nested models, expected posterior priors (intrinsic priors) for 145–146
neural network 189, *189*
non-invertible functional model
general 99–100
simple 98–99
non-invertible models 98
general non-invertible functional model 99–100
simple non-invertible functional model 98–99
variations 99
non-nested models, empirical expected posterior prior for 148–149
nonparametric fiducial inference, Fisher 368
prediction, 364, 368
non-partitionable model, continuous 97
non-simple functional models 95–96
non-simple models
ancillary information 94–95
examples 96–97
non-simple functional models 95–96
non-transitive group models 95
recent developments 97–98
non-transitive group models 95
normal coefficient of variation 90
normal model
with group structure 87
pivotal inference 85–86
notation for model uncertainty 138–139
nuisance parameters 49, 137–138, 230
precise hypothesis testing with common 137–138
null hypothesis 133, 166–167
basics and notation 133–134, **134,** *135*
choice of the prior 136–137
erroneous use of improper or vague proper priors 136
formulating a problem as a precise hypothesis test 135–136
intrinsic prior for testing a precise hypothesis 136–137
precise hypothesis testing with common nuisance parameters 137–138

objective Bayes
inference 58–79
philosophical 64
priors 58, 60
objective Bayesian analysis 58, 63
constant prior and inverse probability 63–64
frequentist matching 67–69
invariance priors 69–71
Jeffreys-rule Prior 64–65
objective Bayesian inference 58
procedure 61
objective epistemic probability 39
objective priors 75
hierarchical normal models 77–79
hierarchical reference approach 76–77
preliminaries 75–76
reference prior as common 76
one-shot methods 246
online hypothesis testing 376
optimal linear prediction 389–390
original fiducial 386–387

parameter generating equation 382
parameter space, truncation of 61
parametric fiducial prediction 365–367
Pareto distribution 73
partitioning difficulties 93–94
perfect-information games 272
pivot 85–86, 167–168, 328–329, 335
pivotal function 83, 86, 94
inference 85–86
validity 86
Plausibility function 98, 302, 303, 305, 347, 354
Poisson distributions 15, 289, 290
posterior predictive distribution 141, 170, 171, 173, 174, 177

posterior predictive p-value
  Bayesian history of 171
pragmatic Bayesian experiment
  Bayesian history of posterior predictive p-value 171
  Bayesian imaginary for posterior predictive p-value 172–173
  posterior predictive p-values 169–171
  prior influence on posterior predictive p-value 173–174
precise hypothesis
  intrinsic prior for testing 136–137
  testing with common nuisance parameters 137–138
precise hypothesis test
  with common nuisance parameters 137–138
  formulating problem as 135–136
predictive distributions
  Bayesian 171, 177
  conformal 364–377
  cross conformal 373–374
  fiducial 177–179, 364–377
  frequentist 177
  full conformal 370–372
  split-conformal 373–374
predictive systems, split-conformal 373
preliminaries, objective priors 75–76
principle component analysis 280
prior
  conjugate prior 137, 138, 149
  data-dependent 42, 43, 232, 355
  improper 41, 62–63
  intrinsic 132, 133–137, 145–148
  invariance 69–71
  Jeffreys' 40–42
  matching 40–42, 60, 67–71
  objective 75–79, 137–138
  reference 71–77
  spike and slab 28
  targeted 122
  vague 41
  weakly informative 41, 42
  Zellner-Siow 142, 143
prior inclusion probability 129
prior probabilities, assigning model 139
probabilistic criterion of efficiency 375
probabilistic inference 165, 304, 346, 367
probability
  aleatory 38, 39, 126
  epistemic 38–40, 118, 325, 338
  empirical 40
probability models and inference goal 324
procedural-based approach 115
procedural frequentism 37
profile
  likelihood 111, 116, **117**, 178
  method 111, **117**
proper priors, vague 61–62
pure likelihood approach 211
purely frequentist concept 48
p-value 161–183
p-value function 45, 111, 112, 114, 115, 119, 212, 243

random effects model 242, 252, 257, 282–284
random set 47, 100, 285, *312*, 317, 347, 354, 355
random variables, CD 118–119
real data applications
  banknote data 203, *203*
  E. coli data 198–200, *199–201*
  user knowledge data 202, *202, 203*
real parameter, reference priors for 72–73
recommended prior 79
rectified linear unit (ReLU) 189
reduced structural model 96
reduction methods, linear dimensionality 190
reference priors 71
  approach 58, 71
  multiple continuous parameters 73–74
  for real parameter 72–73
regularization 18, 28, 189, 191, 252
relational pivotal models (RPMs) 328–329
  taxonomy for 335
relevance 29–32, *31*
right Haar
  measure 87
  prior 70, 71, 75, 391
roadmap of models **101–102**
Robbins' formula 9–10
Robbins' monumental insight 8
robust **143**, 143, 151–152

saddle point approximation 46, 210, 215, 221, 231
Sample Space 2
scaled uniform model 387–389
scale models 65
second-order inference 216
SFM *see* simple functional model (SFM)

shrinkage 18, 30, 191
significance functions 45–47, 212–214, *214*
simple functional model (SFM) 88–89
simple group model 87–88
simple non-invertible functional model 98
  belief functions 98
  monotonic model 98–99
simulation study for hyper-sphere
  model and sampling details 195–196, *196*
  results 196–197, *197*, *198*
Skovgaard's approximation 277
slice-and-dice argument 323, 334
slice-and-dice fiducial distribution 334
software implementations 254
  integration of dependent data sources 254–255
split-conformal predictive distributions 373–374
split-conformal predictive systems 373
split-conformity measure 373
standard deviation, normal model conditioned on 92
STAN software 280, 282
statistical estimator **115**
statistical inference 35, 116, 345
  matching condition for 118
statistical practice 238
strategies for Alice, Bayesian and Non-Bayesian 269–270
structural
  equation 120, 218, 304, 316–317, 393
  model 83, 88, 91, 96, **102**, 329
structural switching principle 331–332
student's taste of t
  null hypothesis 166–167
  pivotal quantity 167–168
subjective Bayes
  inference 36, 38
  prior 36
summary-level information, integration of 245–246
switching principle arguments 332
  fiducial sets, projections, and slices 332–333
  slice-and-dice fiducial argument 333–334
switching principles 329
  direct switching principle 330–331
  Fisher's fiducial switching principle 329–330
  structural switching principle 331–332

tangent
  exponential model 210–233
  plane 216, *217*, 232, 236
tangent exponential model, derivation of
  no nuisance parameters 228–230
  nuisance parameters 230
  preliminary remarks 227–228
taxonomy for relational pivotal models (RPMs) 335
A Theory of Mathematical Evidence (Shafer) 300
truncated model 96–97
truncation of parameter space 61
Tweedie's formulas 20

uncertainty, types of 324–325
uniform distribution 44, 72, 85, 107, 112, 139, 213, 295, 310, 388
upper and lower probabilities 47

vague proper priors 61–62, 136
validity, pivotal inference 86
variational auto-encoders (VAEs) 190
variational Bayes algorithms 190
variational inference 191

weakened simultaneous inference, with measurement errors 310–313, *312*
Wishart distribution 87, 279
Wishart model 90–91
  with lower triangular group 87–88

Zellner-Siow priors 142